Advanced Analysis of Variance

WILEY SERIES IN PROBABILITY AND STATISTICS

Established by *Walter A. Shewhart and Samuel S. Wilks*

Editors: *David J. Balding, Noel A. C. Cressie, Garrett M. Fitzmaurice, Geof H. Givens, Harvey Goldstein, Geert Molenberghs, David W. Scott, Adrian F. M. Smith, Ruey S. Tsay*

Editors Emeriti: *J. Stuart Hunter, Iain M. Johnstone, Joseph B. Kadane, Jozef L. Teugels*

The ***Wiley Series in Probability and Statistics*** is well established and authoritative. It covers many topics of current research interest in both pure and applied statistics and probability theory. Written by leading statisticians and institutions, the titles span both state-of-the-art developments in the field and classical methods.

Reflecting the wide range of current research in statistics, the series encompasses applied, methodological and theoretical statistics, ranging from applications and new techniques made possible by advances in computerized practice to rigorous treatment of theoretical approaches.

This series provides essential and invaluable reading for all statisticians, whether in academia, industry, government, or research.

A complete list of titles in this series can be found at http://www.wiley.com/go/wsps

Advanced Analysis of Variance

Chihiro Hirotsu

WILEY

This edition first published 2017

Registered Office
John Wiley & Sons, Inc., 111 River Street, Hoboken, NJ 07030, USA

Editorial Office
111 River Street, Hoboken, NJ 07030, USA

For details of our global editorial offices, customer services, and more information about Wiley products visit us at www.wiley.com.

Wiley also publishes its books in a variety of electronic formats and by print-on-demand. Some content that appears in standard print versions of this book may not be available in other formats.

Library of Congress Cataloging-in-Publication Data

Names: Hirotsu, Chihiro, 1939– author.
Title: Advanced analysis of variance / by Chihiro Hirotsu.
Description: Hoboken, NJ : John Wiley & Sons, 2017. | Series: Wiley series in probability and statistics | Includes bibliographical references and index. |
Identifiers: LCCN 2017014501 (print) | LCCN 2017026421 (ebook) | ISBN 9781119303343 (pdf) | ISBN 9781119303350 (epub) | ISBN 9781119303336 (cloth)
Subjects: LCSH: Analysis of variance.
Classification: LCC QA279 (ebook) | LCC QA279 .H57 2017 (print) | DDC 519.5/38–dc23
LC record available at https://lccn.loc.gov/2017014501

Cover Design: Wiley
Cover Image: © KTSDESIGN/SCIENCE PHOTO LIBRARY/Gettyimages; Illustration Courtesy of the Author

Set in 10/12pt Times by SPi Global, Pondicherry, India

Printed in the United States of America

10 9 8 7 6 5 4 3 2 1

Contents

Preface

Scheffé's old book (*The Analysis of Variance*, Wiley, 1959) still seems to be best for the basic ANOVA theory. Indeed, his interpretation of the identification conditions on the main and interaction effects in a two-way layout is excellent, while some textbooks give an erroneous explanation even of this. Miller's book *Beyond ANOVA* (BANOVA; Chapman & Hall/CRC, 1998) intended to go beyond this a long time after Scheffé and succeeded to some extent in bringing new ideas into the book – such as multiple comparison procedures, monotone hypothesis, bootstrap methods, and empirical Bayes. He also gave detailed explanations of the departures from the underlying assumptions in ANOVA – such as non-normality, unequal variances, and correlated errors. So, he gave very nicely the basics of applied statistics. However, I think this would still be insufficient for dealing with real data, especially with regard to the points below, and there is a real need for an advanced book on ANOVA (**AANOVA**). Thus, this book intends to provide some new technologies for data analysis following the precise and exact basic theory of ANOVA.

A Unifying Approach to the Shape and Change-point Hypotheses

The shape hypothesis (e.g., monotone) is essential in dose–response analysis, where a rigid parametric model is usually difficult to assume. It appears also when comparing treatments based on ordered categorical data. Then, the isotonic regression is the most well-known approach to the monotone hypothesis in the normal one-way layout model. It has been, however, introduced rather intuitively and has no obvious optimality for restricted parameter spaces like this. Further, the restricted maximum likelihood approach employed in the isotonic regression is too complicated to extend to non-normal distributions, to the analysis of interaction effects, and also to other shape constraints such as convexity and sigmoidicity. Therefore, in the BANOVA book by Miller, a choice of Abelson and Tukey's maximin linear contrast test is recommended for isotonic inference to escape from the complicated calculations of the isotonic regression. However, such a one-degree-of-freedom contrast test cannot keep high power against the wide range of the monotone hypothesis, even by a careful choice of the contrast. Instead, the author's approach is robust against the wide range of the monotone hypothesis and can be extended in a systematic way to various interesting problems, including analysis of the two-way interaction effects. It starts from a complete class lemma for the tests against the general restricted alternative, suggesting the use of singly, doubly, and triply accumulated statistics as the basic statistics for the monotone, convexity, and sigmoidicity hypotheses, respectively. It also suggests two-way accumulated statistics for two-way data with natural ordering in rows and columns. Two promising statistics derived from these basic statistics are the cumulative chi-squared statistics and the maximal contrast statistics. The cumulative chi-squared is very robust and nicely characterized as a directional goodness-of-fit test statistic. In contrast, the maximal contrast statistic is characterized as an efficient score test for the change-point hypothesis. It should be stressed here that there is a close relationship between the monotone hypothesis and the step change-point model. Actually, each component of the step change-point model is a particular monotone contrast, forming the basis of the monotone hypothesis in the sense that every monotone contrast can be expressed by a unique and positive linear combination of the step change-point contrasts. The unification of the monotone and step change-point hypotheses is also important in practice, since in monitoring the spontaneous reporting of the adverse events of a drug, for example, it is interesting to detect a change point as well as a general increasing tendency of reporting. The idea is extended to convexity and slope change-point models, and sigmoidicity and inflection point models, thus giving a unifying approach to the shape and change-point hypotheses generally. The basic statistics of the newly proposed approach are very simple and have a nice Markov property for elegant and exact probability calculation, not only for the normal distribution but also for the Poisson and multinomial distributions. This approach is of so simple a structure that many of the procedures for a one-way layout model can be extended in a systematic way to two-way data, leading to the two-way accumulated statistics. These approaches have been shown repeatedly to have excellent power (see Chapters 6 to 11 and 13 to 15).

The Analysis of Two-way Data

One of the central topics of data science is the analysis of interactions in the generalized sense. In a narrow sense, interactions are a departure from the additive effects of two factors. However, in the one-way layout the main effects of a treatment also become the interaction effects between the treatment and the response if the response is given by a categorical response instead of quantitative measurements. In this case the data y_{ij} are the frequency of cell (i, j) for the ith treatment and the jth categorical response. If we denote the probability of cell (i, j) by p_{ij}, the treatment effect is a change of the profile $(p_{i1}, p_{i2}, \ldots, p_{ib})$ of the ith treatment, and the interaction effects in terms of p_{ij} are concerned. In this case, however, the naïve additive model is often inappropriate and a log linear model

$$\log p_{ij} = \mu + \alpha_i + \beta_j + (\alpha\beta)_{ij}$$

is assumed. Then, the interaction factor $(\alpha\beta)_{ij}$ denotes the treatment effects. In this sense the regression analysis is also a sort of interaction analysis between the explanation and the response variables. Further, the logit model, the probit model, the independence test of a contingency table, and the canonical correlation analysis are all regarded as a sort of interaction analysis. In previous books, however, interaction analysis has been paid less attention than it deserves, and mainly an overall F- or χ^2- test has been described in the two-way ANOVA. Now, there are several immanent problems in the analysis of two-way data which are not described everywhere.

1. The characteristics of the rows and columns – such as controllable, indicative, variational, and response – should be taken into consideration.

2. The degrees of freedom are often so large that an overall analysis can tell almost nothing about the details of the data. In contrast, the multiple comparison procedures based on one-degree-of-freedom contrasts as taken in BANOVA (1998) are too lacking in power and also the test result is usually unclear.

3. There is often natural ordering in the rows and/or columns, which should be taken into account in the analysis. The isotonic regression is, however, too complicated for the analysis of two-way interaction effects.

In the usual two-way ANOVA with controllable factors in the rows and columns, the purpose of the experiment will be to determine the best combination of the two factors that gives the highest productivity. However, let us consider an example of the international adaptability test of rice varieties, where the rows represent the 44 regions [e.g., Niigata (Japan), Seoul, Nepal, Egypt, and Mexico] and the columns represent the 18 varieties of rice [e.g., Rafaelo, Koshihikari, Belle Patna, and Hybrid]. Then the columns are controllable but the rows are indicative, and the problem is by no means to choose the best combination of row and column as in the usual ANOVA. Instead, the purpose should be to assign an optimal variety to each region. Then, the row-wise multiple comparison procedures for grouping rows with a similar response profile to columns and assigning a common variety to those regions in the same group should be an attractive approach. As another example, let us consider a dose–response analysis based on the ordered categorical data in a phase II clinical trial. Then, the rows represent dose levels and are controllable. The columns are the response variables and the data are characterized by the ordinal rows and columns. Of course, the purpose of the trial is to choose an optimal dose level based on the ordered categorical responses. Then, applying the step change-point contrasts to rows should be an attractive approach to detecting the effective dose. There are several ideas for dealing with the ordered columns, including the two-way accumulated statistics. The approach should be regarded as a sort of profile analysis and can also be applied to the analysis of repeated measurements. These examples show that each of the two-way data requires its own analysis. Indeed, the analysis of two-way data is a rich source of interesting theories and applications (see Chapters 10, 11, 13, and 14).

Multiple Decision Processes

Unification of non-inferiority, equivalence, and superiority tests

Around the 1980s there were several serious problems in the statistical analysis of clinical trials in Japan, among which two major problems were the multiplicity problem and non-significance regarded as equivalence. These were also international problems. The outline of the latter problem is as follows.

In a phase III trial for a new drug application in Japan, the drug used to be compared with an active control instead of a placebo, and admitted for publication if it was evaluated as equivalent to the control in terms of efficacy and safety. Then the problem was that the non-significance by the usual t or Wilcoxon test had long been regarded as proof of equivalence in Japan. This was stupid, since non-significance can so easily be achieved by an imprecise clinical trial with a small sample size. The author (and several others) fought against this, and introduced a non-inferiority test which requires rejecting the handicapped null hypothesis

$$H_{\text{non}0} : p_1 \le p_0 - \Delta$$

against the one-sided alternative

$$H_{\text{non}1} : p_1 > p_0 - \Delta,$$

where p_1 and p_0 are the efficacy rates of the test and control drugs, respectively.

Further, the author found that usually $\Delta = 0.10$, with one-sided significance level 0.05, would be appropriate in the sense that the approximately equal observed efficacy proportions of two drugs will clear the non-inferiority criterion by the usual sample sizes employed in Japanese phase III clinical trials. Actually, the Japanese Statistical Guideline employed the procedure six years in advance of the International Guideline (ICH E9), which employed it in 1998. However, there still remains the problem of how to justify the usual practice of superiority testing after proving non-inferiority. This has been overcome by a unifying approach to non-inferiority and superiority tests based on multiple decision processes. It nicely combines the one- and two-sided tests, replacing the usual simple confidence interval for normal means by a more useful confidence region. It does not require a pre-choice of the non-inferiority or superiority test, or the one- or two-sided test. The procedure gives essentially the power of the one-sided test, keeping the two-sided statistical inference without any prior information (see Chapter 4 and Section 5.4).

Mixed and Random Effects Model

In the factorial experiments, if all the factors except error are fixed effects, it is called a fixed effects model. If the factors are all random except for a general mean, it is called a random effects model. If both types of factor are involved in the experiment, it is called a mixed effects model. In this book mainly fixed effects models are described, but there are cases where it is better to consider the effects of a factor to be random; we discuss basic ideas regarding mixed and random effects models in Chapter 12. In particular, the recent development of the mixed effects model in the engineering field profile analysis is introduced in Chapter 13. There is a factor like the variation factor which is dealt with as fixed in the laboratory, but acts as if it were random in the extension to the real world. Therefore, this is a problem of interpretation of data rather than of mathematics (see Chapters 12 and 13).

Software and Tables

The algorithms for calculating the p-value of the maximal contrast statistics introduced in this book have been developed widely and extensively by my colleague and I decided to support some of them on my website. They are based on Markov properties of the component statistics. As described in the text, they are simple in principle; the reader is also recommended to develop their own algorithms. Presently, the probabilities of popular distributions such as the normal, t, F, and chi-squared are obtained very easily on the Internet (see keisan.casio.com, for example), so only a few tables are given in the Appendix, which are not available everywhere. Among them, Tables A and B are original ones calculated by the proposed algorithm.

Examples

Finally it should be stressed that all the newly introduced methods have originated from real problems which the author experienced in his activities in the real field of statistical quality control, clinical trials, and the evaluation of the New Drug Application from the regulatory side. There are naturally plenty of real examples supplied in this book, compared with previous books. Also, this book is not restricted to ANOVA in the narrow sense, but extends these methodologies to discrete data (including contingency tables). Thus, the book intends to provide some advanced techniques for applied statistics beyond the previous elementary books for ANOVA.

Acknowledgments

I would like to thank first the late Professor Tetsuichi Asaka for inviting me to take an interest in applied statistics through the real field of quality control. I would also like to thank Professor Kei Takeuchi for inviting me to study statistical methods based on rigid mathematical statistics. I would also like to thank Sir Professor David Cox for sharing his interest in the wide range of statistical methods available. In particular, my stay at Imperial College in 1978 when visiting him was stimulating and had a significant impact on my later career. The publication of this book is itself due to his encouragement. I would like to thank my research colleagues in foreign countries – Muni Srivastava, Fortunate Pesarin, Ludwig Hothorn, and Stanislaw Mejza in particular – for long-term discussions and also some direct comments on the draft of this book.

I must not forget to thank my students at the University of Tokyo, including Tetsuhisa Miwa, Hiroe Tsubaki, and Satoshi Kuriki, but they are too many to mention one by one. The long and heated discussions with them at seminars were indeed helpful for me to widen and deepen my interest in both theoretical and applied statistics. In particular, as seen in the References, most of my papers published after 2000 are co-authored with these students.

My research would never have been complete without the help of my colleagues who developed various software supporting the newly proposed statistical methods. They include Kenji Nishihara, Shoichi Yamamoto, and most recently Harukazu Tsuruta, who succeeded and extended these algorithms which had been developed for a long time. He also read carefully an early draft of this book and gave many valuable comments, as well as a variety of technical support in preparing this book. For technical support, I would also like to thank Yasuhiko Nakamura and Hideyasu Karasawa.

Financially, my research has been supported for a long time by a grant-in-aid for scientific research of the Japan Society for Promotion of Science. My thanks are also due to Meisei University who provided me with a laboratory and managerial support for a long time, and even after my retirement from the faculty. My thanks are also due to the Wiley Executive Statistics Editor Jon Gurstelle, Project Editor Divya Narayanan, Production Editor Vishnu Priya and other staff for their help and useful suggestions in publishing this book. Finally, thanks to my wife Mitsuko who helped me in calculating Tables 10.10 and 12.4 a long time ago and for continuous support of all kinds since then.

Chihiro Hirotsu
Tokyo, March 2017

Notation and Abbreviations

Notation

Asterisks on the number (e.g., 2.23* or 3.12**):	statistical significance at level 0.05 or 0.01
Column vector:	bold lowercase italic letter, $\boldsymbol{v}$
Matrix:	bold uppercase italic letter: $\boldsymbol{M}$
Transpose of vector and matrix:	$\boldsymbol{v}'$, $\boldsymbol{M}'$

Observation vector: one-way,

$$\boldsymbol{y} = (y_{11}, y_{12}, \ldots, y_{1n_1}, y_{21}, \ldots, y_{2n_2}, \ldots, y_{a1}, \ldots, y_{an_a})'$$
$$= (\boldsymbol{y}_1', \boldsymbol{y}_2', \ldots, \boldsymbol{y}_a')', \; \boldsymbol{y}_i = (y_{i1}, \ldots, y_{in_i})'$$

two-way,

$$\boldsymbol{y} = (y_{111}, y_{112}, \ldots, y_{11n}, y_{121}, \ldots, y_{12n}, \ldots, y_{ab1}, \ldots, y_{abn})'$$
$$= (\boldsymbol{y}_{11}', \boldsymbol{y}_{12}', \ldots, \boldsymbol{y}_{ab}')', \; \boldsymbol{y}_{ij} = (y_{ij1}, \ldots, y_{ijn})'$$

Dot and bar notation: one-way,

$$y_{i\cdot} = \sum_{j=1}^{n_i} y_{ij}, \; \bar{y}_{i\cdot} = \sum_{j=1}^{n_i} y_{ij}/n_i$$

two-way,

$$y_{i\cdot k} = \sum_{j=1}^{b} y_{ijk}, \; \bar{y}_{i\cdot k} = \sum_{j=1}^{b} y_{ijk}/b$$
$$y_{i\cdot\cdot} = \sum_{j=1}^{b}\sum_{k=1}^{n} y_{ijk}, \; \bar{y}_{i\cdot\cdot} = \sum_{j=1}^{b}\sum_{k=1}^{n} y_{ijk}/(bn)$$

Dot and bar notation in vectors: one-way, $\boldsymbol{y}_{(i)\cdot} = (y_{1\cdot}, y_{2\cdot}, \ldots, y_{a\cdot})'$, $\bar{\boldsymbol{y}}_{(i)\cdot} = (\bar{y}_{1\cdot}, \bar{y}_{2\cdot}, \ldots, \bar{y}_{a\cdot})'$

two-way,

$$\boldsymbol{y}_{(i)(j)\cdot} = (y_{11\cdot}, y_{12\cdot}, \ldots, y_{ab\cdot})', \bar{\boldsymbol{y}}_{(i)(j)\cdot} = (\bar{y}_{11\cdot}, \bar{y}_{12\cdot}, \ldots, \bar{y}_{ab\cdot})',$$

$$\boldsymbol{y}_{(i)\cdot\cdot} = (y_{1\cdot\cdot}, y_{2\cdot\cdot}, \ldots, y_{a\cdot\cdot})', \bar{\boldsymbol{y}}_{(i)\cdot\cdot} = (\bar{y}_{1\cdot\cdot}, \ldots, \bar{y}_{a\cdot\cdot})',$$

$$\boldsymbol{y}_{i\cdot} = (y_{i1\cdot}, y_{i2\cdot}, \ldots, y_{ib\cdot})' = \boldsymbol{y}_{i(j)\cdot}$$

$\boldsymbol{0}_n$: a zero vector of size n, the suffix is omitted when it is obvious

$\boldsymbol{j}_n$: a vector of size n with all elements unity, the suffix is omitted when it is obvious

$\boldsymbol{I}_n$: an identity matrix of size n, the suffix is omitted when it is obvious

$\boldsymbol{P}'_a$: an $(a-1) \times a$ orthonormal matrix satisfying $\boldsymbol{P}'_a\boldsymbol{P}_a = \boldsymbol{I}_{a-1}$, $\boldsymbol{P}_a\boldsymbol{P}'_a = \boldsymbol{I}_a - \boldsymbol{j}_a\boldsymbol{j}'_a$

$|\boldsymbol{A}|$: determinant of a matrix $\boldsymbol{A}$

$\mathrm{tr}(\boldsymbol{A})$: trace of a matrix $\boldsymbol{A}$

$\|\boldsymbol{v}\|^2$: squared norm of a vector $\boldsymbol{v} = (v_1, \ldots, v_n)'$: $\|\boldsymbol{v}\|^2 = v_1^2 + \cdots + v_n^2$

$D = \mathrm{diag}(\lambda_i)$, $i = 1, \ldots, n$ and D^ν: a diagonal matrix with diagonal elements $\lambda_1, \ldots, \lambda_n$ arranged in dictionary order and $D^\nu = \mathrm{diag}(\lambda_i^\nu)$

$A \otimes B$: direct (Kronecker's) product of two matrices

$$\boldsymbol{A} = \begin{bmatrix} a_{11} \cdots a_{1l} \\ \vdots \\ a_{k1} \cdots a_{kl} \end{bmatrix} \text{ and } \boldsymbol{B} = [b_{ij}] : \boldsymbol{A} \otimes \boldsymbol{B} = \begin{bmatrix} a_{11}\boldsymbol{B} \cdots a_{1l}\boldsymbol{B} \\ \vdots \\ a_{k1}\boldsymbol{B} \cdots a_{kl}\boldsymbol{B} \end{bmatrix}$$

Kronecker's delta δ_{ij}: $\delta_{ij} = 1$ if $i = j$, 0 otherwise

$a \cong b$: a is nearly equal to b

$N(\mu, \sigma^2)$: normal distribution with mean μ and variance σ^2

$N(\boldsymbol{\mu}, \boldsymbol{\Omega})$: multivariate normal distribution with mean $\boldsymbol{\mu}$ and variance–covariance matrix $\boldsymbol{\Omega}$

z_α: upper α point of standard normal distribution $N(0, 1)$

$t_\nu(\alpha)$: upper α point of t-distribution with degrees of freedom ν

$\chi^2_\nu(\alpha)$: upper α point of χ^2-distribution with degrees of freedom ν

$F_{\nu_1,\nu_2}(\alpha)$: upper α point of F-distribution with degrees of freedom (ν_1, ν_2)

$q_{a,\,\nu}(\alpha)$: upper α point of Studentized range

$B(n, p)$: binomial distribution

$M(n, \boldsymbol{p})$: multinomial distribution

$H(y \mid R_1, C_1, N)$:	hypergeometric distribution
$MH(y_{ij} \mid y_{i\cdot}, y_{\cdot j})$, $MH(y_{ij} \mid R_i, C_j, N)$:	multivariate hypergeometric distribution for two-way data
$f(\mathbf{y}, \boldsymbol{\theta})$ and $p(\mathbf{y}, \boldsymbol{\theta})$:	density function and probability function
$\Pr(A), \Pr\{A\}$:	probability of event A
$L(\mathbf{y}, \boldsymbol{\theta}), L(\boldsymbol{\theta})$:	likelihood function
$E(y)$ and $E(\mathbf{y})$:	expectation
$E(y \mid B)$ and $E(\mathbf{y} \mid B)$:	conditional exception given B
$V(y)$ and $V(\mathbf{y})$:	variance and variance–covariance matrix
$V(y \mid B)$ and $V(\mathbf{y} \mid B)$:	conditional variance and variance- covariance matrix given B
$Cov(x, y)$:	covariance
$Cor(x, y)$:	correlation
$I_n(\theta)$:	Fisher's amount of information
$\boldsymbol{I}_n(\boldsymbol{\theta})$:	Fisher's information matrix

Abbreviations

ANOVA:	analysis of variance
BIBD:	balanced incomplete block design
BLUE:	best linear unbiased estimator
BLUP:	best linear unbiased predictor
df:	degrees of freedom
FDA:	Food and Drug Administration (USA)
ICH E9:	statistical principles for clinical trials by International Conference on Harmonization
LS:	least squares
MLE:	maximum likelihood estimator
MSE:	mean square error
PMDA:	Pharmaceutical and Medical Device Agency of Japan
REML:	residual maximum likelihood
SD:	standard deviation
SE:	standard error
SLB:	simultaneous lower bound
SN ratio:	signal-to-noise ratio
WLS:	weighted least squares

1

Introduction to Design and Analysis of Experiments

1.1 Why Simultaneous Experiments?

Let us consider the problem of estimating the weight μ of a material W using four measurements by a balance. The statistical model for this experiment is written as

$$y_i = \mu + e_i,\ i = 1, 2, 3, 4,$$

where the e_i are uncorrelated with expectation zero (unbiasedness) and equal variance σ^2. Then, a natural estimator

$$\hat{\mu} = \bar{y}_{\cdot} = (y_1 + y_2 + y_3 + y_4)/4$$

is an unbiased estimator of μ with minimum variance $\sigma^2/4$ among all the linear unbiased estimators of μ. Further, if the normal distribution is assumed for the error e_i, then $\hat{\mu}$ is the minimum variance unbiased estimator of μ among all the unbiased estimators, not necessarily linear.

In contrast, when there are four unknown means $\mu_1, \mu_2, \mu_3, \mu_4$, we can estimate all the μ_i with variance $\sigma^2/4$ and unbiasedness simultaneously by the same four measurements. This is achieved by measuring the total weight and the differences among the μ_i's according to the following design, where $\pm$ means putting the material on the right or left side of the balance:

$$\begin{aligned} y_1 &= \mu_1 + \mu_2 + \mu_3 + \mu_4 + e_1, \\ y_2 &= \mu_1 + \mu_2 - \mu_3 - \mu_4 + e_2, \\ y_3 &= \mu_1 - \mu_2 + \mu_3 - \mu_4 + e_3, \\ y_4 &= \mu_1 - \mu_2 - \mu_3 + \mu_4 + e_4. \end{aligned} \tag{1.1}$$

Advanced Analysis of Variance, First Edition. Chihiro Hirotsu.

Then, the estimators

$$\hat{\mu}_1 = (y_1 + y_2 + y_3 + y_4)/4,$$
$$\hat{\mu}_2 = (y_1 + y_2 - y_3 - y_4)/4,$$
$$\hat{\mu}_3 = (y_1 - y_2 + y_3 - y_4)/4,$$
$$\hat{\mu}_4 = (y_1 - y_2 - y_3 + y_4)/4.$$

are the best linear unbiased estimators (BLUE; see Section 2.1), each with variance $\sigma^2/4$. Therefore, a naïve method to replicate four measurements for each μ_i to achieve variance $\sigma^2/4$ is a considerable waste of time. More generally, when the number of measurements n is a multiple of 4, we can form the unbiased estimator of all n weights with variance σ^2/n. This is achieved by applying a Hadamard matrix for the coefficients of μ_i's on the right-hand side of equation (1.1) (see Section 15.3 for details, as well as the definition of a Hadamard matrix).

1.2 Interaction Effects

Simultaneous experiments are not only necessary for the efficiency of the estimator, but also for detecting interaction effects. The data in Table 1.1 show the result of 16 experiments (with averages in parentheses) for improving a printing machine with an aluminum plate. The measurements are fixing time (s); the shorter, the better. The factor F is the amount of ink and G the drying temperature. The plots of averages are given in Fig. 1.1.

From Fig. 1.1, (F_2, G_1) is suggested as the best combination. On the contrary, if we compare the amount of ink first, fixing at the drying temperature 280°C (G_2), we shall erroneously choose F_1. Then we may fix the ink level at F_1 and try to compare the drying temperature. We may reach the conclusion that (F_1, G_2) should be an optimal combination without trying the best combination, (F_2, G_1). In this example the optimal level of ink is reversed according to the levels G_1 and G_2 of the other factor. If there is such an interaction effect between the two factors, then a one-factor-at-a-time

Table 1.1 Fixing time of special aluminum printing.

	Temperature							
Ink supply	G_1 : 170°C				G_2 : 280°C			
F_1 : large	5.9	3.7	4.6	4.4	5.7	5.0	4.9	2.1
		(4.65)				(4.43)		
F_2 : small	4.7	3.3	4.5	1.0	8.2	5.9	10.7	8.5
		(3.38)				(8.33)		

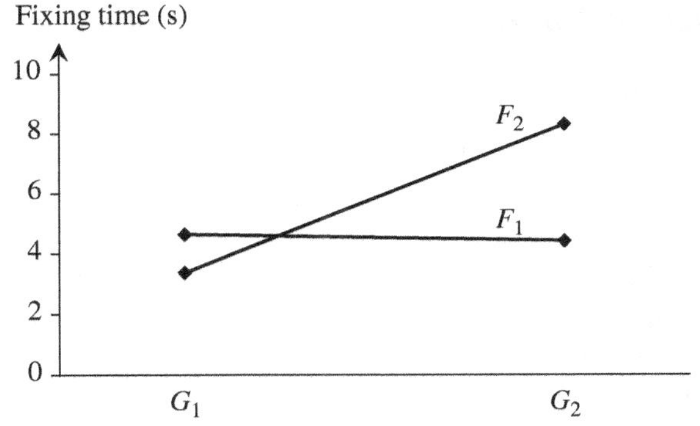

Figure 1.1 Average plots at (F_i, G_j).

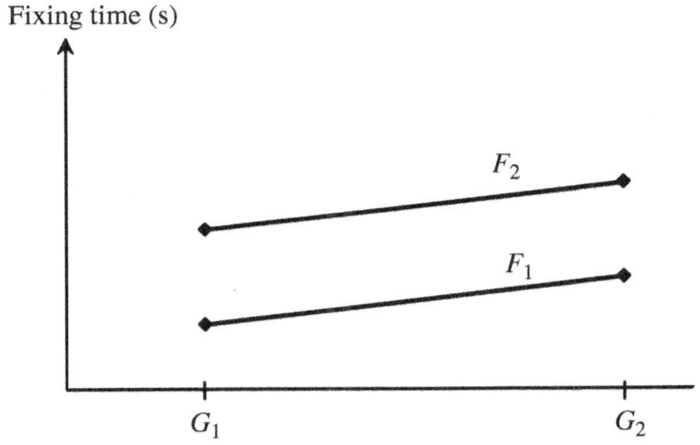

Figure 1.2 No interaction.

experiment will fail to find the optimal combination. In contrast, if there is no such interaction effect, then the effects of the two factors are called additive. In this case, denoting the mean for the combinations (F_i, G_j) by μ_{ij}, the equation

$$\mu_{ij} = \bar{\mu}_{i\cdot} + \bar{\mu}_{\cdot j} - \bar{\mu}_{\cdot\cdot} \tag{1.2}$$

holds, where the dot and overbar denote the sum and average with respect to the suffix replaced by the dot throughout the book. Therefore, $\bar{\mu}_{\cdot\cdot}$ implies the overall average (general mean), for example. If equation (1.2) holds, then the plot of the averages becomes like that in Fig. 1.2. Although in this case a one-factor-at-a-time experiment will also reach the correct decision, simultaneous experiments to detect the interaction effects are strongly recommended in the early stage of the experiment.

1.3 Choice of Factors and Their Levels

A cause affecting the target value is called a factor. Usually, there are assumed to be many affecting factors at the beginning of an experiment. To write down all those factors, a 'cause-and-effect diagram' like in Fig. 1.3 is useful. This uses the thick and thin bones of a fish to express the rough and detailed causes, arranged in order of operation. In drawing up the diagram it is necessary to collect as many opinions as possible from the various participants in the different areas. However, it is impossible to include all factors in the diagram at the very beginning of the experiment, so it is necessary to examine the past data or carry out some preliminary experiments. Further, it is essential to obtain as much information as possible on the interaction effects among those factors. For every factor employed in the experiment, several levels are set up – such as the place of origin of materials $A_1, A_2, \cdots$ and the reaction temperature 170°C, 280°C, $\cdots$. The levels of the nominal variable are naturally determined by the environment of the experiment. However, choosing the levels of the quantitative factor is rather arbitrary. Therefore, sometimes sequential experiments are required first to outline the response surface roughly then design precise experiments near the suggested optimal points. In Fig. 1.1, for example, the optimal level of temperature G with respect to F_2 is unknown – either below G_1 or between G_1 and G_2. Therefore, in the first stage of the experiment, it is desirable to design the experiment so as to obtain an outline of the response curve. The choice of factors and their levels are discussed in more detail in Cox (1958).

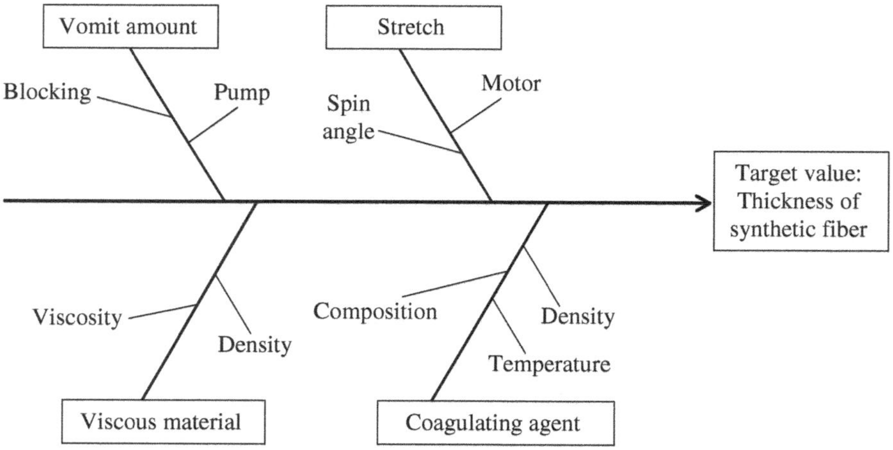

Figure 1.3 Cause-and-effect diagram.

1.4 Classification of Factors

This topic is discussed more in Japan than in other countries, and we follow here the definition of Takeuchi (1984).

(1) **Controllable factor.** The level of the controllable factor can be determined by the experimenter and is reproducible. The purpose of the experiment is often to find the optimal level of this factor.

(2) **Indicative factor.** This factor is reproducible but not controllable by the experimenter. The region in the international adaptability test of rice varieties is a typical example, while the variety is a controllable factor. In this case the region is not the purpose of the optimal choice, and the purpose is to choose an optimal variety for each region – so that an interaction analysis between the controllable and indicative factors is of major interest.

(3) **Covariate factor.** This factor is reproducible but impossible to define before the experiment. It is known only after the experiment, and used to enhance the precision of the estimate of the main effects by adjusting its effect. The covariate in the analysis of covariance is a typical example.

(4) **Variation (noise) factor.** This factor is reproducible and possible to specify only in laboratory experiments. In the real world it is not reproducible and acts as if it were noise. In the real world it is quite common for users to not follow exactly the specifications of the producer. For example, a drug for an infectious disease may be used before identifying the causal germ intended by the producer, or administered to a subject with some kidney difficulty who has been excluded in the trial. Such a factor is called a noise factor in the Taguchi method.

(5) **Block factor.** This factor is not reproducible but can be introduced to eliminate the systematic error in fertility of land or temperature change with passage of time, for example.

(6) **Response factor.** This factor appears typically as a categorical response to a contingency table and there are two important cases: nominal and ordinal. The response is usually not called a factor, but mathematically it can be regarded and dealt with as a factor, with categories just like levels.

One should also refer to Cox (1958) for a classification of the factors from another viewpoint.

1.5 Fixed or Random Effects Model?

Among the factors introduced in Section 1.4, the controllable, indicative and covariate factors are regarded as fixed effects. The variation factor is dealt with as fixed in the laboratory but dealt with as random in extending laboratory results to the real world.

Therefore, the levels specified in the laboratory should be wide enough to cover the wide range of real applications. The block is premised to have no interaction with other factors, so that the treatment either as fixed or random does not affect the result. However, it is necessarily random in the recovery of inter-block information in the incomplete block design (see Section 9.2).

The definition of fixed and random effects models was first introduced by Eisenhart (1947), but there is also the comment that these are mathematically equivalent and the definitions are rather misleading. Although it is a little controversial, the distinction of fixed and random still seems to be useful for the interpretation and application of experimental results, and is discussed in detail in Chapters 12 and 13.

1.6 Fisher's Three Principles of Experiments vs. Noise Factor

To compare the treatments in experiments, Fisher (1960) introduced three principles: (1) randomization, (2) replication and (3) local control.

To explain randomization, Fisher introduced the sensory test of tasting a cup of tea made with milk. The problem then is to know whether it is true or not that a lady can declare correctly whether the milk or the tea infusion was added to the cup first. The experiment consists of mixing eight cups of tea, four in one way and four in the other, and presenting them to the subject for judgment. There are, however, numerous uncontrollable causes which may influence the result: the requirement that all the cups are exactly alike is impossible; the strength of the tea infusion may change between pouring the first and last cup; and the temperature at which the tea is tasted will change in the course of the experiment. One procedure that is used to escape from such systematic noise is to randomize the order of the eight cups for tasting. This process converts the systematic noise to random error, giving the basis of statistical inference.

Secondly, it is necessary to replicate the experiments to raise the sensitivity of comparison. It is also necessary to separate and evaluate the noise from treatment effects, since the outcomes of experiments under the same experimental conditions can vary due to unknown noise. The treatment effects of interest should be beyond such random fluctuations, and to ensure this several replications of experiments are necessary to evaluate the effects of noise.

Local control is a technique to ensure homogeneity within a small area for comparing treatments by splitting the total area with large deviations of noise. In field experiments for comparing a plant varieties, the whole area is partitioned into n blocks so that the fertility becomes homogeneous within each block. Then, the precision of comparisons is improved compared with randomized experiments of all an treatments.

Fisher's idea to enhance the precision of comparisons is useful in laboratory experiments in the first stage of research development. However, in a clinical trial for comparing antibiotics, for example, too rigid a definition of the target population and the causal germs may not coincide with real clinical treatment. This is because, in the real world, antibiotics may be used by patients with some kidney trouble who might be excluded from the trial, by older patients beyond the range of the trial, before identifying the causal germ exactly, or with poor compliance of the taking interval. Therefore, in the final stage of research development it is required to introduce purposely variations in users and environments in the experiments to achieve a robust product in the real world. It should be noted here that the purpose of experiments is not to know all about the sample, but to know all about the background population from which the sample is taken – so the experiment should be designed to simulate or represent well the target population.

1.7 Generalized Interaction

A central topic of data science is the analysis of interaction in a generalized sense. In a narrow sense, it is the departure from the additive effects of two factors. If the effect of one factor differs according to the levels of the other factor, then the departure becomes large (as in the example of Section 1.2).

In the one-way layout also, the main effects of a treatment become the interaction between the treatment and the response if the response is given by a categorical response instead of quantitative measurements. In this case, the data y_{ij} are the frequency of the (i, j) cell for the ith treatment and the jth categorical response. If we denote the probability of cell (i, j) by p_{ij}, then the treatment effect is a change in the profile $(p_{i1}, p_{i2}, \dots, p_{ib})$ of the ith treatment and the interaction effects in terms of p_{ij} are concerned. In this case, however, a naïve additive model like (1.2) is often inappropriate, and the log linear model

$$\log p_{ij} = \mu + \alpha_i + \beta_j + (\alpha\beta)_{ij}$$

is assumed. Then, the factor $(\alpha\beta)_{ij}$ denotes the ith treatment effect. In this sense, the regression analysis is also a sort of interaction analysis between the explanation and the response variables. Further, the logit model, probit model, independence test of a contingency table, and canonical correlation analysis are all regarded as a sort of interaction analysis. One should also refer to Section 7.1 regarding this idea.

1.8 Immanent Problems in the Analysis of Interaction Effects

In spite of its importance, the analysis of interaction is paid much less attention than it deserves, and often in textbooks only an overall F- or χ^2-test is described. However, the degrees of freedom for interaction are usually large, and such an overall test cannot tell any detail of the data – even if the test result is highly significant. The degrees of freedom are explained in detail in Section 2.5.5. In contrast, the multiple comparison procedure based on one degree of freedom statistics is far less powerful and the interpretation of the result is usually unclear. Usually in the text books it is recommended to estimate the combination effect μ_{ij} by the cell mean $\bar{y}_{ij.}$, if the interaction exists. However, it often occurs that only a few degrees of freedom can explain the interaction very well, and in this case we can recover information for μ_{ij} from other cells and improve the naïve estimate $\bar{y}_{ij.}$ of μ_{ij}. This also implies that it is possible to separate the essential interaction from the noisy part without replicated experiments. Further, the purpose of the interaction analysis has many aspects – although the textbooks usually only describe how to find an optimal combination of the controllable factors. In this regard the classification of factors plays an essential role (see Chapters 10, 11, 13, and 14).

1.9 Classification of Factors in the Analysis of Interaction Effects

In case of a two-factor experiment, one factor should be controllable since otherwise the experiment cannot result in any action. In case of controllable vs. controllable, the purpose of the experiment will be to specify the optimal combination of the levels of those two factors for the best productivity. Most of the textbooks describe this situation. However, the usual F test is not useful in practice, and the simple interaction model derived from the multiple comparison approach would be more useful.

In case of controllable vs. indicative, the indicative factor is not the object of optimization but the purpose is to specify the optimal level of the controllable factor for each level of the indicative factor. In the international adaptability test of rice varieties, for example, the purpose is obviously not to select an overall best combination but to specify an optimal variety (controllable) for each region (indicative). Then, it should be inconvenient to hold an optimal variety for each of a lot of regions in the world, and the multiple comparison procedure for grouping regions with similar response profiles is required.

The case of controllable vs. variation is most controversial. If the purpose is to maximize the characteristic value, then the interaction is a sort of noise in extending the laboratory result to the real world, where the variation factor cannot be specified rigidly and may take diverse levels. Therefore, it is necessary to search for a robust level of the controllable factor to give a large and stable output beyond the random fluctuations of the variation factor. Testing main effects by interaction effects in the mixed effects model of controllable vs. variation factors is one method in this line (see Section 12.3.5).

1.10 Pseudo Interaction Effects (Simpson's Paradox) in Categorical Data

In case of categorical responses, the data are presented as the number of subjects satisfying a specified attribute. Binary (1, 0) data with or without the specified attribute are a typical example. In such cases it is controversial how to define the interaction

Table 1.2 Simpson's paradox.

	Young ($j=1$)		Old ($j=2$)		Young + old	
	($k=1$)	($k=2$)	($k=1$)	($k=2$)	($k=1$)	($k=2$)
Drug ($i=1$)	120	40	10	30	130	70
Drug ($i=2$)	30	10	40	120	70	130

effects, see Darroch (1974). In most cases an additive model is inappropriate, and is replaced by a multiplicative model. The numerical example in Table 1.2 will explain well how the additive model is inappropriate, where ($k=1$) denotes useful and ($k=2$) useless. In Table 1.2 it is obvious that drug 1 and drug 2 are equivalent in usefulness for each of the young and old patients, respectively. Therefore, it seems that the two drugs should be equivalent for (young + old) patients. However, the collapsed subtable for all the subjects apparently suggests that drug 1 is better than drug 2. This contradiction is known as Simpson's paradox (Simpson, 1951), and occurs by additive operation according to the additive model of the drug and age effects. The correct interpretation of the data is that both drugs are equally useful for young patients and equally useless for old patients. Drug 1 is employed more frequently for young patients (where the useful cases are easily obtained) than old patients, and as a result the useful cases are seen more in drug 1 than drug 2. By applying the multiplicative model we can escape from this erroneous conclusion (Fienberg, 1980) – see Section 14.3.2 (1).

1.11 Upper Bias by Statistical Optimization

As a simple example, suppose we have random samples $y_{11}, \ldots, y_{1n}$ and $y_{21}, \ldots,$ y_{2n} from the normal population $N(\mu_1, \sigma^2)$ and $N(\mu_2, \sigma^2)$, respectively, where $\mu_1 = \mu_2 = \mu$. Then, if we select the population corresponding to the maximum of $\bar{y}_{1\cdot}$ and $\bar{y}_{2\cdot}$, and estimate the population mean by the maximal sample, an easy calculation leads to

$$E\{\max(\bar{y}_{1\cdot}, \bar{y}_{2\cdot})\} = \mu + \sigma/\sqrt{n\pi}$$

showing the upper bias as an estimate of the population mean μ. The bias is induced by treating the sample employed for selection (optimization) as if it were a random sample for estimation; this is called selection bias.

A similar problem inevitably occurs in variable selection in the linear regression model, see Copas (1983), for example. It should be noted here again that the purpose of the data analysis is not to explain well the current data, but to predict what will happen in the future based on the current data. The estimation based on the data employed for optimization is too optimistic to predict the future. Thus, the Akaike's information criterion (AIC) approach or penalized likelihood is justified. One should also refer to Efron and Tibshirani (1993) for the bootstrap as a non-parametric method of model validation.

1.12 Stage of Experiments: Exploratory, Explanatory or Confirmatory?

Finally, most important in designing experiments is to define the target of the experiments clearly. For this purpose it is useful to define the three stages of experiments. The first stage is exploratory, whose purpose is to discover a promising hypothesis in the actual science – such as industry and clinical medicine. At this stage the exploring data analysis, analysis of variance, regression analysis, and many other statistical methods are applied. Data dredging is allowed to some extent, but it is most inappropriate to take the result as a conclusion. This stage only proposes some interesting hypotheses, which should be confirmed in the following stages. The second stage is explanatory, whose purpose is to clarify the hypothesis under rigid experimental conditions. The design and analysis of experiments following Fisher's principle will be successfully applied. The third stage is confirmatory, whose purpose is to confirm that the result of laboratory experiments is robust enough in the actual world. The robust design of Taguchi is useful here. It should be noted that in these stages of experiments, the essence of the statistical method for summarizing and analyzing data does not change; the change is in the interpretation and degree of confidence of the analytical results. Finally, follow-up analysis of the post-market data is inevitable, since it is impossible to predict all that will happen in the future by pre-market research, even if the most precise and detailed experiments were performed.

References

Copas, J. B. (1983) Regression, prediction, shrinkage. *J. Roy. Statist. Soc.* **B45**, 311–354.

Cox, D. R. (1958) *Planning of experiments*. Wiley, New York.

Darroch, J. N. (1974) Multiplicative and additive interaction in contingency tables. *Biometrika* **61**, 207–214.

Efron, B. and Tibshirani, R. (1993) *An introduction to bootstrap*. Chapman & Hall, New York.

Eisenhart, C. (1947) The assumptions underlying the analysis of variance. *Biometrics* **3**, 1–21.

Fienberg, S. E. (1980) *The analysis of cross classified data*. MIT Press, Boston, MA.

Fisher, R. A. (1960) *The design of experiments*, 7th edn. Oliver & Boyd, Edinburgh.

Simpson, E. H. (1951) The interpretation of interaction in contingency tables. *J. Roy. Statist. Soc.* **B13**, 238–241.

Takeuchi, K. (1984) Classification of factors and their analysis in the factorial experiments. *Kyoto University Research Information Repository* **526**, 1–12 (in Japanese).

2

Basic Estimation Theory

Methods for extracting some systematic variation from noisy data are described. First, some basic theorems are given. Then, a linear model to explain the systematic part and the least squares (LS) method for analyzing it are introduced. The principal result is the best linear unbiased estimator (BLUE). Other important topics are the maximum likelihood estimator (MLE) for a generalized linear model and sufficient statistics.

2.1 Best Linear Unbiased Estimator

Suppose we have a simple model for estimating a weight μ by n experiments,

$$y_i = \mu + e_i,\ i = 1, \ldots, n. \tag{2.1}$$

Then μ is a systematic part and the e_i represent random error. It is the work of a statistician to specify μ out of the noisy data. Maybe most people will intuitively take the sample mean $\bar{y}_.$ as an estimate for μ, but it is by no means obvious for $\bar{y}_.$ to be a good estimator in any sense. Of course, under the assumptions (2.4) ~ (2.6) of unbiasedness, equal variance and uncorrelated error, $\bar{y}_.$ converges to μ in probability by the law of large numbers. However, there are many other estimators that can satisfy such a consistency requirement in large data.

There will be no objection to declaring that the estimator $T_1(\boldsymbol{y})$ is a better estimator than $T_2(\boldsymbol{y})$ if, for any $\gamma_1, \gamma_2 \geq 0$,

$$\Pr\{\mu - \gamma_1 \leq T_1(\boldsymbol{y}) \leq \mu + \gamma_2\} \geq \Pr\{\mu - \gamma_1 \leq T_2(\boldsymbol{y}) \leq \mu + \gamma_2\} \tag{2.2}$$

Advanced Analysis of Variance, First Edition. Chihiro Hirotsu.

holds, where $\boldsymbol{y}=(y_1, \ldots, y_n)'$ denotes an observation vector and the prime implies a transpose of a vector or a matrix throughout this book. A vector is usually a column vector and expressed by a bold-type letter. However, there exists no estimator which is best in this criterion uniformly for any unknown value of μ. Suppose, for example, a trivial estimator $T_3(\boldsymbol{y}) \equiv \mu_0$ that specifies $\mu=\mu_0$ for any observation $\boldsymbol{y}$. Then it is a better estimator than any other estimator when μ is actually μ_0, but it cannot be a good estimator when actually μ is not equal to μ_0. Therefore, let us introduce a criterion of mean squared error (MSE):

$$E\{T(\boldsymbol{y})-\mu\}^2.$$

This is a weaker condition than (2.2), since if equation (2.2) holds, then we obviously have $E\{T_1(\boldsymbol{y})-\mu\}^2 \le E\{T_2(\boldsymbol{y})-\mu\}^2$. However, in this criterion too the trivial estimator $T_3(\boldsymbol{y}) \equiv \mu_0$ becomes best, attaining MSE $=0$ when $\mu=\mu_0$. Therefore, we further request the estimator to be unbiased:

$$E\{T(\boldsymbol{y})\}=\mu \tag{2.3}$$

for any μ, and consider minimizing the MSE under the unbiased condition (2.3). Then, the MSE is nothing but a variance. If such an estimator exists, we call it a minimum variance (or best) unbiased estimator. If we restrict to the linear estimator $T(\boldsymbol{y})=\boldsymbol{l}'\boldsymbol{y}$, the situation becomes easier. Let us assume

$$E(e_i)=0,\ i=1, \ldots, n\,(\text{unbiased}), \tag{2.4}$$

$$V(e_i)=\sigma^2,\ i=1, \ldots, n\,(\text{equal variance}), \tag{2.5}$$

$$Cov(e_i, e_{i'})=0,\ i, i'=1, \ldots, n; i \ne i'\,(\text{uncorrelated}) \tag{2.6}$$

naturally for the error, then the problem of the BLUE is formulated as minimizing $V(\boldsymbol{l}'\boldsymbol{y})$ under the condition $E(\boldsymbol{l}'\boldsymbol{y})=\mu$. Mathematically, it reduces to minimizing $\boldsymbol{l}'\boldsymbol{l}=\sum_i l_i^2$ subject to $\boldsymbol{l}'\boldsymbol{j}_n=\sum_i l_i=1$, where $\boldsymbol{j}_n=(1, \ldots, 1)'$ is an n -dimensional column vector of unity throughout this book and the suffix is omitted if it is obvious. This can be solved at once, giving $\boldsymbol{l}=n^{-1}\boldsymbol{j}$. Namely, $\bar{y}_.$ is a BLUE of μ. The BLUE is obtained generally by the LS method of Section 2.3, without solving the respective minimization problem.

2.2 General Minimum Variance Unbiased Estimator

If μ is a median in model (2.1), then there are many non-linear estimators, like the sample median $\tilde{y}$ and the Hodges–Lehman estimator, the median of all the combinations $(y_i+y_j)/2$, and it is still not obvious in what sense $\bar{y}_.$ is a good estimator. If we assume a normal distribution of error in addition to the conditions (2.4) ~ (2.6), then the sample mean $\bar{y}_.$ is a minimum variance unbiased estimator among all unbiased estimators, called the best unbiased estimator. There are various ways to prove this, and we apply Rao's theorem here. Later, in Section 2.5, another proof based on sufficient statistics will be given.

Theorem 2.1. Rao's theorem. Let $\boldsymbol{\theta}$ be an unknown parameter vector of the distribution of a random vector $\boldsymbol{y}$. Then a necessary and sufficient condition for an unbiased estimator $\hat{g}$ of a function $g(\boldsymbol{\theta})$ of $\boldsymbol{\theta}$ to be a minimum variance unbiased estimator is that $\hat{g}$ is uncorrelated with every unbiased estimator $h(\boldsymbol{y})$ of zero.

Proof

Necessity: For any unbiased estimator $h(\boldsymbol{y})$ of zero, a linear combination $\hat{g}+\lambda h$ is also an unbiased estimator of $g(\boldsymbol{\theta})$. Since its variance is

$$V(\hat{g}+\lambda h)=V(\hat{g})+2\lambda Cov(\hat{g}, h)+\lambda^2 V(h),$$

we can choose λ so that $V(\hat{g}+\lambda h)\le V(\hat{g})$, improving the variance of $\hat{g}$ unless $Cov(\hat{g}, h)$ is zero. This proves that $Cov(\hat{g}, h)=0$ is a necessary condition.

Sufficiency: Suppose that $\hat{g}$ is uncorrelated with any unbiased estimator h of zero. Let $\hat{g}^*$ be any other unbiased estimator of g. Since $\hat{g}-\hat{g}^*$ becomes an unbiased estimator of zero, an equation

$$Cov(\hat{g},\hat{g}-\hat{g}^*)= V(\hat{g})- Cov(\hat{g},\hat{g}^*) =0$$

holds. Then, since an inequality

$$0\le V(\hat{g}-\hat{g}^*)= V(\hat{g})-2\,Cov(\hat{g},\hat{g}^*) + V(\hat{g}^*)= V(\hat{g}^*)-V(\hat{g})$$

holds, $\hat{g}$ is a minimum variance unbiased estimator of $g(\boldsymbol{\theta})$.

Now, assuming the normality of the error e_i in addition to (2.4) ~ (2.6), the probability density function of $\boldsymbol{y}$ is given by

$$f(\boldsymbol{y})=\left(2\pi\sigma^2\right)^{-n/2}\ \exp-\frac{\sum_i(y_i-\mu)^2}{2\sigma^2}.$$

This is a density function of the normal distribution with mean μ and variance σ^2. If $h(\boldsymbol{y})$ is an unbiased estimator of zero, we have

$$\int h(\boldsymbol{y})\times f(\boldsymbol{y})d\boldsymbol{y}=0. \tag{2.7}$$

By the partial derivation of (2.7) with respect to μ, we have

$$\int\left\{\sum_i(y_i-\mu)/\sigma^2\right\}\times h(\boldsymbol{y})\times f(\boldsymbol{y})d\boldsymbol{y}=\left(n/\sigma^2\right)\int(\bar{y}_.-\mu)\times h(\boldsymbol{y})\times f(\boldsymbol{y})d\boldsymbol{y}=0.$$

This equation suggests that $\bar{y}_.$ is uncorrelated with $h(\mathbf{y})$, that is, $\bar{y}_.$ is a minimum variance unbiased estimator of its expectation μ.

On the contrary, for $\bar{y}_.$ to be a minimum variance unbiased estimator, the distribution of e_i in (2.1) must be normal (Kagan *et al.*, 1973).

2.3 Efficiency of Unbiased Estimator

To consider the behavior of sample mean $\bar{y}_.$ under non-normal distributions, it is convenient to consider the t-distribution (Fig. 2.1) specified by degrees of freedom ν:

$$f_\nu(y) = \frac{1}{\nu^{1/2} B\left(2^{-1}, \nu/2\right)} \left\{1 + \frac{(y-\mu)^2}{\nu}\right\}^{-(\nu+1)/2}, \tag{2.8}$$

where $B\left(2^{-1}, \nu/2\right)$ is a beta function. At $\nu = \infty$ this coincides with the normal distribution, and when $\nu = 1$ it is the Cauchy distribution representing a long-tailed distribution with both mean and variance divergent. Before comparing the estimation efficiency of sample mean $\bar{y}_.$ and median $\tilde{y}$, we describe Cramér–Rao's theorem, which gives the lower bound of variance of an unbiased estimator generally.

Theorem 2.2. Cramér–Rao's lower bound. Let the density function of $\mathbf{y} = (y_1, \ldots, y_n)'$ be $f(\mathbf{y}, \theta)$. Then the variance of any unbiased estimator $T(\mathbf{y})$ of θ satisfies an inequality

$$V(T) \geq I_n^{-1}(\theta), \tag{2.9}$$

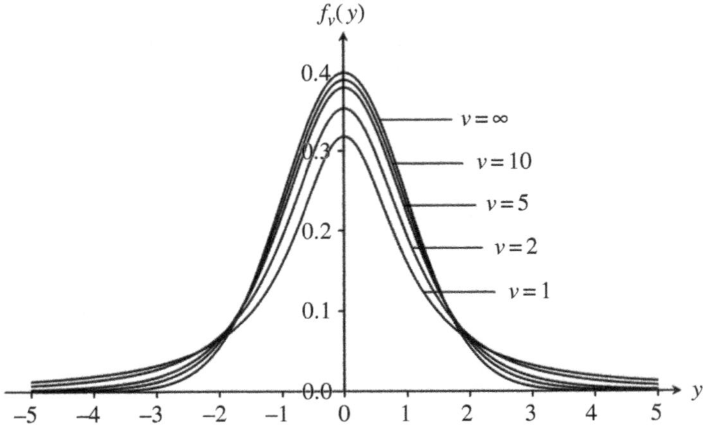

Figure 2.1 t-distribution $f_\nu(y)$.

where

$$I_n(\theta) = -E\left\{\frac{\partial^2 \log f(y, \theta)}{\partial \theta^2}\right\} = E\left[\left\{\frac{\partial \log f(y, \theta)}{\partial \theta}\right\}^2\right] \tag{2.10}$$

is called Fisher's amount of information. In the case of a discrete distribution $P(y, \theta)$, we can simply replace $f(y, \theta)$ by $P(y, \theta)$ in (2.10).

Proof. Since $T(y)$ is an unbiased estimator of θ, the equation

$$\int T(y) \times f(y, \theta) \mathrm{d}y = \theta \tag{2.11}$$

holds. Under an appropriate regularity condition such as exchangeability of derivation and integration, the derivation of (2.11) with respect to θ is obtained as

$$\int T(y) \times \frac{\partial \log f(y, \theta)}{\partial \theta} \times f(y, \theta) \mathrm{d}y = 1. \tag{2.12}$$

Further, by the derivation of $\int f(y, \theta)\mathrm{d}y = 1$ by θ, we have

$$\int \frac{\partial f(y, \theta)}{\partial \theta} \mathrm{d}y = \int \frac{\partial \log f(y, \theta)}{\partial \theta} \times f(y, \theta) \mathrm{d}y = 0. \tag{2.13}$$

Then, equations (2.12) and (2.13) imply

$$E\left[\{T(y) - \theta\} \times \frac{\partial \log f(y, \theta)}{\partial \theta}\right] = 1. \tag{2.14}$$

In contrast, for any random variable g, h and a real number λ, the equation

$$E\left\{(g + \lambda h)^2\right\} = E\left(g^2\right) + 2\lambda E(g \times h) + \lambda^2 E\left(h^2\right) \geq 0 \tag{2.15}$$

holds generally. If inequality (2.15) holds for any real number λ, we have

$$E\left(g^2\right) \times E\left(h^2\right) - \{E(g \times h)\}^2 \geq 0, \tag{2.16}$$

which is Schwarz's inequality. Applying (2.16) to (2.14), we get

$$E\{T(y) - \theta\}^2 \times E\left\{\frac{\partial \log f(y, \theta)}{\partial \theta}\right\}^2 \geq 1$$

and this is one form of (2.9) and (2.10), since $V(T) = E\{T(y) - \theta\}^2$. Next, since we have

$$\frac{\partial^2 \log f(y, \theta)}{\partial \theta^2} = \frac{\partial}{\partial \theta}\left\{\frac{1}{f(y, \theta)} \frac{\partial f(y, \theta)}{\partial \theta}\right\} = \frac{1}{f(y, \theta)} \frac{\partial^2 f(y, \theta)}{\partial \theta^2} - \frac{1}{f^2(y, \theta)}\left\{\frac{\partial f(y, \theta)}{\partial \theta}\right\}^2$$

and $E\left\{\frac{1}{f(y,\theta)}\frac{\partial^2 f(y,\theta)}{\partial\theta^2}\right\} = \int \frac{\partial^2 f(y,\theta)}{\partial\theta^2}dy = 0$, by the derivation of (2.13) we get

$$E\left\{\frac{\partial^2 \log f(y,\theta)}{\partial\theta^2}\right\} = -E\left[\left\{\frac{\partial \log f(y,\theta)}{\partial\theta}\right\}^2\right],$$

which gives another form of (2.10).

If the elements of $\mathbf{y} = (y_1, \ldots, y_n)'$ are independent following the probability density function $f(y_i, \theta)$, $I_n(\theta)$ can be expressed as $I_n(\theta) = nI_1(\theta)$, where

$$I_1(\theta) = -E\left\{\frac{\partial^2 \log f(y_i,\theta)}{\partial\theta^2}\right\} = E\left[\left\{\frac{\partial \log f(y_i,\theta)}{\partial\theta}\right\}^2\right]$$

is Fisher's amount of information per one datum.

An unbiased estimator which satisfies Cramér–Rao's lower bound is called an efficient estimator. When y is distributed as the normal distribution $N(\mu, \sigma^2)$, it is obvious that $I_1(\mu) = \sigma^{-2}$. Therefore, the lower bound of the variance of an unbiased estimator based on n independent samples is σ^2/n. Since $V(\bar{y}.) = \sigma^2/n$, $\bar{y}.$ is not only a minimum variance unbiased estimator but also an efficient estimator. An efficient estimator is generally a minimum variance unbiased estimator, but the reverse is not necessarily true. As a simple example, when $y_1, \ldots, y_n$ are distributed independently as $N(\mu, \sigma^2)$, the so-called unbiased variance

$$\hat{\sigma}^2 = \sum_{i=1}^{n} (y_i - \bar{y}.)^2/(n-1) \tag{2.17}$$

is a minimum variance unbiased estimator of σ^2 but it is not an efficient estimator (see Example 2.2 of Section 2.5.2).

When $y_1, \ldots, y_n$ are distributed independently as a t-distribution of (2.8), we have $I_1(\mu) = (\nu+1)/(\nu+3)$ and therefore the lower bound of an unbiased estimator of μ is

$$\frac{1}{n} \times \frac{\nu+3}{\nu+1}. \tag{2.18}$$

On the contrary, the variance of sample mean $\bar{y}.$ is

$$\frac{1}{n} \times \frac{\nu}{\nu-2}. \tag{2.19}$$

For the sample median $\tilde{y}$, the asymptotic variance

$$nV(\tilde{y}) \cong \frac{f^{-2}(\mu)}{4} = \frac{\nu}{4}B^2\left(\frac{1}{2}, \frac{\nu}{2}\right) \quad (n \to \infty) \tag{2.20}$$

Table 2.1 The efficiency of sample mean $\bar{y}_.$ and median $\tilde{y}$.

ν	1	2	3	4	5	10	∞
$\text{Eff}(\bar{y}_.)$	0	0	0.5	0.7	0.8	0.945	1
$\text{Eff}(\tilde{y})$	0.811	0.833	0.811	0.788	0.769	0.916	0.637

is known. Then the ratios of (2.18) to (2.19) and (2.20), namely

$$\text{Eff}(\bar{y}_.) = \frac{\nu+3}{\nu+1} \times \frac{\nu-2}{\nu}$$

and

$$\text{Eff}(\tilde{y}) = \frac{\nu+3}{\nu+1} \times \frac{4}{\nu B^2\left(\frac{1}{2}, \frac{\nu}{2}\right)}$$

are called the efficiency of $\bar{y}_.$ and $\tilde{y}$, respectively. The inverse of the efficiency implies the necessary sample size to attain Cramér–Rao's lower bound by the respective estimators. The efficiencies are given in Table 2.1.

From Table 2.1 we see that $\bar{y}_.$ behaves well for $\nu \geq 5$ but its efficiency decreases below 5, and in particular the efficiency becomes zero at $\nu = 1$ and 2. In contrast, $\tilde{y}$ keeps relatively high efficiency and is particularly useful at $\nu = 1$ and 2. Actually, for the Cauchy distribution an extremely large or small datum occurs from time to time, and $\bar{y}_.$ is directly affected by it whereas $\tilde{y}$ is quite stable against such disturbance. This property of stability is called robustness in statistics. There are various proposals for the robust estimator when a long-tailed distribution is expected or no prior information regarding error is available at all in advance. However, a simple and established method is not available, except for a simple estimation problem of a population mean. Also, the real data may not follow exactly the normal distribution, but still it will be rare to have to assume such a long-tailed distribution as Cauchy. Therefore, it is actually usual to base the inference on the linear model and BLUE by checking the model very carefully and with an appropriate transformation of data if necessary.

The basis of normality is the central limit theorem, which ensures normality for the error if it consists of infinitely many casual errors. In contrast, for the growth of creatures, the amount of growth is often proportional to the present size, inviting a product model instead of an additive model. In this case, the logarithm of the data fits the normal distribution better. Masuyama (1976) examined widely the germination age of teeth, the time to produce cancer from X-ray irradiation, and so on, and reported that the lognormal distribution generally fitted well the time to appearance of effects. Concentrations of chemical agents in blood have also been said to fit well to a lognormal distribution, and we have employed this in proving bio-equivalence in

Section 5.3.6. Power transformation including square and cube roots is also applied quite often, but in choosing transformations some rationale is desired in addition to apparent fitness.

As another sort of transformation, an arc sine transformation $\sin^{-1}\sqrt{y/n}$ of the data from the binomial distribution $B(n, p)$ is conveniently used for normal approximation, with mean $\sin^{-1}\sqrt{p}$ and stabilized variance $1/(4n)$.

2.4 Linear Model

In the analysis of variance (ANOVA) and regression analysis also, it is usual to assume a linear model

$$\boldsymbol{y}_n = \boldsymbol{X}_{n\times p}\boldsymbol{\theta}_p + \boldsymbol{e}_n, \tag{2.21}$$

where $\boldsymbol{y}_n$ is an n-dimensional observation vector, $\boldsymbol{\theta}_p$ a p -dimensional unknown parameter vector, $\boldsymbol{e}$ an error vector, and $\boldsymbol{X}$ a design matrix of experiments which gives the relationship between the observation vector $\boldsymbol{y}$ and the parameter vector $\boldsymbol{\theta}$. We assume

$$E(\boldsymbol{e}) = \boldsymbol{0}_n, \tag{2.22}$$

$$V(\boldsymbol{e}) = \sigma^2\boldsymbol{I}_n, \tag{2.23}$$

where $\boldsymbol{0}$ denotes a zero vector of an appropriate size, $\boldsymbol{I}_n$ an identity matrix of order n, and the suffix will be omitted if it is obvious. The difference is obvious from Fisher's information matrix $I_n(\theta)$, which is a function of a relevant parameter θ. Equation (2.22) corresponds to the assumption (2.4), and (2.23) to (2.5) and (2.6).

The simple model (2.1) of n repetitions is expressed as, for example,

$$\boldsymbol{y} = \boldsymbol{j}_n\mu + \boldsymbol{e}.$$

As a special case of a 1-dimensional parameter, $\boldsymbol{j}_n\mu$ can also be expressed as $\mu\boldsymbol{j}_n$, with μ a scalar multiplier of vector $\boldsymbol{j}_n$. Model (1.1) is also a linear model, and can be expressed as

$$\boldsymbol{y} = \begin{bmatrix} y_1 \\ y_2 \\ y_3 \\ y_4 \end{bmatrix} = \begin{bmatrix} 1 & 1 & 1 & 1 \\ 1 & 1 & -1 & -1 \\ 1 & -1 & 1 & -1 \\ 1 & -1 & -1 & 1 \end{bmatrix} \begin{bmatrix} \mu_1 \\ \mu_2 \\ \mu_3 \\ \mu_4 \end{bmatrix} + \begin{bmatrix} e_1 \\ e_2 \\ e_3 \\ e_4 \end{bmatrix}.$$

Also, a regression model

$$y_i = \beta_0 + \beta_1 x_i + \beta_2 x_i^2 + e_i,\ i = 1, \ldots, n$$

can be expressed as

$$y = \begin{bmatrix} 1 & x_1 & x_1^2 \\ 1 & x_2 & x_2^2 \\ & \vdots & \\ 1 & x_n & x_n^2 \end{bmatrix} \begin{bmatrix} \beta_0 \\ \beta_1 \\ \beta_2 \end{bmatrix} + \begin{bmatrix} e_1 \\ e_2 \\ \vdots \\ e_n \end{bmatrix}.$$

The name 'linear model' comes from the fact that the systematic part is expressed as a linear combination of the unknown parameters, and therefore it is no problem to include a non-linear term x_i^2 in the explanation variables.

Now, the problem of a minimum variance linear unbiased estimator (BLUE) $\boldsymbol{l}'\boldsymbol{y}$ of a linear combination $\boldsymbol{L}'\boldsymbol{\theta}$ is formulated as a problem of finding $\boldsymbol{l}$ so as to minimize the variance $\boldsymbol{l}'\boldsymbol{l}(\sigma^2)$ subject to $\boldsymbol{l}'\boldsymbol{X} = \boldsymbol{L}'$ for given $\boldsymbol{X}$ and $\boldsymbol{L}'$. It should be noted that in the example at the end of Section 2.1, $\boldsymbol{X}$ was $\boldsymbol{j}$ and $\boldsymbol{L}' = 1$. However, it is very time-consuming to solve this minimization problem each time. Instead, by the LS method we can obtain BLUE very easily.

2.5 Least Squares Method

2.5.1 LS method and BLUE

Let us define the LS estimator $\hat{\boldsymbol{\theta}}$ by

$$\left(\hat{\boldsymbol{\theta}} \,|\, \|\boldsymbol{y} - \boldsymbol{X}\hat{\boldsymbol{\theta}}\|^2 \le \|\boldsymbol{y} - \boldsymbol{X}\boldsymbol{\theta}\|^2 \text{ for any } \boldsymbol{\theta}\right), \tag{2.24}$$

where $\|\boldsymbol{v}\|^2$ denotes a squared norm of a vector $\boldsymbol{v}$. Then, for any linear estimable function $\boldsymbol{L}'\boldsymbol{\theta}$, the BLUE is uniquely obtained from $\boldsymbol{L}'\hat{\boldsymbol{\theta}}$ even when the $\hat{\boldsymbol{\theta}}$ that satisfies (2.24) is not unique. Here, the linear estimable function $\boldsymbol{L}'\boldsymbol{\theta}$ implies that there is at least one linear unbiased estimator for it. The necessary and sufficient condition for the estimability is that $\boldsymbol{L}'$ can be expressed by a linear combination of rows of $\boldsymbol{X}$ by the requirement $E(\boldsymbol{l}'\boldsymbol{y}) = \boldsymbol{l}'\boldsymbol{X}\boldsymbol{\theta} = \boldsymbol{L}'\boldsymbol{\theta}$. Therefore, if rank$(\boldsymbol{X})$ is p, then every linear function $\boldsymbol{L}'\boldsymbol{\theta}$ and $\boldsymbol{\theta}$ itself is estimable. When rank$(\boldsymbol{X})$ is smaller than p, every element of $\boldsymbol{\theta}$ is not estimable and $\hat{\boldsymbol{\theta}}$ of (2.24) cannot be determined uniquely. It is important that even for this case, $\boldsymbol{L}'\hat{\boldsymbol{\theta}}$ is uniquely determined for the estimable function $\boldsymbol{L}'\boldsymbol{\theta}$.

Example 2.1. One-way ANOVA model. Let us consider the one-way ANOVA model of Chapter 5:

$$y_{ij} = \mu_i + e_{ij},\ i = 1, \ldots, a,\ j = 1, \ldots, m. \tag{2.25}$$

This model is expressed in the form of (2.21), taking

$$X_{n\times a}=\begin{bmatrix} j_m & 0 & \cdots & 0 \\ 0 & j_m & \cdots & 0 \\ & & \ddots & \\ 0 & 0 & \cdots & j_m \end{bmatrix},\ \theta_a=\begin{bmatrix} \mu_1 \\ \mu_2 \\ \vdots \\ \mu_a \end{bmatrix}$$

with $n=am$ and $p=a$. Obviously, rank(X) is a and coincides with the number of unknown parameters. Therefore, all the parameters μ_i are estimable. However, the model (2.25) is often rewritten as

$$y_{ij}=\mu+\alpha_i+e_{ij},\ i=1,\ \ldots,\ a, j=1,\ \ldots,\ m, \tag{2.26}$$

factorizing μ_i to a general mean μ and main treatment effect α_i. Then, the linear model is expressed in matrix form as

$$y_n=[j_n\ X_a]\begin{bmatrix} \mu \\ \alpha \end{bmatrix}+e_n,\ \alpha=(\alpha_1,\ \ldots,\alpha_a)',$$

where X_α is equivalent to $X_{n\times a}$ and $p=a+1$. Since rank[j_n X_α] is a, this is the case where the design matrix is not full rank and every unknown parameter is not estimable. The estimable functions are obviously $\mu+\alpha_i$, $i=1,\ \ldots,\ a$, and their linear combinations. Therefore, $\alpha_i-\alpha_{i'}$ is estimable but μ and α_i themselves are not estimable. The linear combination in α with sum of coefficients equal to 0, like $\alpha_i-\alpha_{i'}$, is called a contrast, which implies a sort of difference among parameters. In a one-way layout all the contrasts are estimable, since then μ vanishes.

Theorem 2.3. Gauss–Markov's theorem. We call the linear model (2.21) under the assumptions (2.22) and (2.23), Gauss–Markov's model. With this model, any $\hat{\theta}$ that satisfies

$$X'X\hat{\theta}=X'y \tag{2.27}$$

is called an LS estimator. Equation (2.27) is obtained by equating the derivation of $\|y-X\theta\|^2$ with respect to θ to zero, called a normal equation. Then, for any estimable function $L'\theta$, the BLUE is obtained simply by substituting the LS estimator $\hat{\theta}$ into θ, as $L'\hat{\theta}$.

Proof. The proof is very simple when the design matrix X is full rank. In this case equation (2.27) is solved at once to give the solution $L'\hat{\theta}=L'(X'X)^{-1}X'y$. This is an unbiased estimator of $L'\theta$, since $E\left(L'\hat{\theta}\right)=E\left\{L'(X'X)^{-1}X'y\right\}=L'(X'X)^{-1}$ $X'X\theta=L'\theta$. Next, suppose $l'y$ to be any linear unbiased estimator of $L'\theta$ and denote the difference from $L'\hat{\theta}$ by $b'y$. Then we have

$$V(\boldsymbol{l}'\boldsymbol{y}) = V\left(\boldsymbol{L}'\hat{\boldsymbol{\theta}} - \boldsymbol{b}'\boldsymbol{y}\right) = V\left(\boldsymbol{L}'\hat{\boldsymbol{\theta}}\right) + V(\boldsymbol{b}'\boldsymbol{y}) - 2Cov\left(\boldsymbol{L}'\hat{\boldsymbol{\theta}}, \boldsymbol{b}'\boldsymbol{y}\right)$$

$$= V\left(\boldsymbol{L}'\hat{\boldsymbol{\theta}}\right) + V(\boldsymbol{b}'\boldsymbol{y}) - 2\boldsymbol{L}'(\boldsymbol{X}'\boldsymbol{X})^{-1}\boldsymbol{X}'\boldsymbol{b}\sigma^2$$

$$= V\left(\boldsymbol{L}'\hat{\boldsymbol{\theta}}\right) + V(\boldsymbol{b}'\boldsymbol{y}) \geq V\left(\boldsymbol{L}'\hat{\boldsymbol{\theta}}\right).$$

The last equality holds since the equation $E\left(\boldsymbol{l}'\boldsymbol{y} - \boldsymbol{L}'\hat{\boldsymbol{\theta}}\right) = E(\boldsymbol{b}'\boldsymbol{y}) = \boldsymbol{b}'\boldsymbol{X}\boldsymbol{\theta} = 0$ holds for any $\boldsymbol{\theta}$, suggesting $\boldsymbol{b}'\boldsymbol{X} = 0$.

There are various methods for proving Theorem 2.3 when $\text{rank}(\boldsymbol{X}) = r \leq p$. We apply here an orthonormal transformation of $\boldsymbol{y}$ to a standard form. Another proof is obtained in Section 2.5.4 with the aid of a generalized inverse of a matrix.

Let the orthonormal eigenvectors for the non-zero eigenvalues $\lambda_1, \ldots, \lambda_r\ (>0)$ of $\boldsymbol{X}'\boldsymbol{X}$ be denoted by $\boldsymbol{P}_{p \times r}$,

$$\boldsymbol{X}'\boldsymbol{X}\boldsymbol{P} = \boldsymbol{P}\boldsymbol{D},$$

where $\boldsymbol{D} = \text{diag}\,(\lambda_i)$ is a diagonal matrix with diagonal elements $\lambda_1, \ldots, \lambda_r$. Next, let us define an orthonormal matrix $\boldsymbol{Q}_{n \times (n-r)}$ that is orthogonal to the columns of $\boldsymbol{X}$. Then

$$\boldsymbol{M}' = \begin{bmatrix} \boldsymbol{D}^{-1/2}\boldsymbol{P}'\boldsymbol{X}' \\ \boldsymbol{Q}' \end{bmatrix} \tag{2.28}$$

is an $n \times n$ orthonormal matrix satisfying $\boldsymbol{M}'\boldsymbol{M} = \boldsymbol{I}_n$, where the notation D^{ν} for a diagonal matrix is used for the diagonal matrix with diagonal elements $\lambda_1^{\nu}, \ldots, \lambda_r^{\nu}$ throughout this book. The matrices $\boldsymbol{P}$ and $\boldsymbol{Q}$ are generally not unique, but we can choose and fix either of them. The column space of $\boldsymbol{Q}$ is nothing but the error space introduced in Section 2.5.2. We put forward a transformation of $\boldsymbol{y}$ by $\boldsymbol{M}'$ as

$$z = \begin{bmatrix} z_1 \\ z_2 \end{bmatrix} = \boldsymbol{M}'\boldsymbol{y} = \begin{bmatrix} \boldsymbol{\eta} \\ \boldsymbol{0} \end{bmatrix} + \boldsymbol{\xi}, \tag{2.29}$$

where z_1 and z_2 are the partitions of z according to the partition of $\boldsymbol{M}'$ (2.28) and $\boldsymbol{\xi}$ satisfies the same condition $E(\boldsymbol{\xi}) = \boldsymbol{0}$ and $V(\boldsymbol{\xi}) = \sigma^2\boldsymbol{I}$ as (2.22) and (2.23). In the standard form (2.29), elements of $\boldsymbol{\eta} = \boldsymbol{D}^{-1/2}\boldsymbol{P}'\boldsymbol{X}'\boldsymbol{X}\boldsymbol{\theta} = \boldsymbol{D}^{1/2}\boldsymbol{P}'\boldsymbol{\theta}$ are linearly independent of each other and it is obvious that the class of estimable functions of $\boldsymbol{\theta}$ is formed by all linear combinations of $\boldsymbol{\eta}$. Since the elements of $\boldsymbol{\eta}$ are linearly independent, any estimable function is uniquely expressed by $\boldsymbol{l}_1'\boldsymbol{\eta}$ and its unbiased estimator is given generally by $\boldsymbol{l}_1'z_1 + \boldsymbol{l}_2'z_2$ with arbitrary vector $\boldsymbol{l}_2'$. Then, since z_1 and z_2 are uncorrelated, we have

$$V\left(\boldsymbol{l}_1'z_1 + \boldsymbol{l}_2'z_2\right) = \boldsymbol{l}_1'\boldsymbol{l}_1\sigma^2 + \boldsymbol{l}_2'\boldsymbol{l}_2\sigma^2$$

and the variance is minimized at $l_2=0$. Therefore, $l_1'z_1=l_1'D^{-1/2}P'X'y$ is a BLUE of $l_1'\eta=l_1'D^{1/2}P'\theta$ and its variance is $l_1'l_1\sigma^2$. Next, multiplying by P' on both sides of the normal equation (2.27), we have

$$DP'\hat{\theta}=P'X'y, \tag{2.30}$$

and therefore we have $l_1'z_1=l_1'D^{-1/2}P'X'y=l_1'D^{1/2}P'\hat{\theta}$. Namely, the BLUE of $l_1'D^{1/2}P'\theta$ is obtained simply by replacing θ with $\hat{\theta}$ that satisfies the normal equation.

2.5.2 Estimation space and error space

In the linear model (2.21), the linear subspace spanned by the columns of X is called an estimation space and its orthogonal complement is called an error space. The error space is spanned by the column vectors of Q in (2.28). In this context, the whole space of observation y is called a sample space. Now, noting equation (2.30), $X\hat{\theta}$ is rewritten as

$$X\hat{\theta}=X(PP')\hat{\theta}=XP\left(D^{-1}P'X'y\right)=\Pi_X y,$$

where $\Pi_X=XPD^{-1}P'X'$ is a projection matrix onto the estimation space since it is symmetric, idempotent and keeps $X\theta$ $(=\Pi_X X\theta)$ unchanged. Then, $X\hat{\theta}$ is unique as a projection of y onto the estimation space. Note that when $\text{rank}(X)=p$, Π_X is simply $X(X'X)^{-1}X'$. This is natural, since $X\hat{\theta}$ is closest to the observation vector y in the sense of squared distance among the vectors expressed by $X\theta$ (see Fig. 2.2).

In contrast, we call $y-X\hat{\theta}$ a residual and it is rewritten as

$$y-X\hat{\theta}=(I-\Pi_X)y=QQ'y.$$

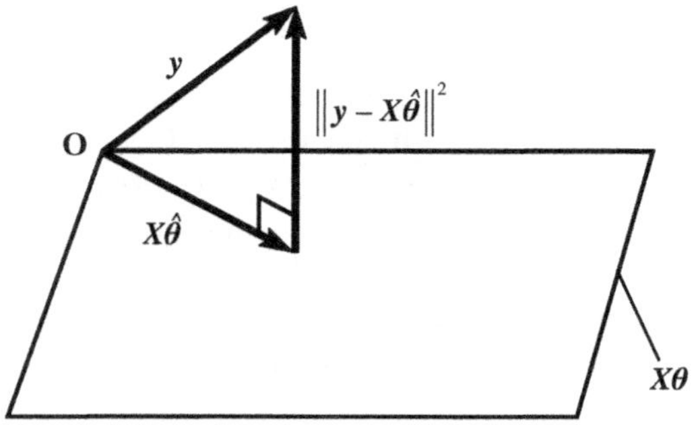

Figure 2.2 Observation vector y and its projection onto the estimation space.

This is obviously a projection of $\boldsymbol{y}$ onto the error space. Its squared norm

$$S(\hat{\boldsymbol{\theta}}) = \|\boldsymbol{y}-\boldsymbol{X}\hat{\boldsymbol{\theta}}\|^2 = \boldsymbol{y}'\boldsymbol{Q}\boldsymbol{Q}'\boldsymbol{y} = \|\boldsymbol{Q}'\boldsymbol{y}\|^2 \tag{2.31}$$

is called a residual sum of squares and generally expressed by S. It should be noted here that $\boldsymbol{X}\hat{\boldsymbol{\theta}}$ and $\boldsymbol{y}-\boldsymbol{X}\hat{\boldsymbol{\theta}}$ are uncorrelated by the orthogonality of estimation space and error space,

$$Cov\left(\boldsymbol{X}\hat{\boldsymbol{\theta}}, \boldsymbol{y}-\boldsymbol{X}\hat{\boldsymbol{\theta}}\right) = \boldsymbol{\Pi}_X V(\boldsymbol{y})(\boldsymbol{I}-\boldsymbol{\Pi}_X) = \sigma^2\boldsymbol{\Pi}_X(\boldsymbol{I}-\boldsymbol{\Pi}_X) = 0.$$

This further implies that $\boldsymbol{X}\hat{\boldsymbol{\theta}}$ and $\boldsymbol{y}-\boldsymbol{X}\hat{\boldsymbol{\theta}}$ are mutually independent under the normality assumption. Since every element of $\boldsymbol{Q}'\boldsymbol{y}$ is distributed with expectation zero, equal variance σ^2 and uncorrelated, we have

$$E\left\{S\left(\hat{\boldsymbol{\theta}}\right)\right\} = (n-r)\sigma^2.$$

Therefore

$$\hat{\sigma}^2 = \|\boldsymbol{y}-\boldsymbol{X}\hat{\boldsymbol{\theta}}\|^2/(n-r) \tag{2.32}$$

is an unbiased estimator of σ^2, called an unbiased variance. For the numerical calculation of $S\left(\hat{\boldsymbol{\theta}}\right)$, the formula below is more convenient:

$$\begin{aligned} S\left(\hat{\boldsymbol{\theta}}\right) &= \boldsymbol{y}'\boldsymbol{y} - \boldsymbol{y}'\boldsymbol{X}\hat{\boldsymbol{\theta}} - \hat{\boldsymbol{\theta}}'\boldsymbol{X}'\boldsymbol{y} + \hat{\boldsymbol{\theta}}'\boldsymbol{X}'\boldsymbol{X}\hat{\boldsymbol{\theta}} \\ &= \boldsymbol{y}'\boldsymbol{y} - \hat{\boldsymbol{\theta}}'\boldsymbol{X}'\boldsymbol{y} \end{aligned} \tag{2.33}$$

The necessary calculation is the sum of squares of y_i and the inner product of the estimate $\hat{\boldsymbol{\theta}}$ and $\boldsymbol{X}'\boldsymbol{y}$, which is the right-hand side of the normal equation (2.27). On the one hand, applying a well-known formula

$$E(U^2) = \{E(U)\}^2 + V(U) \tag{2.34}$$

for any random variable U to $\|\boldsymbol{y}\|^2$ element-wise and noting equation (2.21), we have $E\left(\|\boldsymbol{y}\|^2\right) = \|\boldsymbol{X\theta}\|^2 + n\sigma^2$. Therefore, from (2.33),

$$\begin{aligned} E\left(\|\boldsymbol{X}\hat{\boldsymbol{\theta}}\|^2\right) &= E(\boldsymbol{y}'\boldsymbol{y}) - E\left\{S\left(\hat{\boldsymbol{\theta}}\right)\right\} \\ &= E\left(\|\boldsymbol{y}\|^2\right) - (n-r)\sigma^2 = \|\boldsymbol{X\theta}\|^2 + r\sigma^2. \end{aligned}$$

If we assume a normal distribution of the error $\boldsymbol{e}$, then the BLUE $\boldsymbol{L}'\hat{\boldsymbol{\theta}}$ is the minimum variance among all the unbiased estimators. This is verified by applying Theorem 2.1 to the partial derivation of

$$H(\boldsymbol{\theta}, \sigma^2) = \int h(\boldsymbol{y}) \times (2\pi\sigma^2)^{-n/2} \exp - \frac{\|\boldsymbol{y}-\boldsymbol{X\theta}\|^2}{2\sigma^2} \mathrm{d}\boldsymbol{y} = 0$$

with respect to $\boldsymbol{\theta}$. Further, from the two expressions $\partial^2 H(\boldsymbol{\theta}, \sigma^2)/\partial\boldsymbol{\theta}^2$ and $\partial H(\boldsymbol{\theta}, \sigma^2)/\partial\sigma^2$, the estimator $\hat{\sigma}^2$ of equation (2.32) is shown to be a minimum variance unbiased estimator (see Example 2.2).

Example 2.2. Application of Theorem 2.1 to model $y = j_n\mu + e$ with normal error. The normal equation (2.27) is simply

$$\hat{\mu} = (j'j)^{-1}j'y = n^{-1}\sum_{i=1}^{n} y_i = \bar{y}_{\cdot}.$$

The residual sum of squares is obtained from (2.31) or (2.33) as

$$S(\hat{\mu}) = \sum_{i=1}^{n}(y_i - \bar{y}_{\cdot})^2 = \sum_{i=1}^{n} y_i^2 - \bar{y}_{\cdot} \times y_{\cdot} = \sum_{i=1}^{n} y_i^2 - y_{\cdot}^2/n.$$

The function $H(\mu, \sigma^2)$ is given as equation (2.7) and it has already been shown that $\bar{y}_{\cdot}$ and $h(\mathbf{y})$ are uncorrelated. Then, from $\partial^2 H(\mu, \sigma^2)/\partial\mu^2$, $y_{\cdot}^2$ is shown to be uncorrelated with $h(\mathbf{y})$. Further, from $\partial H(\mu, \sigma^2)/\partial\sigma^2$, $\sum y_i^2$ is shown to be uncorrelated with $h(\mathbf{y})$. Combining these results, the unbiased variance (2.17),

$$\hat{\sigma}^2 = \frac{S(\hat{\mu})}{n-1} = \frac{1}{n-1}\left(\sum_{i=1}^{n} y_i^2 - y_{\cdot}^2/n\right),$$

is shown to be a minimum variance unbiased estimator. Under the normality assumption, elements of $\boldsymbol{Q}'y/\sigma$ are distributed independently as the standard normal distribution and therefore $S(\hat{\mu})/\sigma^2$ is distributed as the chi-squared distribution χ_ν^2 with degrees of freedom $\nu = n-1$ (see Section 2.5.5 for an explanation of degrees of freedom and Lemma 2.1 for the chi-squared distribution). Since the variance of χ_ν^2 is 2ν, the variance of $\hat{\sigma}^2$ is $V(\hat{\sigma}^2) = 2\sigma^4/(n-1)$. In contrast, the Cramér–Rao's lower bound for σ^2 is $2\sigma^4/n$ from $I_1(\sigma^2) = (2\sigma^4)^{-1}$. Therefore, $\hat{\sigma}^2$ is not an efficient estimator. By the general theory of this section, $\hat{\mu}$ and $\hat{\sigma}^2$ are mutually independent and are utilized in Section 3.1.4 to derive Student's t-distribution.

2.5.3 Linear constraints on parameters for solving the normal equation

By Gauss–Markov's theorem the BLUE for an estimable function $\boldsymbol{L}'\boldsymbol{\theta}$ is obtained simply and uniquely by $\boldsymbol{L}'\hat{\boldsymbol{\theta}}$ with any $\hat{\boldsymbol{\theta}}$ satisfying the normal equation. If $\text{rank}(\boldsymbol{X}) = r$ is equal to the number of unknown parameters p, then the normal equation is solved uniquely and there is no problem. However, if $r < p$, then $\hat{\boldsymbol{\theta}}$ is not unique and it is a problem how to find $\boldsymbol{L}'\hat{\boldsymbol{\theta}}$ for the estimable function. We give three ways to obtain $\boldsymbol{L}'\hat{\boldsymbol{\theta}}$ in the following.

(1) **Direct method.** Since $\boldsymbol{L}'\boldsymbol{\theta}$ is estimable, $\boldsymbol{L}'\hat{\boldsymbol{\theta}}$ can be obtained directly by linear combination of the left-hand side of the normal equation $\boldsymbol{X}'\boldsymbol{X}\hat{\boldsymbol{\theta}} = \boldsymbol{X}'\boldsymbol{y}$. Then, the same linear combination of the right-hand side of the normal equation is nothing but the BLUE as shown in Example 2.3.

(2) **Orthonormal transformation.** Transform the linear model to a standard form. This looks difficult, but for many designs of experiments such a transform is known and given in later sections.

(3) **Linear constraints.** Add condition $\boldsymbol{K}\boldsymbol{\theta} = \boldsymbol{0}$ to the normal equation so that it can be solved uniquely. In this case $\hat{\boldsymbol{\theta}}$ will change according to the conditions, but $\boldsymbol{L}'\hat{\boldsymbol{\theta}}$ is uniquely determined. For our purpose the conditions should satisfy:

(a) Intersection of the subspaces spanned by the rows of $\boldsymbol{X}$ and $\boldsymbol{K}$ is $\boldsymbol{0}'$.

(b) $\operatorname{rank}\begin{bmatrix} \boldsymbol{X} \\ \boldsymbol{K} \end{bmatrix} = r.$

Condition (a) is for the condition $\boldsymbol{K}\boldsymbol{\theta} = \boldsymbol{0}$ not to restrict any of the estimable function. Condition (b) is for the joint equations $\boldsymbol{X}'\boldsymbol{X}\hat{\boldsymbol{\theta}} = \boldsymbol{X}'\boldsymbol{y}$ and $\boldsymbol{K}\hat{\boldsymbol{\theta}} = \boldsymbol{0}$ to be solved uniquely for $\hat{\boldsymbol{\theta}}$. Condition (a) is also for the column space of $\boldsymbol{X}$ to be unchanged if we restrict the coefficient $\boldsymbol{\theta}$ of $\boldsymbol{X}\boldsymbol{\theta}$ to satisfy $\boldsymbol{K}\boldsymbol{\theta} = \boldsymbol{0}$. Further, it is a condition that the minimization problem of

$$S(\boldsymbol{\theta}, \boldsymbol{\gamma}) = \|\boldsymbol{y} - \boldsymbol{X}\boldsymbol{\theta}\|^2 + \boldsymbol{\gamma}'\boldsymbol{K}\boldsymbol{\theta},\ \boldsymbol{K}\boldsymbol{\theta} = \boldsymbol{0}$$

with respect to $\boldsymbol{\theta}$ and the Lagrange multiplier $\boldsymbol{\gamma}$ has a consistent solution $\boldsymbol{\gamma} = \boldsymbol{0}$, see Hirotsu (1971) for details.

Example 2.3. Estimation of one-way layout model (2.26). The estimable functions of interest are $\mu + \alpha_i$, $i = 1, \ldots, a$ and the contrast $\boldsymbol{L}'\boldsymbol{\alpha}$, $\boldsymbol{L} = (L_1, \ldots, L_a)'$, $\boldsymbol{\alpha} = (\alpha_1, \ldots, \alpha_a)'$. Since $\boldsymbol{X} = [\boldsymbol{j}_n \boldsymbol{X}_\alpha]$, the normal equation is obtained as

$$\boldsymbol{X}'\boldsymbol{X}\hat{\boldsymbol{\theta}} = \begin{bmatrix} n & m & m & \cdots & m \\ m & m & 0 & \cdots & 0 \\ m & 0 & m & \cdots & 0 \\ & & & \ddots & \\ m & 0 & 0 & \cdots & m \end{bmatrix} \begin{bmatrix} \hat{\mu} \\ \hat{\boldsymbol{\alpha}} \end{bmatrix} = \begin{bmatrix} y_{\cdot\cdot} \\ y_{1\cdot} \\ \vdots \\ y_{a\cdot} \end{bmatrix}. \tag{2.35}$$

Equation (2.35) cannot be solved since $\operatorname{rank}(\boldsymbol{X}) = a$, whereas there are $a+1$ unknown parameters. Therefore, we apply methods (1), (2), and (3).

(1) **Direct method.** The $i+1$st row of (2.35) is

$$m(\hat{\mu}+\hat{\alpha}_i)=m(\widehat{\mu+\alpha_i})=y_{i\cdot}.$$

Therefore, $(\widehat{\mu+\alpha_i})=\bar{y}_{i\cdot}$ is BLUE for its expectation $\mu+\alpha_i$. Since a BLUE is unique, it suffices to find one solution. Then, by the condition $\boldsymbol{L}'\boldsymbol{j}=0$, we have

$$\textstyle\sum_{i=1}^{a}\left\{L_i(\widehat{\mu+\alpha_i})\right\}=(\sum_{i=1}^{a}L_i)\hat{\mu}+\sum_{i=1}^{a}(L_i\hat{\alpha}_i)=\sum_{i=1}^{a}(L_i\hat{\alpha}_i)$$

and therefore $\sum L_i\bar{y}_{i\cdot}$ is BLUE for the contrast $\boldsymbol{L}'\boldsymbol{\alpha}$.

(2) **Orthogonal transformation.** We can take $a-1$ orthonormal vectors $\boldsymbol{p}_1, \ldots, \boldsymbol{p}_{a-1}$ which are orthogonal to each other and also to $\boldsymbol{j}_a$. Let us define

$$\boldsymbol{P}_a=[\boldsymbol{p}_1, \ldots, \boldsymbol{p}_{a-1}], \tag{2.36}$$

where this definition of $\boldsymbol{P}_a$ is used throughout the book. Then, since $\left[a^{-1/2}\boldsymbol{j}_a\,\boldsymbol{P}_a\right]$ is an $a\times a$ orthonormal matrix, we have

$$\boldsymbol{P}_a\boldsymbol{P}_a'=\boldsymbol{I}_a-a^{-1}\boldsymbol{j}_a\boldsymbol{j}_a',\quad \boldsymbol{P}_a'\boldsymbol{P}_a=\boldsymbol{I}_{a-1}. \tag{2.37}$$

Now, since

$$\boldsymbol{X}'\boldsymbol{X}\begin{bmatrix}0\\ \boldsymbol{p}_i\end{bmatrix}=m\begin{bmatrix}0\\ \boldsymbol{p}_i\end{bmatrix},$$

$\left[0\ \boldsymbol{p}_i'\right]'$, $i=1,\ldots,a-1$ are the eigenvectors of $\boldsymbol{X}'\boldsymbol{X}$ with respect to the eigenvalue m. Also, since

$$\boldsymbol{X}'\boldsymbol{X}\begin{bmatrix}a\\ \boldsymbol{j}_a\end{bmatrix}=m(a+1)\begin{bmatrix}a\\ \boldsymbol{j}_a\end{bmatrix},$$

$\left[a\ \boldsymbol{j}_a'\right]'$ is an eigenvector with respect to the eigenvalue $m(a+1)$. Therefore, every column of

$$\boldsymbol{P}=\begin{bmatrix}\{a(a+1)\}^{-1/2}a & \boldsymbol{0}'\\ \{a(a+1)\}^{-1/2}\boldsymbol{j}_a & \boldsymbol{P}_a\end{bmatrix}$$

forms orthonormal eigenvectors for a non-zero eigenvalues of $\boldsymbol{X}'\boldsymbol{X}$ and we get

$$\boldsymbol{P}'\boldsymbol{X}'\boldsymbol{X}\boldsymbol{P}=\boldsymbol{D}=\begin{bmatrix}m(a+1) & \boldsymbol{0}'\\ \boldsymbol{0} & m\boldsymbol{I}_{a-1}\end{bmatrix}.$$

Then we have an orthonormal transformation $\boldsymbol{M}'$ (2.28) of $\boldsymbol{y}$ as

$$\boldsymbol{M}' = \begin{bmatrix} \boldsymbol{D}^{-1/2}\boldsymbol{P}'\boldsymbol{X}' \\ \boldsymbol{Q}' \end{bmatrix} = \begin{bmatrix} n^{-1/2}\boldsymbol{j}_n' \\ m^{-1/2}\boldsymbol{P}_a'\boldsymbol{X}_\alpha' \\ \boldsymbol{Q}' \end{bmatrix}. \tag{2.38}$$

Finally, we get a standard form (2.29) as

$$\boldsymbol{z} = \boldsymbol{M}'\boldsymbol{y} = \begin{bmatrix} n^{-1/2}y_{\cdot\cdot} \\ m^{-1/2}\boldsymbol{P}_a' \begin{bmatrix} y_{1\cdot} \\ \vdots \\ y_{a\cdot} \end{bmatrix} \\ \boldsymbol{Q}'\boldsymbol{y} \end{bmatrix} = \begin{bmatrix} n^{1/2}(\mu + \bar{\alpha}_\cdot) \\ m^{1/2}\boldsymbol{P}_a'\alpha \\ \boldsymbol{0} \end{bmatrix} + \boldsymbol{\xi}. \tag{2.39}$$

Every row of $\boldsymbol{P}_a'\boldsymbol{\alpha}$ represents an orthonormalized contrast in $\boldsymbol{\alpha}$, and $\boldsymbol{P}_a'(\bar{y}_{1\cdot},\ldots,\bar{y}_{a\cdot})'$ gives its BLUE. Further, from $\boldsymbol{P}_a\boldsymbol{P}_a'\hat{\boldsymbol{\alpha}} = (\boldsymbol{I} - a^{-1}\boldsymbol{jj}')\hat{\boldsymbol{\alpha}} = (\boldsymbol{I} - a^{-1}\boldsymbol{jj}')(\bar{y}_{1\cdot},\ldots,\bar{y}_{a\cdot})'$ we have $\widehat{(\alpha_i - \bar{\alpha}_\cdot)} = \bar{y}_{i\cdot} - \bar{y}_{\cdot\cdot}$. In contrast, from the first row of (2.39) we have $\widehat{(\mu + \bar{\alpha}_\cdot)} = \bar{y}_{\cdot\cdot}$ and combining them we get $\widehat{(\mu + \alpha_i)} = \bar{y}_{i\cdot}$. Thus, an orthonormal transformation of $\boldsymbol{y}$ works nicely for taking out estimable functions and their BLUE.

(3) **Linear constraints.** It is obvious that only one condition is necessary and sufficient. Among various conditions to satisfy the requirements, we give two examples below.

(a) Add $\boldsymbol{K} = (1\ \boldsymbol{0}')$ to the normal equation (2.35), which is obviously not expressible by a linear combination of rows of $\boldsymbol{X}$. This is equivalent to setting $\mu = 0$, and the solution is obtained at once as

$$\hat{\mu} = 0,\ \hat{\alpha}_i = \bar{y}_{i\cdot},\ i = 1, \ldots, a. \tag{2.40}$$

(b) Add $\boldsymbol{K} = (0\ \boldsymbol{j}')$, which is equivalent to setting $\sum \alpha_i = 0$, and the solution is obtained easily as

$$\hat{\mu} = \bar{y}_{\cdot\cdot},\ \hat{\alpha}_i = \bar{y}_{i\cdot} - \bar{y}_{\cdot\cdot},\ i = 1, \ldots, a. \tag{2.41}$$

Since μ and α_i are non-estimable, (2.40) and (2.41) give different solutions. However, both solutions give the same BLUE for the estimable functions:

$$\widehat{(\mu + \alpha_i)} = \hat{\mu} + \hat{\alpha}_i = \bar{y}_{i\cdot},\ \widehat{\alpha_i - \alpha_{i'}} = \hat{\alpha}_i - \hat{\alpha}_{i'} = \bar{y}_{i\cdot} - \bar{y}_{i'\cdot}.$$

2.5.4 Generalized inverse of a matrix

Another method of proving Gauss–Markov's theorem is the application of the generalized inverse of a matrix. When there is a consistent equation

$$\boldsymbol{Ax} = \boldsymbol{v} \tag{2.42}$$

and if $\boldsymbol{A}^{-}\boldsymbol{v}$ is a solution to it, $\boldsymbol{A}^{-}$ is called a generalized inverse. Of course, $\boldsymbol{A}^{-}$ is not necessarily unique. We introduce several theorems regarding the generalized inverse.

Theorem 2.4. Generalized inverse of a matrix. $\boldsymbol{A}^{-}$ is a generalized inverse of a matrix $\boldsymbol{A}$ if and only if

$$\boldsymbol{AA}^{-1}\boldsymbol{A} = \boldsymbol{A}. \tag{2.43}$$

Proof. Let $\boldsymbol{A}$ be $(\boldsymbol{a}_1, \ldots, \boldsymbol{a}_n)$. If $\boldsymbol{A}^{-}$ is a generalized inverse of $\boldsymbol{A}$, then $\boldsymbol{x} = \boldsymbol{A}^{-}\boldsymbol{a}_i$ is a solution of the consistent equation $\boldsymbol{Ax} = \boldsymbol{a}_i$, namely $\boldsymbol{AA}^{-}\boldsymbol{a}_i = \boldsymbol{a}_i$ holds for $i = 1, \ldots, a$. This proves at once $\boldsymbol{AA}^{-1}\boldsymbol{A} = \boldsymbol{A}$. Next, suppose there is a consistent equation (2.42), then

$$\boldsymbol{v} = \boldsymbol{Ax} = \boldsymbol{AA}^{-}\boldsymbol{Ax} = \boldsymbol{AA}^{-}\boldsymbol{v}$$

holds for any $\boldsymbol{A}^{-}$ that satisfies $\boldsymbol{AA}^{-}\boldsymbol{A} = \boldsymbol{A}$. Namely, $\boldsymbol{A}^{-}\boldsymbol{v}$ is a solution of (2.42). Finally, it should be noted that $\text{rank}(\boldsymbol{A}^{-}) \geq \text{rank}(\boldsymbol{AA}^{-}\boldsymbol{A}) = \text{rank}(\boldsymbol{A})$.

It is verified that a generalized inverse that satisfies (2.43) exists as follows. Let $\text{rank}(\boldsymbol{A}) = r$ in the following, then there exist non-singular matrices $\boldsymbol{B}(m \times m)$ and $\boldsymbol{C}(n \times n)$ which satisfy

$$\boldsymbol{BAC} = \Delta = \begin{bmatrix} \Delta_r & \boldsymbol{0} \\ \boldsymbol{0} & \boldsymbol{0} \end{bmatrix}_{m \times n},$$

where $\Delta_r = \text{diag}(\delta_1, \ldots, \delta_r)$, $\delta_i \neq 0$, $i = 1, \ldots, r$ is a diagonal matrix. Then it is easily shown that

$$\Delta^{-} = \begin{bmatrix} \Delta_r^{-1} & \boldsymbol{0} \\ \boldsymbol{0} & \boldsymbol{0} \end{bmatrix}_{n \times m}$$

and $\boldsymbol{A}^{-} = \boldsymbol{C}\Delta^{-}\boldsymbol{B}$ satisfies equation (2.43), proving Δ^{-} and $\boldsymbol{A}^{-}$ to be the generalized inverses of Δ and $\boldsymbol{A}$, respectively.

Theorem 2.5. A matrix $\boldsymbol{H} = \boldsymbol{A}^{-}\boldsymbol{A}$ is an $n \times n$ idempotent matrix and rank $(\boldsymbol{H}) = r$.

Proof. It is obvious that the following two equations hold:

$$\boldsymbol{H}^2 = \boldsymbol{A}^{-}\boldsymbol{AA}^{-}\boldsymbol{A} = \boldsymbol{A}^{-}\boldsymbol{A} = \boldsymbol{H}$$

$$r = \text{rank}\ (\boldsymbol{A}) = \text{rank}\ (\boldsymbol{AH}) \leq \text{rank}\ (\boldsymbol{H}) \leq \text{rank}\ (\boldsymbol{A}) = r.$$

Theorem 2.6. General solution of $Ax=0$. A general solution of $Ax=0$ is expressed as

$$x=(I-H)w. \tag{2.44}$$

Proof. It is obvious that $Ax=0$ for x satisfying (2.44). Since $I-H$ is also an idempotent matrix, its rank is equal to $\mathrm{tr}\,(I-H)=\mathrm{tr}\,(I)-\mathrm{tr}\,(H)=n-r$ and therefore equation (2.44) expresses all the solutions of $Ax=0$.

Theorems 2.7 and 2.8 follow at once from Theorems 2.5 and 2.6.

Theorem 2.7. General solution of a consistent equation. A general solution of a consistent equation $Ax=v$ is expressed as

$$x=A^{-}v+(I-H)w,$$

where A^{-} is any generalized inverse of the coefficient matrix A.

Theorem 2.8. Unique solution in a consistent equation. A necessary and sufficient condition for the solution $L'x$ to be defined uniquely for x that satisfies a consistent equation $Ax=v$ is

$$L'H=L'.$$

Now we can apply these theorems to the normal equation (2.27) by taking $A=X'X, x=\hat{\theta}$, $v=X'y$. It is obvious that the normal equation is consistent, since the left and right sides of the equation belong to the same linear subspace. Defining any generalized inverse $(X'X)^{-}$, we obtain

$$\hat{\theta}=(X'X)^{-}X'y. \tag{2.45}$$

For any estimable function $L'\theta$, L' can be expressed as $L'=l'X'X$ and then

$$L'H=l'X'X(X'X)^{-}X'X=l'X'X=L'$$

holds, where it is easy to verify that all the linear combinations of $X'X$ and X coincide each other. Namely, $L'\hat{\theta}$ is unique by Theorem 2.8. Also, once $\hat{\theta}$ has been expressed as (2.45), the minimum variance of $L'\hat{\theta}$ can be proved in exactly the same way as in the case when X is full rank.

2.5.5 Distribution theory of the LS estimator

The variance of BLUE $L'\hat{\theta}$ of an estimable function $L'\theta$ is given by

$$L'(X'X)^{-}X'(\sigma^2 I)X(X'X)^{-}L=L'(X'X)^{-}L\sigma^2. \tag{2.46}$$

If we express $\boldsymbol{L}'\boldsymbol{\theta}$ as $\boldsymbol{l}_1'\boldsymbol{\eta}$ in the standard form of (2.29), then the variance of BLUE $\boldsymbol{l}_1'\hat{\boldsymbol{\eta}}$ is given by $\boldsymbol{l}_1'\boldsymbol{l}_1\sigma^2$. Since $\boldsymbol{l}_1'$ is expressed as $\boldsymbol{L}'\boldsymbol{P}\boldsymbol{D}^{-1/2}$, we have

$$V\left(\boldsymbol{L}'\hat{\boldsymbol{\theta}}\right) = \boldsymbol{L}'\boldsymbol{P}\boldsymbol{D}^{-1}\boldsymbol{P}'\boldsymbol{L}\sigma^2. \tag{2.47}$$

$\boldsymbol{P}\boldsymbol{D}^{-1}\boldsymbol{P}'$ is nothing but the Moore–Penrose generalized inverse of $\boldsymbol{X}'\boldsymbol{X}$.

If we assume the normality of error $\boldsymbol{e}$, $\boldsymbol{L}'\hat{\boldsymbol{\theta}}$ is distributed as normal with expectation $\boldsymbol{L}'\boldsymbol{\theta}$ and variance given by (2.46) or (2.47).

The distribution of $S\left(\hat{\boldsymbol{\theta}}\right)$ (2.31) is easily obtained by the standard form of (2.29). Namely, if the error $\boldsymbol{e}$ is distributed as normal, the elements of $\boldsymbol{Q}'\boldsymbol{y}$ are distributed independently as $N(0, \sigma^2)$ and therefore $S\left(\hat{\boldsymbol{\theta}}\right)/\sigma^2$ is distributed as a chi-squared distribution with degrees of freedom $n-r$ as the sum of squares of $n-r$ independent standard normal variables. Let us denote the chi-squared variable with degrees of freedom ν by χ_ν^2, then $S\left(\hat{\boldsymbol{\theta}}\right)$ is distributed as $\sigma^2\chi_{n-r}^2$. The degrees of freedom (df) have already appeared in Sections 2.3 and 2.5.2 as the parameter to define t- and the chi-squared distribution, respectively. Also, they are explained as follows.

(1) The chi-squared variable with df ν is characterized as the sum of squares of independent ν standard normal variables.

(2) When we express the linear model (2.21) in the standard form (2.29), the number of linearly independent parameters is r. Therefore, $n-r$ is the number obtained from the sample size n subtracting the number of essential parameters r.

(3) Since $\boldsymbol{X}'\left(\boldsymbol{y}-\boldsymbol{X}\hat{\boldsymbol{\theta}}\right)=0$, the elements of $\boldsymbol{y}-\boldsymbol{X}\hat{\boldsymbol{\theta}}$ are not linearly independent. The number of linearly independent elements is only $n-r$. $S\left(\hat{\boldsymbol{\theta}}\right)$ is formally a sum of squares of n elements. However, it can be rewritten as the sum of squares of $n-r$ elements and one expression of it is $\|\boldsymbol{Q}'\boldsymbol{y}\|^2$. In short, the degrees of freedom are the dimension of error space in this context.

Regarding the chi-squared distribution, Lemma 2.1 is useful.

Lemma 2.1. Suppose $\boldsymbol{y}$ is distributed as a multivariate normal distribution $N(\boldsymbol{\mu}, \boldsymbol{\Omega})$. Then the necessary and sufficient condition for a quadratic form $\boldsymbol{y}'\boldsymbol{A}\boldsymbol{y}$ to be distributed as chi-squared is that $\boldsymbol{A}\boldsymbol{\mu}=\boldsymbol{0}$ and $\boldsymbol{A}\boldsymbol{\Omega}$ is an idempotent matrix. In this case the degrees of freedom is $\mathrm{tr}(\boldsymbol{A}\boldsymbol{\Omega})$. If $\boldsymbol{A}\boldsymbol{\mu}\neq\boldsymbol{0}$, it is a non-central chi-squared distribution with non-centrality parameter $\boldsymbol{\mu}'\boldsymbol{A}\boldsymbol{\mu}$.

Proof. These conditions are easily verified by calculating the characteristic function of the quadratic form and omitted.

Example 2.4. $S(\hat{\theta})$ of (2.31) is a quadratic form of $\boldsymbol{y}$ by $\boldsymbol{A} = \boldsymbol{I} - \boldsymbol{\Pi}_X$ and $V(\boldsymbol{y}) = \sigma^2 \boldsymbol{I}$. Since $\boldsymbol{A}E(\boldsymbol{y}) = (\boldsymbol{I} - \boldsymbol{\Pi}_X)\boldsymbol{X\theta} = 0$ and $\sigma^{-2}\boldsymbol{A}V(\boldsymbol{y}) = \boldsymbol{I} - \boldsymbol{\Pi}_X$ is idempotent with tr $(\boldsymbol{I} - \boldsymbol{\Pi}_X) = n - r$, $S(\hat{\theta})/\sigma^2$ is distributed as chi-squared with df $n - r$.

Finally, under the normality assumption, $\hat{\boldsymbol{\theta}}$ and $\boldsymbol{Q}'\boldsymbol{y}$ are mutually independent by the orthogonality of estimation space and error space. Therefore, an estimator $\boldsymbol{L}'\hat{\boldsymbol{\theta}}$ of the estimable function and $S(\hat{\theta})$ are mutually independent.

2.6 Maximum Likelihood Estimator

The linear model is suitable for modeling the mean of the normal distribution and is not necessarily good for a general distribution. As already mentioned in Sections 1.7 and 1.10, a log linear model is more suitable for the categorical data. A well-known independence model $p_{ij} = p_{i\cdot} \times p_{\cdot j}$ is a typical example. As another example, we consider a dose–response model in a bio-assay, where data are obtained as y_i deaths out of n_i experimental animals at dose x_i of a poison. Then, as a model of expectation $E(y_i/n_i)$ a simple regression model $p_i = \beta_0 + \beta_1 x_i$ for death rate is obviously inappropriate, since it is easily beyond unity for large x_i when β_1 is positive. In this case a logit linear model

$$\log\{p_i/(1 - p_i)\} = \beta_0 + \beta_1 x_i \tag{2.48}$$

can escape from the restriction on the range of p_i, since the range of $\log\{p_i/(1 - p_i)\}$ for $0 \le p_i \le 1$ is $-\infty \sim \infty$.

The model (2.48) can also be interpreted as follows. Suppose that every experimental animal has its own threshold and a death occurs when the dose x goes beyond the threshold. Let us assume a logistic distribution

$$F(x) = \exp(\beta_0 + \beta_1 x)/\{1 + \exp(\beta_0 + \beta_1 x)\}$$

for the threshold over the population of experimental animals. Then the death rate p_i for dose x_i is $p_i = \Pr(X \le x_i) = F(x_i)$ and this is nothing but the logit linear model (2.48). A linear model for some function of the expectation parameter like (2.48) is generally called a generalized linear model, and the function is called a link function. The link function employed in (2.48) is called a logit function. One should refer to McCullagh and Nelder (1989) and Agresti (2012) for this type of modeling. A statistical method employed for the general non-linear model is the maximum likelihood method, instead of the LS method for the linear model.

We call the probability density function of $\boldsymbol{y}$ a likelihood function of parameter $\boldsymbol{\theta}$ and express it as $L(\boldsymbol{y}, \boldsymbol{\theta})$ or simply $L(\boldsymbol{\theta})$. The value $\hat{\boldsymbol{\theta}}(\boldsymbol{y})$ that maximizes $L(\boldsymbol{\theta})$ is called a maximum likelihood estimator (MLE), and intuitively expected to be a good estimator. Actually, when the sample size is large and under an appropriate regularity condition it asymptotically coincides with $\boldsymbol{\theta}$ (consistency), attains Cramér–Rao's lower bound (efficiency), distributed as normal, and is called a best asymptotically normal estimator. In particular, when the underlying distribution is normal in linear model (2.21) the MLE coincides with the LS estimator.

Example 2.5. MLE for the normal distribution. Let $y_1, \ldots, y_n$ be distributed independently as the normal distribution $N(\mu, \sigma^2)$. Then the likelihood function is

$$L(\mu, \sigma^2) = \left(2\pi\sigma^2\right)^{-n/2} \exp - \frac{\sum_i (y_i - \mu)^2}{2\sigma^2}.$$

The maximum likelihood equations are obtained by maximizing $\log L$ with respect to μ and σ^2 as follows:

$$\frac{\partial \log L}{\partial \mu} = \frac{1}{\sigma^2} \sum_i (y_i - \mu)^2 = 0 \tag{2.49}$$

$$\frac{\partial \log L}{\partial \sigma^2} = -\frac{n}{2} \times \frac{1}{\sigma^2} + \frac{1}{2\sigma^4} \sum_i (y_i - \mu)^2 = 0 \tag{2.50}$$

From equations (2.49) and (2.50), the MLE are obtained as

$$\hat{\mu} = \bar{y}_{\cdot}, \hat{\sigma}^2 = \frac{1}{n} \sum_i (y_i - \mu)^2.$$

Here, $\hat{\mu}$ is an efficient estimator and $\hat{\sigma}^2$ is a consistent and asymptotically efficient estimator.

In the following we prove the asymptotic consistency, efficiency and normality of the MLE $\hat{\theta}$ in the simplest case that $y_1, \ldots, y_n$ are independently and identically distributed following one parameter probability density function $f(y, \theta)$, not necessarily normal, where we assume some regularity conditions such as an exchange of the order of integration and derivation. The log likelihood function and its first partial derivation with respect to θ are obtained as follows:

$$\log L(\theta) = \sum_i \log f(y_i, \theta)$$

$$\frac{\partial \log L(\theta)}{\partial \theta} = \sum_i \frac{\partial \log f(y_i, \theta)}{\partial \theta}.$$

Now, $n^{-1}\partial \log L(\theta)/\partial\theta$ is an average of independent and identical random variables so that with n large it converges in probability to its expectation

$$E_{\theta_0}\left\{\frac{\partial \log f(y,\theta)}{\partial\theta}\right\} = \xi_{\theta_0}(\theta)$$

by the law of large numbers, where we put the true value of θ as θ_0. Then, ξ_{θ_0} at $\theta=\theta_0$ is calculated as

$$\xi_{\theta_0}(\theta_0) = \int \frac{\partial f(y,\theta_0)/\partial\theta_0}{f(y,\theta_0)} \times f(y,\theta_0)\mathrm{d}y = \int \{\partial f(y,\theta_0)/\partial\theta_0\}\mathrm{d}y. \tag{2.51}$$

The last equation of (2.51) is just the derivation of $\int f(y,\theta_0)\mathrm{d}y = 1$ with respect to θ_0, which reduces to zero. Therefore, if θ_0 is a unique solution of $\xi_{\theta_0}(\theta) = 0$, $\hat\theta$ converges to θ_0 in probability. This is the proof of consistency. Next we expand the log likelihood function at $\theta=\hat\theta$ around $\theta=\theta_0$:

$$\frac{1}{\sqrt{n}}\frac{\partial \log L(\theta)}{\partial\theta}\Big|_{\theta=\hat\theta} = \frac{1}{\sqrt{n}}\frac{\partial \log L(\theta)}{\partial\theta}\Big|_{\theta=\theta_0} + \sqrt{n}\left(\hat\theta-\theta_0\right)\cdot\frac{1}{n}\frac{\partial^2 \log L(\theta)}{\partial\theta^2}\Big|_{\theta=\theta^*} = 0, \tag{2.52}$$

where θ^* is some value of θ that satisfies $\left(\theta^*-\hat\theta\right)(\theta^*-\theta_0) < 0$. The first term of the middle equation of (2.52) is

$$\frac{1}{\sqrt{n}}\sum\nolimits_{i=1}^{n}\frac{\partial \log f(y_i,\theta_0)}{\partial\theta_0}$$

and converges in law by the central limit theorem to the normal distribution with expectation zero and variance:

$$V\left\{\frac{\partial \log f(y_i,\theta_0)}{\partial\theta_0}\right\} = E\left\{\frac{\partial \log f(y_i,\theta_0)}{\partial\theta_0}\right\}^2 = I_1(\theta_0),$$

where $I_1(\theta_0)$ is Fisher's amount of information evaluated at $\theta=\theta_0$. In contrast, the coefficient of $\sqrt{n}\left(\hat\theta-\theta_0\right)$ in the second term converges in probability to

$$E\left\{\frac{\partial^2 \log f(y_i,\theta_0)}{\partial{\theta_0}^2}\right\} = -I_1(\theta_0)$$

by the law of large numbers. Therefore, $\sqrt{n}\left(\hat\theta-\theta_0\right)$ is asymptotically distributed as normal, $N\left\{0, I_1^{-1}(\theta_0)\right\}$.

More generally, suppose an observation vector $\boldsymbol{y} = (y_1, \ldots, y_n)'$ has a likelihood function $L(\boldsymbol{y}, \boldsymbol{\theta})$ then, under an appropriate regularity condition, the asymptotic distribution of the MLE $\hat{\boldsymbol{\theta}}$ is shown to be a multivariate normal distribution

$$\sqrt{n}\left(\hat{\boldsymbol{\theta}}-\boldsymbol{\theta}_0\right) \cong N\{\boldsymbol{0}, \boldsymbol{I}_1^{-1}(\boldsymbol{\theta}_0)\},$$

where $\boldsymbol{I}_1(\boldsymbol{\theta}_0)$ is Fisher's information matrix per datum.

2.7 Sufficient Statistics

In statistical inference, the concept of sufficient statistics has an essential role. As a simple example, let us consider the Bernoulli sequence $\boldsymbol{y} = (y_1, \ldots, y_n)'$ where y_i takes value 0 or 1 independently with probability

$$\Pr(Y_i = 1) = p \text{ and } \Pr(Y_i = 0) = 1 - p,$$

respectively. Then, the likelihood function of $\boldsymbol{y}$ is

$$\Pi_i\left\{p^{y_i}(1-p)^{1-y_i}\right\} = p^{\sum_i y_i}(1-p)^{n-\sum_i y_i}. \tag{2.53}$$

If the parametric inference is made through equation (2.53), it is reasonable to consider the data $\boldsymbol{y}$ to contribute to the inference only through the statistic $T(\boldsymbol{y}) = \sum_i y_i$. More exactly, it can be shown that the conditional distribution given $T(\boldsymbol{y})$ does not depend on the parameter p, that is, knowing more details about $\boldsymbol{y}$ beyond $T(\boldsymbol{y})$ contributes no information regarding p. Since the marginal distribution of $T(\boldsymbol{y})$ is

$$\Pr(T = t) = \sum_{y_1} \cdots \sum_{y_n} p^t(1-p)^{n-t} = \binom{n}{t} p^t(1-p)^{n-t}, \tag{2.54}$$

we obtain the conditional distribution of $\boldsymbol{y}$ given $T(\boldsymbol{y})$ as

$$\Pr(\boldsymbol{Y} = \boldsymbol{y} \mid t = \sum y_i) = \frac{p^t(1-p)^{n-t}}{\left\{\binom{n}{t} p^t(1-p)^{n-t}\right\}} = 1\Big/\binom{n}{t}.$$

The summation in equation (2.54) is with respect to $y_1, \ldots, y_n$ subject to $\sum y_i = t$. The result shows that there are $\binom{n}{t}$ sequences of y_i subject to $\sum y_i = t$ and each of them occurs with equal probability. In other words, knowing $t = \sum y_i$, knowing further each y_i gives no extra information regarding p. Such a statistic $T(\boldsymbol{y})$ is called a sufficient statistic.

Definition 2.1. Sufficient statistics. Regarding the probability density function $f(\boldsymbol{y}, \boldsymbol{\theta})$, if the conditional distribution given a set of statistics $\boldsymbol{t} = (t_1, \ldots, t_k)'$ does not depend on parameter $\boldsymbol{\theta}$, then $(t_1, \ldots, t_k)$ are called sufficient statistics.

Theorem 2.9 gives a necessary and sufficient condition for $\boldsymbol{t}$ to be sufficient statistics.

Theorem 2.9. Factorization theorem. The necessary and sufficient condition for $\boldsymbol{t} = (t_1, \ldots, t_k)'$ to be sufficient statistics is that $f(\boldsymbol{y}, \boldsymbol{\theta})$ is factorized as

$$f(\boldsymbol{y}, \boldsymbol{\theta}) = c(\boldsymbol{\theta}) \times g(\boldsymbol{y}) \times h(\boldsymbol{t}, \boldsymbol{\theta}) \tag{2.55}$$

Proof. The proof is given for the probability function $P(\boldsymbol{y}, \boldsymbol{\theta})$ of a discrete distribution.

Necessity. For the conditional distribution to be free from parameter $\boldsymbol{\theta}$, the following equation necessarily holds:

$$\frac{P(\boldsymbol{y}, \boldsymbol{\theta})}{\sum_{T(\boldsymbol{y}) = \boldsymbol{t}} P(\boldsymbol{y}, \boldsymbol{\theta})} = g(\boldsymbol{y}), \tag{2.56}$$

where $g(\boldsymbol{y})$ is free from $\boldsymbol{\theta}$. The denominator on the left-hand side of (2.56) is a summation with respect to $\boldsymbol{y}$ satisfying $T(\boldsymbol{y}) = \boldsymbol{t}$ so that it should be a function of $\boldsymbol{t}$ and can be written as $c(\boldsymbol{\theta}) \times h(\boldsymbol{t}, \boldsymbol{\theta})$. Thus it is necessary for $P(\boldsymbol{y}, \boldsymbol{\theta})$ to be factorized as $c(\boldsymbol{\theta})g(\boldsymbol{y})h(\boldsymbol{t}, \boldsymbol{\theta})$.

Sufficiency. If $P(\boldsymbol{y}, \boldsymbol{\theta})$ is factorized as (2.55), we have

$$\sum_{T(\boldsymbol{y}) = \boldsymbol{t}} P(\boldsymbol{y}, \boldsymbol{\theta}) = c(\boldsymbol{\theta}) \times g^*(\boldsymbol{t}) \times h(\boldsymbol{t}, \boldsymbol{\theta}).$$

Then the conditional distribution becomes

$$c(\boldsymbol{\theta}) \times g(\boldsymbol{y}) \times h(\boldsymbol{t}, \boldsymbol{\theta}) / \{c(\boldsymbol{\theta}) \times g^*(\boldsymbol{t}) \times h(\boldsymbol{t}, \boldsymbol{\theta})\} = g(\boldsymbol{y}) / g^*(\boldsymbol{t})$$

and does not depend on $\boldsymbol{\theta}$.

Example 2.6. If $y_1, \ldots, y_n$ are distributed independently as the normal distribution $N(\mu, \sigma^2)$, then the density function is factorized as

$$f(\boldsymbol{y}, \boldsymbol{\theta}) = \left(2\pi\sigma^2\right)^{-n/2} \exp -\frac{1}{2\sigma^2}\left\{n(\bar{y}_{\cdot} - \mu)^2 + \sum_i (y_i - \bar{y}_{\cdot})^2\right\}. \tag{2.57}$$

Therefore, $\boldsymbol{t}=\left\{\bar{y}_{\cdot},\ \sum_i(y_i-\bar{y}_{\cdot})^2\right\}$ is a set of sufficient statistics.

Example 2.7. If $y_1, \ldots, y_n$ are distributed independently as a uniform distribution in the interval $[0,\ \theta]$, then the density function can be written as

$$f(\boldsymbol{y}, \boldsymbol{\theta}) = \begin{cases} \dfrac{1}{\theta^n},\ 0 \le y_1, \cdots, y_n \le \theta, \\ \quad 0,\ \text{otherwise}, \end{cases}$$

$$= \frac{1}{\theta^n} h_v(\theta - y_{\max}) h_v(y_{\min}),$$

where

$$h_v(x) = \begin{cases} 0,\ x<0, \\ 1,\ x \ge 0 \end{cases}$$

is the Heaviside step function. Therefore, $y_{\max}$ is a sufficient statistic.

Example 2.8. Let $\boldsymbol{y}=(y_1, \ldots, y_n)'$ be distributed independently according to an unknown density function $f(\boldsymbol{y})$. This is regarded as the case of infinite number of parameters. Let us denote the ordered statistics by

$$y_{(1)} \le \cdots \le y_{(n)}.$$

Then the joint density function of the ordered statistics $y_{(1)}, \ldots, y_{(n)}$ is

$$g\left(y_{(1)}, \ldots, y_{(n)}\right) = n! \times \Pi_i f\left\{y_{(i)}\right\}.$$

Actually, the following equation is easily verified:

$$\int \cdots \int g\left(y_{(1)}, \ldots, y_{(n)}\right) \Pi_i \mathrm{d}y_{(i)} = 1,$$

where the integration is subject to $y_{(1)} \le \cdots \le y_{(n)}$. This proves that $y_{(1)}, \ldots, y_{(n)}$ are sufficient statistics because of the factorization

$$f(\boldsymbol{y}) = \frac{1}{n!} \times g\left(y_{(1)}, \cdots, y_{(n)}\right).$$

The conditional distribution of $\boldsymbol{y}$ given the sufficient statistics gives equal probability $1/\ n\,!$ to all the permutations of $(y_{(1)}, \ldots, y_{(n)})$. This distribution is utilized later to evaluate the p-value of the non-parametric permutation test.

The following theorem tells how sufficient the sufficient statistics are for statistical inference. It also gives a method of constructing a minimum variance unbiased estimator.

Theorem 2.10. Rao–Blackwell. For simplicity, we describe the case of one parameter but the theorem holds more generally. Let t be a sufficient statistic for the probability function $P(y, \theta)$ or $f(y, \theta)$. Let $\hat{\theta}$ be an arbitrary unbiased estimator of θ and consider the conditional expectation of $\hat{\theta}$ given t:

$$\hat{\theta}^* = E(\hat{\theta}|t).$$

Then $\hat{\theta}^*$ is also an unbiased estimator and the variance is improved as

$$V\left(\hat{\theta}^*\right) \le V(\hat{\theta}). \tag{2.58}$$

Proof. Since t is a sufficient statistic, $\hat{\theta}^*$ is a function of t not depending on the unknown parameter. Then its expectation with respect to t is equivalent to the overall expectation of $\hat{\theta}$:

$$E_T\left\{\hat{\theta}^*(T)\right\} = E_T\left\{E(\hat{\theta}|T)\right\} = E(\hat{\theta}) = \theta.$$

Similarly, we have

$$\begin{aligned} V(\hat{\theta}) &= E\left\{\left(\hat{\theta} - \hat{\theta}^* + \hat{\theta}^* - \theta\right)^2\right\} \\ &= E\left\{\left(\hat{\theta} - \hat{\theta}^*\right)^2\right\} + 2E\left\{\left(\hat{\theta} - \hat{\theta}^*\right)\left(\hat{\theta}^* - \theta\right)\right\} + E\left\{\left(\hat{\theta}^* - \theta\right)^2\right\}. \end{aligned} \tag{2.59}$$

The second expectation of (2.59) is found to be zero by taking the conditional expectation given t first. The first expectation is non-negative and the last expectation is the variance of $\hat{\theta}^*$, thus proving the inequality (2.58).

This theorem tells us that for any unbiased estimator it is possible to construct an unbiased estimator based on the sufficient statistics whose variance is equal to or smaller than the original unbiased estimator. In other words, to search for the minimum variance unbiased estimator we can restrict to the function of sufficient statistics. This theorem is not restricted to the unbiased estimator. For any $\hat{\theta}$ there is a function of sufficient statistics improving the MSE with the same expectation as $\hat{\theta}$.

Example 2.9. Assume a Bernoulli sequence $\boldsymbol{y}=(y_1,\ldots,y_n)'$ which is distributed as (2.53). Then $\hat{p}=y_1$ is an unbiased estimator, since it satisfies

$$E(\hat{p})=E(Y_1)=1\times p+0\times(1-p)=p$$

with variance

$$V(\hat{p})=E(\hat{p}^2)-\{E(\hat{p})\}^2=(1\times p+0\times p)-p^2=p(1-p).$$

Therefore, we take a conditional expectation of $\hat{p}$ given the sufficient statistic $T=\sum y_i$. The conditional distribution of $\boldsymbol{y}$ takes one of the sequences with total number of unity equal to t at equal probability, that is, $1/\binom{n}{t}$. The conditional expectation is therefore

$$\hat{p}^*=E(Y_1\mid t)=1\times\frac{\binom{n-1}{t-1}}{\binom{n}{t}}+0\times\frac{\binom{n-1}{t}}{\binom{n}{t}}=\frac{t}{n},$$

giving an unbiased estimator again with a smaller variance $V(\hat{p}^*)=p(1-p)/n$.

Example 2.10. Suppose $\boldsymbol{y}=(y_1,\ldots,y_n)'$ is distributed as normal $N(\mu,1)$. In this case it is obvious that $\bar{y}.$ is a sufficient statistic and y_1 is an unbiased estimator of μ. Therefore,we take a conditional expectation of y_1 given the sufficient statistic $\bar{y}.$. To obtain the conditional distribution it is convenient to transform the data $\boldsymbol{y}$ to $\bar{y}.$ and a set of statistics which are independent of $\bar{y}.$. As one such method, we take an orthonormal transformation

$$\boldsymbol{u}=\begin{bmatrix} n^{-1/2}\boldsymbol{j}_n' \\ \boldsymbol{P}_n' \end{bmatrix}\boldsymbol{y},$$

where we already defined P_n in (2.36) and (2.37). The first element of $\boldsymbol{u}$ is $u_1=\sqrt{n}\bar{y}.$ and the last $(n-1)$ elements $\boldsymbol{u}_2$ are distributed as $N(\boldsymbol{0},\boldsymbol{I}_{n-1})$ independently of u_1. Then the conditional distribution of

$$\boldsymbol{P}_n\boldsymbol{u}_2=\boldsymbol{P}_n\boldsymbol{P}_n'\boldsymbol{y}=(\boldsymbol{I}-n^{-1}\boldsymbol{jj}')\boldsymbol{y}=\boldsymbol{y}-\bar{y}.\boldsymbol{j}$$

given $\bar{y}.$ is $N(\boldsymbol{0},\boldsymbol{P}_n\boldsymbol{P}_n')$. Therefore, the conditional distribution of $\boldsymbol{y}=\boldsymbol{P}_n\boldsymbol{u}_2+\bar{y}.\boldsymbol{j}$ given $\bar{y}.$ is known to be

$$N(\bar{y}.\boldsymbol{j}_n,\ \boldsymbol{I}_n-n^{-1}\boldsymbol{j}_n\boldsymbol{j}_n'). \tag{2.60}$$

The distribution (2.60) is degenerate, since $\bar{y}.$ is fixed. However, in $(n-1)$ dimensions it is a usual multivariate normal distribution. In particular, the conditional distribution of y_1 is $N(\bar{y}.,\ 1-n^{-1})$ and its conditional expectation is $\bar{y}.$.

By definition the observation vector $\boldsymbol{y}$ is itself a sufficient statistic. The discussion up to here does not contradict this, but we have been implicitly assuming a minimal dimension of the sufficient statistics. This is called a minimal sufficient statistic and one should refer to Lehman and Scheffé (1950) for a more rigid definition.

Definition 2.2. Complete sufficient statistics. If a function of sufficient statistics whose expectation is zero is only zero, the sufficient statistics are called complete.

Suppose there are two unbiased estimators T_1 and T_2 composed of the complete sufficient statistics for a parameter, then T_1 and T_2 should be equal, since

$$E(T_1 - T_2) = E(T_1) - E(T_2) = 0.$$

Namely, the unbiased estimator based on the complete sufficient statistics is a unique and minimum variance unbiased estimator by the Rao–Blackwell theorem. In Examples 2.9 and 2.10, t and $\bar{y}_.$ are complete sufficient statistics so that t/n and $\bar{y}_.$ are minimum variance unbiased estimators of p and μ, respectively. This gives another proof from the Cramér–Rao's lower bound or Rao's Theorem 2.1. More generally, if the y_i are distributed independently as an exponential family

$$C(\boldsymbol{\theta})h(y_i)\ \exp-\{\theta_1 t_1(y_i) + \cdots + \theta_k t_k(y_i)\},$$

where $\theta_1, \ldots, \theta_k$ are functionally independent and the range of y for the density to be positive does not depend on $\boldsymbol{\theta}$, then

$$t_j = \sum\nolimits_i t_j(y_i),\ j = 1, \ldots, k$$

gives a set of complete sufficient statistics.

Example 2.11. Example 2.5 continued. The density function $f(y_i, \mu, \sigma^2)$ of a normal distribution is written as

$$f\left(y_i, \mu, \sigma^2\right) \propto \ \exp-\frac{1}{2}\left(\frac{1}{\sigma^2}y_i^2 - \frac{2\mu}{\sigma^2}y_i\right).$$

Therefore, $\left(\sum y_i^2, \sum y_i\right)$ is a set of complete sufficient statistics.

References

Agresti, A. (2012) *Categorical data analysis*, 3rd edn. Wiley, New York.

Hirotsu, C. (1971) Zero and non-zero Lagrange multipliers in the least squares method. *Rep. Statist. Appl. Res., JUSE* **18**, 105–114.

Kagan, K. M., Linigue, Yu. V. and Rao, C. R. (1973) *Characterization problems in mathematical statistics*. Wiley, New York.

Lehman, E. L. and Scheffé, H. (1950) Completeness, similar regions and unbiased estimation. *Sankhyā* **10**, 305; 15, 219.

Masuyama, M. (1976) On stochastic models for quasi-constancy of biochemical individual variability. *Jap. J. Applied Statist.* **5**, 95–114 (in Japanese).

McCullagh, P. and Nelder, J. A. (1989) *Generalized linear models*, 2nd edn. Chapman & Hall/ CRC, London.

3

Basic Test Theory

A test is a statistical method to judge whether a specified value of a parameter is true or not. The specification is called a null hypothesis, H_0. By the nature of the test, a rejection of H_0 is useful information whereas an acceptance does not make sense unless the sample size is appropriately chosen for the purpose. When constructing a test it is important to specify the direction of departure from the null hypothesis, which is called the alternative hypothesis H_1. The null hypothesis H_0 can be a single value, a range, or a simplified structure of parameters with lower dimension. Reduction of the estimation space of a linear model is an example of the last, and the first is considered a special case of the last.

3.1 Normal Mean

3.1.1 Setting a null hypothesis and a rejection region

Suppose we have the simple model of Section 2.1:

$$y_i = \mu + e_i,\ i = 1, \ldots, n,$$

with e_i distributed independently as $N(0, \sigma^2)$, where we assume for a while σ^2 to be known for simplicity. Now, the problem is to test the null hypothesis

$$H_0 : \mu = \mu_0 \tag{3.1}$$

against the alternative hypothesis

$$H_1 : \mu = \mu_1, \tag{3.2}$$

Advanced Analysis of Variance, First Edition. Chihiro Hirotsu.

where μ_0 is a past process mean and μ_1 is the value after making some change in the process. A test specifies a region R in the sample space which supports H_1 more than H_0 and rejects H_0 if the data $\boldsymbol{y}$ belong to R. The region R is therefore called a rejection region. The rejection region R is characterized by two kinds of probability:

(1) Probability of error of the first kind $\Pr\left(\boldsymbol{y} \in R \mid H_0\right)$. This is the error of rejecting H_0 when it is true.

(2) Probability of error of the second kind $\Pr\left(\boldsymbol{y} \notin R \mid H_1\right)$: This is the error of accepting H_0 when it is not true.

These two probabilities are contrary to each other. For example, it is possible to make the second kind of error have probability zero by taking R equal to the whole sample space (always rejecting H_0, irrespective of the value of $\boldsymbol{y}$), but then the first kind of error has probability unity. In contrast, we can make the first kind of error have probability zero by setting R equal to the empty set, but then the second kind of error has probability unity. One approach to avoid such a nonsense rejection region is to set an upper bound α for the first kind of error probability and to minimize the second kind of error probability under this condition.

The upper bound α is called a significance level and usually taken as 0.05 or 0.01. We define the power of a test by

$$\Pr(\boldsymbol{y} \in R \mid H_1) = 1 - \Pr(\boldsymbol{y} \notin R \mid H_1).$$

This approach is then formulated as an optimization problem of maximizing

$$\Pr(\boldsymbol{y} \in R \mid H_1) \text{ subject to } \Pr(\boldsymbol{y} \in R \mid H_0) \le \alpha.$$

The result is called a rejection region of the most powerful test. Let us define an indicator function of R as

$$\varphi(\boldsymbol{y}) = \begin{cases} 1, & y \in R, \\ 0, & y \notin R, \end{cases}$$

then defining a rejection region is equivalent to determining the function $\varphi(\mathbf{y})$, which is called a test function. The most powerful test is given by Neyman–Pearson's fundamental lemma.

Theorem 3.1. Neyman–Pearson's fundamental lemma. When $\boldsymbol{y}$ is distributed according to a density function $f(\boldsymbol{y}, \mu)$, the rejection region of the most powerful test for testing H_0 (3.1) against H_1 (3.2) is given by

$$R : f(\boldsymbol{y}, \mu_1)/f(\boldsymbol{y}, \mu_0) > c, \tag{3.3}$$

where the constant c is determined by

$$\Pr(\mathbf{y} \in R \mid H_0) = \alpha. \tag{3.4}$$

Proof. Let c be an unknown constant then, for any test $\varphi(\mathbf{y})$ that satisfies the condition $\Pr(y \in R \mid H_0) \leq \alpha$, we have

$$\int \varphi(\mathbf{y})\{f(\mathbf{y}, \mu_1) - cf(\mathbf{y}, \mu_0)\}\mathrm{d}\mathbf{y} \geq \int \varphi(\mathbf{y})f(\mathbf{y}, \mu_1)\mathrm{d}\mathbf{y} - c\alpha.$$

Define a particular test function $\varphi^*(\mathbf{y})$ by

$$\varphi^*(\mathbf{y}) = \begin{cases} 1, & f(\mathbf{y}, \mu_1) - cf(\mathbf{y}, \mu_0) > 0, \\ 0, & f(\mathbf{y}, \mu_1) - cf(\mathbf{y}, \mu_0) \leq 0. \end{cases}$$

Then we have

$$\int \varphi^*(\mathbf{y})\{f(\mathbf{y}, \mu_1) - cf(\mathbf{y}, \mu_0)\}\mathrm{d}\mathbf{y} \geq \int \varphi(\mathbf{y})f(\mathbf{y}, \mu_1)\mathrm{d}\mathbf{y} - c\alpha.$$

Therefore, if we can define c so as to satisfy $E\{\varphi^*(\mathbf{y}) \mid H_0\} = \alpha$, we have

$$E\{\varphi^*(\mathbf{y}) \mid H_1\} \geq E\{\varphi(\mathbf{y}) \mid H_1\}$$

and $\varphi^*(\mathbf{y})$ is the most powerful test function since $\varphi(\mathbf{y})$ is an arbitrary test function with significance level α.

Equation (3.3) implies that the data $\mathbf{y}$ support μ_1 more than μ_0, and equation (3.4) tells us that the probability of the error of the first kind is equal to α. The probability of the error of the first kind that the test actually has is called the size or risk rate of the test, whereas the significance level α implies the upper bound of the risk in constructing a test. To raise the power of a test, the size of the test should be set at the upper bound α, but this is generally impossible for discrete data.

If we apply equation (3.3) to the normal density $N(\mu, \sigma^2)$, we have

$$R: (\mu_1 - \mu_0)\bar{y}_. > c'.$$

If we assume $\mu_1 - \mu_0 > 0$, then a rejection region is obtained as

$$R: \bar{y}_. > c''.$$

Namely, a larger $\bar{y}_.$ supports H_1 more. The value of c'' is determined from (3.4). Under H_0, $\bar{y}_.$ is distributed as $N(\mu, \sigma^2/n)$ and the standardized form $u = (\bar{y}_. - \mu_0)/\sqrt{\sigma^2/n}$ is distributed as $N(0, 1)$. We generally denote the upper 100 α percentile of $N(0, 1)$ as z_α, as shown in Fig. 3.1.

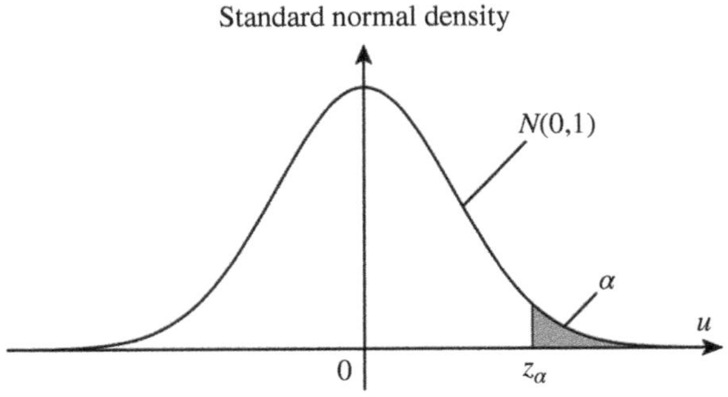

Figure 3.1 An upper 100 α percentile of $N(0, 1)$.

Then, from

$$\Pr\left\{(\bar{y}_{\cdot}-\mu_0)/\sqrt{\sigma^2/n}>z_\alpha\right\}=\alpha$$

we have

$$c''=\mu_0+\left(\sigma/\sqrt{n}\right)z_\alpha.$$

Namely, the rejection region of the most powerful test is given by

$$R:\bar{y}_{\cdot}>\mu_0+\left(\sigma/\sqrt{n}\right)z_\alpha. \quad (3.5)$$

It rejects H_0 when $\bar{y}_{\cdot}$ exceeds μ_0 more than $(\sigma/\sqrt{n})z_\alpha$. It should be noted that the rejection region (3.5) does not depend on μ_1, except for the condition $\mu_1-\mu_0>0$. That is, it is most powerful against every μ_1 larger than μ_0. Therefore, the rejection region (3.5) is uniformly most powerful against the alternative hypothesis

$$H_1':\mu>\mu_0. \quad (3.6)$$

Similarly, against the alternative hypothesis

$$H_1'':\mu<\mu_0 \quad (3.7)$$

the rejection region

$$R':\bar{y}_{\cdot}<\mu_0-\left(\sigma/\sqrt{n}\right)z_\alpha \quad (3.8)$$

gives the uniformly most powerful test. The hypotheses H_1' (3.6) and H_1'' (3.7) are called right and left one-sided alternative hypothesis, respectively. In contrast, for the two-sided hypothesis

$$H_1''':\mu\neq\mu_0 \quad (3.9)$$

there is obviously no uniformly most powerful test, since it is obvious that the rejection region R (3.5) is uniformly most powerful against the alternative hypothesis

$\mu > \mu_0$ but the power of such a rejection region against $\mu < \mu_0$ becomes less than α. However, there certainly exists a rejection region R' that is more powerful than R against $\mu < \mu_0$. Under the one- and two-sided hypotheses (3.6), (3.7), and (3.9), the assumed distribution is not unique and they are called a composite hypothesis. In contrast, under H_0 (3.1) and H_1 (3.2) the distribution is uniquely determined and they are called a simple hypothesis. Another type of composite hypothesis will be introduced in Section 3.1.4.

An approach to finding a reasonable test against the two-sided alternative is to request for the power of the test not to go down under the significance level α. A test which satisfies this property is called an unbiased test. The rejection region of the most powerful level α unbiased test

$$R_U : \bar{y}_. - \mu_0 > a \quad \text{or} \quad \bar{y}_. - \mu_0 < b$$

is obtained by determining a and b to satisfy the requirements that the power function defined in the next section should take a minimum value α at $\mu = \mu_0$ (see Fig. 3.3). Then we reach the two-sided rejection region

$$R_2 : |\bar{y}_. - \mu_0| > \left(\sigma/\sqrt{n}\right) z_{\alpha/2}. \tag{3.10}$$

The rejection region R_2 does not depend on the particular value of $\mu\ (\neq \mu_0)$ and is called a uniformly most powerful unbiased test.

As mentioned above, a rejection region is often expressed by the amount of test statistic $T(\mathbf{y})$, like $T > c$ or $|T| > c$. In this case we can evaluate the significance by the probability above the observed value $T = t$ evaluated at the null hypothesis H_0. This probability, $\Pr(T \ge t \mid H_0)$, is called the p-value. To declare significance when the p-value is less than α is equivalent to the significance test by rejection region with significance level α. Generally, having the p-value is a more informative procedure.

3.1.2 Power function

The power of the one-sided test (3.5) at a particular value $\mu = \mu_1$ is calculated, noting that $u = (\bar{y}_. - \mu_1)/\sqrt{\sigma^2/n}$ is distributed as $N(0, 1)$. By a simple calculation we have

$$\begin{aligned}\Pr(\mathbf{y} \in R \mid \mu = \mu_1) &= \Pr\left\{\bar{y}_. > \mu_0 + \left(\sigma/\sqrt{n}\right) z_\alpha \mid \mu = \mu_1\right\} \\ &= \Pr\left\{u + (\mu_1 - \mu_0)/\left(\sigma/\sqrt{n}\right) > z_\alpha\right\}.\end{aligned}$$

In Fig. 3.2 the power is the upper-tail probability of the shaded area by oblique lines with respect to the standard normal distribution shifted to the right by the amount of $\gamma = (\mu_1 - \mu_0)/\left(\sigma/\sqrt{n}\right)$. The left one-sided rejection region R' (3.8) should be considered symmetrically. Similarly, for the two-sided rejection region R_2 (3.10) we get

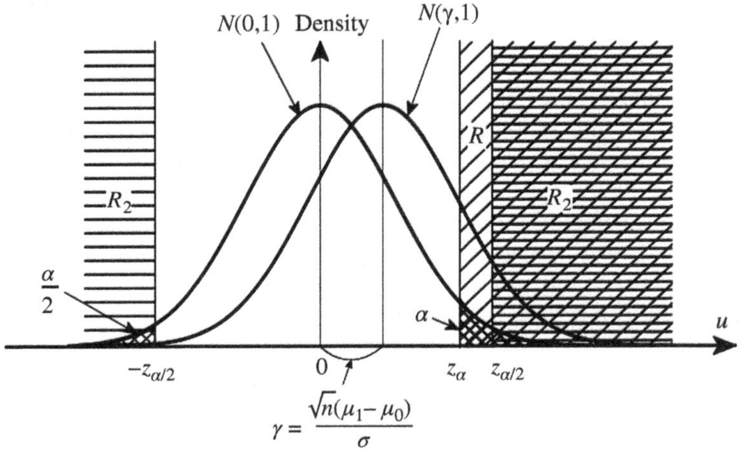

Figure 3.2 Rejection region R and R_2.

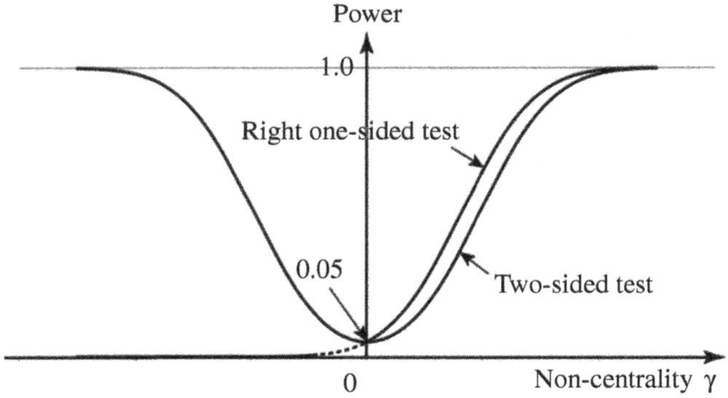

Figure 3.3 Power function.

$$\Pr\left(y \in R_2 \mid \mu = \mu_1\right) = \Pr\left\{u + \frac{(\mu_1 - \mu_0)}{(\sigma/\sqrt{n})} > z_{\alpha/2}\right\} + \Pr\left\{u + \frac{(\mu_1 - \mu_0)}{(\sigma/\sqrt{n})} < -z_{\alpha/2}\right\}.$$

The power is a function of

$$\gamma = \sqrt{n}(\mu_1 - \mu_0)/\sigma$$

and increases with it or the absolute value of it. In other words, the power is large when the departure of μ_1 from μ_0 is large, n is large, and the variance is small. The power as a function of γ is called the power function. The outline of the power function for the rejection regions R_1 and R_2 is as in Fig. 3.3.

In the introduction of the test above, the treatments of the two hypotheses H_0 and H_1 are not symmetrical. If we take small values 0.01 or 0.05 for α as usual and $\mathbf{y}$ belongs to the rejection region, this strongly supports the negation of H_0. However, if $\mathbf{y}$ does not belong to the rejection region this does not necessarily support the fact that H_0 is true. In the two-sided rejection region R_2 with $\alpha = 0.05$, for example, power reach 0.5 at $\gamma = \sqrt{n}(\mu_1 - \mu_0)/\sigma = z_{0.05/2} = 1.96$, and for γ smaller than 1.96, the probability of acceptance is larger than the probability of rejection of the null hypothesis. This should occur quite often when the departure $\delta = (\mu_1 - \mu_0)/\sigma$ is moderate and the sample size is not large enough. Therefore, it is not appropriate to declare the acceptance of H_0 by not rejecting H_1 unless the sample size has been determined to have a sufficient power against the amount of departure δ of interest. We often call the complement of a rejection region an acceptance region, but careful interpretation is strongly recommended. The equivalence test of Section 5.3 is required in this context.

3.1.3 Sample size determination

Suppose we wish to detect a departure $\mu_1 - \mu_0 = k\sigma$ at power $1-\beta$, where β is usually taken as 0.10 or 0.20.

(1) **Right one-sided alternative.** The power requirement at $\mu_1 = \mu_0 + k\sigma$ is formulated as follows:

$$\Pr\left\{u > z_\alpha - \sqrt{n}(\mu_1 - \mu_0)/\sigma\right\} = \Pr\left(u > z_\alpha - \sqrt{n}k\right) \geq 1-\beta.$$

It requests

$$z_\alpha - \sqrt{n}k \leq z_{1-\beta} = -z_\beta \Rightarrow n \geq \left(z_\alpha + z_\beta\right)^2/k^2.$$

The left one-sided alternative is dealt with similarly, and the same result obtained.

(2) **Two-sided alternative.** If $\mu_1 = \mu_0 + k\sigma \ (k>0)$ and $1-\beta$ is not small, the probability that the observation $\mathbf{y}$ belongs to the left part of the rejection region R_2 is negligible. Therefore, the required sample size is

$$n \geq \left(z_{\alpha/2} + z_\beta\right)^2/k^2.$$

For the same α and β, the required sample size is larger for the two-sided alternative than the one-sided alternative. This is quite reasonable, since the one-sided alternative could restrict the direction of departure by prior information.

The one-sided alternative is employed when the direction of departure is clear, for example, the defective percentage should decrease with improvement of a production process or the freezing point should go down with contamination. The multiple decision processes approach of Chapter 4 is attractive in this context, as it unifies nicely the right and left one-sided and two-sided tests.

3.1.4 Nuisance parameter

It was for convenience that we assumed σ^2 to be known in Section 3.1.1. In this section we consider a composite null hypothesis

$$K_0 : \mu = \mu_0,\ \sigma^2 \text{ is an arbitrary positive number,}$$

where the distribution under the null hypothesis is not unique. An unknown parameter like σ^2 here, which is not the object of statistical inference but should be taken into account in the test procedure, is called a nuisance parameter. The usual approach in this case is to consider a test which keeps the significance level α irrespective of the value of the nuisance parameter. Such a test is called a similar test and, if there exists a most powerful test among similar tests, it is called a most powerful similar test. A similar test is usually constructed by a test statistic whose distribution is free from the nuisance parameter under the null hypothesis. Therefore, it is generally constructed based on the conditional distribution given the sufficient statistics under the null hypothesis.

In this problem the statistic

$$t = (\bar{y}_{\cdot} - \mu_0)/\sqrt{\hat{\sigma}^2/n} \tag{3.11}$$

has been obtained from the standardized statistic $(\bar{y}_{\cdot} - \mu_0)/\sqrt{\sigma^2/n}$ by replacing σ^2 by the unbiased variance $\hat{\sigma}^2$ (2.32). If we rewrite t (3.11) as

$$t = \frac{(\bar{y}_{\cdot} - \mu_0)/\sqrt{\sigma^2/n}}{\sqrt{\{S(\hat{\mu})/\sigma^2\}/(n-1)}},$$

then the numerator is distributed as a standard normal distribution and the denominator is distributed as $\sqrt{\chi^2_{n-1}/(n-1)}$, independently of the numerator (see Example 2.2). This statistic is called the Student's t-statistic, and its distribution is Student's, or simply the t-distribution with df $n-1$. The test based on t is called Student's t, or simply the t -test. We have already shown the outline of the density function of the t-distribution in Fig. 2.1. If y_i is distributed as $N(\mu, \sigma^2)$, namely under the alternative hypothesis, it is called a non-central t-distribution with non-centrality parameter

$$\gamma = \sqrt{n}(\mu - \mu_0)/\sigma. \tag{3.12}$$

When $\gamma = 0$ it is a usual t-distribution.

For the right one-sided composite alternative

$$K_1 : \ \mu > \mu_0,\ \sigma^2 \text{ is an arbitrary positive number} \tag{3.13}$$

the rejection region

$$R_t : t > t_{n-1}(\alpha) \tag{3.14}$$

gives a uniformly most powerful similar test with significance level α, where $t_{n-1}(\alpha)$ is the upper α point of the t-distribution. Similarly, for the left one-sided alternative K_1' we get a uniformly most powerful similar test

$$R_t' : t < -t_{n-1}(\alpha).$$

For the two-sided alternative

$$K_2: \ \mu \neq \mu_0, \ \sigma^2 \text{is an arbitrary positive number,}$$

we consider the class of unbiased tests as in Section 3.1.1 to obtain

$$R_{2t}: \ |t| > t_{n-1}(\alpha/2).$$

This is a uniformly most powerful unbiased test.

The power of the t-test is calculated as a function of the non-centrality parameter γ (3.12). The t-distribution is asymptotically distributed as normal. Therefore, the outline of the power function is rather similar to Fig. 3.3 when the sample size is moderately large.

3.1.5 Non-parametric test for median

There are various test procedures for the normal mean other than the t-test of the previous section. Suppose that six data points $y_1, \ldots, y_6$ are all larger than μ_0 in testing the null hypothesis H_0 (3.1) against H_1' (3.6) or K_1 (3.13), for example. Such an outcome is an extreme value supporting the right one-sided hypothesis obtained at probability $1/2^6 \doteqdot 0.016$ under H_0. In other words, the one-sided p-value is 0.016. Therefore, we can conclude that the result is significant at level 0.05. This test uses only the sign of $y_i - \mu_0$ and is called a sign test. It does not require the normality assumption of the error $\boldsymbol{e}$ and gives a non-parametric test procedure for the null hypothesis of the median

$$K_0': \Pr(y > \mu_0) = \Pr(y < \mu_0) \tag{3.15}$$

against the right one-sided alternative hypothesis

$$K_1'': \Pr(y > \mu_0) > \Pr(y < \mu_0). \tag{3.16}$$

However, it is natural that the power of the sign test is considerably lower than the most powerful tests R (3.5) and R_t (3.14) if the distribution of the error is actually normal. In contrast, if the distribution of the error is symmetric then the permutation test and the signed rank sum test below can give a reasonable power.

(1) **Permutation test.** If we assume symmetry of the error distribution, then the distribution of $y - \mu_0$ under K_0' (3.15) becomes symmetric at both sides of zero. Let us define $x_i = y_i - \mu_0$, $i = 1, \ldots, n$. Then the sum $x.$ should be around zero under K_0' and take a large positive value under the right one-sided alternative K_1'' (3.16). The significance of $x.$ can be evaluated by the conditional distribution given $|x_1|, \ldots, |x_n|$. The conditional distribution of x_i takes $\pm |x|_i$ at equal probability 1/2 under K_0' by the symmetry assumption. More exactly, the joint conditional distribution of $\boldsymbol{x} = (x_1, \ldots, x_n)'$ given $|x|_1, \ldots, |x|_n$ takes all the 2^n combinations $(\pm |x_1|, \ldots, \pm |x_n|)'$ at equal probability $1/2^n$. Then the p-value of $x.$ is given by $n_x/2^n$, where n_x is the number

of sample points with sum $X.$ exceeding or equal to the observed value $x.$. For the left one-sided alternative we may count the number of $X.$ below or equal to $x.$. It is also equal to the number of $X.$ exceeding or equal to the observed value $-x.$ For the two-sided test we count the number n_x of $|X.|$ exceeding or equal to the observed value $|x.|$, and then the p-value is $2n_x/2^n$.

If n is moderately large, it is time-consuming to count the number n_x. In this case a normal approximation of the conditional distribution is available. We can calculate the conditional expectation and variance of

$$Z = X_1 + \cdots + X_n \tag{3.17}$$

given $|X_i| = |x_i|$, $i = 1, \ldots, n$, under K_0' as follows. Since X_i takes $\pm |x_i|$ at equal probability 1/2, we have

$$E(X_i) = \frac{1}{2} \times |x_i| + \frac{1}{2} \times (-|x_i|) = 0$$

and $V(X_i) = E(X_i^2) - \{E(X_i)\}^2 = \frac{1}{2} \times |x_i|^2 + \frac{1}{2} \times (-|x_i|)^2 - 0^2 = x_i^2$.

Assuming $X_1, \ldots, X_n$ are mutually independent, we have

$$E(Z) = 0,\ V(Z) = \sum_{i=1}^{n} x_i^2.$$

Therefore, when n is large the null distribution of the standardized variable

$$u = (x_1 + \cdots + x_n) \big/ \sqrt{\sum_{i=1}^{n} x_i^2} \tag{3.18}$$

is a standard normal $N(0, 1)$. Therefore, the rejection regions are

$$u > z_\alpha \text{ for the right one-sided alternative hypothesis,}$$
$$u < -z_\alpha \text{ forthe left one-sided alternative hypothesis,}$$
$$|u| > z_{\alpha/2} \text{ forthe two-sided alternative hypothesis.}$$

We call this test simply the u-test. This approximation is useful roughly for $n \geq 10$.

It is interesting to note the relationship between the t (3.11) and u (3.18) tests. It is easy to see that

$$t = u \big/ \left\{ \left(n - u^2\right) / (n-1) \right\}^{1/2}. \tag{3.19}$$

Namely, the u-test is equivalent to the t-test employing the critical value of (3.19) at $u = z_\alpha$ or $z_{\alpha/2}$ instead of $t_{n-1}(\alpha)$ or $t_{n-1}(\alpha/2)$. We denote the value of (3.19) at $u = z_\alpha$ by u_α. Then, as shown in Table 3.1, there is not much difference between the two critical values and at $n = \infty$ they coincide precisely. Namely, the t-test is an asymptotically

Table 3.1 Comparing the critical values of the t (3.11) and u (3.18) tests.

n	10		20		30		50		∞	
α	0.025	0.05	0.025	0.05	0.025	0.05	0.025	0.05	0.025	0.05
$t_{n-1}(\alpha)$	2.262	1.833	2.093	1.729	2.045	1.699	2.010	1.677	1.960	1.645
u_α	2.369	1.827	2.125	1.724	2.064	1.696	2.019	1.674	1.960	1.645

non-parametric test and gives an asymptotically exact critical value under the assumption of symmetry. This property is called the criterion robustness of the t-test against a departure from the normal distribution. In other words, the permutation test has asymptotically equivalent power to the t-test.

(2) **Signed rank sum test.** In the permutation test we may use the rank of $|x_i|$ instead of using $|x_i|$ as it is. Namely, instead of z (3.17) we use

$$W = (\text{Sum of ranks of } |x_i| \text{ for } x_i > 0) - (\text{Sum of ranks of } |x_i| \text{ for } x_i < 0). \qquad (3.20)$$

Then we obtain

$$E(W) = 0,\ V(W) = \sum_1^n i^2 = n(n+1)(2n+1)/6,$$

assuming there is no tie. The standardized statistic

$$W/\sqrt{n(n+1)(2n+1)/6}$$

can be evaluated asymptotically as a standard normal distribution $N(0, 1)$. When there are ties a modification is necessary, but for the quantitative data the chance of ties is small. The formulae when ties exist are given in Sections 5.2.2 and 5.2.3. When the sample size is not too large, an exact enumeration is possible, as shown in Example 3.1.

Example 3.1. The data in Table 3.2 are the extension strength of stainless steel for 10 samples (Moriguti, 1976). The samples are for testing whether the average $\mu = 70(\text{kgf/mm}^2)$ has changed with a change of materials. Since the direction of change is not specified, we apply the two-sided test. We give three procedures here, but actually one method should be selected before seeing the data.

(1) ***t*-test.** The necessary calculations are as follows:

$$\bar{y}_\cdot = \sum_i y_i/n = 717.0/10 = 71.7,$$

$$\sum_i y_i^2 = 72.8^2 + 69.7^2 + \cdots + 70.5^2 = 51453.60,$$

$$S(\hat{\mu}) = \sum_i y_i^2 - y_\cdot^2/n = 51453.60 - 717^2/10 = 44.70,$$

$$\hat{\sigma}^2 = S(\hat{\mu})/(n-1) = 44.70/9 = 4.97.$$

Table 3.2 Preparation for calculating statistics.

Sample	1	2	3	4	5	6	7	8	9	10	Total
y_i	72.8	69.7	77.4	73.3	71.2	73.4	69.8	68.2	75.7	70.5	117.0
$x_i = y_i - 70$	2.8	−0.3	2.4	3.3	1.2	3.4	−0.2	−1.8	5.7	0.5	17.0
x_i^2	7.84	0.09	5.76	10.89	1.44	11.56	0.04	3.24	32.49	0.25	73.6
Rank of $\lvert x_i \rvert$	7	2	6	8	4	9	1	5	10	3	
Signed rank	7	−2	6	8	4	9	−1	−5	10	3	39

From these statistics we obtain

$$|t| = \left| (\bar{y}_{\cdot} - \mu_0) / \sqrt{\hat{\sigma}^2 / n} \right| = (71.7 - 70.0) / \sqrt{4.97/10} = 2.41^*.$$

Since $t_9(0.05/2) = 2.23$, the result is significant at the two-sided level 0.05. Namely, a mean shift is suggested.

(2) **Permutation test.** First we prepare Table 3.2 to calculate the test statistics for non-parametric tests.
From $z = 17.0$ and $\sum_{i=1}^{10} x_i^2 = 73.60$ we obtain

$$u = 17.0 / \sqrt{73.60} = 1.98^*.$$

Since this is larger than 1.96, the permutation test also gives a significant result at the two-sided level 0.05. The result is very similar to that of the t-test.

(3) **Signed rank sum test.** From Table 3.2 we obtain the statistic $W = 39$ (3.20) and by standardization

$$W / \sqrt{\frac{n(n+1)(2n+1)}{6}} = 39 / \sqrt{385} = 1.99^*.$$

This value is very close to $u = 1.98$, and the same conclusion is obtained. In this case it is possible to count the number of signed rank sums which are equal to or larger than 39 among all possible outcomes. This is equivalent to counting the rank sum less than or equal to 8. It can be counted as follows.

Number of minus ranks equal to 1:	1, 2, 3, 4, 5, 6, 7 or 8	8 cases
Number of minus ranks equal to 2:	1 and 2, 3, 4, 5, 6 or 7	6 cases
	2 and 3, 4, 5 or 6	4 cases
	3 and 4 or 5	2 cases
Number of minus ranks equal to 3:	1, 2 and 3, 4 or 5	3 cases
	1, 3 and 4	1 case

Therefore, the total number of cases is 24 and the two-sided p-value is obtained as

$$2\times 24/2^{10}=0.047. \tag{3.21}$$

The normal approximation to $W=1.99$ gives a p-value 0.0469, which is very close to (3.21).

In the above example all the tests give similar results and this is a general tendency, unless the underlying distribution is considerably long tailed or skewed. For a long-tailed distribution the rank sum test behaves better than others. For the skewed distribution the rank sum test also has a poor power. The test with df 1, like the permutation test or the signed rank sum test in this section, can keep the significance level correct under the null hypothesis and is called robust. However, the test with df 1 is not necessarily robust regarding power. Actually, we can choose various scores to form a linear score test, each of which is specialized to a particular alternative, but this is not powerful enough against the other alternatives out of its target. Multiple degrees of freedom will be necessary against the long-tailed or skewed distributions (see also the remarks after Examples 5.8 and 5.10). Refer to Pesarin (2001) or Pesarin and Salmaso (2010) for extensive theory and applications of the permutation test.

3.2 Normal Variance

3.2.1 Setting a null hypothesis and a rejection region

Let us go back to a simple model again,

$$y_i=\mu+e_i,\ i=1,\ \ldots,\ n,$$

with e_i distributed independently as $N(0, \sigma^2)$, where σ^2 is a parameter of interest while μ is a nuisance parameter. Now, the problem is to test the null hypothesis

$$H_0:\ \sigma^2=\sigma_0^2$$

against one of the alternative hypotheses:

$$\begin{aligned}
&\text{right one-sided} && H_1:\ \sigma^2>\sigma_0^2,\\
&\text{left one-sided} && H_1':\ \sigma^2<\sigma_0^2,\\
&\text{two-sided} && H_2:\ \sigma^2\neq\sigma_0^2.
\end{aligned}$$

When μ is known, the sufficient statistic $S=\sum(y_i-\mu)^2/\sigma_0^2$ under H_0 is distributed as chi-squared with df n, and gives a uniformly most powerful test:

$$R_1 \text{ against } H_1: S>\chi_n^2(\alpha), \tag{3.22}$$

$$\text{and }\ R_1' \text{ against } H_1':\ S<\chi_n^2(1-\alpha), \tag{3.23}$$

Table 3.3 Critical values of the uniformly most powerful unbiased test.

	$\alpha = 0.05$		$\alpha = 0.01$	
df	Upper bound	Lower bound	Upper bound	Lower bound
2	9.530	0.0847	13.286	0.0175
3	11.192	0.296	15.127	0.101
4	12.802	0.607	16.901	0.264
5	14.368	0.989	18.621	0.496
6	15.897	1.425	20.296	0.786
7	17.392	1.903	21.931	1.122
8	18.860	2.414	23.533	1.498
9	20.305	2.953	25.106	1.907
10	21.729	3.516	26.653	2.344
11	23.135	4.100	28.178	2.807
12	24.525	4.701	29.683	3.291
13	26.900	5.319	31.171	3.795
14	27.263	5.948	32.641	4.316
15	28.614	6.591	34.097	4.883
16	29.955	7.245	35.540	5.404
17	31.286	7.910	36.971	5.968
18	32.607	8.584	38.390	6.544
19	33.921	9.267	39.798	7.131
20	35.227	9.958	41.197	7.729

respectively. The problem is not so simple for two-sided inference. Usually we assign $\alpha/2$ to both sides and employ

$$R_2 \text{ against } H_2 : S > \chi_n^2(\alpha/2) \text{ or } S < \chi_n^2(1-\alpha/2), \tag{3.24}$$

but this is not unbiased. The critical values of the uniformly most powerful unbiased test are given in Table 3.3 (Takeuchi, 1963), and are obtained as follows.

Let the rejection region of the most powerful unbiased test be

$$R_U : S > a \text{ or } S < b$$

with the corresponding level α test function

$$\varphi(S) = \begin{cases} 1, & S \in R_U, \\ 0, & S \notin R_U. \end{cases}$$

Since

$$S = \sum_{i=1}^{n} (y_i - \mu)^2 / \sigma_0^2 = \left(\sigma^2/\sigma_0^2\right) \times \left\{\sum_{i=1}^{n} (y_i - \mu)^2/\sigma^2\right\} = \left(\sigma^2/\sigma_0^2\right)\chi_n^2$$

is a multiple of chi-squared variable and depends only on the unknown parameter $\delta = \sigma^2/\sigma_0^2$, we denote the power function by $P(\delta) = E\{\varphi(S)\}$. Then, a and b are determined by solving the two equations

$$P(\delta)|_{\delta=1} = \alpha \text{ and } \partial P(\delta)/\partial\delta|_{\delta=1} = 0,$$

which implies that $P(\delta)$ takes a minimum value α at the null hypothesis $\sigma^2 = \sigma_0^2$ (see also Ramachandran, 1958).

When μ is unknown, the statistic $S = \sum_{i=1}^{n} (y_i - \bar{y}_.)^2/\sigma_0^2$ is distributed under H_0 as chi-squared with df $n-1$ (see Example 2.2). The rejection regions against H_1, H_1', and H_2 are obtained from (3.22), (3.23), and (3.24) by replacing df n by $n-1$:

$$R_1 \text{ against } H_1 : S > \chi_{n-1}^2(\alpha),$$
$$R_1' \text{ against } H_1' : S < \chi_{n-1}^2(1-\alpha),$$
$$R_2 \text{ against } H_2 : S > \chi_{n-1}^2(\alpha/2) \text{ or } S < \chi_{n-1}^2(1-\alpha/2).$$

R_1 and R_1' are uniformly most powerful similar tests. The uniformly most powerful unbiased test against H_2 is obtained from Table 3.3 at df $n-1$.

3.2.2 Power function

The difference between the μ known and unknown case lies only in the degrees of freedom, so we describe only the unknown case here. We have already defined the power function $P(\delta)$ in a previous section regarding the unbiased test, which is obviously calculated for R_1 and R_2 by the following formulae.

Right one-sided alternative:

$$P(\delta) = \Pr\{\chi_{n-1}^2 > \delta^{-1}\chi_{n-1}^2(\alpha)\}.$$

Two-sided alternative:

$$P(\delta) = \Pr\{\chi_{n-1}^2 > \delta^{-1}\chi_{n-1}^2(\alpha/2)\} + \Pr\{\chi_{n-1}^2 < \delta^{-1}\chi_{n-1}^2(1-\alpha/2)\}.$$

We give the power of the right one-sided test (df 6, $\alpha = 0.05$) in Table 3.4. It is seen from this table that the power reaches 0.90 at the ratio $\delta = 6$.

The behavior of the two-sided test at $\delta \gg 1$ or $\delta \ll 1$ is similar to that of the one-sided test, so we examine the power at δ around 1 taking $n = 5$ and $\alpha = 0.05$ with

Table 3.4 Power of the right one-sided test (df 6, $\alpha = 0.05$).

δ	1.2	1.5	2.0	2.5	3.0	4.0	5.0	6.0	8.0	10.0	12.0	15.0
$P(\delta)$	0.105	0.211	0.391	0.539	0.650	0.790	0.866	0.910	0.954	0.974	0.984	0.991

Table 3.5 Power of the two-sided test ($n=5$, $\alpha=0.05$).

δ	$\Pr\{\chi_4^2 < 0.48442/\delta\}$	$\Pr\{\chi_4^2 > 11.143/\delta\}$	$P(\delta)$
0.5	0.0855	0.0002	0.0857
0.6	0.0625	0.0010	0.0635
0.7	0.0477	0.0031	0.0508
0.8	0.0376	0.0075	0.0451
0.9	0.0303	0.0147	0.0451
1.0	0.0250	0.0250	0.0500
1.1	0.0210	0.0383	0.0593
1.2	0.0178	0.0543	0.0722
1.3	0.0153	0.0727	0.0881
1.4	0.0133	0.0931	0.1064
1.5	0.0117	0.1149	0.1266

$\chi_4^2(0.025) = 11.143$, $\chi_4^2(0.975) = 0.48442$, and give the result in Table 3.5. It is seen that the power goes down below 0.05 at $\delta = 0.8$ and 0.9. That is, this test is not unbiased and to make it unbiased we have to use the critical value of Table 3.3.

3.3 Confidence Interval

3.3.1 Normal mean

The point estimation of μ has been given in detail in Chapter 2. Actually, however, it is important to know how the point estimator is close to the true value of μ. The variance of an estimator is one such measure. Further, by finding appropriate upper and lower bounds, $l(\mathbf{y})$ and $u(\mathbf{y})$, we often define the interval

$$l(\mathbf{y}) \le \mu \le u(\mathbf{y}), \tag{3.25}$$

which includes the true value of μ at a specified probability $1-\alpha$. This procedure is called interval estimation. Such an interval is usually constructed by inversion of the acceptance region of a test.

Let $y_1, \ldots, y_n$ be distributed independently as normal $N(\mu, \sigma^2)$ and assume σ^2 to be known. We have already introduced a rejection region $|\bar{y}. - \mu_0| > (\sigma/\sqrt{n})z_{\alpha/2}$ (3.10) with significance level α. By inverting it we have an acceptance region

$$\Pr\left\{|\bar{y}. - \mu_0| \le \left(\sigma/\sqrt{n}\right)z_{\alpha/2}\right\} = 1-\alpha.$$

Solving this equation with respect to μ, we have

$$\bar{y}. - \left(\sigma/\sqrt{n}\right)z_{\alpha/2} \le \mu \le \bar{y}. + \left(\sigma/\sqrt{n}\right)z_{\alpha/2} \tag{3.26}$$

in the form of (3.25). The probability that the interval (3.26) includes μ is $1-\alpha$. In this context $1-\alpha$ is called a confidence coefficient and α is usually taken as 0.05 or 0.10. We often abbreviate (3.26) as

$$\bar{y}_{\cdot} \pm \left(\sigma/\sqrt{n}\right) z_{\alpha/2}. \tag{3.27}$$

It should be noted that in interpreting this interval estimation, the expression 'μ is included in the interval (3.26)' is not correct, since the random variables are the upper and lower bounds (3.27) and not μ. The correct expression is 'the interval (3.26) includes μ at confidence coefficient $1-\alpha$'.

A narrower interval is preferable if the confidence coefficient is the same. The confidence interval constructed from a uniformly most powerful test like (3.26) is uniformly optimal in this sense. If the probability that the confidence interval with confidence coefficient $1-\alpha$ does not include a non-true value of μ is larger than or equal to α, then the interval is called an unbiased confidence interval. The confidence interval constructed from a uniformly most powerful unbiased test is uniformly optimum among the unbiased confidence intervals.

When σ^2 is unknown, the confidence interval obtained by inverting the t-test,

$$\bar{y}_{\cdot} \pm \left(\hat{\sigma}/\sqrt{n}\right) t_{n-1}(\alpha/2),$$

is the best unbiased confidence interval with confidence coefficient $1-\alpha$.

Since a confidence interval is constructed by inverting a test, the rejection of the null hypothesis $H_0 : \mu = \mu_0$ at significance level α is equivalent for a confidence interval not to include μ_0 at confidence coefficient $1-\alpha$. Therefore, a confidence interval can work as a test procedure. Further, if the null hypothesis is rejected by a large experiment, the departure can be small and may not be important. A confidence interval generally has more information than a binary decision by a statistical test. Therefore, construction of a confidence interval is strongly recommended.

3.3.2 Normal variance

When μ is unknown, the best unbiased estimator of σ^2 is given by

$$\hat{\sigma}^2 = \sum_{i=1}^{n} (y_i - \bar{y}_{\cdot})^2/(n-1).$$

The confidence interval is obtained by inversion of the two-sided test in the form

$$\Pr\left\{ a \le \sum_{i=1}^{n} \frac{(y_i - \bar{y}_{\cdot})^2}{\sigma^2} \le b \right\} = 1-\alpha. \tag{3.28}$$

In equation (3.28), $\sum (y_i - \bar{y}_{\cdot})^2/\sigma^2$ is distributed as χ^2_{n-1} and there are innumerable combinations of a, b satisfying equation (3.28). A solution is $a = \chi^2_{n-1}(1-\alpha/2)$, $b = \chi^2_{n-1}(\alpha/2)$, for example, but taking upper and lower bounds of an unbiased test is preferable to obtain an unbiased confidence interval. If we choose the values a, b from an unbiased test of σ^2, then the unbiased confidence interval is expressed as

$$\left[\frac{\sum(y_i-\bar{y}.)^2}{b}, \frac{\sum(y_i-\bar{y}.)^2}{a}\right].$$

If μ is known, simply replace $\bar{y}.$ by μ and search a, b for the df n.

3.4 Test Theory in the Linear Model

3.4.1 Construction of F-test

As already stated in Chapter 2, many statistical models in the design of experiments can be expressed as a linear model

$$\boldsymbol{y}_n = \boldsymbol{X}_{n\times p}\boldsymbol{\theta}_p + \boldsymbol{e}_n, \text{ rank } (\boldsymbol{X}) = r. \tag{3.29}$$

In this section we consider a linear hypothesis $\boldsymbol{H\theta} = \boldsymbol{0}$, generally assuming an independent normal distribution $N(\boldsymbol{0}, \sigma^2\boldsymbol{I})$ for the error $\boldsymbol{e}$.

We start from the null hypothesis on the single estimable function,

$$H_0 : \boldsymbol{L}'\boldsymbol{\theta} = 0.$$

Since the least squares estimator $\boldsymbol{L}'\hat{\boldsymbol{\theta}}$ is distributed as $N\{\boldsymbol{L}'\boldsymbol{\theta}, \boldsymbol{L}'(\boldsymbol{X}'\boldsymbol{X})^-\boldsymbol{L}\sigma^2\}$, the t-statistic

$$t = \boldsymbol{L}'\hat{\boldsymbol{\theta}}/\sqrt{\boldsymbol{L}'(\boldsymbol{X}'\boldsymbol{X})^-\boldsymbol{L}\hat{\sigma}^2}, \tag{3.30}$$

$$\hat{\sigma}^2 = \|\boldsymbol{y}-\boldsymbol{X}\hat{\boldsymbol{\theta}}\|^2/(n-r) = \left(\boldsymbol{y}'\boldsymbol{y}-\hat{\boldsymbol{\theta}}'\boldsymbol{X}'\boldsymbol{y}\right)/(n-r) \tag{3.31}$$

is obtained with df $n-r$. The t-test based on (3.30) gives a uniformly most powerful similar test for the one-sided alternative and a uniformly most powerful unbiased test for the two-sided alternative, respectively.

Suppose, more generally, the linear hypothesis of estimable functions

$$H_0' : \boldsymbol{H}_{q\times p}\boldsymbol{\theta}_p = 0. \tag{3.32}$$

In this case, since the least squares estimator $\boldsymbol{H}\hat{\boldsymbol{\theta}}$ is distributed as a multivariate normal distribution $N\{\boldsymbol{H\theta}, \boldsymbol{H}(\boldsymbol{X}'\boldsymbol{X})^-\boldsymbol{H}'\sigma^2\}$, we have a chi-squared variable

$$\chi^2 = \left(\boldsymbol{H}\hat{\boldsymbol{\theta}}\right)'\{\boldsymbol{H}(\boldsymbol{X}'\boldsymbol{X})^-\boldsymbol{H}'\}^-\left(\boldsymbol{H}\hat{\boldsymbol{\theta}}\right)/\sigma^2 \tag{3.33}$$

(see Lemma 2.1). Then, replacing σ^2 by $\hat{\sigma}^2$ (3.31), we can construct an F-test as the ratio of two independent chi-squared variables.

Definition 3.1. *F*-statistic. Suppose there are two independent chi-squared variables χ_1^2 and χ_2^2 with df f_1 and f_2, respectively, then the ratio

$$F = \frac{\chi_1^2/f_1}{\chi_2^2/f_2}$$

is called the F-statistic and its distribution is the F-distribution with df (f_1, f_2). When $f_1 = 1$ this is equivalent to the square of the t-statistic with df f_2.

Then, for the chi-squared χ^2 (3.33),

$$F = \left[\chi^2/\mathrm{tr}(\{\boldsymbol{H}(\boldsymbol{X}'\boldsymbol{X})^-\boldsymbol{H}'\}^-\{\boldsymbol{H}(\boldsymbol{X}'\boldsymbol{X})^-\boldsymbol{H}'\})\right]/\hat{\sigma}^2 \tag{3.34}$$

is distributed as F with df $[\mathrm{tr}(\{\boldsymbol{H}(\boldsymbol{X}'\boldsymbol{X})^-\boldsymbol{H}'\}^-\{\boldsymbol{H}(\boldsymbol{X}'\boldsymbol{X})^-\boldsymbol{H}'\}), n-r]$, where $\hat{\sigma}^2$ is the unbiased variance (3.31) with df $n-r$. When rank $(\boldsymbol{H}) = 1$, namely $q = 1$ in (3.32), F (3.34) is equal to the square of t (3.30). However, including the case where $\boldsymbol{H}$ is not full rank, it is recommended to rewrite the null model as

$$\boldsymbol{y} = \boldsymbol{X\theta} + \boldsymbol{e},\ \boldsymbol{H\theta} = \boldsymbol{0} \Rightarrow \boldsymbol{y} = \boldsymbol{X}_0\boldsymbol{\theta}_0 + \boldsymbol{e}. \tag{3.35}$$

In Example 3.2 below, equation (3.40) is in the form of H_0' (3.32) and equation (3.41) is in the form of (3.35). Then, we rewrite the original model (3.29) as

$$\boldsymbol{y} = \begin{bmatrix} \boldsymbol{X}_0 & \boldsymbol{\Pi}_{X_0}^{\perp}\boldsymbol{X}_1 \end{bmatrix} \begin{bmatrix} \boldsymbol{\theta}_0 \\ \boldsymbol{\theta}_1 \end{bmatrix} + \boldsymbol{e} \tag{3.36}$$

where $\boldsymbol{\Pi}_{X_0}^{\perp} = \boldsymbol{I} - \boldsymbol{\Pi}_{X_0}$ and $\boldsymbol{\Pi}_{X_0}^{\perp}\boldsymbol{X}_1\boldsymbol{\theta}_1$ expresses the restricted estimation space by the null hypothesis in a form orthogonal to the null space $\boldsymbol{X}_0\boldsymbol{\theta}_0$. (3.29) and (3.36) are constructed to be equivalent as a linear model. Since $\boldsymbol{X}_0$ and $\boldsymbol{\Pi}_{X_0}^{\perp}\boldsymbol{X}_1$ are orthogonal to each other, the normal equation is obtained simply as

$$\begin{bmatrix} \boldsymbol{X}_0'\boldsymbol{X}_0 & \boldsymbol{0} \\ \boldsymbol{0} & \boldsymbol{X}_1'\boldsymbol{\Pi}_{X_0}^{\perp}\boldsymbol{X}_1 \end{bmatrix} \begin{bmatrix} \hat{\boldsymbol{\theta}}_0 \\ \hat{\boldsymbol{\theta}}_1 \end{bmatrix} = \begin{bmatrix} \boldsymbol{X}_0'\boldsymbol{y} \\ \boldsymbol{X}_1'\boldsymbol{\Pi}_{X_0}^{\perp}\boldsymbol{y} \end{bmatrix}.$$

That is, the LS estimator $\hat{\boldsymbol{\theta}}_1 = \left(\boldsymbol{X}_1'\boldsymbol{\Pi}_{X_0}^{\perp}\boldsymbol{X}_1\right)^-\boldsymbol{X}_1'\boldsymbol{\Pi}_{X_0}^{\perp}\boldsymbol{y}$ is distributed as normal with variance $\left(\boldsymbol{X}_1'\boldsymbol{\Pi}_{X_0}^{\perp}\boldsymbol{X}_1\right)^-\sigma^2$. Therefore, we have another expression of χ^2 (3.33),

$$\chi^2 = \hat{\boldsymbol{\theta}}_1'\left(\boldsymbol{X}_1'\boldsymbol{\Pi}_{X_0}^{\perp}\boldsymbol{X}_1\right)\hat{\boldsymbol{\theta}}_1/\sigma^2, \tag{3.37}$$

with df $\{r - \mathrm{rank}(\boldsymbol{X}_0)\}$ (see Lemma 2.1). The F-statistic is constructed as

$$F = \frac{\hat{\boldsymbol{\theta}}_1'\left(\boldsymbol{X}_1'\boldsymbol{\Pi}_{X_0}^{\perp}\boldsymbol{X}_1\right)\hat{\boldsymbol{\theta}}_1/\{r - \mathrm{rank}(\boldsymbol{X}_0)\}}{\hat{\sigma}^2}.$$

It should be noted that $r - \mathrm{rank}(\boldsymbol{X}_0) = \mathrm{tr}\left\{\left(\boldsymbol{X}_1'\boldsymbol{\Pi}_{X_0}^{\perp}\boldsymbol{X}_1\right)\left(\boldsymbol{X}_1'\boldsymbol{\Pi}_{X_0}^{\perp}\boldsymbol{X}_1\right)^-\right\} = \mathrm{rank}\left(\boldsymbol{X}_1'\boldsymbol{\Pi}_{X_0}^{\perp}\boldsymbol{X}_1\right)$.

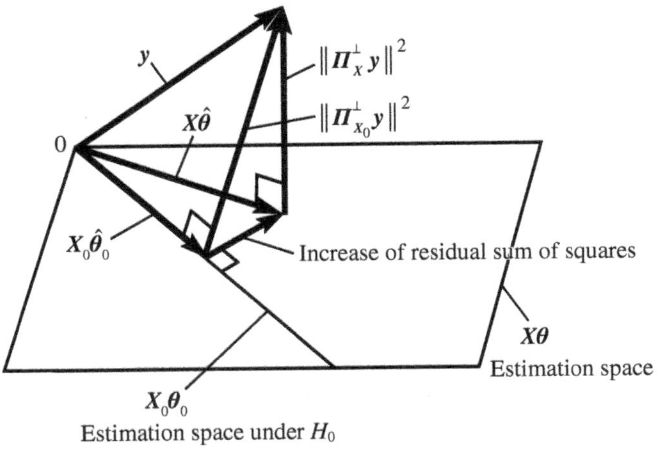

Figure 3.4 Construction of F-test.

Now, it is very important to understand the chi-squared component (3.33) or (3.37) corresponding to the alternative hypothesis as the partition of a residual sum of squares. From (3.36) we have

$$\|\Pi_X^{\perp} y\|^2 = \|y - X_0\hat{\theta}_0 - \Pi_{X_0}^{\perp} X_1\hat{\theta}_1\|^2 = \|\Pi_{X_0}^{\perp} y - \Pi_{X_0}^{\perp} X_1\hat{\theta}_1\|^2$$
$$= \|\Pi_{X_0}^{\perp} y\|^2 - \hat{\theta}_1'\left(X_1'\Pi_{X_0}^{\perp} X_1\right)\hat{\theta}_1.$$

Therefore, the chi-squared component of (3.37) is expressed as

$$\hat{\theta}_1'\left(X_1'\Pi_{X_0}^{\perp} X_1\right)\hat{\theta}_1 = \|\Pi_{X_0}^{\perp} y\|^2 - \|\Pi_X^{\perp} y\|^2,$$

which is the increase in residual sum of squares from the original model by imposing the restriction of the alternative hypothesis. Thus, the F-statistic is interpreted as

$$F = \frac{(\text{Increase of residual sum of squares by imposing restriction})/(\text{Decrease of df by the restriction})}{(\text{Residual sum of squares for the original model})/(\text{Corresponding df})}.$$

This relationship is illustrated in Fig. 3.4. It should be noted that this figure is an extension of Fig. 2.2 in Section 2.5.2. It is recommended that the reader fit full and null models of Example 3.2 to this figure.

The example of this procedure is given in Example 3.2, and also in detail in Example 10.4 (2) for the analysis of two-way unbalanced data where the usual ANOVA table cannot be applied as it is.

3.4.2 Optimality of F-test

In constructing an F-test in the previous section we did not refer to any particular alternative hypothesis. This is because a negation of the null hypothesis is implicitly considered as an alternative. Then, it is naturally expected that the F-test will have equal power against various directions of departure from the null hypothesis in case of a multi-dimensional hypothesis. On the contrary, it is obvious that there is no uniformly most powerful test for the multi-dimensional hypothesis since, if we choose one direction in the alternative hypothesis, the t-test should be most powerful. Now, the power of the F-test is determined only by the degrees of freedom and the non-centrality parameter

$$\gamma = \boldsymbol{\theta}_1' \left(\boldsymbol{X}_1' \boldsymbol{\Pi}_{X_0}^{\perp} \boldsymbol{X}_1 \right) \boldsymbol{\theta}_1 / \sigma^2, \tag{3.38}$$

since χ^2 (3.37) is distributed as chi-squared with non-centrality (3.38) under the alternative hypothesis. The non-null distribution is called a non-central F-distribution. Namely, the F-test has the same power against various $\boldsymbol{\theta}_1$ if γ is the same. As an example, we give γ of the ANOVA in a one-way layout.

Example 3.2. ANOVA in one-way layout. Let us consider the one-way ANOVA model,

$$y_{ij} = \mu_i + e_{ij},\ i = 1, \ldots, a,\ j = 1, \ldots, m, \tag{3.39}$$

where the errors e_{ij} are assumed to be distributed independently as $N(0, \sigma^2)$. The matrix expression of (3.39) is given in Example 2.1. Then, the null hypothesis of the homogeneity,

$$H_0 : \mu_1 = \cdots = \mu_a$$

is a linear hypothesis, since it is rewritten as

$$H_0 : \boldsymbol{P}_a' \boldsymbol{\mu} = 0. \tag{3.40}$$

It is also obvious that a linear model under H_0 is expressed as

$$H_0 : \boldsymbol{y} = \boldsymbol{j}\mu + \boldsymbol{e}. \tag{3.41}$$

We already have the LS estimate $\hat{\mu}_i = \bar{y}_{i\cdot}$ for the original model (3.39), and the residual sum of squares is obtained as

$$S_e = \sum_i \sum_j (y_{ij} - \bar{y}_{i\cdot})^2 = \sum_i \sum_j y_{ij}^2 - \sum_i \bar{y}_{i\cdot} y_{i\cdot} = \sum_i \sum_j y_{ij}^2 - \sum_i y_{i\cdot}^2 / m.$$

In contrast, under H_0 (3.41) we obtain $\hat{\mu} = \bar{y}_{\cdot\cdot}$ and the residual sum of squares

$$S_0 = \sum_i \sum_j y_{ij}^2 - \bar{y}_{\cdot\cdot} y_{\cdot\cdot} = \sum_i \sum_j y_{ij}^2 - y_{\cdot\cdot}^2 / n,\ n = am.$$

In calculating the residual sum of squares, formula (2.33) has been utilized. Then, the increase in residual sum of squares is obtained as

$$S_H = S_0 - S_e = \left(\sum_i \sum_j y_{ij}^2 - y_{\cdot\cdot}^2/n\right) - \left(\sum_i \sum_j y_{ij}^2 - \sum_i y_{i\cdot}^2/m\right) = \sum_i y_{i\cdot}^2/m - y_{\cdot\cdot}^2/n.$$

In this case, $\mathrm{rank}(\boldsymbol{X}) = a$, rank $(\boldsymbol{X}_0) = 1$, and we have finally

$$F = \frac{S_H/(a-1)}{S_e/(n-a)} \tag{3.42}$$

for the homogeneity test of H_0 (3.40). Since S_e represents variation within the same level of the factor and S_H represents variation between different levels, the F-statistic (3.42) is called a between vs. within variance ratio. By the general theory in Chapter 2, we already know $E(S_e) = (n-a)\sigma^2$. Here we can calculate $E(S_H)$ by applying a well-known formula (2.34):

$$E(S_H) = \left\{\sum_i (m\mu_i)^2/m + \sum_i m\sigma^2/m\right\} - \left\{(n\bar{\mu}.)^2/n + n\sigma^2/n\right\}$$
$$= m\sum(\mu_i - \bar{\mu}.)^2 + (a-1)\sigma^2.$$

Then, the non-centrality parameter γ (3.38) is obtained as

$$\gamma = m\sum(\mu_i - \bar{\mu}.)^2/\sigma^2.$$

This suggests that the F-test has a uniform power on the $(a-1)$-dimensional sphere centered at $\bar{\mu}_{\cdot}$. It is described in Scheffé (1959) that the F-test is most powerful among tests with uniform power on the sphere, and the maximal loss in power compared with the most powerful test in every direction is minimum. The latter property is called most stringent.

3.5 Likelihood Ratio Test and Efficient Score Test

3.5.1 Likelihood ratio test

Let the likelihood function of the observation vector $\boldsymbol{y} = (y_1, \ldots, y_n)'$ be $L(\boldsymbol{\theta})$ and consider a simple hypothesis on the parameter $\boldsymbol{\theta} = (\theta_1, \ldots, \theta_k)'$,

$$H_0 : \boldsymbol{\theta} = \boldsymbol{\theta}_0.$$

The likelihood ratio at $\boldsymbol{\theta} = \boldsymbol{\theta}_0$ and $\boldsymbol{\theta} = \hat{\boldsymbol{\theta}}$ (MLE),

$$\lambda = L(\boldsymbol{\theta}_0)/L\left(\hat{\boldsymbol{\theta}}\right)$$

satisfies $0 < \lambda \le 1$ and takes a value close to 1 when H_0 is true and close to 0 when H_0 is false. Therefore, a rejection region like

$$R : \lambda < c \tag{3.43}$$

seems reasonable. If the form of likelihood $L(\boldsymbol{\theta})$ is given explicitly, we can determine c so as to satisfy the requirement of significance level α. In contrast, we can generally determine an asymptotic rejection region as follows.

Expand $\log L(\boldsymbol{\theta})$ around $\boldsymbol{\theta}=\hat{\boldsymbol{\theta}}$ up to second order, since the first order vanishes due to the nature of MLE,

$$\begin{aligned}\log L(\boldsymbol{\theta}) &= \log L\left(\hat{\boldsymbol{\theta}}\right) + \left(\frac{\partial \log L(\boldsymbol{\theta})}{\partial \theta_1}, \ldots, \frac{\partial \log L(\boldsymbol{\theta})}{\partial \theta_k}\right)_{\boldsymbol{\theta}=\hat{\boldsymbol{\theta}}}\left(\boldsymbol{\theta}-\hat{\boldsymbol{\theta}}\right) \\ &+ \frac{1}{2}\left(\boldsymbol{\theta}-\hat{\boldsymbol{\theta}}\right)'\left[\frac{\partial^2 \log L(\boldsymbol{\theta})}{\partial \theta_i \partial \theta_j}\right]_{\boldsymbol{\theta}=\boldsymbol{\theta}^*,\,\left(\boldsymbol{\theta}^*-\hat{\boldsymbol{\theta}}\right)'(\boldsymbol{\theta}^*-\boldsymbol{\theta})<0}\left(\boldsymbol{\theta}-\hat{\boldsymbol{\theta}}\right),\end{aligned} \tag{3.44}$$

where $\boldsymbol{\theta}^*$ is some value of $\boldsymbol{\theta}$ that satisfies the inequality in (3.44). Let the true value of $\boldsymbol{\theta}$ be $\boldsymbol{\theta}_0$, then (3.44) becomes

$$2\log\left\{L(\boldsymbol{\theta}_0)/L\left(\hat{\boldsymbol{\theta}}\right)\right\} = \sqrt{n}\left(\hat{\boldsymbol{\theta}}-\boldsymbol{\theta}_0\right)'\frac{1}{n}\left[\frac{\partial^2 \log L(\boldsymbol{\theta})}{\partial \theta_i \partial \theta_j}\right]_{\boldsymbol{\theta}=\boldsymbol{\theta}^*}\sqrt{n}\left(\hat{\boldsymbol{\theta}}-\boldsymbol{\theta}_0\right). \tag{3.45}$$

Since asymptotically $\boldsymbol{\theta}^*$ converges to $\boldsymbol{\theta}_0$, the matrix in (3.45) converges to $-\boldsymbol{I}_1(\boldsymbol{\theta}_0)$ in probability. Then we have

$$-2\log\left\{L(\boldsymbol{\theta}_0)/L\left(\hat{\boldsymbol{\theta}}\right)\right\} \cong \sqrt{n}\left(\hat{\boldsymbol{\theta}}-\boldsymbol{\theta}_0\right)'\boldsymbol{I}_1(\boldsymbol{\theta}_0)\sqrt{n}\left(\hat{\boldsymbol{\theta}}-\boldsymbol{\theta}_0\right)$$

and the right-hand side of this equation is asymptotically distributed as a chi-squared distribution with df k under H_0 (see Section 2.6 and also Lemma 2.1). Therefore, we have a rejection region

$$-2\log\lambda > \chi_k^2(\alpha).$$

This test is called a likelihood ratio test.

3.5.2 Test based on the efficient score

The partial derivation of the log likelihood function is called an efficient score and we denote it by $\boldsymbol{\nu}(\boldsymbol{\theta})$,

$$\boldsymbol{\nu}(\boldsymbol{\theta}) = \left(\frac{\partial \log L(\boldsymbol{\theta})}{\partial \theta_1}, \ldots, \frac{\partial \log L(\boldsymbol{\theta})}{\partial \theta_k}\right)'.$$

We expand each element of $\nu(\boldsymbol{\theta})$ around $\boldsymbol{\theta}=\hat{\boldsymbol{\theta}}$ to obtain

$$\frac{1}{\sqrt{n}}\boldsymbol{\nu}(\boldsymbol{\theta}) = \mathbf{0} + \frac{1}{n}\begin{bmatrix} \frac{\partial^2 \log L(\boldsymbol{\theta})}{\partial \theta_1 \partial \theta_1}, \ldots, \frac{\partial^2 \log L(\boldsymbol{\theta})}{\partial \theta_1 \partial \theta_k} \\ \cdots \\ \frac{\partial^2 \log L(\boldsymbol{\theta})}{\partial \theta_i \partial \theta_1}, \ldots, \frac{\partial^2 \log L(\boldsymbol{\theta})}{\partial \theta_i \partial \theta_k} \\ \cdots \\ \frac{\partial^2 \log L(\boldsymbol{\theta})}{\partial \theta_k \partial \theta_1}, \ldots, \frac{\partial^2 \log L(\boldsymbol{\theta})}{\partial \theta_k \partial \theta_k} \end{bmatrix}_{\boldsymbol{\theta}=\boldsymbol{\theta}^*}\sqrt{n}\left(\boldsymbol{\theta}-\hat{\boldsymbol{\theta}}\right). \tag{3.46}$$

When $\boldsymbol{\theta}=\boldsymbol{\theta}_0$, the coefficient matrix in (3.46) converges to $-I_1(\boldsymbol{\theta}_0)$ in probability and

$$n^{-1/2}\boldsymbol{\nu}(\boldsymbol{\theta}_0) \cong I_1(\boldsymbol{\theta}_0)\times\sqrt{n}\left(\hat{\boldsymbol{\theta}}-\boldsymbol{\theta}_0\right)$$

is asymptotically distributed as $N\{\mathbf{0}, \boldsymbol{I}_1(\boldsymbol{\theta}_0)\}$. Therefore

$$n^{-1}\boldsymbol{\nu}(\boldsymbol{\theta}_0)'I_1^{-1}(\boldsymbol{\theta}_0)\boldsymbol{\nu}(\boldsymbol{\theta}_0) > \chi_k^2(\alpha)$$

is a rejection region with significance level α. The test based on efficient score is a locally most powerful test. That is, it is most powerful in the neighborhood of H_0 (see Cox and Hinkley, 2000).

3.5.3 Composite hypothesis

Consider testing a composite hypothesis that $\boldsymbol{\theta}$ is expressed by a parameter $\boldsymbol{\beta}$ with lower dimension k_0 than $\boldsymbol{\theta}$,

$$H_0: \boldsymbol{\theta}=\boldsymbol{\theta}_0=\boldsymbol{\theta}(\boldsymbol{\beta}).$$

The likelihood ratio test and the efficient score test are extended to the composite hypothesis by MLE $\hat{\boldsymbol{\theta}}_0=\boldsymbol{\theta}\left(\hat{\boldsymbol{\beta}}\right)$ under H_0 as follows.

(1) **Likelihood ratio test:**

$$R: -2\log\left\{L\left(\hat{\theta}_0\right)/L\left(\hat{\theta}\right)\right\} = -2\log\left[L\left\{\theta\left(\hat{\beta}\right)\right\}/L\left(\hat{\theta}\right)\right] > \chi_{k-k_0}^2(\alpha). \qquad (3.47)$$

(2) **Efficient score test:**

$$R: n^{-1}\boldsymbol{\nu}\left\{\boldsymbol{\theta}\left(\hat{\boldsymbol{\beta}}\right)\right\}'\boldsymbol{I}_1^{-1}\left\{\boldsymbol{\theta}\left(\hat{\boldsymbol{\beta}}\right)\right\}\boldsymbol{\nu}\left\{\boldsymbol{\theta}\left(\hat{\boldsymbol{\beta}}\right)\right\} > \chi_{k-k_0}^2(\alpha). \qquad (3.48)$$

The efficient score test is easier to apply, since it needs MLE only under H_0 (see Examples 3.3 and 3.4).

Example 3.3. ANOVA in one-way layout (likelihood ratio test). Let us consider the one-way ANOVA model

$$y_{ij}=\mu_i+e_{ij},\ i=1,\cdots,a, j=1,\ldots,m,$$

where the error e_{ij} are assumed to be distributed independently as $N(0, \sigma^2)$. Then, the likelihood is expressed as

$$L(\boldsymbol{\mu}, \sigma^2) = \Pi_{i=1}^a\left\{\left(2\pi\sigma^2\right)^{-m/2}\exp-\frac{\sum_{j=1}^m\left(y_{ij}-\mu_i\right)^2}{2\sigma^2}\right\}$$

$$= \left(2\pi\sigma^2\right)^{-n/2}\exp-\frac{1}{2\sigma^2}\left\{m\sum_i\left(\bar{y}_{i\cdot}-\mu_i\right)^2+\sum_i\sum_j\left(y_{ij}-\bar{y}_{i\cdot}\right)^2\right\}, n=am. \qquad (3.49)$$

From equation (3.49) we get the MLE

$$\hat{\mu}_i = \bar{y}_{i\cdot}, \hat{\sigma}^2 = \frac{1}{n}\sum_i\sum_j \left(y_{ij} - \bar{y}_{i\cdot}\right)^2, \tag{3.50}$$

where it should be noted that in (3.50), $\hat{\sigma}^2$ is different from the usual unbiased variance. Finally, we have

$$L(\hat{\mu}, \hat{\sigma}^2) = \left(2\pi\hat{\sigma}^2\right)^{-n/2} \exp - \left(\frac{n}{2}\right)$$

Under the null hypothesis of homogeneity in μ_i's,

$$H_0 : \mu_1 = \cdots = \mu_a = \mu,$$

the likelihood is obtained as

$$L(\mu, \sigma^2) = \left(2\pi\sigma^2\right)^{-n/2} \exp - \frac{1}{2\sigma^2}\left\{m\sum_i (\bar{y}_{i\cdot} - \bar{y}_{\cdot\cdot})^2 + \sum_i\sum_j \left(y_{ij} - \bar{y}_{i\cdot}\right)^2\right\}.$$

In this case we have MLE

$$\hat{\mu} = \bar{y}_{\cdot\cdot}, \hat{\sigma}_0^2 = \frac{1}{n}\sum_i\sum_j \left(y_{ij} - \bar{y}_{\cdot\cdot}\right)^2$$

and $L(\hat{\mu}, \hat{\sigma}_0^2) = \left(2\pi\hat{\sigma}_0^2\right)^{-n/2} \exp - \left(\frac{n}{2}\right)$.

The rejection region of the likelihood ratio test is obtained in the form of (3.43), but it is equivalent to

$$F = \frac{m\sum_i (\bar{y}_{i\cdot} - \bar{y}_{\cdot\cdot})^2 / (a-1)}{\sum_i\sum_j \left(y_{ij} - \bar{y}_{i\cdot}\right)^2 / \{a(m-1)\}} > c',$$

which has already been introduced as the F-test in Example 3.2. Therefore, an exact test is available in this case.

Example 3.4. Goodness-of-fit test in multinomial distribution (efficient score test). Let $\boldsymbol{y} = (y_1, \ldots, y_a)'$ be distributed as a multinomial distribution $M(y_\cdot, \boldsymbol{p})$,

$$L(\boldsymbol{p}) = \frac{n!}{y_1! \cdots y_a!} p_1^{y_1} \times \cdots \times p_a^{y_a} \quad (p_\cdot = 1, y_\cdot = n). \tag{3.51}$$

In this case, since $p_\cdot = 1$, the dimension of $\boldsymbol{p}$ is $a-1$. Therefore, the likelihood (3.51) is considered as a function of $p_1, \ldots, p_{a-1}$ and p_a is replaced by $p_a = 1 - p_1 - \cdots - p_{a-1}$. The efficient score is then calculated as

$$\frac{\partial \log L(\boldsymbol{p})}{\partial p_i} = \frac{y_i}{p_i} + \frac{y_a}{p_a}(-1), \; i = 1, \cdots, a-1.$$

The (i, i) and (i, j) elements of Fisher's information matrix are obtained as

$$-E\left\{\frac{\partial^2 \log L(\boldsymbol{p})}{\partial p_i^2}\right\} = -E\left\{-\frac{y_i}{p_i^2} + \frac{y_a}{p_a^2}(-1)\right\} = n\left(p_i^{-1} + p_a^{-1}\right),$$

$$-E\left\{\frac{\partial^2 \log L(\boldsymbol{p})}{\partial p_i \partial p_j}\right\} = n/p_a (i \neq j),$$

respectively. Then, Fisher's information matrix per datum is expressed as

$$\boldsymbol{I}_1(\boldsymbol{p}) = -E\left\{\frac{1}{n}\left[\frac{\partial^2 \log L(\boldsymbol{p})}{\partial p_i \partial p_j}\right]_{(a-1)\times(a-1)}\right\} = \text{diag}\left(p_i^{-1}\right) + p_a^{-1}\boldsymbol{j}\boldsymbol{j}'.$$

We consider a composite null hypothesis

$$H_0 : \boldsymbol{p} = \boldsymbol{p}_0 = \boldsymbol{p}(\boldsymbol{\beta})$$

and the MLE under H_0 is denoted by $\hat{\boldsymbol{p}}_0 = \boldsymbol{p}\left(\hat{\boldsymbol{\beta}}\right)$, where the dimension of $\boldsymbol{\beta}$ is assumed to be $k_0(< a-1)$. Then, the test statistic based on the efficient score is constructed according to (3.48) as

$$\chi^2 = n^{-1}\boldsymbol{\nu}(\hat{\boldsymbol{p}}_0)'\boldsymbol{I}_1^{-1}(\hat{\boldsymbol{p}}_0)\boldsymbol{\nu}(\hat{\boldsymbol{p}}_0)$$

$$= n^{-1}\left(\begin{bmatrix} \hat{p}_1^{-1} & 0 & 0\cdots 0 & 0 - \hat{p}_a^{-1} \\ 0 & \hat{p}_2^{-1} & 0\cdots 0 & 0 - \hat{p}_a^{-1} \\ & & \cdots & \\ 0 & 0 & 0\cdots 0 & \hat{p}_{a-1}^{-1} - \hat{p}_a^{-1} \end{bmatrix}\boldsymbol{y}\right)'\left(\begin{bmatrix} \hat{p}_1^{-1} & \cdots & 0 \\ & \ddots & \\ 0 & \cdots & \hat{p}_{a-1}^{-1} \end{bmatrix} + \hat{p}_a^{-1}\boldsymbol{j}\boldsymbol{j}'\right)^{-1} \times$$

$$\left(\begin{bmatrix} \hat{p}_1^{-1} & 0 & 0\cdots 0 & 0 & -\hat{p}_a^{-1} \\ 0 & \hat{p}_2^{-1} & 0\cdots 0 & 0 & -\hat{p}_a^{-1} \\ & & \cdots & & \\ 0 & 0 & 0\cdots 0 & \hat{p}_{a-1}^{-1} & -\hat{p}_a^{-1} \end{bmatrix}\boldsymbol{y}\right), \tag{3.52}$$

where we omit the suffix 0 of MLE $\hat{p}_{0i}$ under H_0 to avoid complex expression. Equation (3.52) is further simplified by the formula $(\boldsymbol{I} + \boldsymbol{A}\boldsymbol{B})^{-1} = \boldsymbol{I} - \boldsymbol{A}(\boldsymbol{I} + \boldsymbol{B}\boldsymbol{A})^{-1}\boldsymbol{B}$ as

$$\begin{aligned}\left\{\text{diag}\left(\hat{p}_i^{-1}\right) + \hat{p}_a^{-1}\boldsymbol{j}\boldsymbol{j}'\right\}^{-1} &= \left\{\boldsymbol{I} + \hat{p}_a^{-1}(\hat{p}_1, \ldots, \hat{p}_{a-1})'\boldsymbol{j}'\right\}^{-1}\text{diag}(\hat{p}_i) \\ &= \text{diag}(\hat{p}_i) - \hat{p}_a^{-1}(\hat{p}_1, \ldots, \hat{p}_{a-1})'\left\{1 + \hat{p}_a^{-1}\boldsymbol{j}'(\hat{p}_1, \ldots, \hat{p}_{a-1})'\right\}^{-1}\boldsymbol{j}'\text{diag}(\hat{p}_i) \\ &= \text{diag}(\hat{p}_i) - (\hat{p}_1, \ldots, \hat{p}_{a-1})'(\hat{p}_1, \ldots, \hat{p}_{a-1}).\end{aligned}$$

Finally, a well-known formula

$$\chi^2 = n^{-1}\mathbf{y}'\mathrm{diag}\left(\hat{p}_i^{-1}\right)_{a\times a}\{\mathrm{diag}(\hat{p}_i) - (\hat{p}_1,\cdots,\hat{p}_a)'(\hat{p}_1,\cdots,\hat{p}_a)\}\mathrm{diag}\left(\hat{p}_i^{-1}\right)\mathbf{y}$$
$$= \sum_{i=1}^{a}\frac{(y_i - n\hat{p}_i)^2}{n\hat{p}_i} \tag{3.53}$$

is obtained, where it should be noted that in equation (3.53) all the cells are dealt with symmetrically to recover p_a. This equation is expressed as

$$\sum_i \frac{\{(\text{Observation of the } i\text{th cell}) - (\text{MLE of the } i\text{th cell frequency under } H_0)\}^2}{\text{MLE of the } i\text{th cell frequency under } H_0} \tag{3.54}$$

and called a goodness-of-fit chi-squared, where the summation is with respect to all the cells of the multinomial distribution. It is asymptotically distributed as a chi-squared distribution with df

$$\nu = a - 1 - k_0.$$

Equation (3.54) is expressed more simply as

$$\sum_i \frac{(\text{observation} - \text{fitted value})^2}{\text{fitted value}}.$$

Example 3.5. Testing independence in a two-way contingency table. Let each item be cross-classified according to two attributes and the occurrence probability of cell (i, j) be denoted by p_{ij}, $i = 1, \ldots, a$, $j = 1, \ldots, b$. Assuming a multinomial distribution $M(y_{..}, \mathbf{p})$ for the cell frequencies y_{ij}, we have a likelihood function

$$L(\mathbf{p}) = n!\Pi_i\Pi_j\frac{1}{y_{ij}!}p_{ij}^{y_{ij}} \quad (p_{..} = 1,\ y_{..} = n). \tag{3.55}$$

The null hypothesis that the classification is independent is defined by

$$H_0: p_{ij} = p_{i\cdot} \times p_{\cdot j}. \tag{3.56}$$

The MLE for p_{ij} of (3.55) is obviously $\hat{p}_{ij} = y_{ij}/n$ and the MLE under H_0 (3.56) is obtained as $\hat{p}_{0ij} = (y_{i\cdot}/n)(y_{\cdot j}/n)$. Therefore, the likelihood ratio test (3.47) is obtained as

$$-2\log\lambda = 2\sum_i\sum_j\left(y_{ij}\log\hat{p}_{ij} - y_{ij}\log\hat{p}_{0ij}\right)$$
$$= 2\left\{\sum_i\sum_j(y_{ij}\log y_{ij}) - \sum_i(y_{i\cdot}\log y_{i\cdot}) - \sum_i(y_{\cdot j}\log y_{\cdot j}) + n\log n\right\}.$$

Table 3.6 2×2 contingency table.

	Response	
Drug	Improved	Not improved
Test	87	25
Control	81	40

In contrast, since the estimate of cell frequency under H_0 is obtained as

$$n \times \left(y_{i\cdot}y_{\cdot j}/n^2\right) = y_{i\cdot}y_{\cdot j}/n$$

we have a goodness-of-fit chi-squared

$$\chi^2 = \sum_i \sum_j \frac{\left(y_{ij} - y_{i\cdot}y_{\cdot j}/n\right)^2}{y_{i\cdot}y_{\cdot j}/n}. \tag{3.57}$$

The degrees of freedom are

$$ab - 1 - (a + b - 2) = (a - 1)(b - 1)$$

because of the constraint $p_{\cdot\cdot} = 1$. In the case of $a = b = 2$, the chi-squared statistic is expressed in a simple single squared form as

$$\chi^2 = \sum_{i=1}^{2} \sum_{j=1}^{2} \frac{\left(y_{ij} - y_{i\cdot}y_{\cdot j}/n\right)^2}{y_{i\cdot}y_{\cdot j}/n} = \frac{y_{\cdot\cdot}\left(y_{11}y_{22} - y_{12}y_{21}\right)^2}{y_{1\cdot}y_{2\cdot}y_{\cdot 1}y_{\cdot 2}}. \tag{3.58}$$

This gives good reasoning that the degree of freedom of the chi-squared (3.58) is unity, although it is composed of four terms.

Example 3.6. Example 3.5 continued. The data of Table 3.6 are from a clinical trial explained in Example 5.13 of Section 5.3.5, where the total of 233 subjects are cross-classified by the drugs which they received and the results of treatment. The chi-squared statistic (3.58) is calculated as

$$\chi^2 = \frac{233(87 \times 40 - 25 \times 81)^2}{112 \times 121 \times 168 \times 65} = 3.33.$$

Since $\chi_1^2(0.05) = 3.84$, the result is not significant at significance level 0.05.

The data of Table 3.6 are reanalyzed in Section 5.3.5 as a comparison of two binomial distributions.

References

Moriguti, S. (ed.) (1976) *Statistical methods* (new edn). Japanese Standards Association, Tokyo (in Japanese).

Pesarin, F. (2001) *Multivariate permutation tests: with applications in biostatistics*. Wiley, New York.

Pesarin, F. and Salmaso, L. (2010) *Permutation tests for complex data: Theory, application and software*. Wiley, New York.

Ramachandran, K. V. (1958) A test of variances. *J. Amer. Statist. Assoc.* **53**, 741–747.

Scheffé, H. (1959) *The analysis of variance*. Wiley, New York.

Takeuchi, K. (1963) *Mathematical statistics*. Toyo-Keizai, Tokyo (in Japanese).

4

Multiple Decision Processes and an Accompanying Confidence Region

4.1 Introduction

To test the null hypothesis $H_0 : \mu = 0$ in the normal model, the two-sided and right or left one-sided tests have been introduced in Section 3.1. The latter tests are directional and more powerful than the two-sided omnibus test, because of the prior information for the directions of change. If the null hypothesis H_0 is rejected by the right one-sided test with significance level α, we conclude that $\mu > 0$ with confidence coefficient $1 - \alpha$. However, in this case, if we actually obtain a sufficiently large negative value supporting $H_2 : \mu < 0$, we will wish to conclude that $\mu < 0$ instead of accepting H_0. Also, applying a one-sided confidence interval after obtaining a significant result with a two-sided test is often seen in practice. These common but erroneous procedures might be justified by the idea of multiple decision processes.

4.2 Determining the Sign of a Normal Mean – Unification of One- and Two-Sided Tests

Suppose σ^2 is known and is unity for simplicity, and assume that the datum y is distributed as $N(\mu, 1)$. If we have n independent observations y_i, $i = 1, \ldots, n$, simply replace y by $n^{1/2}\bar{y}.$ in the following equations. In case of unknown σ^2, we employ

Advanced Analysis of Variance, First Edition. Chihiro Hirotsu.

the usual unbiased estimator of variance $\hat{\sigma}^2$ (2.17) and replace y by $n^{1/2}\bar{y}./\hat{\sigma}$. Then, the critical value is replaced by $t_{n-1}(\alpha)$, where $t_{n-1}(\alpha)$ is the upper α point of the t-distribution with degrees of freedom $(n-1)$.

Let us partition the sample space into three regions:

$$K_1 : \mu < 0,$$
$$K_2 : \mu = 0,$$
$$K_3 : \mu > 0.$$

Since they are disjoint, only one of the three K's is true. Therefore, by testing each hypothesis with significance level α, without any adjustment, the probability of rejecting the true hypothesis is at most α (Takeuchi, 1973). We can therefore apply a one-sided level-α test for K_1 and K_3, and a two-sided level-α test for K_2, keeping the overall type I error rate at α. Then, the rejection region of each test is given by

$$\begin{aligned} R_{K_1} &: y \geq z_\alpha, \\ R_{K_2} &: y > z_{\alpha/2} \text{ or } y < -z_{\alpha/2}, \\ R_{K_3} &: y \leq -z_\alpha. \end{aligned} \tag{4.1}$$

Inverting the rejection region (4.1), the acceptance region is obtained as

$$A_{K_1} : y < z_\alpha,$$
$$A_{K_2} : -z_{\alpha/2} \leq y \leq z_{\alpha/2},$$
$$A_{K_3} : y > -z_\alpha.$$

These regions are sketched in Fig. 4.1, where the black and white circles denote the inclusion and exclusion of the point, respectively. From Fig. 4.1, the accepted hypotheses are K_3 for $y > z_{\alpha/2}$, K_2 and K_3 for $z_\alpha \leq y \leq z_{\alpha/2}$, K_1, K_2 and K_3 for $-z_\alpha < y < z_\alpha$, K_1 and K_2 for $-z_{\alpha/2} \leq y \leq -z_\alpha$, and only K_1 for $y < -z_{\alpha/2}$.

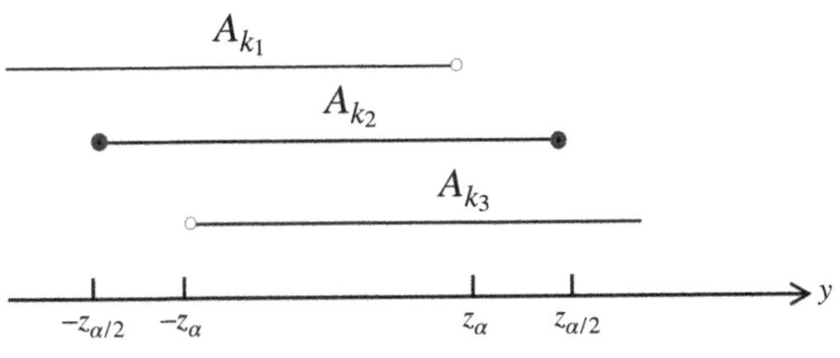

Figure 4.1 The acceptance region.

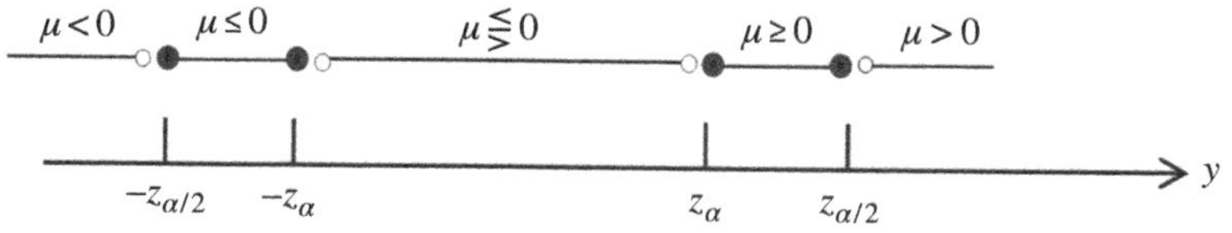

Figure 4.2 The accepted hypotheses.

In other words, we can conclude

$$
\begin{aligned}
&\mu>0 \text{ for } y>z_{\alpha/2},\\
&\mu\geq 0 \text{ for } z_\alpha \leq y \leq z_{\alpha/2},\\
&\mu\leq 0 \text{ for } -z_{\alpha/2} \leq y \leq -z_\alpha,\\
&\mu<0 \text{ for } y<-z_{\alpha/2},
\end{aligned}
$$

and the sign cannot be determined for $-z_\alpha<y<z_\alpha$, as shown in Fig. 4.2.

Thus, the proposed method can give two-sided inference, keeping essentially the power of a one-sided test. It should be noted that the direction of departure from the null hypothesis $K_2 : \mu=0$ need not be declared in advance.

4.3 An Improved Confidence Region

Extending the idea of determining the sign, an interesting confidence region is obtained. Define the acceptance region

$$
\begin{aligned}
&A'_{K_1} : y-\mu<z_\alpha \text{ for } \mu<0,\\
&A'_{K_2} : -z_{\alpha/2}\leq y-\mu \leq z_{\alpha/2} \text{ for } \mu=0,\\
&A'_{K_3} : y-\mu>-z_\alpha \text{ for } \mu>0.
\end{aligned}
\tag{4.2}
$$

Inverting the acceptance region (4.2), the confidence region with confidence coefficient $1-\alpha$ is obtained as

$$
\begin{aligned}
&y-z_\alpha<\mu<0 \text{ for } y<-z_{\alpha/2}\\
&y-z_\alpha<\mu\leq 0 \text{ for } -z_{\alpha/2}\leq y\leq -z_\alpha,\\
&(0>)y-z_\alpha<\mu<y+z_\alpha\ (>0) \text{ for} -z_\alpha<y<z_\alpha,\\
&0\leq\mu<y+z_\alpha \text{ for } z_\alpha\leq y\leq z_{\alpha/2}\\
&0<\mu<y+z_\alpha \text{ for } z_{\alpha/2}<y.
\end{aligned}
$$

This confidence region is sketched in Fig. 4.3, where the shaded area including the solid line is the acceptance region and the dotted line denotes the exclusion. This confidence region holds the conclusion of Section 4.2, and simultaneously gives an upper

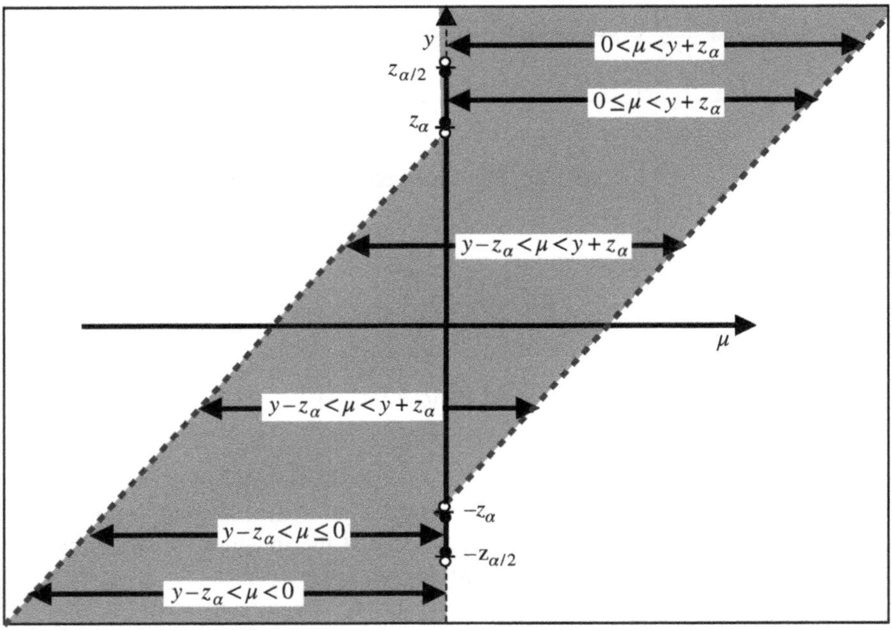

Figure 4.3 An improved confidence region.

bound for $\mu > 0$, a lower bound for $\mu < 0$, and lower and upper bounds when the sign of μ cannot be determined. Further, compared with the naïve confidence interval

$$y - z_{\alpha/2} \leq \mu \leq y + z_{\alpha/2} \tag{4.3}$$

the confidence region gives stricter upper bounds $y + z_\alpha$ and 0 at $y > -z_\alpha$ and $-z_{\alpha/2} \leq y \leq -z_\alpha$, respectively. It also gives stricter lower bounds $y - z_\alpha$ and 0 at $y < z_\alpha$ and $z_\alpha \leq y \leq z_{\alpha/2}$, respectively. This is with the compensation of lowering the lower bound to 0 at $y > z_{\alpha/2}$ and raising the upper bound to 0 at $< -z_{\alpha/2}$, but the merits of multiple decisions will surpass. Therefore, this method detects the sign of μ very efficiently and simultaneously gives very sharp lower and upper bounds. Thus, this method is useful for proving bio-equivalence, where stricter upper and lower bounds are required (see Section 5.3.6), and also for the unifying approach to prove superiority, equivalence or non-inferiority of a new drug against an active control (see Section 5.3.2). It is recommended that the reader writes the naïve confidence interval (4.3) on Fig. 4.3.

Reference

Takeuchi, K. (1973) *Methodological basis of mathematical statistics*. Toyo-Keizai Shinposha, Tokyo (in Japanese).

5

Two-Sample Problem

In this chapter the problem of comparing two populations, two distributions, or two parameters defining the distributions is discussed. First we introduce in Section 5.1.1 the comparison of two means, assuming equal variance for the two normal distributions. This is a simple application of the LS method to a linear model in Chapters 2 and 3. It is also a special case of a one-way layout model of next section with number of levels two. The method is essentially a t-test and the confidence interval is obtained as an inversion of the t-test. In Section 5.3 it is extended to a more useful confidence region by the method of multiple decision processes introduced in Chapter 4. It is also extended to comparing two binomial populations, including the case where there are stratifications by a factor influential on the outcome. A paired t-test is introduced in Section 5.1.3, and Section 5.1.4 is for comparison of variances. Some non-parametric approaches are introduced in Section 5.2.

5.1 Normal Theory

5.1.1 Comparison of normal means assuming equal variances

We assume a linear model

$$y_{ij} = \mu_i + e_{ij},\ i = 1, 2,\ j = 1, \ldots, n_i, \tag{5.1}$$

where the e_{ij} are distributed independently of each other as $N(0, \sigma^2)$. This is a special case of model (2.25) with $a = 2$ and unequal number of repetitions n_i, expressed in matrix form as

Advanced Analysis of Variance, First Edition. Chihiro Hirotsu.

$$\boldsymbol{y} = \boldsymbol{X\mu} + \boldsymbol{e} = \begin{bmatrix} \boldsymbol{j}_{n_1} & 0 \\ 0 & \boldsymbol{j}_{n_2} \end{bmatrix} \begin{bmatrix} \mu_1 \\ \mu_2 \end{bmatrix} + \begin{bmatrix} \boldsymbol{e}_1 \\ \boldsymbol{e}_2 \end{bmatrix}.$$

We apply the test theory of Section 3.4 to the null hypothesis of interest

$$H_0 : \boldsymbol{L}'\boldsymbol{\mu} = \mu_1 - \mu_2 = 0,$$

where $\boldsymbol{\mu} = (\mu_1, \mu_2)'$, $\boldsymbol{L} = (1, -1)'$. Then the LS estimator is obtained as

$$\boldsymbol{L}'\hat{\boldsymbol{\mu}} = \bar{y}_{1\cdot} - \bar{y}_{2\cdot} \text{ with variance} \left(n_1^{-1} + n_2^{-1}\right)\sigma^2.$$

The residual sum of squares is obtained as

$$\mathrm{S}(\hat{\boldsymbol{\mu}}) = \|\boldsymbol{y} - \boldsymbol{X}\hat{\boldsymbol{\mu}}\|^2 = \sum_{i=1}^{2} \sum_{j=1}^{n_i} y_{ij}^2 - \sum_{i=1}^{2} \frac{y_{i\cdot}^2}{n_i},$$

by formulae (2.31) and (2.33) with df $n_1 + n_2 - 2$. Therefore, the t-statistic

$$t = (\bar{y}_{1\cdot} - \bar{y}_{2\cdot}) / \sqrt{\left(n_1^{-1} + n_2^{-1}\right)\hat{\sigma}^2}, \tag{5.2}$$

with unbiased variance

$$\hat{\sigma}^2 = \mathrm{S}(\hat{\boldsymbol{\mu}}) / (n_1 + n_2 - 2),$$

is obtained. The tests against

two-sided alternative hypothesis $H_1 : \mu_1 - \mu_2 \neq 0$,

one-sided alternative hypothesis $H_1' : \mu_1 - \mu_2 > 0$,

based on t (5.2) are the uniformly most powerful unbiased test and uniformly most powerful similar test, respectively. The two-sided test is equivalent to the F-test of (3.42) at $a = 2$. It will be a matter of simple algebra to verify $F = t^2$.

A naïve confidence interval

$$\bar{y}_{1\cdot} - \bar{y}_{2\cdot} \pm \sqrt{\left(n_1^{-1} + n_2^{-1}\right)\hat{\sigma}^2} t_{n_1 + n_2 - 2}(\alpha/2) \tag{5.3}$$

is obtained as an inversion of the two-sided t-test. However, a more useful confidence region is given in Section 5.3, unifying the one- and two-sided inference.

Example 5.1. Half-life of NFLX (antibiotic). In Table 5.1 we are interested in whether there is an extension of the half-life according to the increase in dose. Therefore, we perform a t-test for the null hypothesis $H_0 : \mu_1 - \mu_2 = 0$ against the one-sided alternative $H_1 : \mu_2 - \mu_1 > 0$. From the table we obtain

$$\bar{y}_{2\cdot} - \bar{y}_{1\cdot} = 9.28/5 - 8.34/5 = 0.188$$

Table 5.1 Half-life of NFLX at two doses.

Dosage (mg/kg/day)			Half-life (h)			Total
$25(i=1)$	1.55	1.63	1.49	1.53	2.14	8.34
$200(i=2)$	1.78	1.93	1.80	2.07	1.70	9.28

and $S(\hat{\mu})=\sum_i\sum_j y_{ij}^2-\sum_i y_{i\cdot}^2/n_i=1.55^2+\cdots+1.70^2-(8.34^2/5+9.28^2/5)=0.37340$. Then, $\hat{\sigma}^2=0.37332/(10-2)=0.046675$ and finally we have

$$t=0.188/\sqrt{(5^{-1}+5^{-1})\times 0.046675}=1.376.$$

The p-value is 0.103 as a t-distribution with df 8, and the result is not significant at one-sided $\alpha=0.05$.

The one-sided lower bound at confidence coefficient 0.95 is obtained from (5.3) as

$$\mu_2-\mu_1>0.188-\sqrt{(5^{-1}+5^{-1})\times 0.046675}\times 1.86=-0.066, \tag{5.4}$$

where 1.86 is the critical value $t_8(0.05)$. The interval (5.4) includes 0, corresponding to the test result that the difference is not significant at one-sided level 0.05.

We notice that the last datum of the first group ($i=1$) in Table 5.1 is too large compared with the others and looks like an outlier. The t-test is asymptotically equivalent to the permutation test as shown in Section 3.1.5 (1), and is robust on the whole. However, it is still sensitive to an outlier, since it causes a bias in estimating the mean and simultaneously an over-estimate of the variance. Therefore, a preliminary check for outliers is recommended by the Smirnov–Grubbs test. For a data set $\boldsymbol{y}=(y_1,\ldots,y_n)$ it calculates

$$\max\{(y_{(n)}-\bar{y}_\cdot)/\hat{\sigma},\ (\bar{y}_\cdot-y_{(1)})/\hat{\sigma}\}, \tag{5.5}$$

where $y_{(i)}$ is the ith largest statistic and $\hat{\sigma}$ is the root of the usual unbiased variance. The meaning of the statistic (5.5) will be obvious, and the critical values are given in Table XIII*a* of Barnett and Lewis (1993). This test can also be used for a one-sided test if the direction of departure is decided before taking data. One should refer to Barnett and Lewis (1993) for more detailed explanations, including multiple outliers. Anyway, it should be noted that an outlier test is not like a usual statistical test. A significant result itself is not of much importance, but it suggests that if we go back to examining the data it is very likely to reach some reasoning of the outlier – such as failure of experiment, miswriting, and so on.

Example 5.2. Example 5.1 continued, checking outlier

Data ($i=1$): 1.55, 1.63, 1.49 $(=y_{(1)})$, 1.53, 2.14 $(=y_{(5)})$, $\bar{y}_{\cdot}=1.668$, $\hat{\sigma}=0.2687$.

Smirnov–Grubbs statistic: $\max\{(2.14-1.668)/\hat{\sigma},\ (1.668-1.49)/\hat{\sigma}\}=1.756^{*}$.

Since the upper 0.01 point is 1.749, the result is significant at level 0.02 two-sided.

Data ($i=2$): 1.78, 1.93, 1.80, 2.07 $(=y_{(5)})$, 1.70 $(=y_{(1)})$, $\bar{y}_{\cdot}=1.856$, $\hat{\sigma}=0.1454$.

Smirnov–Grubbs statistic: $\max\{(2.07-1.856)/\hat{\sigma},\ (1.856-1.70)/\hat{\sigma}\}=1.472$.

Since the upper 0.10 point is 1.602, there is no evidence for an outlier in this data set.

Example 5.3. Example 5.1 continued, eliminating the outlier. After an outlier check and discussion with the medical doctor, we decided to eliminate the datum 2.14. Then we obtain

$$\bar{y}_{2\cdot}-\bar{y}_{1\cdot}=9.28/5-6.20/4=0.306,$$

$$S(\hat{\boldsymbol{\mu}})=\sum_i\sum_j y_{ij}^2-\sum_i y_{i\cdot}^2/n_i=1.55^2+\cdots+1.70^2-\left(6.20^2/4+9.28^2/5\right)=0.09492$$

$$\hat{\sigma}^2=0.09484/(9-2)=0.01356$$

Finally we have

$$t=0.306/\sqrt{\left(4^{-1}+5^{-1}\right)\times 0.01355}=3.92^{**}$$

with p-value 0.003 (one-sided) as a t-distribution with df 7. That is, an increase of half-life is observed for dose 200 (mg/kg/day) over dose 25 (mg/kg/day). In this case it is seen that the outlier 2.14 induces a bias in the estimate of difference of means, and also a large variance within group 1.

The data of Table 5.1 are part of Table 6.9 in Example 6.7, and a full analysis is given there.

5.1.2 Remark on the unequal variances

We assume the linear model (5.1) and consider the case where the e_{ij} are distributed independently of each other as $N\left(0,\sigma_i^2\right)$ with possible different variances σ_1^2 and σ_2^2. Then $\bar{y}_{1\cdot}-\bar{y}_{2\cdot}$ is still a minimum variance unbiased estimator of $\mu_1-\mu_2$ with variance $\sigma_1^2/n_1+\sigma_2^2/n_2$. In contrast,

$$\hat{\sigma}_i^2=\sum_{j=1}^{n_i}\left(y_{ij}-\bar{y}_{i\cdot}\right)^2/(n_i-1),\ i=1,2 \tag{5.6}$$

give the minimum variance unbiased estimator of σ_i^2 and intuitively a test statistic

$$t = (\bar{y}_{1\cdot} - \bar{y}_{2\cdot}) / \sqrt{\hat{\sigma}_1^2/n_1 + \hat{\sigma}_2^2/n_2} \tag{5.7}$$

is suggested. However, the null distribution of t (5.7) is different from the t-distribution and depends on σ_2^2/σ_1^2. If we apply formally the t-distribution with df $(n_1 + n_2 - 2)$, it is known to enhance the type I error. By any other method we cannot make the test statistic free from σ_i^2, and this situation is called the Behrens–Fisher problem.

In practice, a chi-squared approximation $d\chi_f^2$ of $\hat{\sigma}_1^2/n_1 + \hat{\sigma}_2^2/n_2$ is often employed, adjusting the first two cumulants. However, the meaning of testing equality of means when there is clearly a difference in variances is not clear. If the mean and variance are simultaneously increased for one population against the other, then the underlying normal assumption is suspicious. A simultaneous test of mean and variance might then be preferable to their separate analyses. Also under this situation an application of a lognormal distribution might be more reasonable, see the remarks at the end of Section 2.3. In contrast, since the t-test is asymptotically non-parametric, it is expected that the unequal variances to some extent may not be so serious (see Section 5.2.1). Therefore, we do not go deep into this subject.

5.1.3 Paired sample

The data of Table 5.2 are measurements of total cholesterol before and after 6 months' treatment for 10 subjects. The problem of interest is to evaluate the effects of treatment, and it looks similar to the problem of Section 5.1.1. The necessary calculation of the left one-sided t-test is given in the table, and we have

$$\begin{aligned} \bar{y}_{2\cdot} - \bar{y}_{1\cdot} &= 2727/10 - 2942/10 = -21.5, \\ \sum_i \left(\sum_j y_{ij}^2 - y_{i\cdot}^2/n_i \right) &= 34009.6 + 19230. = 53239.7, \\ \hat{\sigma}^2 &= 53239.7/(20-2) = 2957.76, \end{aligned} \tag{5.8}$$

and finally

$$t = -21.5 / \sqrt{(10^{-1} + 10^{-1}) \times 2957.76} = -0.884.$$

This result is not significant at one-sided $\alpha = 0.05$. However, the data in Table 5.2 are the paired data for each subject and we can also apply a t-test to the differences before and after the treatment instead of analyzing the averages. Now the data are differences, given in the fourth column of Table 5.2, which we denote by x_j, $j = 1, \ldots, 10$.

Table 5.2 Cholesterol measurements before and after treatment.

Subject j	Before ($i=1$)	After ($i=2$)	Difference	Average
1	333	338	5	335.5
2	240	229	−11	234.5
3	364	305	−59	334.5
4	337	301	−36	319.0
5	326	279	−47	302.5
6	279	239	−40	259.0
7	188	210	22	199.0
8	371	339	−32	355.0
9	273	242	−31	257.5
10	231	245	14	238.0
$y_{i\cdot}=\sum_j y_{ij}$	2942	2727	−215	
$\sum_j y_{ij}^2$	899546	762883	11397	
$S=\sum_j y_{ij}^2 - y_{i\cdot}^2/n_i$	34009.6	19230.1	6774.5	

The null and alternative hypotheses are $H_0 : \mu_x = 0$ and $H_1 : \mu_x < 0$, where μ_x denotes the expectation of x_j. Then, the t-statistic is obtained as

$$t = \sqrt{10}(-21.5)/\sqrt{6774.5/(10-1)} = -2.478^*,$$

with p-value 0.0175 and significant at level 0.05.

The difference of two kinds of t-test lies in the evaluation of the error variance. The sum of squares (5.8) for error employed for the usual t-test is dissolved into two parts:

$$\sum_{i=1}^{2}\left(\sum_{j=1}^{n} y_{ij}^2 - y_{i\cdot}^2/n\right) = 2^{-1}S_x + S_B, \tag{5.9}$$

where

$$S_x = \sum_{j=1}^{n}(x_j - \bar{x}_\cdot)^2 = \sum_{j=1}^{n}\{y_{2j} - y_{1j} - (\bar{y}_{2\cdot} - \bar{y}_{1\cdot})\}^2$$

denotes the error sum of squares of the difference x_j. In contrast,

$$S_B = 2\sum_{j=1}^{n}(\bar{y}_{\cdot j} - \bar{y}_{\cdot\cdot})^2$$

denotes the variation of averages among the subjects. Now, suppose a model

$$y_{ij} = \mu_i + \beta_j + e_{ij},$$

adding the effect β_j of potential amount of cholesterol of each subject to the original model (5.1). Then it is easy to see the expectations

$$E(S_x) = 2(n-1)\sigma^2$$
$$E(S_B) = 2\sum_j (\beta_j - \bar{\beta}_.)^2 + (n-1)\sigma^2$$

Without effects β_j, both components on the right-hand side of (5.9) equally represent the $\sigma^2\chi^2_{n-1}$ but S_B suffers from the variation of the β_j if they exist. Under the assumption that the potential amount of cholesterol will not affect the treatment, the variation of β_j can be removed from the error variation and the analysis of difference based on x_j is justified. The ratio of $2^{-1}S_x$ to S_B is 0.068, and the variation between averages is much larger than the measurement errors in this example. That is, the reduction of variation among subjects is of larger effect than the reduction of degrees of freedom of the error. The subject in this experiment is a sort of block, discussed in Chapter 9, where it is assumed there is no interaction effect between the block and the treatment.

Example 5.4. The difference of skill between the expert and a beginner. The data of Table 5.3 are for evaluating the bias of measurements between the expert and a newly employed beginner (Moriguti, 1976). If there is a bias, the training of the beginner should be continued. We can apply the paired t-test, eliminating the variation in materials and the two-sided alternative hypothesis is employed. From the table we have

$$t = \sqrt{10}(2.2/10)/\sqrt{43.856/(10-1)} = 0.315.$$

This is smaller than $t_9(0.025)$, and the difference between the beginner and the expert is not significant at level 0.05. In this case the difference was not detected even by the paired t-test, but still the ratio of $2^{-1}S_x$ to S_B is 0.14, suggesting an effect of eliminating the variation in materials.

5.1.4 Comparison of normal variances

In this section we consider testing the equality of variances of two normal populations $N(\mu_i, \sigma_i^2)$, $i = 1, 2$. This test is used also for a preliminary check of the assumption in comparisons of means in Section 5.1.1, not only for the interest in variances. Now the estimates of variance given by (5.6) are distributed independently as $\sigma_i^2\chi^2_{n_i-1}/(n_i-1)$, $i = 1, 2$. Therefore

$$F = \hat{\sigma}_1^2/\hat{\sigma}_2^2 \tag{5.10}$$

Table 5.3 Measurements in the chemical analysis.

Material j	Expert ($i=1$)	Beginner ($i=2$)	Difference	Average
1	47.9	48.2	0.3	48.05
2	50.9	51.8	0.9	51.35
3	51.9	51.8	−0.1	51.85
4	54.0	53.8	−0.2	53.9
5	49.3	46.9	−2.4	48.1
6	49.1	46.6	−2.5	47.85
7	47.1	45.5	−1.6	46.3
8	50.0	50.4	0.4	50.2
9	50.3	53.2	2.9	51.75
10	53.2	57.7	4.5	55.45
$x_{\cdot}=\sum_j x_j$			2.2	
$\sum_j x_j^2$			44.34	
$S=\sum_j x_j^2 - x_{\cdot}^2/n$			43.856	

is distributed as $(\sigma_1^2/\sigma_2^2)\times F_{n_1-1,n_2-1}$, where F_{f_1,f_2} denotes a random variable distributed as an F-distribution with df (f_1, f_2). Therefore, under the null hypothesis

$$H_0: \sigma_1^2 = \sigma_2^2,$$

F (5.10) is distributed as F_{n_1-1,n_2-1}. Then, against each of the alternative hypotheses

right one-sided $H_1: \sigma_1^2 > \sigma_2^2$,
two-sided $H_2: \sigma_1^2 \neq \sigma_2^2$,

the rejection region is obtained as

$$R_1: F > F_{n_1-1,n_2-1}(\alpha),$$
$$R_2: F > F_{n_1-1,n_2-1}(\alpha/2) \text{ or } F < F_{n_1-1,n_2-1}(1-\alpha/2),$$

respectively. Usually, tabulation for the F-distribution is made for upper tail probability up to 0.50 and the following relation is applied:

$$F < F_{n_1-1,n_2-1}(1-\alpha/2) \Leftrightarrow F < F^{-1}_{n_2-1,n_1-1}(\alpha/2)$$

for tail probability beyond 0.50.

The confidence interval for σ_1^2, σ_2^2 and also the common variance when σ_1^2 and σ_2^2 are regarded as equal are constructed as described in Section 3.3.2. We describe here the confidence interval for the ratio $\gamma=\sigma_1^2/\sigma_2^2$. Since F of (5.10) is distributed as $\gamma F_{n_1-1,n_2-1}$, we have

$$\Pr\{F_{n_1-1,n_2-1}(1-\alpha/2)\le\gamma^{-1}F\le F_{n_1-1,n_2-1}(\alpha/2)\}=1-\alpha$$

and therefore the confidence interval for γ with confidence coefficient $1-\alpha$ is given by

$$F/F_{n_1-1,n_2-1}(\alpha/2)\le\gamma\le F\times F_{n_2-1,n_1-1}(\alpha/2).$$

Example 5.5. Comparing the densities of food between two companies. Two companies A and B are competitors in some food market. The density is an important quality of the food, and some data are taken by company B as shown in Table 5.4. It is a serious problem for company B if the density is lower than that of company A. Therefore, they planned to test the null hypothesis $H_0:\mu_A=\mu_B$ against the right one-sided alternative $H_1:\mu_A>\mu_B$.

Since the variance is also an important quality for a product in a large market, they planned first a two-sided test at significance level 0.05 for variance. From Table 5.4 the unbiased variances are obtained as

$$S_1=765.21-87.3^2/10=3.081\Rightarrow\hat\sigma_1^2=3.081/9=0.3423$$
$$S_2=675.37-82.1^2/10=1.329\Rightarrow\hat\sigma_2^2=1.329/9=0.1477$$

The F-statistic is obtained as

$$F=\frac{0.3423}{0.1477}=2.32,$$

not significant compared with $F_{9,9}(0.05/2)=4.03$. Then, they proceeded to apply a t-test to the means, assuming equality of variance. The t-statistic is calculated as

$$t=\left(\frac{87.3}{10}-\frac{82.1}{10}\right)\Big/\sqrt{\frac{(10^{-1}+10^{-1})(3.081+1.329)}{(20-2)}}=2.35^*.$$

Table 5.4 Density of food (Moriguti, 1976).

Company	Density										$\sum_j y_{ij}$	$\sum_j y_{ij}^2$
$A\,(i=1)$	9.1	8.1	9.1	9.0	7.8	9.4	8.2	9.1	8.2	9.3	87.3	765.21
$B\,(i=2)$	8.2	8.6	7.8	7.6	8.4	8.6	8.0	8.1	8.8	8.0	82.1	675.37

The p-value is 0.015 as a t-distribution with df 18. Since a significant difference has been suggested, company B should make improvements to raise the density of the food product.

5.2 Non-parametric Tests

5.2.1 Permutation test

We test the equality of two populations

$$H_0:\ \Pr(Y_1 \le y_0) = \Pr(Y_2 \le y_0) \tag{5.11}$$

against the alternative hypothesis

$$H_1:\ \Pr(Y_1 \le y_0) < \Pr(Y_2 \le y_0),$$

where y_0 is any constant and Y_i is a random variable from population $i = 1$ or 2. The alternative hypothesis H_1 implies that population 1 is statistically larger than population 2. Let $y_{11}, \ldots, y_{1n_1}$ and $y_{21}, \ldots, y_{2n_2}$ be random samples from two populations. Then, without any distributional assumption, we can test the hypothesis by evaluating the statistic

$$z = y_{11} + \cdots + y_{1n_1} \tag{5.12}$$

as a random sample from the given finite population

$$(y_{11}, \ldots, y_{1n_1}; y_{21}, \ldots, y_{2n_2}). \tag{5.13}$$

In Example 5.3 we were interested in whether population 2 is statistically larger than population 1 or not. Instead, we may test the statistical smallness of population 1. Then, the observed $z = 6.20$ is the smallest among all the random samples of size four from a population

$$(1.55,\ 1.63,\ 1.49,\ 1.53; 1.78,\ 1.93,\ 1.80,\ 2.07,\ 1.70).$$

Therefore the one-sided p-value is evaluated as $1/\binom{9}{4} = 0.008$.

In the general case it would be difficult to enumerate all samples of size four with sum larger than or equal to z. In such a case, a normal approximation is useful just as in Section 3.1.5 (1). The mean, variance, and covariance of the random sample from population (5.13) are calculated as follows:

$$E(Y_{1i}) = \mu = (n_1 + n_2)^{-1}(y_{11} + \cdots + y_{1n_1} + y_{21} + \cdots + y_{2n_2}), \tag{5.14}$$

$$\begin{aligned} V(Y_{1i}) &= \sigma^2 = E\left(Y_{1i}^2\right) - \{E(Y_{1i})\}^2 \\ &= (n_1 + n_2)^{-1}\left(y_{11}^2 + \cdots + y_{1n_1}^2 + y_{21}^2 + \cdots + y_{2n_2}^2\right) - \mu^2, \end{aligned} \tag{5.15}$$

$$Cov(Y_{1i}, Y_{1j}) = E(Y_{1i} \times Y_{1j}) - E(Y_{1i}) \times E(Y_{1j})$$

$$= (y_{11}y_{12} + y_{11}y_{13} + \cdots + y_{2n_2-1}y_{2n_2}) \Big/ \binom{n_1+n_2}{2} - \mu^2$$

$$= \left[\frac{\{(n_1+n_2)\mu\}^2 - \left(y_{11}^2 + \cdots + y_{1n_1}^2 + y_{21}^2 + \cdots + y_{2n_2}^2\right)}{2} - \binom{n_1+n_2}{2}\mu^2 \right] \Big/ \binom{n_1+n_2}{2}$$

$$= -\sigma^2/(n_1+n_2-1). \tag{5.16}$$

It should be noted that a negative correlation is induced between two random samples from a finite population. Thus, the expectation and variance of z (5.12) are obtained as follows:

$$E(z) = E(\textstyle\sum Y_{1i}) = n_1\mu$$

$$V(z) = V(\textstyle\sum Y_{1i}) = \sum V(Y_{1i}) + 2\sum_{i<j} Cov(Y_{1i}, Y_{1j})$$

$$= n_1\sigma^2 + 2 \times \binom{n_1}{2} \frac{-\sigma^2}{n_1+n_2-1} = \frac{n_1 n_2}{n_1+n_2-1}\sigma^2$$

Finally, we have a normal approximation

$$u = \frac{z - n_1\mu}{\sqrt{n_1 n_2 \sigma^2/(n-1)}} \sim N(0, 1) \tag{5.17}$$

under H_0 (5.11), where $n = n_1 + n_2$. Just as in Section 3.1.5 (1), u of (5.17) is related to t of (5.2) by the equation

$$t = u/\{(n-1-u^2)/(n-2)\}^{1/2}.$$

Therefore, the t-test is asymptotically equivalent to the non-parametric permutation test in this case, too.

Example 5.6. Example 5.1 continued

(1) **Eliminating outlier.** We have

$$\hat{\mu} = (6.20 + 9.28)/9 = 1.720,$$

$$\hat{\sigma}^2 = (9.6204 + 17.3082)/9 - 1.720^2 = 0.033667$$

$$u = (6.20 - 4 \times 1.720) \Big/ \sqrt{4 \times 5 \times \frac{0.033667}{(9-1)}} = -2.34^{**}$$

This value corresponds to a lower tail probability 0.0095 of the standard normal distribution. It is slightly above the exact value 0.008 but does not differ much, leading to the same conclusion. The normal approximation is improved for non-extreme cases unlike this example, and moderate sample size.

(2) **Full data.** We have

$$\hat{\mu} = (8.34 + 9.28)/10 = 1.762$$

$$\hat{\sigma}^2 = (14.2000 + 17.3082)/10 - 1.762^2 = 0.046176$$

$$u = (8.34 - 5 \times 1.762)/\sqrt{5 \times 5 \times \frac{0.046176}{(10-1)}} = -1.312.$$

This value corresponds to a lower tail probability 0.095 of the standard normal distribution, and is close to the t-test result in Example 5.1.

Example 5.7. Example 5.5 continued. From Table 5.4 we have

$$\hat{\mu} = (87.3 + 82.1)/20 = 8.47,$$

$$\hat{\sigma}^2 = (765.21 + 675.37)/20 - 8.47^2 = 0.2881,$$

$$u = (87.3 - 10 \times 8.47)/\sqrt{10 \times 10 \times \frac{0.2881}{(20-1)}} = 2.11^*.$$

The one-sided p-value is 0.017, and close to the p-value 0.016 of the$t$-test.

5.2.2 Rank sum test

The idea of the permutation test is applied to the rank data as it is. If we replace the data by rank in Example 5.6 (1), we have

$$3, 4, 1, 2 \text{ for } i = 1 \text{ and } 6, 8, 7, 9, 5 \text{ for } i = 2.$$

Then, the ranks for $i = 1$ are the four smallest among all the ranks and the conclusion is the same as in Example 5.6 (1).

For a general case, let r_{ij} denote the rank of the jth datum of the ith group. We employ the rank sum

$$W_i = r_{i\cdot}, \; i = 1, 2$$

as the test statistic and usually put the same average rank for the ties. Then we get the expectation and variance under the null hypothesis (5.11) as

$$E(W_i) = n_i(n_1 + n_2 + 1)/2,$$

and

$$V(W_i) = \frac{n_1 n_2 \sigma^2}{n_1 + n_2 - 1}, \quad \sigma^2 = \left\{ \frac{1}{n_1 + n_2} \sum_i \sum_j r_{ij}^2 - \left(\frac{n_1 + n_2 + 1}{2} \right)^2 \right\}, \tag{5.18}$$

just like *E(z)* and *V(z)* of Section 5.2.1. The test statistic is given by

$$\{W_i - n_i(n_1 + n_2 + 1)/2\} / \sqrt{V(W_i)},$$

which corresponds to (5.17). If there is no tie, we have simply

$$V(W_i) = n_1 n_2 (n_1 + n_2 + 1)/12.$$

This test is called the Wilcoxon–Mann–Whitney rank sum or simply Wilcoxon ranks test.

Example 5.8. Example 5.1 continued. For the full data we have ranks

$$3, 4, 1, 2, 10 \text{ for } i = 1 \text{ and } 6, 8, 7, 9, 5 \text{ for } i = 2.$$

Therefore we have

$$\frac{\{W_2 - n_2(n_1 + n_2 + 1)/2\}}{\sqrt{n_1 n_2 (n_1 + n_2 + 1)/12}} = \frac{35 - 5 \times 11/2}{\sqrt{5 \times 5 \times 11/12}} = 1.5667.$$

The upper tail probability as a standard normal distribution is 0.059. By enumeration, there are 19 cases which give the rank sum equal to or larger than 35 and exact p-value $19/\binom{10}{5} = 0.075$. It is seen that the normal approximation is not so bad for sample size as small as five per group. Compared with the permutation test, the rank test is less affected by the outlier. Also, for a long-tailed distribution the rank test is slightly less affected than the permutation test. However, on the whole they give similar results for a logistic distribution or a t-distribution with df around 10, for example.

The permutation test can be applied to any score which is monotonically related to y_i. Further, we can choose the score so as to achieve an asymptotic efficiency of unity against the most powerful test if the underlying distribution is known. The Wilcoxon score (rank) is optimum for a logistic distribution. The Fisher–Yates and Van der Waerden scores are optimum for the normal distribution and for the extreme distribution, the Savage score is obtained. However, actually the underlying distribution is unknown and we cannot choose an optimum score. Some adaptive method has been proposed to choose an asymptotically optimum score, but we need a large sample size for the method to work well. Also, any score test with df 1 cannot keep high power against the wide range of the alternative hypothesis, except for the target alternative. We therefore introduce some multi-df tests in part (2) of the next section.

5.2.3 Methods for ordered categorical data

(1) **Linear score test.** The data of Table 5.5 are taken in a phase II clinical trial for comparing the two doses of a drug, AF3mg and AF6mg, based on the degree of general improvement after administration.

The data can be regarded as a sort of rank data of Section 5.2.2 with many ties. However, we employ here the notation of Section 11.2 for a general a-sample problem, where we denote the frequency data as in Table 5.5 by $\boldsymbol{y}_i = (y_{i1}, \ldots, y_{iJ})'$, $R_i = y_{i\cdot}$, $C_j = y_{\cdot j}$, $N = y_{\cdot\cdot}$, $i = 1, 2, j = 1, \ldots, J, J = 6$. The C_j subjects in the jth response category are given the same score w_j, of which a typical example is the Wilcoxon averaged rank score

$$w_j = C_1 + \cdots + C_{j-1} + (C_j + 1)/2, j = 1, \ldots, J.$$

Then, the statistics of a linear score test are

$$W_i = \sum\nolimits_j w_j y_{ij}, \ i = 1, 2.$$

Under the null hypothesis (5.11) of equality of two populations the expectation and variance of y_{ij} can be calculated as a random sample of size R_i from a finite population $(w_1, \ldots, w_1; \ldots; w_J, \ldots, w_J)$, where the number of w_j is C_j. The calculation is the same as given in (5.14) ~ (5.16) in Section 5.2.1. We can also calculate them regarding the y_{ij} as random variables from the multivariate hypergeometric distribution

$$MH(y_{ij} \mid R_i, C_j, N) \sim \frac{\Pi_i R_i! \Pi_j C_j!}{N! \Pi_i \Pi_j y_{ij}!}.$$

Table 5.5 Degree of general improvement in a phase II trial.

Drug	General improvement						Total
	Least favorable	Unfavorable	No good	Slightly favorable	Favorable	Most favorable	
AF3mg	7	4	33	21	10	1	$R_1 = 76$
AF6mg	5	6	21	16	23	6	$R_2 = 77$
Total	12	10	54	37	33	7	153

This is a special case of the general case of $a \geq 2$ in Section 6.3, and detailed and sophisticated calculations in this line are given by equations (11.3) and (11.4) of Section 11.2. Anyway, we have

$$E(W_i) = (R_i/N)\sum_j w_j C_j,\ i = 1, 2$$

$$V(W_1) = V(W_2) = \frac{R_1 R_2}{(N-1)}\sigma_w^2,$$

where

$$\sigma_w^2 = N^{-1}\left\{\sum_j w_j^2 C_j - \frac{1}{N}\left(\sum_j w_j C_j\right)^2\right\}$$

is the same as σ^2 in (5.15) and (5.18) but indexed by w here to express a particular score system $\{w_j\}$. Then the test statistic is given by

$$W = \left\{\frac{N-1}{R_1 R_2 \sigma_w^2}\right\}^{1/2}\left\{W_2 - \frac{R_2}{N}\sum_j w_j C_j\right\}$$
$$= \left(\frac{N-1}{N}\right)^{1/2}\frac{1}{\sigma_w}\left(\frac{1}{R_1} + \frac{1}{R_2}\right)^{-1/2}\left(\frac{W_2}{R_2} - \frac{W_1}{R_1}\right),$$

which is asymptotically distributed as $N(0, 1)$ under the null hypothesis H_0 (5.11).

Example 5.9. Wilcoxon rank test for Table 5.5. For calculation, it is convenient to make Table 5.6.

Table 5.6 Calculation of Wilcoxon rank test statistic.

	General improvement j						
Drug i	1	2	3	4	5	6	Total
1	7	4	33	21	10	1	76
2	5	6	21	16	23	6	77
C_j	12	10	54	37	33	7	153
w_j	6.5	17.5	49.5	95	130	150	
$w_j y_{1j}$	45.5	70	1633.5	1995	1300	150	5194
$w_j y_{2j}$	32.5	105	1039.5	1520	2990	900	6587
$w_j C_j$	78	175	2673	3515	4290	1050	11781
$w_j^2 C_j$	507	3062.5	132313.5	333925	557700	157500	1185008

$$\sigma_w^2 = 153^{-1}\left\{1185008 - 153^{-1}(11781)^2\right\} = 1816.150$$

From this table we obtain

$$W=\left\{\frac{153-1}{76\times77\times1816.150}\right\}^{1/2}\left\{6587-\frac{77}{153}\times11781\right\}=2.49^{*}$$

and the related two-sided p-value 0.0128.

(2) **Max acc.t1 and cumulative chi-squared χ^{*2}.** In contrast, there is the problem of comparing two multinomial distributions $M(n_i, \boldsymbol{p}_i)$, $\boldsymbol{p}_i=(p_{i1}, \ldots, p_{i6})'$, $i=1, 2$, where the hypothesis of interest is

$$\frac{p_{21}}{p_{11}}\le\frac{p_{22}}{p_{12}}\le\ldots\le\frac{p_{26}}{p_{16}}. \tag{5.19}$$

This inequality implies that the distribution of AF6mg is shifted to the right compared with AF3mg, or simply that AF6mg is statistically larger than AF3mg, while the null hypothesis H_0 is given by all the equalities in (5.19) corresponding to (5.11). Now, the data are very similar to the transpose of the data of Table 7.1 for comparing five ordered binomial distributions, as in Table 5.7. For the latter case a monotone hypothesis H_{mon}(6.23) in the logit $\theta_i=\text{logit}\,p_i=\log\{p_i/(1-p_i)\}$ of the improvement rate p_i is assumed. However, this is mathematically equivalent to the inequality (5.19) if we consider $p_{1j}+p_{2j}=p_{\cdot j}$ as a fixed nuisance parameter. Further, in both cases the analyses are based on fixed marginal totals $R_i=y_{i\cdot}$, $C_j=y_{\cdot j}$, and $N=y_{\cdots}$. Therefore, max acc. t1 and the cumulative chi-squared χ^{*2} of Section 7.3.1 are applied as they are to the problem here.

Example 5.10. Max acc. t1 and cumulative chi-squared χ^{*2} for Table 5.5. Following the procedures explained in Section 7.3.1 (2) or 11.5.2 (2), we make the 2×2 collapsed sub-tables for all the five cut points in the columns of Table 5.5 and calculate the goodness-of-fit chi-squares as given in Table 5.8. Then, the maximal statistic is $t_4^2=10.651$ and its two-sided p-value as the maximal statistic is 0.0033. The calculation of the p-value for this type of maximal statistic is not so easy. However, a very elegant and efficient formula is obtained in Section 7.3.1 (1) (c). The cumulative chi-squared χ^{*2} is obtained as $\sum_{i=1}^{5}t_i^2=18.80$ as the sum of chi-squares

Table 5.7 Transpose of the data in Table 7.1.

	Dosage (mg)					
Improvement	100	150	200	225	300	Total
No	16	18	9	9	5	57
Yes	20	23	27	26	9	105

Table 5.8 Collapsing columns of Table 5.5 at five cut points.

	General improvement j									
Drug i	1	2 ~ 6	1, 2	3 ~ 6	1 ~ 3	4 ~ 6	1 ~ 4	5, 6	1 ~ 5	6
1	7	69	11	65	44	32	65	11	75	1
2	5	72	11	66	32	45	48	29	71	6
	⇓ $t_1^2=0.3906$		⇓ $t_2^2=0.0011$		⇓ $t_3^2=4.0832$		⇓ $t_4^2=10.6514$		⇓ $t_5^2=3.6746$	

from the sub-tables. Its approximate p-value is 0.01 two-sided, see Section 7.3.1 (2) for the calculation. The χ^{*2} is inevitably two-sided because of the sum of squares, while the max acc. $t1$ and linear score test can be applied both to one- and two-sided tests. The data of Table 5.5 are part of Table 11.8, and a full analysis is given in Section 11.5.

The non-parametric tests, such as the permutation and linear score test, are robust in the sense that they keep the nominal significance level against various underlying distributions. However, such non-parametric tests with df 1 cannot keep high power against the wide range of the alternative hypotheses. In contrast, the kth sub-table of Table 5.8 is understood as yielding a score test statistic of a particular score system

$$w_j = 0 \text{ for } j = 1, \ldots, k; w_j = 1 \text{ for } j = k+1, \ldots, J, k = 1, \ldots, J-1.$$

Thus, max acc. $t1$ and the cumulative chi-squared χ^{*2} come from the multiple score systems and are characterized as multi-df statistics. In the analysis of Table 5.5, the p-values are 0.0128 for the Wilcoxon test, 0.0033 for max acc. $t1$, and 0.01 for the cumulative chi-squared χ^{*2}. Of course they vary according to the response pattern, but max acc. $t1$ and the cumulative chi-squared χ^{*2} are characterized to keep high power against the wide range of the monotone hypotheses. In Table 5.5 there is a step change in the ratio p_{2j}/p_{1j} between $j = 4$ and 5, and max acc. $t1$ responded reasonably well to this pattern. However, it can also respond well to a linear trend, as demonstrated in later sections.

5.3 Unifying Approach to Non-inferiority, Equivalence and Superiority Tests

5.3.1 Introduction

Around the 1980s there were several serious problems in the statistical analysis of clinical trials for new drug applications in Japan, among which two major problems were the multiplicity problem and the non-significance regarded as equivalence. The outline of the latter problem is as follows.

In a confirmatory phase III trial for a new drug application, a drug used to be compared with an active control instead of a placebo in Japan, and be admitted for publication if it was evaluated as equivalent to the control in efficacy. This is because drugs have multiple characteristics, and equivalence in efficacy is considered to be satisfactory if there is any advantage such as mildness or ease of administration. Then, the problem was that the non-significance by the usual t- or Wilcoxon test had long been regarded as proof of equivalence in Japan. This was stupid, since equivalence can so easily be achieved by an imprecise clinical trial with a small sample size. In contrast, it is true that it is too severe to request proof of superiority of a new drug over an established active control by the usual significance test. This requires too many subjects and too much time to prove superiority, even for a good drug, which results in a delay in introducing good drugs to the market. To overcome this situation, a non-inferiority test has been introduced, which requires rejecting the handicapped null hypothesis in the normal model

$$H_{\text{non0}}: \mu_1 - \mu_2 \le -\delta$$

against the one-sided alternative

$$H_{\text{non1}}: \mu_1 - \mu_2 > -\delta,$$

in favor of the test drug, where μ_1 and μ_2 are the normal means of the test and control drugs, respectively, and δ a non-inferiority margin. This is a positive proof instead of the classical negative proof, and never cleared by an imprecise clinical trial. It also never means that an inferior drug in the amount of $-\delta$ in the mean difference actually goes out through the screening process, and gives a very nice practical procedure. As stated later, we can make the outgoing mean difference approximately equal to or slightly above 0 by adjusting the non-inferiority margin. However, there still remains a multiplicity problem of how to justify the usual practice of superiority testing after proving non-inferiority. This has been overcome by a unifying approach to non-inferiority and superiority tests based on the multiple decision processes of Hirotsu (2007); it nicely combines the one- and two-sided tests and the handicapped test, replacing the usual simple confidence interval for normal means by a more useful confidence region. This is an application of the confidence region introduced in Section 4.3.

5.3.2 Unifying approach via multiple decision processes

(1) **Method.** Let the response of treatments be denoted by y_{ij} and assume them to be independently distributed as $N(\mu_i, \sigma^2)$, $j=1,\ldots,n_i$, where $i=1$ stands for the test drug, $i=2$ for the control, and n_i is the repetition number. We partition the parameter space into three regions:

$$K_1: \mu_1 - \mu_2 < 0$$
$$K_2: \mu_1 - \mu_2 = 0$$
$$K_3: \mu_1 - \mu_2 > 0.$$

Then, for K_1, K_2, and K_3, we can take the acceptance regions with confidence coefficient $1-\alpha$ as

$$A_{K_1}: \bar{y}_{1\cdot} - \bar{y}_{2\cdot} - (\mu_1 - \mu_2) < T_\alpha \text{ for } \mu_1 - \mu_2 < 0,$$
$$A_{K_2}: -T_{\alpha/2} \le \bar{y}_{1\cdot} - \bar{y}_{2\cdot} - (\mu_1 - \mu_2) \le T_{\alpha/2} \text{ for } \mu_1 - \mu_2 = 0,$$
$$A_{K_3}: \bar{y}_{1\cdot} - \bar{y}_{2\cdot} - (\mu_1 - \mu_2) > -T_\alpha \text{ for } \mu_1 - \mu_2 > 0,$$

respectively, where $\bar{y}_{i\cdot}$, $i=1, 2$, is the sample mean and

$$T_\alpha = \hat{\sigma}\sqrt{\frac{1}{n_1}+\frac{1}{n_2}}t_{n_1+n_2-2}(\alpha),$$

with $\hat{\sigma}^2$ the usual unbiased estimate of variance σ^2 employed in (5.3) and $t_f(\alpha)$ the upper α point of the t-distribution with df f. Since the partition is disjoint, the overall significance level of this procedure is α, so that the union of these acceptance regions form an acceptance region with confidence coefficient exactly $1-\alpha$. Then, by inverting the acceptance region into the regions for $\mu_1-\mu_2$, we have the confidence region for $\mu_1-\mu_2$ with confidence coefficient $1-\alpha$ as

$$\begin{aligned}
0<\mu_1-\mu_2<\bar{y}_{1\cdot}-\bar{y}_{2\cdot}+T_\alpha &\text{ for } T_{\alpha/2}<\bar{y}_{1\cdot}-\bar{y}_{2\cdot},\\
0\le\mu_1-\mu_2<\bar{y}_{1\cdot}-\bar{y}_{2\cdot}+T_\alpha &\text{ for } T_\alpha\le\bar{y}_{1\cdot}-\bar{y}_{2\cdot}\le T_{\alpha/2},\\
\bar{y}_{1\cdot}-\bar{y}_{2\cdot}-T_\alpha<\mu_1-\mu_2<\bar{y}_{1\cdot}-\bar{y}_{2\cdot}+T_\alpha &\text{ for } -T_\alpha<\bar{y}_{1\cdot}-\bar{y}_{2\cdot}<T_\alpha\\
\bar{y}_{1\cdot}-\bar{y}_{2\cdot}-T_\alpha<\mu_1-\mu_2\le 0 &\text{ for } -T_{\alpha/2}\le\bar{y}_{1\cdot}-\bar{y}_{2\cdot}\le -T_\alpha,\\
\bar{y}_{1\cdot}-\bar{y}_{2\cdot}-T_\alpha<\mu_1-\mu_2<0 &\text{ for } \bar{y}_{1\cdot}-\bar{y}_{2\cdot}<-T_{\alpha/2},
\end{aligned} \tag{5.20}$$

see also Fig. 5.1, where the shaded area is the acceptance region. It is seen that the upper and lower bounds of this confidence region are essentially the same as those

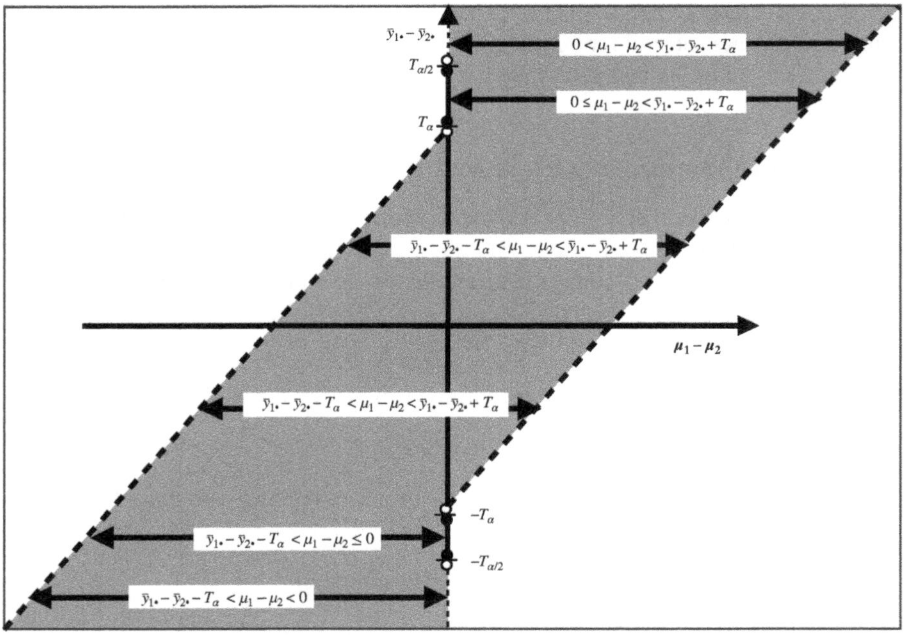

Figure 5.1 A confidence region with confidence coefficient $1-\alpha$.

of the conventional confidence interval with confidence coefficient $1-2\alpha$, which is given by the two parallel dotted lines

$$\bar{y}_{1\cdot}-\bar{y}_{2\cdot}-T_\alpha \le \mu_1-\mu_2 \le \bar{y}_{1\cdot}-\bar{y}_{2\cdot}+T_\alpha,$$

but raise the inclusion probability of $\mu_1-\mu_2=0$ up to $1-\alpha$ by widening the region at $|\bar{y}_{1\cdot}-\bar{y}_{2\cdot}|>T_\alpha$.

For the non-inferiority test we further introduce two regions

$$K_{-\delta}: \mu_1-\mu_2=-\delta,$$
$$K_{-\delta}^-: \mu_1-\mu_2<-\delta.$$

It should be noted that $K_{-\delta}^-$ and $K_{-\delta}$ are disjoint and both are included in K_1. We have therefore

$$\begin{aligned} &K_1 \cap K_{-\delta}^- = K_{-\delta}^-, \quad K_1 \cap K_{-\delta} = K_{-\delta}, \quad K_{-\delta}^- \cap K_{-\delta} = \phi, \\ &K_1 \cap K_2 = K_{-\delta}^- \cap K_2 = K_{-\delta} \cap K_2 = \phi, \end{aligned} \tag{5.21}$$

where ϕ is an empty set. Then we can test $K_{-\delta}^-$ and $K_{-\delta}$ before K_1 without adjusting the significance level α according to the closed test procedure (Section 6.4.4).

If they are rejected, we can also test K_1 and K_2 without adjusting the significance level. Therefore, we first apply the one-sided α test for the null hypothesis $K_{-\delta}^-$ and if it is not rejected, the non-inferiority at the margin δ cannot be achieved. If it is rejected, we proceed to testing for $K_{-\delta}$ by a two-sided α test. If it is not rejected, we can declare only the non-inferiority: $\mu_1-\mu_2 \ge -\delta$. If it is rejected, we proceed to testing for K_1 by a one-sided α test. If it is not rejected, we can declare only one step above the non-inferiority: $\mu_1-\mu_2>-\delta$. If it is rejected, we proceed to testing for K_2 by a two-sided α test. If it is not rejected, we can declare up to the equivalence: $\mu_1-\mu_2 \ge 0$ and if it is rejected, then we can declare the superiority: $\mu_1-\mu_2>0$. The multiple decision processes are summarized below by setting α at 0.05.

1. If $\bar{y}_{1\cdot}+\delta-\bar{y}_{2\cdot}<T_{0.05}$, then non-inferiority is not achieved.
2. If $T_{0.05} \le \bar{y}_{1\cdot}+\delta-\bar{y}_{2\cdot} \le T_{0.025}$, then declare weak non-inferiority: $\mu_1-\mu_2 \ge -\delta$.
3. If $\bar{y}_{1\cdot}+\delta-\bar{y}_{2\cdot}>T_{0.025}$ and $\bar{y}_{1\cdot}-\bar{y}_{2\cdot}<T_{0.05}$, then declare strong non-inferiority: $\mu_1-\mu_2>-\delta$.
4. If $T_{0.05} \le \bar{y}_{1\cdot}-\bar{y}_{2\cdot} \le T_{0.025}$, then declare at least equivalence: $\mu_1-\mu_2 \ge 0$.
5. If $\bar{y}_{1\cdot}-\bar{y}_{2\cdot}>T_{0.025}$, then declare superiority: $\mu_1-\mu_2>0$.

This could include the non-inferiority test of ICH E9 (the International Statistical Guideline; Lewis, 1999) via one-sided level 0.025 as its step 3 and superiority test as its step 5, while having an intermediate decision: at least equivalent between those two steps. Further, the non-inferiority has been made precise in two parts with and without equality to the non-inferiority margin, and could include the old Japanese practice via one-sided level 0.05 as weak non-inferiority at step 2. Thus it could give a very good unifying approach to the ICH E9 and the old Japanese Guideline (Koseisho, 1992).

Here we recommend requesting

$$\bar{y}_{1\cdot}-\bar{y}_{2\cdot} \geq -T_{0.025} \tag{5.22}$$

for the non-inferiority of steps 2 and 3, since otherwise the confidence region is totally below 0 and seems not to deserve the name of non-inferiority (even if it clears the non-inferiority margin). It should be noted that $\bar{y}_{1\cdot}-\bar{y}_{2\cdot} < -T_{0.025}$ is the region where the null hypothesis K_2 is rejected in the negative direction by the usual two-sided 0.05 significance test. This is therefore a device to avoid asserting the non-inferiority by a very large trial when actually the inequality $-\delta \leq \mu_1-\mu_2 < 0$ holds and $\bar{y}_{1\cdot}-\bar{y}_{2\cdot}$ is significantly different from zero in the negative direction. By this device we can hold essentially the same power with the one-sided 0.05 test while having even less consumer's risk than the one-sided 0.025 test for large n, as described in (2) and (3) below.

In some areas the non-inferiority margin may be decided to be a clinically relevant quantity. However, there are also many cases where it is difficult to decide such a quantity and the non-inferiority margin is introduced just as an operating parameter for approving a slightly better drug in terms of efficacy compared with the standard. We discuss this point further in Section 5.3.5.

In the non-inferiority trial it has been recommended to include a placebo for assessing the sensitivity of the trial. Let the response of the placebo be denoted by μ_0, and assume

$$A: \mu_0 < \mu_2 - \delta.$$

Then we can apply the closed test procedure starting from testing the null hypothesis

$$K_p: \mu_1 \leq \mu_0$$

against

$$K_p^+ \mu_1 > \mu_0$$

since, by the assumption A, K_p is included in $K_{-\delta}^-$. The assumption A should be satisfied by taking the non-inferiority margin δ sufficiently smaller than the true difference $\mu_2-\mu_0$ between the active control and the placebo. It should be noted that the assumption A assures that the negation of $K_{-\delta}^-$ implies also $\mu_1 > \mu_0$.

(2) **Consideration of producer's risk.** For convenience, we consider (1−producer's risk) and call it power. It is defined as the probability that the test drug can clear the

non-inferiority criterion when $\mu_1 - \mu_2 = 0$. Then, for the conventional confidence interval with confidence coefficient $1-2\alpha$, the power is simply the probability

$$\Pr\{\bar{y}_{1\cdot} - \bar{y}_{2\cdot} > T_\alpha - \delta \mid \mu_1 - \mu_2 = 0\}$$

with $\alpha = 0.05$ for the old Japanese practice and 0.025 for the ICH E9. For simplicity, we assume σ to be known (or n_i sufficiently large), and then it is

$$\Pr\left\{u > z_\alpha - \left(\frac{1}{n_1} + \frac{1}{n_2}\right)^{-1/2} \delta/\sigma\right\}, \tag{5.23}$$

where u is distributed as $N(0, 1)$.

For the proposed method, the probability is obtained after some calculations as

$$\Pr\left[u > \max\left\{-z_{0.025}, z_{0.05} - \left(\frac{1}{n_1} + \frac{1}{n_2}\right)^{-1/2} \delta/\sigma\right\}\right]. \tag{5.24}$$

The non-inferiority margin δ/σ is usually taken around 1/3, and if we assume $n_1 = n_2 = n$ then we have

$$z_{0.05} - \left(\frac{2}{n}\right)^{-1/2} \delta/\sigma > -z_{0.025}$$

for $n \le 233$, so that region (5.24) is equivalent to region (5.23) with $\alpha = 0.05$. On the contrary, if n is larger than 233 then the power is at least 0.975, so that regarding the power, the proposed method is essentially equivalent to a one-sided 0.05 test (the old Japanese practice). More precisely, for unknown σ

$$\Pr\left\{T_{0.05} - \left(\frac{1}{n_1} + \frac{1}{n_2}\right)^{-1/2} \frac{\delta}{\hat{\sigma}} > -T_{0.025} \,\middle|\, \frac{\delta}{\sigma} = \frac{1}{3}\right\}$$

is 0.99 at $n = 198$, 0.54 at $n = 233$, and 0.01 at $n = 269$, so that the normal approximation doesn't differ much from the exact theory. We compare the exact power of the proposed method with one-sided 0.05 and 0.025 tests for several n's in Table 5.9. It is seen that the power of the proposed method is essentially equal to a one-sided 0.05 test unless n is extremely large, as suggested by the normal theory. In contrast, if n is extremely large, the power of all the tests approaches unity and the differences are very small.

(3) **Consideration of consumer's risk.** The consumer's risk is defined as the probability that the test drug may clear the non-inferiority criterion when actually $\mu_1 - \mu_2 = -\delta$. It is obviously 0.05 and 0.025 for the classical one-sided 0.05 and 0.025 level tests, respectively. Since it has been verified that the normal

Table 5.9 Power comparison at $\delta/\sigma = 1/3$.

	$n_1 = n_2$			
Test	100	198	233	269
0.05 test	0.7593	0.9522	0.9743	0.9867
0.025 test	0.6501	0.9113	0.9485	0.9712
Proposed test	0.7593	0.9521	0.9718	0.9750

approximation works well for the proposed method in the previous section, we employ it again here for simplicity. Then, by definition of the consumer's risk, we have $\Pr\{(T_{0.05} \le \bar{y}_{1\cdot} + \delta - \bar{y}_{2\cdot}) \cap (\bar{y}_{1\cdot} - \bar{y}_{2\cdot} \ge -T_{0.025})\}$. After some calculations the consumer's risk is obtained as

$$\Pr\left[u > \max\left\{\left(\frac{1}{n_1} + \frac{1}{n_2}\right)^{-1/2} \delta/\sigma - z_{0.025},\ z_{0.05}\right\}\right]. \tag{5.25}$$

Again we assume $\delta/\sigma = 1/3$, then up to $n_1 = n_2 = 233$ this probability is 0.05 but after this point it begins to decrease according to n and at $n = 277$ it becomes even less than 0.025, which is the consumer's risk of the one-sided 0.025 test. If δ/σ is taken at 1/2, the probability goes below 0.05 and 0.025 at n as small as 104 and 124, respectively. It is seen that the lower bound of (5.25) goes up with increasing δ and n, and this is a very useful property, avoiding setting an inappropriately large margin. It also implies that a trial which intends to prove the non-inferiority of a test drug that is actually inferior to the control by an extremely large number of subjects should be unsuccessful. This is the effect of requesting inequality (5.22) for non-inferiority.

5.3.3 Extension to the binomial distribution model

The test for non-inferiority was introduced originally for the binomial model and asymptotically it can be dealt with similarly to the normal model. Now, suppose we are comparing two efficacy rates p_1, p_2 of the test and control drugs based on the samples y_1, y_2 for which we assume the binomial distributions $B(n_i, p_i)$, $i = 1, 2$. Just like with the normal theory in Section 5.3.2, we first partition the space of $p_1 - p_2$ into three regions:

$$K_1 : p_1 - p_2 < 0,$$

$$K_2 : p_1 - p_2 = 0,$$

$$K_3 : p_1 - p_2 > 0.$$

We further introduce two regions

$$K_{-\delta}: p_1 - p_2 = -\delta$$
$$K_{-\delta}^{-}: p_1 - p_2 < -\delta$$

with δ some pre-specified non-inferiority margin. We have the same relationship among the hypotheses K_i as in (5.21), and by the asymptotic theory we can proceed similarly to Section 5.3.2. The only difficulty arising differently from the previous section is that the variance estimators for standardization are different for the null hypothesis K_1 and the handicapped hypothesis $K_{-\delta}$. This arises from the characteristic of the binomial distribution that the variance depends on the mean whereas the mean and variance are independent parameters in the normal model. Fortunately, however, it is seen that the naïve variance estimator

$$\hat{V}_0 = \left(\frac{1}{n_1} + \frac{1}{n_2}\right) \frac{y_1 + y_2}{n_1 + n_2} \left(1 - \frac{y_1 + y_2}{n_1 + n_2}\right)$$

for $(y_1/n_1 - y_2/n_2)$ under K_2 is also valid under the handicapped hypothesis $K_{-\delta}$. More exactly, Dunnett and Gent (1977) show that the variance estimator

$$\hat{V}_\delta = \frac{1}{n_1}(\hat{p}_{2\delta} - \delta)(1 - \hat{p}_{2\delta} + \delta) + \frac{1}{n_2}\hat{p}_{2\delta}(1 - \hat{p}_{2\delta}) \tag{5.26}$$

with

$$\hat{p}_{2\delta} = (y_1 + y_2 + n_1\delta)/(n_1 + n_2) \tag{5.27}$$

is appropriate for the handicapped non-inferiority test if the sample sizes are nearly balanced. Then, substituting $\hat{p}_{2\delta}$ of (5.27) into (5.26) and expanding it with respect to δ, we have

$$\hat{V}_\delta \cong \hat{V}_0 + \delta \times \frac{n_1 - n_2}{n_1 n_2}\left(1 - 2\frac{y_1 + y_2}{n_1 + n_2}\right) - \delta^2 \times \frac{n_1^2 - n_1 n_2 + n_2^2}{n_1 n_2 (n_1 + n_2)}.$$

Further, putting $\hat{p}_{20} = (y_1 + y_2)/(n_1 + n_2)$ we have

$$\frac{\left|\sqrt{\hat{V}_\delta} - \sqrt{\hat{V}_0}\right|}{\sqrt{\hat{V}_\delta}} \cong \frac{\delta}{8\hat{p}_{20}(1 - \hat{p}_{20})}\left\{2(2\hat{p}_{20} - 1)\frac{n_1 - n_2}{n_1 + n_2} + \delta\right\}. \tag{5.28}$$

That is, the relative error of using $\sqrt{\hat{V}_0}$ instead of $\sqrt{\hat{V}_\delta}$ is the order of δ^2 if n_1 and n_2 are nearly equal. Usually, δ is taken around 0.1 and the error of using $\sqrt{\hat{V}_0}$ instead of $\sqrt{\hat{V}_\delta}$ is practically negligible; therefore we can simply use $\sqrt{\hat{V}_0}$ throughout all the steps of this unifying procedure. Actually, in Example 5.8 the value of (5.28) is only 0.002. By this approximation, we can summarize the multiple decision processes as follows, where $\hat{p}_i = y_i/n_i$, $i = 1, 2$.

1. If $(\hat{p}_1+\delta-\hat{p}_2)/\sqrt{\hat{V}_0}<z_{0.05}$, then non-inferiority is not achieved.

2. If $z_{0.05}\le(\hat{p}_1+\delta-\hat{p}_2)/\sqrt{\hat{V}_0}\le z_{0.025}$, then declare weak non-inferiority: $p_1-p_2\ge-\delta$.

3. If $(\hat{p}_1+\delta-\hat{p}_2)/\sqrt{\hat{V}_0}>z_{0.025}$ and $(\hat{p}_1-\hat{p}_2)/\sqrt{\hat{V}_0}<z_{0.05}$, then declare strong non-inferiority: $p_1-p_2>-\delta$.

4. If $z_{0.05}\le(\hat{p}_1-\hat{p}_2)/\sqrt{\hat{V}_0}\le z_{0.025}$, then declare at least equivalence: $p_1-p_2\ge 0$.

5. If $(\hat{p}_1-\hat{p}_2)/\sqrt{\hat{V}_0}>z_{0.025}$, then declare superiority: $p_1-p_2>0$.

In the non-inferiority tests (2. and 3.), it is also useful to impose the requirement

$$(\hat{p}_1-\hat{p}_2)/\sqrt{\hat{V}_0}>-z_{0.025},$$

which corresponds to (5.22). By this alteration, the power for weak and strong non-inferiority is kept essentially the same and the consumer's risk is reduced just as in Section 5.3.2.

5.3.4 Extension to the stratified data analysis

As a typical example from a phase III clinical trial, the test and control drugs are often compared in several different classes. The data of Table 5.10 are for comparing a new drug against an active control in the infectious disease of respiratory organs. The result is known to be highly affected by the existence of pseudomonas. Therefore,

Table 5.10 A phase III trial in the infectious disease of respiratory organs.

Pseudomonas	Drug	Effectiveness		Total
i	j	$k=1$ (−)	$k=2$ (+)	
1 (Detected)	1 (Active)	15	21	36
	2 (Control)	13	10	23
Total		28	31	59
2 (No)	1 (Active)	7	20	27
	2 (Control)	9	34	43
Total		16	54	70
Total	1 (Active)	22	41	63
	2 (Control)	22	44	66

achievements are shown in each class of pseudomonas detected or not. The analysis ignoring this classification is quite misleading, often suffering from Simpson's paradox as described in Section 1.10. One standard approach to avoiding Simpson's paradox is testing interaction effects assuming a log linear model (see Section 14.3). Instead, in this section we first test the constancy of the differences in efficacy rates between test and control drugs through stratification. Then, if the constancy model is not rejected, we proceed to estimating the constant difference.

(1) **Model with a constant difference of efficacy rates through the stratification.** Let y_{ij} be the number of successful patients (denoted by $k=2$ in the table) for the jth drug in the ith class, where $j=1$ for the test drug and 2 for the control, and $i=1, \ldots, a$, $a=2$, in Table 5.10. We assume a binomial distribution $B(n_{ij}, p_{ij})$ for y_{ij} and test the null hypothesis of constant difference

$$H_0 : p_{i1}-p_{i2} \equiv \Delta, \; i=1, \ldots, a. \tag{5.29}$$

Under the full model, the likelihood is given by

$$L_1 = \Pi_{i=1}^{a} \Pi_{j=1}^{2} \left\{ \binom{n_{ij}}{y_{ij}} p_{ij}^{y_{ij}} (1-p_{ij})^{n_{ij}-y_{ij}} \right\}$$

and the MLE is obtained as $\hat{p}_{ij} = y_{ij}/n_{ij}$, $i=1, \ldots, a, j=1, 2$. Under the composite hypothesis H_0 (5.29), the MLE is obtained from the partial derivations of the log-likelihood

$$\log L_0 = \log \left[\Pi_{i=1}^{a} \left\{ \binom{n_{i1}}{y_{i1}} p_{i1}^{y_{i1}} (1-p_{i1})^{n_{i1}-y_{i1}} \binom{n_{i2}}{y_{i2}} (p_{i1}-\Delta)^{y_{i2}} (1-p_{i1}+\Delta)^{n_{i2}-y_{i2}} \right\} \right]$$

with respect to p_{i1} and Δ. Then, the likelihood ratio test statistic is given by

$$2\log\left(\hat{L}_1/\hat{L}_0\right)$$

with df $a-1$.

Example 5.11. Analysis of Table 5.10. An algorithm for calculating the MLE and the likelihood ratio test statistic is given in Hirotsu *et al.* (1997). Then, $2\log\left(\hat{L}_1/\hat{L}_0\right) = 1.39$ with df $= a-1 = 1$ is obtained by the algorithm. The p-value is 0.24 and non-significant at $p=0.15$, which is the probability usually employed for model checking.

(2) **Estimation of efficacy difference with skewness correction.** First, a naïve weighted mean for estimating Δ under the constant difference model (5.29) is given in Hirotsu *et al.* (1997) as

$$\hat{\Delta} = \frac{V^{-1}(\hat{\Delta}_1)\hat{\Delta}_1 + \cdots + V^{-1}(\hat{\Delta}_a)\hat{\Delta}_a}{V^{-1}(\hat{\Delta}_1) + \cdots + V^{-1}(\hat{\Delta}_a)}$$

$$\hat{\Delta}_i = \frac{y_{i1}}{n_{i1}} - \frac{y_{i2}}{n_{i2}},\ i=1,\ \ldots,\ a,\ V(\hat{\Delta}_i) = \left(\frac{1}{n_{i1}} + \frac{1}{n_{i2}}\right) \cdot \frac{y_{i\cdot}}{n_{i\cdot}} \cdot \frac{n_{i\cdot} - y_{i\cdot}}{n_{i\cdot}},\ i=1,\ \ldots,\ a \tag{5.30}$$

with variance

$$V(\hat{\Delta}) = \left\{V^{-1}(\hat{\Delta}_1) + \cdots + V^{-1}(\hat{\Delta}_a)\right\}^{-1}.$$

Then, a naïve confidence interval with confidence coefficient $1-2a$ is obtained as

$$\hat{\Delta} - V^{1/2}(\hat{\Delta})z_\alpha \le \Delta \le \hat{\Delta} + V^{1/2}(\hat{\Delta})z_\alpha. \tag{5.31}$$

As shown in Table 5.11, this naïve method (NAI) gives a reasonable result if the sample size is around 50 in each class. However, in the stratified analysis there will

Table 5.11 Consumer's risk at non-inferiority margin $\delta=0.\ 1$.

(p_{12}, p_{22}, p_{32})	Method	$n_{i1}=(10,10,10)$ $n_{i2}=(10,10,10)$	$n_{i1}=(10,20,10)$ $n_{i2}=(20,10,20)$	$n_{i1}=(100,100,100)$ $n_{i2}=(100,100,100)$
(0.7, 0.7, 0.7)	YTH	0.0540	0.0529	0.0516
	NAI	0.0569	0.0566	0.0515
	ES1	0.0522	0.0534	0.0517
	ES2	0.0518	0.0529	0.0516
(0.5, 0.5, 0.5)	YTH	0.0517	0.0567	0.0541
	NAI	0.0527	0.0551	0.0514
	ES1	0.0506	0.0564	0.0539
	ES2	0.0501	0.0548	0.0532
(0.3, 0.3, 0.3)	YTH	0.0579	0.0545	0.0509
	NAI	0.0675	0.0599	0.0526
	ES1	0.0513	0.0523	0.0511
	ES2	0.0522	0.0518	0.0510
(0.5, 0.6, 0.7)	YTH	0.0528	0.0549	0.0517
	NAI	0.0545	0.0554	0.0511
	ES1	0.0516	0.0547	0.0518
	ES2	0.0510	0.0541	0.0516
(0.3, 0.4, 0.5)	YTH	0.0559	0.0541	0.0510
	NAI	0.0575	0.0541	0.0507
	ES1	0.0527	0.0535	0.0506
	ES2	0.0525	0.0524	0.0504

often be a class of size as small as 10. Under these circumstances, the normal approximation based on the efficient score

$$T(\Delta) = \partial \log L_0 / \partial \Delta$$

with skewness correction by Bartlett (1953) works better. Since the nuisance parameters $p_{i1}, i = 1, \ldots, a$, are included in the efficient score $T(\Delta)$, we substitute the solution $p_{i1}(\Delta)$ of the likelihood equation

$$\frac{\partial \log \mathrm{L}_0}{\partial p_{i1}} = \frac{(y_{i1} - n_{i1} p_{i1})}{p_{i1}(1 - p_{i1})} + \frac{y_{i2} - n_{i2}(p_{i1} - \Delta)}{(p_{i1} - \Delta)(1 - p_{i1} + \Delta)} = 0. \tag{5.32}$$

This substitution causes a bias in the expectation of $T(\Delta)$. Although the bias of order $n^{-1/2}$ vanishes, the skewness of that order comes out. For the skewness correction we actually replace $T(\Delta)$ by

$$T^*(\Delta) = \left(\sigma_\Delta^2\right)^{-1/2} T(\Delta) - 6^{-1} \gamma \left\{ \left(\sigma_\Delta^2\right)^{-1} T^2(\Delta) - 1 \right\} \tag{5.33}$$

$$\sigma_\Delta^2 = \textstyle\sum_i \left\{ n_{i1}^{-1} p_{i1}(1 - p_{i1}) + n_{i2}^{-1}(p_{i1} - \Delta)(1 - p_{i1} + \Delta) \right\}^{-1}$$

$$\gamma = \left(\sigma_\Delta^2\right)^{-3/2} \left[\textstyle\sum_i b_i^3 \left\{ \frac{n_{i1}}{p_{i1}^2} - \frac{n_{i1}}{(1 - p_{i1})^2} \right\} + \textstyle\sum_i (1 - b_i)^3 \left\{ \frac{n_{i2}}{(1 - p_{i1} + \Delta)^2} - \frac{n_{i2}}{(p_{i1} - \Delta)^2} \right\} \right]$$

$$b_i = \frac{n_{i2} p_{i1}(1 - p_{i1})}{n_{i1}(p_{i1} - \Delta)(1 - p_{i1} + \Delta) + n_{i2} p_{i1}(1 - p_{i1})}$$

Since $T^*(\Delta)$ is a monotone decreasing function of Δ, we can solve two equations

$$T^*(\Delta_L) = z_\alpha \text{ and } T^*(\Delta_U) = -z_\alpha \tag{5.34}$$

to obtain lower and upper confidence bounds for Δ. In the numerical calculation we use the naïve estimator (5.30) for the initial value of Δ to solve equation (5.32) for $p_{i1}, i = 1, \ldots, a$. Then, we substitute the solution into (5.33) to obtain the function $T^*(\Delta)$ and solve the two equations in (5.34). We go back to (5.32) to calculate $p_{i1}, i = 1, \ldots, a$, by the renewed Δ, respectively, by the two equations of (5.34). The computer program for this is also given in Hirotsu *et al.* (1997). This program can be used to compare two binomial distributions by setting $a = 1$.

We can apply the confidence interval to the simultaneous tests of non-inferiority and superiority. It is an extension of the procedure of Section 5.3.3. Let $\Delta_L(\alpha)$ and $\Delta_U(\alpha)$ be the lower and upper confidence bounds defined by (5.34). Then we have the following multiple decision processes at significance level 0.05.

1. If $\Delta_L(0.05) < -\delta$, then non-inferiority is not achieved.
2. If $\Delta_L(0.05) \geq -\delta$ and $\Delta_L(0.025) \leq -\delta$, then declare weak non-inferiority: $p_1 - p_2 \geq -\delta$.

3. If $\Delta_L(0.025) > -\delta$ and $\Delta_L(0.05) < 0$, then declare strong non-inferiority: $p_1 - p_2 > -\delta$.
4. If $\Delta_L(0.025) \le 0$ and $\Delta_L(0.05) \ge 0$, then declare at least equivalence: $p_1 - p_2 \ge 0$.
5. If $\Delta_L(0.025) > 0$, then declare superiority: $p_1 - p_2 > 0$.

In the non-inferiority tests (2. and 3.), it is also useful to impose the requirement

$$\Delta_U(0.025) \ge 0$$

or

$$\hat{\Delta}/V^{1/2}(\hat{\Delta}) \ge -z_{0.025},$$

which corresponds to (5.22). By this alteration the power for weak and strong non-inferiority is kept essentially the same, and the consumer's risk is reduced just as in Section 5.3.2.

Example 5.12. Example 5.11 continued. Since a constant difference model was not rejected, we calculate Δ_L from $T^*(\Delta_L) = z_{\alpha/2}$ at $\alpha = 0.05$ to obtain $\Delta_L = -0.1078$. Since it is below the non-inferiority margin -0.10, non-inferiority has not been proved and the analysis is finished here. The lower bound by the naïve estimate (5.31) is –0.1098 and leads to the same conclusion. The naïve method is not as bad with this size of clinical trial.

In Table 5.11 we compare the risk of declaring non-inferiority when actually $p_{i1} - p_{i2} \equiv -0.1$ by simulation of 10^5 repetitions for each. The methods compared are the handicapped test of Yanagawa *et al.* (1994) (YTH), the naïve confidence interval (NAI), the normal approximation of efficient score (ES1), and the efficient score with skewness correction (ES2). The YTH tests the null hypothesis $H_0: p_{i1} - p_{i2} \equiv \Delta_i$, $i = 1, \ldots, a$ against $H_1: p_{i1} - p_{i2} > \Delta_i$, $i = 1, \ldots, a$ by Mantel–Haenszel's method, and we apply it with $\Delta_i \equiv \Delta$. As a result, it becomes a normal approximation method for $\sum_i y_{i2}$ and similar to ES1. However, it is different from ES1, which is based on the weighted sum of y_{i2}. From Table 5.11 it is seen that ES2 works well for all situations. The naïve method also works well when the sample size is moderately large.

5.3.5 Meaning of non-inferiority test and a rationale of switching to superiority test

There is an objection to the procedure of switching to a superiority test after obtaining favorable data which were not expected in the planning stage. In this section we discuss the problem, assuming a very simple but typical situation of non-inferiority

Table 5.12 Required sample size per arm ($p_2=0.7, \beta=0.1$).

	K_0	$K_{-0.05}$		$K_{-0.10}$	
Efficacy of test drug	0.025	0.025	0.05	0.025	0.05
$p_1=0.7$	–	1766	1439	442	360
$p_1=0.7+0.05$	1658	418	341	$186^{\dagger 2}$	$152^{\dagger 2}$
$p_1=0.7+0.10$	389	$173^{\dagger 2}$	$141^{\dagger 2}$	$98^{\dagger 2}$	$80^{\dagger 2}$

testing of $K_{-\delta}$ against $K_{-\delta}^{+}: p_1-p_2>-\delta$ and superiority testing of K_2 against K_3, denoted by K_0 in Table 5.12. We assume $p_2=0.7$ and power of the non-inferiority test $1-\beta=0.9$, and summarize the required sample size per arm for some combinations of α and δ in Table 5.12.

We should consider $p_1=0.75$ and 0.80 as quite acceptable against $p_2=0.7$, and the required sample size for a successful trial in Table 5.12 seems too great except for those cases marked by a dagger (†). Especially, such a scale of the superiority test ($\delta=0$) is usually unfeasible in Japan, and led to introducing non-inferiority tests with $\delta=0.1$ and $\alpha=0.05$ one-sided in the old Japanese Guideline (Koseisho, 1992) (see also Hirotsu, 1986). The change of α from 0.05 to 0.025 increases the required sample size by more than 20%, and seems unnecessary. Now, suppose the situation $p_1=0.80$, $p_2=0.70$, $\delta=0.1$, $\alpha=0.05$, and $n_1=n_2=80$, then the acceptance region for $K_{-0.1}$ at step 2 of Section 5.3.3 given by replacing $\hat{V}_0$ by $(2/80)\times 0.75\times 0.25$ becomes $\hat{p}_{10}-\hat{p}_{20}\geq\sqrt{\hat{V}_0}\times z_{0.05}-0.10=0.013$ (slightly above 0). Interestingly, in a trial around $p_1=p_2=0.75$, $n=100$ per arm with $\delta=0.1$ and $\alpha=0.05$ the approximately zero outcome ($\hat{p}_{10}-\hat{p}_{20}=0$) just clears the non-inferiority test and this was also a reason why we introduced a non-inferiority test with $\delta=0.1$ and $\alpha=0.05$ in Koseisho (1992) (see Hirotsu, 1986, 2004; Hirotsu and Hothorn, 2003 for more details). Mathematically, the non-inferiority margin δ is defined as the maximal tolerable difference but the above discussion suggests that it is simply a device to approve the new drug with slightly better efficacy rate compared with the standard drug. It should be noted that the outgoing quality level p_1 passing through the non-inferiority test of $K_{-\delta}$ is by no means $p_2-\delta$, but expected to be much higher. In contrast, in the normal model there may be some cases where δ is interpreted as a clinically relevant quantity.

To discuss the switching procedure from non-inferiority to superiority test, it is crucial that $K_{-\delta}^{-}$ and K_1 are not disjoint hypotheses but $K_{-\delta}^{-}$ implies K_1 $\left(K_{-\delta}^{-}\subset K_1\right)$. Now suppose we have $p_1=0.80, p_2=0.70$. Then, to get approval for the new drug, the required sample size 389 for K_2 is too large, so we employ the non-inferiority test with $\delta=0.1, \beta=0.1$, which requires $n=98$ per arm under the current rule of $\alpha=0.025$ one-sided (Table 5.12). Then the test has approximate power

$$P = \Pr\left\{u > 1.96 - \frac{0.1}{(0.75\times 0.25\times 2/98)^{1/2}}\right\} = 0.366 \text{ for superiority,}$$

$$P = \Pr\left\{u > 1.645 - \frac{0.1}{(0.75\times 0.25\times 2/98)^{1/2}}\right\} = 0.489 \text{ for at least equivalence,}$$

respectively. So, it does not happen that a trial planned as a non-inferiority test could clear the superiority or at least the equivalence criterion, but it occurs quite often. Further, since $K_{-0.10}$: $p_1 - p_2 = -0.1$ is just an operating hypothesis and the true hypothesis (estimate) that the experimenter has is $p_1 - p_2 = 0.1$, the switching procedure should be justified.

Example 5.13. An example of the switching procedure in the binomial model. The data in Table 5.13 are from a phase III randomized, double-blind clinical trial comparing the efficacy of the test and control drugs in chronic urticaria (Nishiyama *et al.*, 2001).

We set δ at 0.1 and apply the procedure introduced in Section 5.3.3. First $\hat{V}_0$ is obtained as

$$\hat{V}_0 = \left(\frac{1}{112} + \frac{1}{121}\right) \times \frac{168}{233} \times \frac{65}{233} = 0.0034583.$$

Then we calculate

$$(\hat{p}_1 + \delta - \hat{p}_2)/\sqrt{\hat{V}_0} = \left(\frac{87}{112} + 0.1 - \frac{81}{121}\right)/\sqrt{0.0034583} = 3.526.$$

This value is larger than $z_{0.025}$, and we proceed to the next step. Since

$$(\hat{p}_1 - \hat{p}_2)/\sqrt{\hat{V}_0} = \left(\frac{87}{112} - \frac{81}{121}\right)/\sqrt{0.0034583} = 1.826 \quad (5.35)$$

is larger than $z_{0.05}$ but smaller than $z_{0.025}$, at least equivalence: $p_1 - p_2 \geq 0$ has been proved. According to ICH E9 and the old Japanese Guideline also, only the

Table 5.13 Phase III trial for chronic urticarial.

	Response		
Drug	Improved	Not improved	Total
Test	87	25	112
Control	81	40	121

non-inferiority can be asserted in this case but at least equivalence seems to be a more appropriate conclusion from the achievement of the trial ($\hat{p}_1 = 0.78$, $\hat{p}_2 = 0.67$). In this case $\hat{V}_\delta$ in (5.26) is 0.003444 and the relative error (5.28) is equal to 0.002, suggesting that the difference is negligible.

Incidentally, it is worthwhile noting that the square of (5.35), $1.826^2 = 3.33$, coincides exactly with the χ^2 obtained in Example 3.6. This is verified by noting that

$$(\hat{p}_1 - \hat{p}_2)/\sqrt{\hat{V}_0} = \frac{(y_1/n_1 - y_2/n_2)}{\sqrt{(n_1^{-1} + n_2^{-1})\{(y_1 + y_2)/(n_1 + n_2)\}\{1 - (y_1 + y_2)/(n_1 + n_2)\}}}$$

$$= \frac{\sqrt{n_1 + n_2}\{y_1(n_2 - y_2) - y_2(n_1 - y_1)\}}{\sqrt{n_1 n_2 (y_1 + y_2)\{n_1 + n_2 - (y_1 + y_2)\}}} \tag{5.36}$$

Now, by the notation of Example 3.6, $y_1 = y_{11}$, $n_2 - y_2 = y_{22}$, $y_2 = y_{21}$, $n_1 - y_1 = y_{12}$ and the square of (5.36) is exactly equal to (3.58).

5.3.6 Bio-equivalence

The topic discussed so far is closely related to bio-equivalence, where the mean levels μ_1, μ_2 of two drugs are required to satisfy

$$0.80 \le \mu_1/\mu_2 \le 1.25. \tag{5.37}$$

This is usually confirmed by showing the 0.90 confidence interval for μ_1/μ_2 to be within the range of (5.37). In this section we assume a lognormal model for the pharmaco-kinetic measurements y_{ij} for two drugs $i = 1, 2$:

$$\log y_{ij} = \log \mu_i + e_{ij},\ i = 1,2; j = 1, \ldots, n_i$$

where the measurement errors e_{ij} are identically and independently distributed as $N(0, \sigma^2)$. Let $x_{ij} = \log y_{ij}$, then the usual approach for assuring bio-equivalence is to require the 0.90 confidence interval

$$\begin{aligned} &\bar{x}_{1\cdot} - \bar{x}_{2\cdot} - \hat{\sigma}\sqrt{\frac{1}{n_1} + \frac{1}{n_2}} t_{n_1+n_2-2}(0.05) \\ &\le \log \mu_1 - \log \mu_2 \le \bar{x}_{1\cdot} - \bar{x}_{2\cdot} + \hat{\sigma}\sqrt{\frac{1}{n_1} + \frac{1}{n_2}} t_{n_1+n_2-2}(0.05) \end{aligned} \tag{5.38}$$

to be totally inside the range (log 0.80, log 1.25). The procedure assures that the consumer's risk, defined in this case as the probability of wrongly declaring the ratio μ_1/μ_2 truly beyond the interval (5.36) to be equivalent, is less than or equal to 0.05 for the respective departures up and down. Now, for this problem we can also apply the confidence region (5.20) introduced in Section 5.3.2. As stated there, the confidence region raises the confidence coefficient up to 0.95 by widening the interval at $\bar{x}_{1\cdot}-\bar{x}_{2\cdot}\geq T_{0.05}$ and $\bar{x}_{1\cdot}-\bar{x}_{2\cdot}\leq -T_{0.05}$. However, widening the confidence interval obviously does not increase the probability of declaring equivalence in any case, so that the consumer's risk is not increased. The producer's risk, defined as the probability of failing to declare equivalence for the ratio μ_1/μ_2 truly inside the interval (5.37), should also be unchanged by this alteration, since the difference is only in the case when $|\bar{x}_{1\cdot}-\bar{x}_{2\cdot}|\geq T_{0.05}$. Suppose $\bar{x}_{1\cdot}-\bar{x}_{2\cdot}\geq T_{0.05}$, for example, then both lower confidence limits of 0 (by the proposed method) and $\bar{x}_{1\cdot}-\bar{x}_{2\cdot}-T_{0.05}$ (by the classical approach) exceed log 0.80 (negative value) with probability 1. A similar argument holds for the upper bound when $\bar{x}_{1\cdot}-\bar{x}_{2\cdot}\leq -T_{0.05}$. That is, the risk of accepting μ_i's beyond the range of (5.37) and also the risk of rejecting μ_i's truly inside the interval (5.37) are exactly the same for both methods (5.20) and (5.38). It should, however, be noted that even if the equivalence criterion has been satisfied by a confidence interval, some discussion would be necessary if it could not include $\mu_1/\mu_2=1$. We therefore propose requesting 0 to be included in the confidence interval for declaring equivalence, in addition to clearing the lower and upper bounds (log 0.80 and log 1.25), respectively. Then, the probability of including 0 in the confidence region when μ_1 and μ_2 are truly equivalent ($\mu_1/\mu_2=1$) is surely 0.95 for the new approach (5.20), while it is 0.90 for the conventional confidence interval (5.38). This could be an advantage of the proposed approach (5.20) over the conventional confidence interval (5.38).

Example 5.14. Proving bio-equivalence between Japanese and Caucasians. As an example, we apply the method to the data taken for a bridging study and given in Table 5.14, which are the $\log_{10} C_{\max}$ for Japanese and Caucasians after prescribing 6 mg of a drug on an empty stomach (Hirotsu, 2004). In a bridging study it is essential to prove that the bio-availability is approximately the same between the Japanese and the Caucasians. We denote the Japanese and Caucasian data by $x_{1j}, j=1,\ldots,n_1$ and $x_{2j}, j=1,\ldots,n_2$, respectively. Then we have the summary statistics

$$\bar{x}_{1\cdot}=1.518,\ \bar{x}_{2\cdot}=1.457,\ \mathrm{S}_1=\sum\left(x_{1j}-\bar{x}_{1\cdot}\right)^2=0.1255,\ \mathrm{S}_2=\sum\left(x_{2j}-\bar{x}_{2\cdot}\right)^2=0.1086.$$

Since the equivalence test is sensitive to outliers, we first check for outliers by the Smirnov–Grubbs test. We have

Table 5.14 Comparison of $\log_{10} C_{\max}$ between Japanese and Caucasians at 6 mg dose.

Japanese ($n=20$):	1.567, 1.515, 1.500, 1.591, 1.624, 1.691, 1.531, 1.456, 1.351, 1.478, 1.461, 1.571, 1.565, 1.586, 1.406, 1.488, 1.500, 1.577, 1.500, 1.407
Caucasians ($n=13$):	1.455, 1.375, 1.474, 1.650, 1.464, 1.375, 1.479, 1.413, 1.423, 1.389, 1.441, 1.650, 1.348

Table 5.15 Comparison of $\log_{10} C_{\max}$ between Japanese and Caucasians at 3 mg dose.

Japanese ($n=12$):	1.206, 1.042, 1.148, 1.198, 1.132, 1.039, 0.848, 1.171, 1.176, 1.101, 1.252, 1.082
Caucasian ($n=13$):	0.964, 0.983, 1.041, 1.083, 1.215, 1.104, 1.041, 1.405, 1.253, 1.228, 1.121, 1.114, 1.199

$$(x_{16}-\bar{x}_{1\cdot})/\hat{\sigma}=(1.691-1.518)/\sqrt{0.0066}=2.13 \text{ for Japanese,}$$
$$(x_{24}-\bar{x}_{2\cdot})/\hat{\sigma}=(1.650-1.457)/\sqrt{0.0091}=2.03 \text{ for Caucasians.}$$

These values are not significant for the two-sided test at level 0.10.

Now we proceed to the equivalence test. The pooled estimate of variance $\hat{\sigma}^2=0.007553$ and the critical values are obtained as $T_{0.05}=0.053$ and $T_{0.025}=0.063$, respectively. Therefore, the mean difference $\bar{x}_{1\cdot}-\bar{x}_{2\cdot}=1.518-1.457=0.061$ is non-significant at two-sided $\alpha=0.05$ but significant at one-sided $\alpha=0.05$, so that we have the confidence interval by (5.20) with confidence coefficient 0.95:

$$0 \le \log_{10}\mu_1-\log_{10}\mu_2 \le 0.061+0.053=0.114$$

or

$$1 \le \mu_1/\mu_2 \le 1.30. \tag{5.39}$$

The interval (5.39) includes 1 (equality of means,) and although the upper limit is slightly over the usual criterion of 1.25 for bio-equivalence, it would be acceptable in considering the data not from a cross-over trial. On the contrary, the conventional confidence interval (5.38): $1.02 \le \mu_1/\mu_2 \le 1.30$ excludes 1 and some discussion will be necessary before asserting the equivalence of $C_{\max}$ between Japanese and Caucasians.

The data at 3 mg dose are also given in Table 5.15 (Hirotsu, 2004). For these data the confidence interval (5.20) at $\alpha=0.05$ is obtained as $0.80 \le \mu_1/\mu_2 \le 1.14$, which coincides with the naïve confidence interval at confidence coefficient 0.90. Thus, the usual criterion of bio-equivalence is satisfied for 3 mg dose. Incidentally, the smallest datum 0.848 in the Japanese data is significant at approximately two-sided level 0.05 by the Smirnov–Grubbs test. If we eliminate this datum, the Japanese and Caucasian data become even closer.

5.3.7 Concluding remarks

The ICH E9 Guideline discusses the non-inferiority and superiority tests separately, and recommends a one-sided $\alpha=0.025$ test for the former and usually a two-sided $\alpha=0.05$ test is applied to the latter. In contrast, according to the old Japanese Statistical Guideline (Koseisho, 1992), Japan used to employ one-sided $\alpha=0.05$ for the non-inferiority test and it had a very big impact when changing α to 0.025. It should be noted, however, that the approaches are not contradictory but lie along the same line with slightly different position, although an explanation will be necessary on why to change α in the superiority and non-inferiority tests under ICH E9. Now, the newly proposed procedure in Sections 5.3.2 ~ 5.3.4 consistently uses $\alpha = 0.05$ for all tests (one- or two-sided) in the process, and as a result keeps the family-wise type I error rate at 0.05. It could include the non-inferiority test of ICH E9 as its step 3 and the superiority test as its step 5, while having an intermediate decision: at least equivalent between those two steps. Further, the non-inferiority test has been made precise in two parts (with and without equality) for the non-inferiority margin, and could include the old Japanese practice as weak non-inferiority at step 2. Thus, it could give a very good compromise justifying simultaneously the ICH E9 and the old Japanese Guideline. Ideally, the significance level α or $\alpha/2$, and also the non-inferiority margin δ in the non-inferiority test, should be chosen optimally to control the outgoing quality level of drugs passing through these testing processes – but this is a very difficult problem. Then, it will be attractive to have such multiple decisions changing the strength of evidence of the relative efficacy of the test drug against the control according to the achievement of the clinical trial. It should be noted that this procedure improves the consumer's risk even against a one-sided 0.025 test in a large trial, while keeping the producer's risk essentially the same as for the one-sided 0.05 test. The confidence region is useful also for proving bio-equivalence.

References

Barnett, V. and Lewis, T. (1993) *Outliers in statistical data*, 3rd edn. Wiley, New York.

Bartlett, M. S. (1953) Approximate confidence intervals II. More than one unknown parameters. *Biometrika* **40**, 306–317.

Dunnett, C. W. and Gent, M. (1977) Significance testing to establish equivalence between treatments with special reference to 2 × 2 tables. *Biometrics* **33**, 593–602.

Hirotsu, C. (1986) Statistical problems in the clinical trial: With particular interest in testing for equivalence. *Clinical Eval.* **14**, 467–475 (in Japanese).

Hirotsu, C. (2004) *Statistical analysis of medical data*. University of Tokyo Press, Tokyo (in Japanese).

Hirotsu, C. (2007) A unifying approach to non-inferiority, equivalence and superiority tests via multiple decision processes. *Pharma. Statist.* **6**, 193–203.

Hirotsu, C. and Hothorn, L. (2003) Impact of ICH E9 Guideline: Statistical principles for clinical trials on the conduct of clinical trials in Japan. *Drug Inf. J.* **37**, 381–395.

Hirotsu, C., Hashimoto, W., Nishihara, K. and Adachi, E. (1997) Calculation of the confidence interval with skewness correction for the difference of two binomial probabilities. *Jap. J. Appl. Statist.* **26**, 83–97 (in Japanese).

Koseisho (1992) *Statistical guideline for clinical trials*. Japanese Ministry for Public Health and Welfare, Tokyo (in Japanese).

Lewis, J. A. (1999) Statistical principles for clinical trials (ICH E9): An introductory note on an international guideline. *Statist. Med.* **18**, 1903–1942.

Moriguti, S. (ed.) (1976) *Statistical methods*, new edn. Japanese Standards Association, Tokyo (in Japanese).

Nishiyama, S., Okamoto, S., Ishibashi, Y. *et al.* (2001) Phase III study of KW-4679 (olopatadine hydrochloride) on chronic urticaria: A double blind study in comparison with Ketifen fumarate. *J. Clin. Therapeut. Med.* **17**, 237–264.

Yanagawa, T., Tango, T. and Hiejima, Y. (1994) Mantel–Haenszel type tests for testing equivalence or more than equivalence in comparative clinical trials. *Biometrics* **50**, 859–864.

6

One-Way Layout, Normal Model

In this chapter we first introduce an overall ANOVA of a one-way layout by the F-test. Then, Section 6.2 presents a homogeneity test of variances. Section 6.3 gives a non-parametric approach. We introduce multiple comparison approaches in Section 6.4, which are more useful in various aspects of applied statistics. Finally, Section 6.5 shows directional tests when there is a natural ordering in the levels of treatment, such as temperature, concentration, and dose levels.

6.1 Analysis of Variance (Overall F-Test)

We extend the t-test in a two-sample problem to comparisons of a treatments. We assume a statistical model

$$y_{ij} = \mu_i + e_{ij},\ i = 1, \ldots, a, j = 1, \ldots, n_i, \tag{6.1}$$

where the error e_{ij} are assumed to be distributed independently as $N(0, \sigma^2)$. We are interested in testing the null hypothesis of the homogeneity in μ_i,

$$H_0 : \mu_1 = \cdots = \mu_a. \tag{6.2}$$

The case of equal number of repetitions, $n_i \equiv m$, has already been discussed in Example 3.2. We use $n(=n_.)$ for the total sample size, as before. The LS estimate for model (6.1) is $\hat{\mu}_i = \bar{y}_{i\cdot}$ again, and the residual sum of squares is obtained as

$$S_e = \sum_{i=1}^{a} \sum_{j=1}^{n_i} (y_{ij} - \bar{y}_{i\cdot})^2 = \sum_{i=1}^{a} \sum_{j=1}^{n_i} y_{ij}^2 - \sum_{i=1}^{a} \bar{y}_{i\cdot} y_{i\cdot} = \sum\sum y_{ij}^2 - \sum y_{i\cdot}^2 / n_i.$$

Advanced Analysis of Variance, First Edition. Chihiro Hirotsu.

Under H_0 (6.2) we obtain $\hat{\mu}=\bar{y}_{..}$ and the residual sum of squares is obtained as

$$S_0=\sum\sum y_{ij}^2-\bar{y}_{..}y_{..}=\sum\sum y_{ij}^2-y_{..}^2/n. \tag{6.3}$$

For the calculation of these residual sums of squares, formula (2.33) has been utilized. Then, the increase in residual sum of squares is obtained as

$$S_H=S_0-S_e=\left(\sum\sum y_{ij}^2-y_{..}^2/n\right)-\left(\sum\sum y_{ij}^2-\sum y_{i\cdot}^2/n_i\right)=\sum y_{i\cdot}^2/n_i-y_{..}^2/n. \tag{6.4}$$

In this case, $\operatorname{rank}(\boldsymbol{X})=a$ and $\operatorname{rank}(\boldsymbol{X}_0)=1$, and we have finally the F-statistic

$$F=\frac{S_H/(a-1)}{S_e/(n-a)}. \tag{6.5}$$

The F-statistic (6.5) is called a between vs. within variance ratio. The expectations of S_H and S_e are calculated similarly to Example 3.2:

$$E(S_e)=(n-a)\sigma^2$$
$$E(S_H)=(a-1)\sigma^2+\sum_i n_i(\mu_i-\bar{\mu}_{.})^2, \bar{\mu}_{.}=\sum_i n_i\mu_i/n.$$

Therefore, the unbiased variance is obtained as

$$\hat{\sigma}^2=S_e/(n-a). \tag{6.6}$$

The non-centrality parameter γ (3.38) is now

$$\gamma=\sum_i n_i(\mu_i-\bar{\mu}_{.})^2/\sigma^2. \tag{6.7}$$

These results are summarized in the ANOVA Table 6.1. It should be noted that actually, S_0 and S_H are calculated first and S_e is calculated by subtracting S_H from S_0 ($S_e=S_0-S_H$). The sum of squares S_0 (6.3) is called a total sum of squares, since it expresses the total variation of the data. In this context, S_0 is written as S_T, denoting the total sum of squares, and hereafter we employ this notation in the ANOVA table. In contrast, S_H (6.4) is the sum of squares for treatments.

Table 6.1 ANOVA table for one-way layout.

Factor	Sum of squares	df	Mean sum of squares	F	Non-centrality
Treatment	S_H (6.4)	$a-1$	$S_H/(a-1)$	F (6.5)	γ (6.7)
Error	S_e	$n-a$	$\hat{\sigma}^2=S_e/(n-a)$ (6.6)		
Total	S_0 (6.3)	$n-1$			

6.2 Testing the Equality of Variances

The F-test in the previous section assumes an equality of variances among treatments. The F-test is asymptotically equivalent to a permutation test, and robust like the t-test against departures from the underlying assumptions. However, if there is a large heterogeneity among variances, it is necessary to control the error in the experiments or reconsider modeling. We mention here three methods of testing the equality of variances in the normal model:

$$\begin{cases} y_{ij} = \mu_i + e_{ij},\ i = 1, \ldots, a, j = 1, \ldots, n_i, \\ e_{ij} \sim N(0, \sigma_i^2) \text{ and mutually independent}, \\ H_\sigma : \sigma_1^2 = \cdots = \sigma_a^2 = \sigma^2. \end{cases}$$

If the null hypothesis H_σ holds, then this model reduces to model (6.1).

6.2.1 Likelihood ratio test (Bartlett's test)

We define the sum of squares for the ith population by $S_i = \sum_j y_{ij}^2 - y_{i\cdot}^2/n_i$. Then the statistics S_i/σ_i^2, $i = 1, \ldots, a$ are distributed independently from each other as a chi-squared distribution with df $(n_i - 1)$. Therefore, the joint distribution is

$$f(S_1, \cdots, S_a) = \left[2^a \Pi_{i=1}^a \left\{ \Gamma\left(\frac{n_i - 1}{2}\right) \sigma_i^2 \right\}\right]^{-1} \Pi_{i=1}^a \left(\frac{S_i}{2\,\sigma_i^2}\right)^{(n_i-3)/2} \exp\left(-\frac{1}{2}\sum_{i=1}^a \frac{S_i}{\sigma_i^2}\right). \tag{6.8}$$

We denote the right-hand side of (6.8) by $L(\sigma_1^2, \ldots, \sigma_a^2)$, which is a likelihood function with respect to σ_i^2. From $\partial \log L(\sigma_1^2, \ldots, \sigma_a^2)\,/\partial\sigma_i^2 = 0$, we obtain the MLE

$$\hat{\sigma}_i^2 = S_i/(n_i - 1),\ i = 1, \ldots, a.$$

The likelihood function L^* under H_σ satisfies

$$\log L^* = \text{const.} - \frac{\sum(n_i - 1)}{2} \log \sigma^2 - \frac{\sum S_i}{2\sigma^2}$$

and we have the MLE

$$\hat{\sigma}^2 = \sum S_i / \sum (n_i - 1).$$

Denoting the likelihood ratio by λ, we obtain the test statistic

$$\begin{aligned} -2\log\lambda &= 2\left\{\log L(\hat{\sigma}_1^2, \ldots, \hat{\sigma}_a^2) - \log L^*(\hat{\sigma}^2)\right\} \\ &= (n-a)\left[\log \frac{\sum\{(n_i - 1)\,\hat{\sigma}_i^2\}}{n-a} - \frac{1}{n-a}\sum\{(n_i - 1)\log \hat{\sigma}_i^2\}\right]. \end{aligned} \tag{6.9}$$

This equation is a logarithmic function of the ratio of the arithmetic and geometric means. When the n_i are moderately large and under H_σ, the test statistic (6.9) is distributed as chi-square with df $a-1$. A better approximation is obtained by Bartlett's correction

$$B=(-2\log\lambda)/\left[1+\left(\sum\frac{1}{n_i-1}-\frac{1}{n-a}\right)/\{3(a-1)\}\right],$$

called Bartlett's test. This test does not need a particular table of critical values, and can be applied to the unbalanced case without difficulty.

6.2.2 Hartley's test

The test statistic is given by

$$F_{\max}=\max_i\hat{\sigma}_i^2/\left(\min_i\hat{\sigma}_i^2\right).$$

This test gives high power when two variances are extreme and the others are rather homogeneous and located in the middle of these two. The critical values are given in Hartley (1950) when the repetition numbers are equal among the groups.

6.2.3 Cochran's test

The test statistic is given by

$$G=\max_i\hat{\sigma}_i^2/\sum_i\hat{\sigma}_i^2.$$

This test gives high power when one population is outlying. The critical values are given in Cochran (1941) when the repetition numbers are equal among the groups.

Example 6.1. Magnetic strength of a ferrite core. The data of Table 6.2 are the measurements of magnetic strength (μ) of a ferrite core by four different treatments. Table 6.3 is prepared for calculation of the sum of squares.

Table 6.2 Magnetic strength (μ) of a ferrite core (Moriguti, 1976).

Treatment			Data			Total	(Total)^2
A_1	10.8	9.9	10.7	10.4	9.7	51.5	2652.25
A_2	10.7	10.6	11.0	10.8	10.9	54.0	2916.00
A_3	11.9	11.2	11.0	11.1	11.3	56.5	3192.25
A_4	11.4	10.7	10.9	11.3	11.7	56.0	3136.00
Total						218.0	11896.50

Table 6.3 Squared data.

Treatment	(Data)^2					Total
A_1	116.64	98.01	114.49	108.16	94.09	531.39
A_2	114.49	112.36	121.00	116.64	118.81	583.30
A_3	141.61	125.44	121.00	123.21	127.69	638.95
A_4	129.96	114.49	118.81	127.69	136.89	627.84
Total						2381.48

(1) **Testing the equality of variances by Bartlett's test**. The sums of squares are obtained as follows:

$$S_1 = 531.39 - 51.5^2/5 = 0.94 \text{ with df } n_1 - 1 = 4,$$
$$S_2 = 583.30 - 54.0^2/5 = 0.10 \text{ with df } n_2 - 1 = 4,$$
$$S_3 = 638.95 - 56.5^2/5 = 0.50 \text{ with df } n_3 - 1 = 4,$$
$$S_4 = 627.84 - 56.0^2/5 = 0.64 \text{ with df } n_4 - 1 = 4.$$

The test statistic (6.9) is obtained as

$$(20-4)\left\{\log\frac{0.94+0.10+0.50+0.64}{20-4} - \frac{1}{20-4}\left(4\log\frac{0.94}{4} + 4\log\frac{0.10}{4} + 4\log\frac{0.50}{4} + 4\log\frac{0.64}{4}\right)\right\} = 4.30.$$

The coefficient for improvement is

$$\left\{1 + \left(\frac{1}{4}\times 4 - \frac{1}{20-4}\right)/(3\times 3)\right\}^{-1} = 0.906,$$

and Bartlett's test statistic B is obtained as 3.90 by the product. This is smaller than $\chi_3^2(0.05) = 7.81$, and the null hypothesis H_σ is not rejected.

For reference, Hartley's test statistic is

$$F_{\max} = (0.94/4)/(0.10/4) = 9.4$$

and it is smaller than the upper 0.05 tail point 20.6.

Cochran's test statistic is

$$G = \frac{0.94/4}{(0.94+0.10+0.50+0.64)/4} = 0.431,$$

which is smaller than the upper 0.05 tail point 0.629.

There is no significant difference suggested among variances by the three methods. Of course, the reader is obliged to choose one method before seeing the data.

Table 6.4 ANOVA table for one-way layout.

Factor	Sum of squares	df	Mean sum of squares	F	Non-centrality
Treatment	$S_H = 3.10$	3	1.03	7.58^{**}	$\sum_i n_i(\mu_i - \bar{\mu}.)^2/\sigma^2$
Error	$S_e = 2.18$	16	$\hat{\sigma}^2 = 0.136$		
Total	$S_T = 5.28$	19			

(2) **ANOVA**. The sums of squares are obtained as follows:

$$S_T = 2381.48 - 218.0^2/20 = 5.28 \text{ by } (6.3),$$
$$S_H = 11896.5/5 - 218.0^2/20 = 3.10 \text{ by } (6.4),$$
$$S_e = S_T - S_H = 2.18.$$

Then, the F-statistic is obtained by (6.5) as

$$F = \frac{3.10/(4-1)}{2.18/(20-4)} = 7.58^{**}.$$

Since $F_{3,\,16}(0.01) = 5.29$, the result is highly significant. These calculations are summarized in the ANOVA Table 6.4.

The maximal strength is obtained by treatment A_3, so we derive a confidence interval for μ_3. However, there is not much difference between A_3 and A_4, and some multiple comparison procedure and related simultaneous confidence interval of the next section will be more appropriate. A naïve confidence interval for μ_3 at confidence coefficient 0.90 is obtained from $\bar{y}_{3.} = 11.3$, $\hat{\sigma}^2 = 0.136$ with df 16 and $t_{16}(0.05) = 1.746$ as

$$\mu_3 \sim 11.3 \pm \sqrt{0.136/5} \times 1.746 = 11.3 \pm 0.3.$$

6.3 Linear Score Test (Non-parametric Test)

In this section we derive a method for a linear score test for the ordered categorical data, which can include permutation test and rank test with or without ties. Following and extending Section 5.2.3 for a two-sample problem, we use the notation $\boldsymbol{y}_i = (y_{i1}, \ldots, y_{ib})'$, $R_i = y_{i\cdot}$, $C_j = y_{\cdot j}$, $N = y_{\cdot\cdot}$, $i = 1, \ldots, a$, $j = 1, \ldots, b$. The C_j subjects in the jth response category are given the same score w_j, a typical example of which is the Wilcoxon averaged rank score

$$w_j = C_1 + \cdots + C_{j-1} + (C_j + 1)/2, j = 1, \ldots, b. \tag{6.10}$$

Then, the statistic of a linear score test is

$$W_i = \sum_j w_j y_{ij},\ i = 1,\ \ldots, a.$$

The expectation and variance of W_i can be calculated as a random sample of size R_i from a finite population of given scores, where the number of w_j is C_j. However, it is more easily calculated by a multivariate hypergeometric distribution

$$MH\left(y_{ij}|R_i, C_j, N\right) \sim \frac{\Pi_i R_i! \Pi_j C_j!}{N! \Pi_i \Pi_j y_{ij}!}.$$

The detailed calculation is shown in Section 11.2. It should be noted that this structure of data is the frequency in the two-way table of the treatment levels and ordered categorical responses, and thus will be treated more generally in Chapter 11. Anyway, we get from (11.3) and (11.4)

$$E(W_i) = (R_i/N)\sum_j (w_j C_j),\ i = 1,\ \ldots, a,$$

$$V(W_i) = \frac{R_i(N - R_i)}{(N-1)} \sigma_w^2,$$

where

$$\sigma_w^2 = N^{-1}\left[\sum_j \left(w_j^2 C_j\right) - N^{-1}\left\{\sum_j (w_j C_j)\right\}^2\right]$$

is the same as σ^2 in (5.15) and (5.18). The covariance of W_i and $W_{i'}$ can also be obtained easily from the multivariate hypergeometric distribution, and we have an overall test statistic for the homogeneity of populations as

$$W^2 = \frac{1}{\sigma_w^2}\left[\sum_i \left(W_i^2/R_i\right) - \frac{1}{N}\left\{\sum_j (w_j C_j)\right\}^2\right] \times \frac{N-1}{N},$$

which is asymptotically distributed as a chi-squared distribution with df $a - 1$ under the null hypothesis of homogeneity of treatments. This test is called the Kruskal–Wallis test when the averaged rank (6.10) is employed as score. The other methods described in Section 5.2.3 (2) – such as max acc. $t1$ and the cumulative chi-squared statistic – are introduced in Section 7.3.

Example 6.2. Kruskal–Wallis test for Table 6.2. We apply the Kruskal–Wallis test for the data of Table 6.2. Rewriting the table as rank data, we prepare Table 6.5 for the necessary calculations. Then we have the test statistic

Table 6.5 Calculation of Kruskal–Wallis test statistic.

Treatment	Rank j														Total
	1	2	3	4	5	6	7	8	9	10	11	12	13	14	
A_1	1	1	1		1	1									5
A_2				1	1	1	1	1							5
A_3								1	1	1	1			1	5
A_4					1		1				1	1	1		5
C_j	1	1	1	1	3	2	2	2	1	1	2	1	1	1	20
w_j (6.15)	1	2	3	4	6	8.5	10.5	12.5	14	15	16.5	18	19	20	
$w_j y_{1j}$	1	2	3	0	6	8.5	0	0	0	0	0	0	0	0	20.5
$w_j y_{2j}$	0	0	0	4	6	8.5	10.5	12.5	0	0	0	0	0	0	41.5
$w_j y_{3j}$	0	0	0	0	0	0	0	12.5	14	15	16.5	0	0	20	78
$w_j y_{4j}$	0	0	0	0	6	0	10.5	0	0	0	16.5	18	19	0	70
$w_j C_j$	1	2	3	4	18	17	21	25	14	15	33	18	19	20	210
$w_j^2 C_j$	1	4	9	16	108	144.5	220.5	312.5	196	225	544.5	324	361	400	2866

$$\sigma_w^2 = 20^{-1}\left\{2866 - 20^{-1}(210)^2\right\} = 33.05$$

$$W^2 = \frac{1}{33.05}\left\{\frac{20.5^2 + 41.5^2 + 78^2 + 70^2}{5} - \frac{1}{20} \times 210^2\right\} \times \frac{19}{20} = 12.08^{**}.$$

Since $\chi_3^2(0.01) = 11.34$, the statistic is highly significant and the result coincides well with the F-test in Example 6.1 (2).

6.4 Multiple Comparisons

6.4.1 Introduction

In the 1980s a sensational event happened, in that the new drug applications from Japan to the FDA (US Food and Drug Administration) were rejected one after another because of the defectiveness of the statistical procedures for dealing with the multiplicity problems. One of these procedures was the multiple comparison procedure developed already in the early 1960s, but this was not known widely among Japanese practitioners and caused considerable confusion. Another problem was that of applying various statistical tests to a set of data and taking the most favorable result. At that time the results of the t and Wilcoxon tests were written simultaneously in the report of the two-sample problem, for example, which of course makes the significance level of a test meaningless. Now it is required to define a statistical method to apply before taking data or before opening the key at latest in a phase III trial.

In a clinical trial, multiple characteristics of a treatment is also a big problem – known as the problem of multiple endpoints. In a simple example of lowering blood pressure, where the systolic or diastolic blood pressure is of interest, is the effect of lowering pressure by 30 mmHg from 200 or 160 mmHg regarded as equivalent and how can we deal with the circadian rhythm, which usually amounts to a 20 mmHg difference between the day and night? To select a target characteristic after taking data obviously causes a false positive. Further, Armitage and Palmer (1986) pointed out stratified analyses as one of the most awkward multiplicity problems. In infectious diseases, for example, the effects of a new drug and an active control can be reversed in acute and chronic patients. This is statistically a problem of interaction between the drug and the status of the disease, and of course it is not allowed to declare the effectiveness of a new drug in a particular status of disease after seeing all the data. A proper approach would be to define such an interaction between the drug and the status of the disease in advance, based on sufficient clinical evidence, and to prove it in the trial.

In contrast, the multiple comparison procedures developed in statistics are the procedures for dealing with the multiple degrees of freedom which arise in comparisons of more than two treatments. In the one-way ANOVA model

$$y_{ij} = \mu_i + e_{ij},\ i = 1, \ldots, a,\ j = 1, \ldots, n_i,$$

where the error e_{ij} are assumed to be distributed independently as $N(0, \sigma^2)$, for example, the overall null hypothesis in μ_i,

$$H_0 : \mu_1 = \cdots = \mu_a, \tag{6.11}$$

can be decomposed into various sub-hypotheses. One of the simplest is to take H_0 as a set of paired comparisons,

$$H_{ij} : \mu_i = \mu_j,\ 1 \le i < j \le a.$$

In this case, testing each H_{ij} at level α obviously causes a false positive and an appropriate procedure for adjustment is required. In Section 6.4.2 we develop statistical methods for the known structure of sub-hypotheses, as in this example. A general approach without any particular structure is given in Section 6.4.3 and in Section 6.4.4 a closed test procedure is given, which can be applied without adjusting the level α.

In this section we assume normal independent errors and the multiple comparisons of binomial probabilities are discussed in Section 7.2. Further, row-wise multiple comparisons in two-way data are developed in Chapters 10, 11, and 13.

6.4.2 Multiple comparison procedures for some given structures of sub-hypotheses

(1) **Tukey's method for paired comparisons.** We consider a set of paired comparisons,

$$H_{ij} : \mu_i = \mu_j,\ 1 \le i < j \le a. \tag{6.12}$$

A test of a simple homogeneity hypothesis like (6.11) cannot bring useful information, such as which of the treatments is recommended, even if the test results are statistically significant. Therefore, comparisons like (6.12) are preferable, but repetitions of the t-test obviously suffer from a large false positive. In this case we can take a maximal t-statistic $\max_{1 \le i < j \le a} t_{i,j}$,

$$t_{i,j} = |\bar{y}_{i\cdot} - \bar{y}_{j\cdot}| / \sqrt{\left(n_i^{-1} + n_j^{-1}\right)\hat{\sigma}^2},$$

$$\hat{\sigma}^2 = \sum_{i=1}^{a} \sum_{j=1}^{n_i} (y_{ij} - \bar{y}_{i\cdot})^2 / (n-a),$$

as a test statistic, where $\hat{\sigma}^2$ is equal to $S_e/(n-a)$ (6.6) in ANOVA Table 6.1. When all the repetition numbers are equal, the distribution of $\sqrt{2}\max_{1 \le i < j \le a} t_{i,j}$ is nothing but that of the Studentized range and the upper tail points $q_{a,\,n-a}(\alpha)$ are tabulated. Also, a simple integration formula for exact calculation of the p-value has been given by Hochberg and Tamhane (1987). When the repetition numbers are not equal, it has been shown by Hayter (1984) that the tail probability for equal number of repetitions gives a very precise and conservative approximation. In contrast, the use of a one-sided Studentized range was suggested in Hirotsu (1976) and extended by Hayter (1990). Although Tukey's method was originally intended for paired comparisons, it can be extended to all the contrasts $\boldsymbol{L}'\boldsymbol{\mu}$ and the simultaneous confidence interval is obtained as follows:

$$\boldsymbol{L}'\boldsymbol{\mu} \sim \boldsymbol{L}'\bar{\boldsymbol{y}}_{(i)\cdot} \pm \left(2\sqrt{m}\right)^{-1}\left(\sum |L_i|\right)\hat{\sigma} q_{a,\, n-a}(\alpha), \tag{6.13}$$

where $\bar{\boldsymbol{y}}_{(i)\cdot} = (\bar{y}_{1\cdot}, \ldots, \bar{y}_{a\cdot})'$ is a vector of sample means, $\boldsymbol{L} = (L_1, \ldots, L_a)'$, and we assume equal number m of repetitions in (6.13). It should be noted that the (i) is not a label of a vector but implies a running variable to specify the vector.

(2) **Scheffé's method for comparing all the contrasts**. The null hypothesis (6.11) is also equivalent to

$$H_L : \boldsymbol{L}'\boldsymbol{\mu} = 0, \ {}^{\forall}\boldsymbol{L}'\boldsymbol{j} = 0. \tag{6.14}$$

The t-statistic corresponding to (6.14) is given by

$$t_L = \boldsymbol{L}'\bar{\boldsymbol{y}}_{(i)\cdot} / \sqrt{\boldsymbol{L}'\mathrm{diag}\left(n_i^{-1}\right)\boldsymbol{L}\hat{\sigma}^2}.$$

Since $\boldsymbol{L}'\boldsymbol{j} = 0$, we have

$$\begin{aligned} \left(\boldsymbol{L}'\bar{\boldsymbol{y}}_{(i)\cdot}\right)^2 &= \left\{\boldsymbol{L}'\mathrm{diag}\left(n_i^{-1/2}\right)\mathrm{diag}\left(n_i^{1/2}\right)(\bar{y}_{i\cdot} - \bar{y}_{\cdot\cdot}\boldsymbol{j})\right\}^2 \\ &\le \left\{\boldsymbol{L}'\mathrm{diag}\left(n_i^{-1}\right)\boldsymbol{L}\right\}\left\{\textstyle\sum_i n_i(\bar{y}_{i\cdot} - \bar{y}_{\cdot\cdot})^2\right\}, \end{aligned}$$

where the inequality is due to Schwarz's inequality. Then we have

$$\frac{t_L^2}{a-1} \le \frac{\sum_i n_i(\bar{y}_{i\cdot} - \bar{y}_{\cdot\cdot})^2/(a-1)}{\hat{\sigma}^2}. \tag{6.15}$$

However, the right-hand side of equation of (6.15) is nothing but an F-statistic with df $(a-1, n-a)$ in the one-way ANOVA, so the statistic t_L for any contrast $\boldsymbol{L}'\boldsymbol{\mu}$ is evaluated by the F-distribution. That is, the simultaneous confidence interval is obtained as

$$\boldsymbol{L}'\boldsymbol{\mu} \sim \boldsymbol{L}'\bar{\boldsymbol{y}}_{(i)\cdot} \pm \sqrt{(a-1)\left\{\boldsymbol{L}'\mathrm{diag}\left(n_i^{-1}\right)\boldsymbol{L}\right\}\hat{\sigma}^2 F_{a-1,\, n-a}(\alpha)}. \tag{6.16}$$

This method is applicable when the sample sizes are not equal. One should refer to Scheffé (1953) for detailed comparisons of Scheffé's and Tukey's methods.

Table 6.6 Simultaneous confidence intervals by Tukey's and Scheffé's methods.

		Tukey		Scheffé	
Contrast	Point estimates	Lower bound	Upper bound	Lower bound	Upper bound
$\mu_2-\mu_1$	0.5	−0.168	1.168	−0.228	1.228
$\mu_3-\mu_1$	1.0	0.332	1.668	0.272	1.728
$\mu_4-\mu_1$	0.9	0.232	1.568	0.172	1.628
$\mu_3-\mu_2$	0.5	−0.168	1.168	−0.228	1.228
$\mu_4-\mu_2$	0.4	−0.268	1.068	−0.328	1.128
$\mu_4-\mu_3$	−0.1	−0.768	0.568	−0.828	0.628
c_1	0.8	0.132	1.468	0.206	1.394
c_2	0.7	0.032	1.368	0.186	1.214

Example 6.3. Example 6.1 continued. We construct simultaneous confidence intervals for the contrasts $\mu_j-\mu_i$, $1\le i<j\le 4$, $c_1=(\mu_2+\mu_3+\mu_4)/3-\mu_1$, and $c_2=(\mu_3+\mu_4)/2-(\mu_1+\mu_2)/2$ by each of Scheffé's and Tukey's methods. Now, $a=4$, $n_i\equiv m=5$, $q_{4,16}(0.05)=4.046$, $F_{3,16}(0.05)=3.24$, and we already have $\hat{\sigma}=\sqrt{0.136}=0.369$ in ANOVA Table 6.5. The calculations of (6.13) and (6.16) are quite easy, and the results are summarized in Table 6.6.

It is seen that the range of intervals for paired comparisons is smaller for Tukey's method than for Scheffé's method, and the reverse is true for general contrasts like c_1 and c_2.

(3) **Dunnett's method for comparing treatments with a standard**. A typical situation often encountered in a clinical trial is comparing treatments with a standard, which is indexed by 1. In this case a set of sub-hypotheses of interest are expressed as

$$H_{i1}: \mu_i=\mu_1, i=2,\ldots,a.$$

The t-statistic corresponding to H_{i1} is given by

$$t_{i,1}=|\bar{y}_{i\cdot}-\bar{y}_{1\cdot}|/\sqrt{\left(n_i^{-1}+n_1^{-1}\right)\hat{\sigma}^2}.$$

When $n_2=\cdots=n_a$, the upper tail points of the maximal statistic, $\max_i t_{i,1}$, are given by Dunnett (1964). Also, a simple integration formula for exact calculation of the p-value is shown by Hochberg and Tamhane (1987).

Example 6.4. From Dunnett's original paper. This example is concerned with the effect of certain drugs on the fat content of the breast muscle in cockerels.

In the experiment, 80 cockerels were divided at random into four treatment groups. The birds in group A were the untreated controls, while groups B, C, and D received, respectively, stilbesterol and two levels of acetyl enheptin in their diets. Birds from each group were sacrificed at specified times (four levels) for the purpose of certain measurements. One of these was the fat content of the breast muscle.

The analysis of variance for the 4×4 two-way data with repetition five at each cell was performed to obtain the unbiased variance $\hat{\sigma}^2=0.1086$ with df 64. The absence of an interaction between treatments and sacrifice times was shown, justifying the comparisons of treatment groups based on the overall means $\bar{y}_{(i)\cdot}=(2.493, 2.398, 2.240, 2.494)'$.

Now, the main comparisons of interest to the experimenter were between each of the three treatments and the control. The one differing most from the control is treatment C, and the standardized difference is obtained as

$$t_{3,1}=|2.240-2.493|/\sqrt{(1/20+1/20)0.1086}=2.43^{*}.$$

To evaluate this value as Student's t obviously suffers from a false positive, since treatment C was selected because of the extreme outcome. By the table given in the paper specific to the comparison of the best treatment with the control, the critical values are 2.41 and 3.02 for two-sided levels 0.05 and 0.01, respectively. Therefore, we can state that treatment C is significantly different from the control at two-sided level 0.05. The other two treatments can be tested by the closed step-down procedures of Section 6.4.4, and it is found that neither of them is significant.

It is commented in the paper, however, that this is a bit surprising, since group D – which received the same drug at twice the dose – does not show any apparent difference from the control. Whether one should conclude in this instance that a real treatment effect has been demonstrated, which for some reason is not manifested at the higher dose level, depends on the experimenter's prior knowledge regarding the properties of this particular drug, together with his assessment of the likelihood of the observed effect's being due to a chance occurrence or a flaw in the conduct of the experiment. Thus, a statistical analysis is by no means a conclusion, but only supports scientifically the real-world decision.

6.4.3 General approach without any particular structure of sub-hypotheses

(1) **Bonferroni's inequality**. Let us consider simultaneous tests of a set of basic hypotheses $H_1, \ldots, H_K$ and any subset of it, namely

$$\cap_{j\in J}H_j, J\subseteq(1, \ldots, K),$$

where we identify H_j and the set of $\boldsymbol{\mu}$ satisfying H_j. We also assume none of the hypotheses $H_1, \ldots, H_K$ can be expressed by the intersection of the other hypotheses. Let the rejection region of each of $H_1, \ldots, H_K$ be $R_1, \ldots, R_K$. Then we should determine $R_1, \ldots, R_K$, so that the probability of wrongly rejecting at least one of $H_1, \ldots, H_K$ when they are true is less than or equal to α. That is, we choose the rejection regions so as to satisfy

$$\Pr(R_1 \cup \cdots \cup R_K \mid H_1 \cap \cdots \cap H_K) \le \alpha.$$

We can choose R_j satisfying

$$\Pr(R_j \mid H_j) \le \alpha/K, \tag{6.17}$$

since then we have

$$\begin{aligned}\Pr(R_1 \cup \cdots \cup R_K \mid H_1 \cap \cdots \cap H_K) &\le \textstyle\sum_{j=1}^{K} \Pr(R_j \mid H_1 \cap \cdots \cap H_K)\\ &\le \textstyle\sum_{j=1}^{K} \Pr(R_j \mid H_j)\\ &\le K \times (\alpha/K) = \alpha.\end{aligned}$$

where the first inequality is due to Bonferroni's inequality.

Next, suppose the rejection region is expressed by a test statistic T_j in the form

$$R_j : T_j \ge c_j$$

as usual. Then define

$$\hat{\alpha}_j(t_j) = \Pr(T_j > t_j \mid H_j)$$

for the observed test statistic t_j, which is called the p-value in the situation of multiple comparisons. When H_j is true we have

$$\begin{aligned}\Pr\{\hat{\alpha}_j(t_j) \le p \mid H_j\} &= \Pr\{\Pr(T_j > t_j \mid H_j) \le p \mid H_j\}\\ &= \Pr\{1 - F(t_j) \le p \mid H_j\} = p,\end{aligned} \tag{6.18}$$

and therefore the rejection region given by (6.17) is equivalent to rejecting the hypothesis when $\hat{\alpha}_j(t_j) \le \alpha/K$ for observed t_j. In equation (6.18) we used the fact that the distribution function $F(T_j)$ of T_j follows a uniform distribution.

Unless the rejection regions R_j are mutually exclusive, the procedure according to Bonferroni's inequality is obviously conservative and should be inefficient when K is moderately large. Further, while the minimal p-value is naturally compared with α/K, the next one seems not necessarily to be compared with α/K. From these view points, several improvements have been made to Bonferroni's procedure. Among them, Holm (1979) and Schaffer (1986) are particularly important.

(2) **Holm's procedure**. Let the ordered p-values be denoted by $\hat{\alpha}_{(1)} \le \cdots \le \hat{\alpha}_{(K)}$ and the corresponding hypotheses by $H_{(1)}, \ldots, H_{(K)}$. Then we have Theorem 6.1 by Holm (1979).

Theorem 6.1. Holm's theorem. Let j^* be the minimal j that satisfies $\hat{\alpha}_{(j)} > \alpha/(K-j+1)$. Then, due to the multiple comparison procedure rejecting $H_{(1)}, \ldots, H_{(j^*-1)}$ and accepting $H_{(j^*)}, \ldots, H_{(K)}$, the probability of rejecting at least one true hypothesis wrongly is less than or equal to α.

Proof. Let the set of true hypotheses be denoted by $J(\subseteq (1, \ldots, K))$ and the number of elements included in J be m. Then it is sufficient to prove that the probability of accepting all the hypotheses $H_{(j)}, j \in J$ is larger than or equal to $1-\alpha$. We have the following inequalities:

$$\begin{aligned}\Pr\left(\hat{\alpha}_j > \frac{\alpha}{m}, {}^\forall j \in J | \cap_{j\in J} H_j\right) &= 1 - \Pr\left(\hat{\alpha}_j \le \frac{\alpha}{m}, {}^\exists j \in J | \cap_{j\in J} H_j\right)\\ &= 1 - \Pr\left\{\cup_{j\in J} H_j \left(\hat{\alpha}_j \le \frac{\alpha}{m}\right) | \cap_{j\in J} H_j\right\}\\ &\ge 1 - \sum_{j\in J} \Pr\left(\hat{\alpha}_j \le \frac{\alpha}{m} | \cap_{j\in J} H_j\right)\\ &\ge 1 - m \times (\alpha/m) = 1-\alpha\end{aligned}$$

That is, it is verified that for all true hypotheses H_j the probability $\Pr(\hat{\alpha}_j > \alpha/m)$ is larger than or equal to $1-\alpha$. However, since the number of elements included in J is m, the event

$$\hat{\alpha}_j \le \frac{\alpha}{m}, {}^\forall j \in J$$

implies that at least m largest p-values – namely $\hat{\alpha}_{(K)}, \ldots, \hat{\alpha}_{(K-m+1)}$ – are larger than α/m. Therefore we have

$$\hat{\alpha}_{(K-m+1)} > \alpha/m = \frac{\alpha}{K-(K-m+1)+1}$$

and the test procedures should have stopped at $K-m+1$ or before. This implies that for any j satisfying $\hat{\alpha}_j > \alpha/m$, the hypothesis H_j should be accepted.

It is obvious that Holm's procedure improves the naïve Bonferroni procedure. When there is an inclusive relationship among the subsets of $(H_1, \ldots, H_K)$, Holm's procedure is further improved by Schaffer (1986).

6.4.4 Closed test procedure

While the procedures of the previous section can be applied widely without any particular structure among sub-hypotheses, we can construct multiple comparison procedures more efficiently for some given structures of sub-hypotheses as mentioned in Section 6.4.2. Then there is a case that we can test sub-hypotheses in an appropriate order without adjusting the significance level. The procedure is given by Theorem 6.2 of Marcus *et al.* (1976).

Theorem 6.2. Marcus *et al.*'s theorem (1976). In the set of sub-hypotheses, let us assign a test φ with significance level α to every intersection of sub-hypotheses. Then a multiple comparison procedure of the significance level α is constructed by testing the null hypothesis H_β if and only if all the null hypotheses included in H_β have been rejected. That is, the probability of rejecting at least one true hypothesis by this procedure is less than or equal to α.

Proof. Define the events A and B as follows:

A – any true hypothesis ω_β is rejected;

B – the intersection ω_τ of all the true hypotheses is rejected by φ_τ.

For ω_β to be rejected, all the null hypotheses included in ω_β have to be rejected by the assumption of the theorem and therefore A implies B. That is, $A \cap B = A$ holds. However, since the intersection ω_τ is one of the true hypotheses and the significance level of φ_τ is α we have

$$P(A) = P(A \cap B) = P(B)P(A|B) \le \alpha \times 1 = \alpha,$$

where $P(A)$ denotes the probability of event A.

An application of a closed test procedure has been given in Section 5.3.2 (2) and will be given also in Section 6.5.3 (1) (g) to define the dose–response pattern. A closed Tukey test procedure and step-down Dunnett procedure are described by Hochberg and Tamhane (1987) and Bretz *et al.* (2011), as examples.

6.5 Directional Tests

6.5.1 Introduction

The shape hypothesis, like the monotone hypothesis, is inevitable in the dose–response analysis where a rigid parametric model is usually difficult to assume. It appears also in comparing treatments based on ordered categorical data (Section 5.2.3). The omnibus F-test is obviously inappropriate against these restricted alternatives. Then, the isotonic regression is the most well-known approach to the monotone hypothesis in the normal one-way layout model (Barlow *et al.*, 1972; Robertson *et al.*, 1988). It was, however, introduced rather intuitively by Bartholomew (1959a,b), and has no obvious optimality for the restricted parameter space like this monotone hypothesis. Further, the restricted maximum likelihood approach employed in the isotonic regression is too complicated to extend to the non-normal distributions, including the discrete distributions, to the analysis of interaction effects, and also to other shape constraints such as convexity and sigmoidicity. Therefore, in the book of BANOVA (Miller, 1998), a choice of Abelson and Tukey's (1963) maximin linear contrast test is recommended for the monotone hypothesis, to escape from the complicated calculations of the isotonic regression. However, such a 1 df contrast test cannot keep high power against the wide range of monotone hypotheses, even by careful choice of contrasts. Actually, it is stated in Robertson *et al.* (1988) that if no prior information is available concerning the location of the true mean vector, and if the likelihood ratio test is viable, then such a contrast test cannot be recommended. Instead, we propose a more robust approach against the wide range of monotone hypotheses, which can be extended in a systematic way to various interesting problems such as convexity and sigmoidicity hypotheses, and also to the order-restricted inference in interaction effects. It starts from a complete class lemma for the tests against the general restricted alternative in Hirotsu (1982). It suggests the use of accumulated statistics as the basic statistics in case of the monotone hypothesis. Two promising statistics derived from the basic cumulative statistics and belonging to the complete class are the maximal contrast statistic (max acc. $t1$) and the cumulative chi-squared statistic χ^{*2}. Although χ^{*2} does not belong to the complete class in the strict sense, it has been verified to have high power widely against the two-sided order restricted hypotheses. It is very robust and nicely characterized as a directional goodness-of-fit test statistic (Hirotsu, 1986). In contrast, max acc. $t1$ strictly belongs to the complete class and is characterized also as an efficient score test for the step change-point hypothesis as shown in Section 6.5.3 (1) (b). The basic statistics are very simple compared with the restricted MLE, and have a very nice Markov property for probability calculation (Hirotsu, 2013; Hirotsu *et al.*, 1992, 1997, 2016). It leads to a very elegant and exact algorithm for probability calculation not only for the normal distribution, but also for the general univariate exponential family including the Poisson and binomial distributions. As shown in Section 6.5.3 (1) (e) and (f), max acc. $t1$ is essential for forming the simultaneous confidence intervals of the monotone contrasts satisfying the inevitable properties of the uniqueness and positivity of the linear combination (see Hirotsu and Srivastava, 2000; Hirotsu *et al.*, 2011). Hirotsu (1997) and Hirotsu and Marumo (2002) proved a close relationship between the monotone hypothesis and the step change-point hypothesis, unifying those two topics developed independently in the two different streams of statistics. Let us consider a set of all the monotone contrasts, which defines a convex polyhedral cone. Then, very interestingly, every corner vector of the polyhedral cone represents a step change-point contrast, suggesting a close relationship between the monotone hypothesis and the step change-point model. Actually, every step change-point contrast is a special case of the monotone contrast and every monotone contrast can be expressed by a unique and positive linear combination of step change-point contrasts (see Example 6.5). This leads to a very interesting approach to defining the dose–response pattern without assuming any parametric model (see Section 6.5.3 (1) (g)).

This unification is also practically important, since in monitoring the spontaneous reporting of the adverse events of a drug, for example, it is interesting to detect a change-point as well as the general increasing tendency of the reportings as shown in Example 8.1 of Section 8.1. This unification approach extends also to the convexity and slope change-point models, and the sigmoid and inflection point models. These basics are the doubly and triply accumulated statistics, respectively. The accumulated statistics have so simple a structure that many of the procedures for a one-way layout model can be extended in a systematic way to two- and three-way data in Chapters 10, 11, 13, and 14. In particular, this leads to the two-way accumulated statistics in Section 11.5. Further, the power of the proposed method has been evaluated repeatedly and proved to be excellent compared with the isotonic regression and some other maximal contrast type test statistics. Thus, this unification is truly unifying in the sense that:

(1) it unifies the approaches to shape and change-point hypotheses, which have been developed independently in the two different streams of statistics;

(2) monotone, convex, and sigmoid hypotheses are approached in a systematic way, corresponding to step, slope, and inflection change-point models, respectively;

(3) a general univariate exponential family is approached in a systematic way, not restricted to the normal distribution;

(4) two-way data are approached in a systematic way, not restricted to one-way data.

The theory and application of singly, doubly, triply, and two-way accumulated statistics are the original work of this book.

6.5.2 General theory for unifying approach to shape and change-point hypotheses

Let $\mathcal{A}$ be a class of tests for testing the alternative hypothesis K on $\boldsymbol{\mu}$ and $P_\varphi(\boldsymbol{\mu})$ the power function of test φ. Then, $\mathcal{A}$ is an essentially complete class of tests if there always exists some $\varphi_2 \in \mathcal{A}$ such that $P_{\varphi_1}(\boldsymbol{\mu}) \le P_{\varphi_2}(\boldsymbol{\mu})$, $\boldsymbol{\mu} \in K$, for any $\varphi_1 \notin \mathcal{A}$. If strict inequality holds for at least one $\boldsymbol{\mu} \in K$, then $\mathcal{A}$ is called a complete class. We first give an essentially complete class for testing the multivariate one-sided alternative in normal means.

Lemma 6.1. Complete class lemma for a normal model (Takeuchi, 1979). Let $\boldsymbol{y} = (y_1, \ldots, y_a)'$ be a sample from a multivariate normal distribution $N(\boldsymbol{\mu}, \boldsymbol{\Omega})$, with $\boldsymbol{\Omega}$ a known non-singular covariance matrix. The prime on the matrix (vector) denotes a transpose of the matrix (vector). Then, an essentially complete class of tests for H_0 : $\boldsymbol{\mu}=\mathbf{0}$ against a multivariate one-sided alternative $K:\boldsymbol{\mu}\geq\mathbf{0}$ with at least one inequality strong is given by all the test functions that are increasing in every element of $\boldsymbol{\Omega}^{-1}\boldsymbol{y}$ and with convex acceptance region, where the inequality for a vector implies the inequality for every element of the vector.

Proof. Let $f(\boldsymbol{y},\boldsymbol{\mu})$ be a probability density function of $\boldsymbol{y}$ and $G(\boldsymbol{\mu})$ a prior distribution defined for $\boldsymbol{\mu}\geq\mathbf{0}$. Then the class of Bayes test

$$R:\int\{f(\boldsymbol{y},\boldsymbol{\mu})/f(\boldsymbol{y},\mathbf{0})\}\mathrm{d}G(\boldsymbol{\mu})>c$$

corresponding to all possible a priori distribution $G(\boldsymbol{\mu})$ forms an essentially complete class (Wald, 1950). For the normal density we have

$$R:\int\exp\left(-2^{-1}\boldsymbol{\mu}'\boldsymbol{\Omega}^{-1}\boldsymbol{\mu}\right)\times\exp\left(\boldsymbol{\mu}'\boldsymbol{\Omega}^{-1}\boldsymbol{y}\right)\mathrm{d}G(\boldsymbol{\mu})>c.$$

Since this is a weighted mean of the convex function $\exp\left(\boldsymbol{\mu}'\boldsymbol{\Omega}^{-1}\boldsymbol{y}\right)$, the acceptance region should be a convex region of $\boldsymbol{y}$ (Birnbaum, 1955). Let $\boldsymbol{y}_1$ and $\boldsymbol{y}_2$ be two points in the sample space satisfying $\boldsymbol{\Omega}^{-1}\boldsymbol{y}_1\geq\boldsymbol{\Omega}^{-1}\boldsymbol{y}_2$, then the inequality $\boldsymbol{\mu}'\boldsymbol{\Omega}^{-1}(\boldsymbol{y}_1-\boldsymbol{y}_2)\geq 0$ holds for any $\boldsymbol{\mu}\in K$. This implies

$$\int\{f(\boldsymbol{y}_1,\boldsymbol{\mu})/f(\boldsymbol{y}_1,\mathbf{0})\}\mathrm{d}G(\boldsymbol{\mu})\geq\int\{f(\boldsymbol{y}_2,\boldsymbol{\mu})/f(\boldsymbol{y}_2,\mathbf{0})\}\mathrm{d}G(\boldsymbol{\mu}).$$

That is, if $\boldsymbol{y}_2\in R$ then $\boldsymbol{y}_1\in R$. In other words, the essentially complete class is given by all the tests that are increasing in every element of $\boldsymbol{\Omega}^{-1}\boldsymbol{y}$ and with a convex acceptance region.

Corollary to Lemma 6.1 (Hirotsu, 1982). Suppose an observation vector $\boldsymbol{y}$ is distributed as an a-variate normal distribution with mean vector $\boldsymbol{\mu}$ and known covariance matrix $\boldsymbol{\Omega}$. Then an essentially complete class of tests for testing the null hypothesis

$$K_0:\boldsymbol{A}'\boldsymbol{\mu}=\mathbf{0}$$

against a restricted alternative

$$K_1:\boldsymbol{A}'\boldsymbol{\mu}\geq\mathbf{0}$$

with $\boldsymbol{A}'$ a $p\times a$ full rank matrix is formed by all the tests that are increasing in every element of

$$(\boldsymbol{A}'\boldsymbol{A})^{-1}\boldsymbol{A}'\boldsymbol{\Omega}^{-1}\left\{\boldsymbol{y}-E_0\left(\boldsymbol{y}|\boldsymbol{A}^{*\prime}\boldsymbol{y}\right)\right\}$$

and with convex acceptance region, where $E_0\left(\boldsymbol{y}|\boldsymbol{A}^{*\prime}\boldsymbol{y}\right)$ denotes the conditional expectation of $\boldsymbol{y}$ given the sufficient statistics $\boldsymbol{A}^{*\prime}\boldsymbol{y}$ under the null hypothesis K_0.

Proof. Let $[\boldsymbol{A} \mid \boldsymbol{A}^*]'$ denote any non-degenerate (full rank) linear transformation of $\boldsymbol{y}$ satisfying $\boldsymbol{A}^{*\prime}\boldsymbol{\Omega}\boldsymbol{A}=\boldsymbol{0}$. Then $\boldsymbol{A}^{*\prime}\boldsymbol{y}$ is a set of sufficient statistics under K_0 and independent of $\boldsymbol{A}'\boldsymbol{y}$. Therefore, a similar test should be formed based on $\boldsymbol{A}'\boldsymbol{y}$ which is distributed as $N(\boldsymbol{A}'\boldsymbol{\mu}, \boldsymbol{A}'\boldsymbol{\Omega}\boldsymbol{A})$. An essentially complete class is then formed by all the tests that are increasing in every element of $(\boldsymbol{A}'\boldsymbol{\Omega}\boldsymbol{A})^{-1}\boldsymbol{A}'\boldsymbol{y}$ and with convex acceptance region. Further, let us introduce an $a \times (a-p)$ matrix $\boldsymbol{B}$ that satisfies $\boldsymbol{A}'\boldsymbol{B}=\boldsymbol{0}$ and $\boldsymbol{A}^{*\prime}\boldsymbol{B}=\boldsymbol{I}_p$. This is easily justified, since $\boldsymbol{B}$ and $\boldsymbol{\Omega}\boldsymbol{A}^*$ form the same linear subspace orthogonal to the column space of $\boldsymbol{A}$. Then we have a relation $A^{*\prime}=\left(\boldsymbol{B}'\boldsymbol{\Omega}^{-1}\boldsymbol{B}\right)^{-1}\boldsymbol{B}'\boldsymbol{\Omega}^{-1}$. By inverting both sides of

$$\begin{bmatrix}\boldsymbol{A}^{*\prime}\\ \boldsymbol{A}'\end{bmatrix}\boldsymbol{\Omega}\left[\boldsymbol{A}^{*\prime} \mid \boldsymbol{A}'\right]=\begin{bmatrix}\boldsymbol{A}^{*\prime}\boldsymbol{\Omega}\boldsymbol{A}^* & \boldsymbol{0}\\ \boldsymbol{0} & \boldsymbol{A}'\boldsymbol{\Omega}\boldsymbol{A}\end{bmatrix}$$

and using the expression

$$\left[\boldsymbol{A}^* \vdots \boldsymbol{A}\right]^{-1}=\begin{bmatrix}\boldsymbol{B}'\\ (\boldsymbol{A}'\boldsymbol{A})^{-1}\boldsymbol{A}'(\boldsymbol{I}-\boldsymbol{A}^*\boldsymbol{B}')\end{bmatrix}$$

we get $(\boldsymbol{A}'\boldsymbol{\Omega}\boldsymbol{A})^{-1}\boldsymbol{A}'\boldsymbol{y}=(\boldsymbol{A}'\boldsymbol{A})^{-1}\boldsymbol{A}'(\boldsymbol{I}-\boldsymbol{A}^*\boldsymbol{B}')\boldsymbol{\Omega}^{-1}\boldsymbol{y}=(\boldsymbol{A}'\boldsymbol{A})^{-1}\boldsymbol{A}'\boldsymbol{\Omega}^{-1}\left(\boldsymbol{y}-\boldsymbol{B}\boldsymbol{A}^{*\prime}\boldsymbol{y}\right)$. On one hand, we have under the null hypothesis

$$\begin{aligned}E_0\left(\boldsymbol{y} \mid \boldsymbol{A}^{*\prime}\boldsymbol{y}\right)&=\begin{bmatrix}\boldsymbol{A}^{*\prime}\\ \boldsymbol{A}'\end{bmatrix}^{-1}E_0\left\{\begin{bmatrix}\boldsymbol{A}^{*\prime}\\ \boldsymbol{A}'\end{bmatrix}\boldsymbol{y} \mid \boldsymbol{A}^{*\prime}\boldsymbol{y}\right\}\\ &=\begin{bmatrix}\boldsymbol{A}^{*\prime}\\ \boldsymbol{A}'\end{bmatrix}^{-1}\begin{bmatrix}\boldsymbol{A}^{*\prime}\boldsymbol{y}\\ \boldsymbol{0}\end{bmatrix}=\boldsymbol{B}\boldsymbol{A}^{*\prime}\boldsymbol{y}\end{aligned}$$

and this completes the proof. By a similar argument we have the conditional variance

$$\begin{aligned}V_0\left(\boldsymbol{y} \mid \boldsymbol{A}^{*\prime}\boldsymbol{y}\right)&=\begin{bmatrix}\boldsymbol{A}^{*\prime}\\ \boldsymbol{A}'\end{bmatrix}^{-1}V_0\left\{\begin{bmatrix}\boldsymbol{A}^{*\prime}\\ \boldsymbol{A}'\end{bmatrix}\boldsymbol{y} \mid \boldsymbol{A}^{*\prime}\boldsymbol{y}\right\}\left[\boldsymbol{A}^* \vdots \boldsymbol{A}\right]^{-1}\\ &=\begin{bmatrix}\boldsymbol{A}^{*\prime}\\ \boldsymbol{A}'\end{bmatrix}^{-1}\begin{bmatrix}\boldsymbol{0} & \boldsymbol{0}\\ \boldsymbol{0} & \boldsymbol{A}'\boldsymbol{\Omega}\boldsymbol{A}\end{bmatrix}\left[\boldsymbol{A}^* \vdots \boldsymbol{A}\right]^{-1}\\ &=\boldsymbol{\Omega}-\boldsymbol{B}\left(\boldsymbol{B}\boldsymbol{\Omega}^{-1}\boldsymbol{B}\right)^{-1}\boldsymbol{B}'\\ &=\boldsymbol{\Omega}-\boldsymbol{\Omega}\boldsymbol{A}^*\left(\boldsymbol{A}^{*\prime}\boldsymbol{\Omega}^{-1}\boldsymbol{A}^*\right)^{-1}\boldsymbol{A}^{*\prime}\boldsymbol{\Omega}.\end{aligned}\tag{6.19}$$

Lemma 6.2. Complete class lemma for a general case (Hirotsu, 1982). More generally, suppose an a-dimensional observation vector $\boldsymbol{y}$ is distributed following the likelihood function $L(\boldsymbol{y}, \boldsymbol{\theta})$ and consider testing the null hypothesis $K_0 : \boldsymbol{A}'\boldsymbol{\theta} = \boldsymbol{0}$ against a restricted alternative $K_1 : \boldsymbol{A}'\boldsymbol{\theta} \geq \boldsymbol{0}$ with $\boldsymbol{A}'$ a $p \times a$ full rank matrix. Then, a complete class of tests is formed by all those tests that are increasing in every element of $(\boldsymbol{A}'\boldsymbol{A})^{-1}\boldsymbol{A}'\boldsymbol{\nu}(\hat{\boldsymbol{\theta}}_0)$ and with convex acceptance region, where $\nu(\hat{\boldsymbol{\theta}}_0)$ is an efficient score vector evaluated at MLE $\hat{\boldsymbol{\theta}}_0$ under K_0.

Proof. Let $\hat{\boldsymbol{\theta}}$ be the MLE of $\boldsymbol{\theta}$ following asymptotically a multivariate normal distribution. Then, by virtue of the Corollary to Lemma 6.1, the essentially complete class is given by all the tests that are increasing in every element of $(\boldsymbol{A}'\boldsymbol{\Gamma}_0^{-1}\boldsymbol{A})^{-1}A'\hat{\boldsymbol{\theta}} = (\boldsymbol{A}'\boldsymbol{A})^{-1}\boldsymbol{A}'\boldsymbol{\Gamma}_0\left\{\hat{\boldsymbol{\theta}} - E_0\left(\hat{\boldsymbol{\theta}} \mid A^{*\prime}\hat{\boldsymbol{\theta}}\right)\right\}$ and with convex acceptance region, where $\boldsymbol{\Gamma}_0$ is Fisher's information matrix under K_0. However, $\boldsymbol{\Gamma}_0\left\{\hat{\boldsymbol{\theta}} - E_0\left(\hat{\boldsymbol{\theta}} \mid A^{*\prime}\hat{\boldsymbol{\theta}}\right)\right\}$ is asymptotically equivalent to the efficient score vector $\nu\left(\hat{\boldsymbol{\theta}}_0\right)$ evaluated at the MLE $\hat{\boldsymbol{\theta}}_0$ under K_0. Therefore, the essentially complete class of tests is formed by all the tests that are increasing in every element of $(\boldsymbol{A}'\boldsymbol{A})^{-1}\boldsymbol{A}'\boldsymbol{\nu}\left(\hat{\boldsymbol{\theta}}_0\right)$ and with a convex acceptance region. It should be noted that the asymptotic null distribution of $\boldsymbol{\nu}\left(\hat{\boldsymbol{\theta}}_0\right)$ is normal with mean zero and covariance matrix $\boldsymbol{\Gamma}_0 - \boldsymbol{\Gamma}_0\boldsymbol{B}(\boldsymbol{B}'\boldsymbol{\Gamma}_0\boldsymbol{B})^{-1}\boldsymbol{B}'\boldsymbol{\Gamma}_0$, by virtue of (6.19). Usually, $\boldsymbol{\Gamma}_0$ is consistently estimated by the value of minus the second derivation of the log likelihood function evaluated at $\hat{\boldsymbol{\theta}}_0$.

In Lemma 6.2 the vector $\boldsymbol{\nu}\left(\hat{\boldsymbol{\theta}}_0\right)$ is essentially the observation vector $\boldsymbol{y}$ for the independent univariate exponential family

$$L(\boldsymbol{y}, \boldsymbol{\theta}) = \Pi_k a(\theta_k) b(y_k) \exp(\theta_k y_k), \tag{6.20}$$

since it is equal to $\boldsymbol{y} - \hat{E}(\boldsymbol{y} \mid \boldsymbol{\theta}_0)$, where $\hat{E}(\boldsymbol{y} \mid \boldsymbol{\theta}_0)$ is the expectation of $\boldsymbol{y}$ evaluated at the MLE $\hat{\boldsymbol{\theta}}_0$ under K_0, and therefore a function of the sufficient statistics under the null model. Those sufficient statistics are the constants in developing a similar test. The distribution (6.20) includes the normal, binomial, Poisson distributions and also a contingency table. We therefore call $(\boldsymbol{A}'\boldsymbol{A})^{-1}\boldsymbol{A}'\boldsymbol{y}$ a key vector generally for testing the restricted alternative $\boldsymbol{A}'\boldsymbol{\theta} \geq \boldsymbol{0}$. Next, we give Lemma 6.3 for the interpretation of the coefficient matrix $(\boldsymbol{A}'\boldsymbol{A})^{-1}\boldsymbol{A}'$.

For contingency tables with ordered categories Cohen and Sackrowitz (1991) give the class of all tests that are simultaneously exact, unbiased and admissible.

Lemma 6.3. Corner vectors of the restricted alternative hypothesis. Every column of the coefficient matrix $\boldsymbol{A}(\boldsymbol{A}'\boldsymbol{A})^{-1}$ represents a corner vector of the polyhedral cone defined by the restricted alternative $A'\theta \geq 0$. In other words, all the θ that satisfy $A'\theta \geq 0$ can be expressed as a positive linear combination of the columns of $A(A'A)^{-1}$, except for the constant term orthogonal to A.

Proof. We consider any θ that satisfies

$$A'\theta \geq 0. \tag{6.21}$$

Since there are additional degrees of freedom in θ, we impose a restriction

$$B'\theta = 0 \tag{6.22}$$

by a non-singular matrix B satisfying $A'B = 0$ without violating the inequality $A'\theta \geq 0$. We can assume the matrix $[A \mid B]$ to be non-singular. Then, all the θ that satisfy (6.21) and (6.22) can be expressed as such θ satisfying

$$\begin{bmatrix} B' \\ A' \end{bmatrix} \theta = \begin{bmatrix} 0 \\ h \end{bmatrix}$$

with some $h \geq 0$. Then, for all those θ we have

$$\theta = \left\{A(A'A)^{-1}A' + B(B'B)^{-1}B'\right\}\theta = A(A'A)^{-1}h$$

since $A(A'A)^{-1}A' + B(B'B)^{-1}B'$ is an identity matrix as the sum of projection matrices orthogonal to each other and of full rank. Thus, all those θ can be expressed by a positive linear combination of the columns of $A(A'A)^{-1}$. Now, excluding the condition (6.22) we have an expression for θ satisfying (6.21):

$$\theta = B\eta + A(A'A)^{-1}h, \; h \geq 0$$

with η an arbitrary vector of coefficients, which completes the proof.

Lemma 6.3 implies that the matrix $A(A'A)^{-1}$ is more directly concerned with the restricted alternative $A'\theta \geq 0$ than the defining matrix A' itself. Thus, the key vector $(A'A)^{-1}A'y$ is interpreted as a projection of the observation vector y onto the corner vector of the cone. For a given model and given A, the key vector can sometimes be very simple and give reasonable test statistics.

Example 6.5. Monotone and step change-point hypotheses in the exponential family (6.20). The simple ordered alternative (monotone hypothesis) in the natural parameter θ,

$$H_{\text{mon}}: \theta_1 \leq \theta_2 \leq \cdots \leq \theta_a \text{ with at least one inequality strong,}$$

can be rewritten in matrix form as

$$H_{\text{mon}}: D_a'\theta \geq 0 \tag{6.23}$$

by the first-order differential matrix

$$\boldsymbol{D}_a' = \begin{bmatrix} -1 & 1 & 0 & 0 & \cdots & 0 & 0 & 0 & 0 \\ 0 & -1 & 1 & 0 & \cdots & 0 & 0 & 0 & 0 \\ 0 & 0 & -1 & 1 & \cdots & 0 & 0 & 0 & 0 \\ & & & & \vdots & & & & \\ 0 & 0 & 0 & 0 & \cdots & 0 & 0 & -1 & 1 \end{bmatrix}_{(a-1)\times a}. \quad (6.24)$$

In this case we have an explicit form of the corner vectors as

$$(\boldsymbol{D}_a'\boldsymbol{D}_a)^{-1}\boldsymbol{D}_a' = \frac{1}{a}\begin{bmatrix} -(a-1) & 1 & 1 & \cdots & 1 \\ -(a-2) & -(a-2) & 2 & \cdots & 2 \\ & & & \vdots & \\ -1 & -1 & -1 & \cdots & (a-1) \end{bmatrix}_{(a-1)\times a}. \quad (6.25)$$

Usually, the first-order differential matrix $\boldsymbol{D}_a'$ is considered essential for defining the monotone hypothesis H_{mon} (6.23), but actually the rows of $(\boldsymbol{D}_a'\boldsymbol{D}_a)^{-1}\boldsymbol{D}_a'$ are the corner vectors of the convex polyhedral cone defined by H_{mon} and more closely related to H_{mon} in the sense that every vector belonging to H_{mon} can be expressed by a unique and positive linear combination of these rows. Also, from equation (6.25) a close relationship between the monotone and step change-point hypotheses is suggested. That is, every element of the step change-point hypothesis represents the corner vector of the polyhedral cone defined by H_{mon} (6.23). Actually, every step change-point contrast is a particular monotone contrast and acts as the base of all the monotone contrasts. Finally, the key vector is explicitly given by

$$(\boldsymbol{D}_a'\boldsymbol{D}_a)^{-1}\boldsymbol{D}_a'\boldsymbol{y} = \begin{bmatrix} (Y_a/a) - Y_1 \\ 2(Y_a/a) - Y_2 \\ \vdots \\ (a-1)(Y_a/a) - Y_{a-1} \end{bmatrix}$$

suggesting the accumulated statistics

$$Y_k = \sum_{i=1}^{k} y_i,\ k = 1, \ldots, a-1 \quad (6.26)$$

as the basic statistics, since the general mean Y_a/a is the sufficient statistic corresponding to a general mean $\bar{\theta}.$ under the null model. Then, max acc. $t1$ and the cumulative chi-squared χ^{*2} are developed based on the accumulated statistics in the following sections.

6.5.3 Monotone and step change-point hypotheses

(1) Maximal contrast method for testing and estimating the dose–response pattern

(a) **Max acc. $t1$**. In a phase II clinical trial, as one of the typical situations, it is simultaneously requested to prove a drug as truly effective by showing a monotone dose response and also to obtain information (estimates) on the recommended dose for ordinary clinical treatments (ICH E9). Then, multiple comparisons of the dose–response patterns of interest are preferable to an overall testing of the null hypothesis or a fitting of a particular parametric model (such as logistic regression). Actually, the most popular practice in a clinical trial is to select the best-fit model approach among the candidate models. For this purpose the maximal accumulated t-test (max acc. $t1$) for the monotone hypothesis is introduced in this section and compared with other maximal contrast tests for estimating the dose–response pattern. The simultaneous lower bounds obtained by the inversion of max acc. $t1$ are shown to be useful for this purpose, and to have some advantage in giving the lower confidence bound of the efficacy difference between the estimated optimal dose and the base level.

Suppose the y_{ij}, $i=1,\ldots,a; j=1,\ldots,n_i$ are distributed independently as $N(\mu_i, \sigma^2)$ and σ^2 is known. The density function is proportional to $\exp\{\mu_i(y_{i\cdot}/\sigma^2)\}$, so that in this case $\boldsymbol{y}$ of the previous section is a vector of $y_{i\cdot}/\sigma^2$, $i=1,\ldots,a$. We therefore replace y_i by $y_{i\cdot}/\sigma^2$ to obtain $Y_k=\sum_{i=1}^{k} y_{i\cdot}/\sigma^2$, $k=1,\ldots,a-1$, in (6.26). Then, after standardization we have

$$t_k=\left\{\left(\frac{1}{N_k}+\frac{1}{N_k^*}\right)\sigma^2\right\}^{-\frac{1}{2}}\left(\bar{Y}_k^*-\bar{Y}_k\right),\quad k=1,\ldots,a-1 \tag{6.27}$$

as the basic statistics, where for convenience we newly define

$$Y_k=y_{1\cdot}+\cdots+y_{k\cdot},\ \bar{Y}_k=Y_k/N_k,\ N_k=n_1+\cdots+n_k,$$
$$Y_k^*=y_{k+1\cdot}+\cdots+y_{a\cdot},\ \bar{Y}_k^*=Y_k^*/N_k^*,\ N_k^*=n_{k+1}+\cdots+n_a.$$

The max acc. $t1$ defined by max $(t_1,\ldots, t_{a-1})$ is the maximal standardized statistic of (6.27). Equation (6.27) is interpreted as the standardized difference between the averaged responses up to k and from k+1 to the end, $k=1,\ldots,a-1$, and max acc. $t1$ is the maximum of them.

(b) **Max acc. $t1$ as an efficient score test of the step change-point hypothesis**. The kth component (6.27) of max acc. $t1$ is nothing but a standardized efficient score for testing the kth step change-point hypothesis, which is defined as

$$M_k : \begin{cases} y_{ij} = \mu + e_{ij}, \ i = 1, \ldots, k, j = 1, \ldots, n_i, \\ y_{ij} = \mu + \Delta + e_{ij}, \ i = k+1, \ldots, a, \ j = 1, \ldots, n_i. \end{cases}$$

Now, assuming the independent normal model, we have the log likelihood function for the kth step change-point hypothesis as

$$\log L = \text{const.} - \frac{\left\{ \sum_{i=1}^{k} \sum_{j=1}^{n_i} (y_{ij} - \mu)^2 + \sum_{i=k+1}^{a} \sum_{j=1}^{n_i} (y_{ij} - \mu - \Delta)^2 \right\}}{(2\sigma^2)}.$$

Then it is easy to verify that the efficient score with respect to Δ evaluated at the null hypothesis is

$$\frac{\partial \log L}{\partial \Delta}\Big|_{\Delta = 0, \ \mu = \bar{y}_{..}} = \left\{ \left(\frac{1}{N_k} + \frac{1}{N_k^*} \right) \sigma^2 \right\}^{-1} \left(\bar{Y}_k^* - \bar{Y}_k \right).$$

Thus, max acc. $t1$ is appropriate also for testing the step change-point hypothesis M_k, $k = 1, \ldots, a-1$ with unknown k.

(c) **Markov property**. Since the component t_k is essentially the cumulative sum of independent variables $y_{i\cdot}$ up to k, the first-order Markov property is obtained for the sequence of t_k, which leads to a very elegant and efficient algorithm for probability calculation. However, we need Lemma 6.5 for an exact proof of the Markov property, since the t_k are standardized by the sufficient statics under the null model. As a preparation, we give Lemma 6.4 on the conditional distribution of a multivariate normal distribution.

Lemma 6.4. Formula for the conditional distribution of a multivariate normal distribution. Let $z = (z_1', z_2')'$ be distributed as a multivariate normal distribution $N\left(\begin{bmatrix} \boldsymbol{\mu}_1 \\ \boldsymbol{\mu}_2 \end{bmatrix}, \begin{bmatrix} \boldsymbol{A} & \boldsymbol{B} \\ \boldsymbol{B}' & \boldsymbol{D} \end{bmatrix}^{-1} \right)$, where the mean vector $\boldsymbol{\mu}$ and the covariance matrix $\boldsymbol{\Omega}$ are partitioned according to the partition of z. Then, the conditional distribution $f(z_1 \mid z_2)$ of z_1 given z_2 is normal, $N\{\boldsymbol{\mu}_1 + \boldsymbol{K}(z_2 - \boldsymbol{\mu}_2), \boldsymbol{A}^{-1}\}$, where $\boldsymbol{K} = -\boldsymbol{A}^{-1}\boldsymbol{B}$.

Proof. By the form of the density function of the multivariate normal distribution, it is obvious that the conditional distribution is also normal. We therefore need only to calculate the conditional mean and variance. We have a formula for the inverse of a partitioned matrix:

$$\boldsymbol{\Omega} = \begin{bmatrix} \boldsymbol{A}^{-1} + \boldsymbol{A}^{-1}\boldsymbol{B}(\boldsymbol{D} - \boldsymbol{B}'\boldsymbol{A}^{-1}\boldsymbol{B})^{-1}\boldsymbol{B}'\boldsymbol{A}^{-1} & -\boldsymbol{A}^{-1}\boldsymbol{B}(\boldsymbol{D} - \boldsymbol{B}'\boldsymbol{A}^{-1}\boldsymbol{B})^{-1} \\ -(\boldsymbol{D} - \boldsymbol{B}'\boldsymbol{A}^{-1}\boldsymbol{B})^{-1}\boldsymbol{B}'\boldsymbol{A}^{-1} & (\boldsymbol{D} - \boldsymbol{B}'\boldsymbol{A}^{-1}\boldsymbol{B})^{-1} \end{bmatrix}.$$

Then, for $\boldsymbol{K} = -\boldsymbol{A}^{-1}\boldsymbol{B}$, we have $Cov\,(\mathbf{z}_1 - \boldsymbol{K}z_2, z_2) = \mathbf{0}$ and therefore $z_1 - Kz_2$ and z_2 are mutually independent. This implies

$$E(z_1 - \boldsymbol{K}z_2 \mid z_2) = E(z_1 - \boldsymbol{K}z_2) = \boldsymbol{\mu}_1 - \boldsymbol{K}\boldsymbol{\mu}_2,$$
$$V(z_1 - \boldsymbol{K}z_2 \mid z_2) = V(z_1 - \boldsymbol{K}z_2) = \boldsymbol{A}^{-1},$$

which completes the proof.

Lemma 6.5. Equivalence theorem of Markov property (Hirotsu *et al.*, 1992). If $z = (z_1, \ldots, z_a)'$ is distributed as a multivariate normal distribution with a non-singular covariance matrix and with no two elements mutually independent, then the following three statements are equivalent.

1. The Markov property
$$f(z_1, \ldots, z_l | z_{l+1}, \ldots, z_n) = f(z_1, \ldots, z_l | z_{l+1}), \; l < n$$
holds for the conditional densities of z_i.

2. The inverse of the covariance matrix of z is a tri-diagonal matrix with no $(i, i \pm 1)$th element equal to zero.

3. The correlation matrix of z has the structure $Cor\,(z_i, z_j) = \gamma_i / \gamma_j,\; i \le j$.

Proof

2.⇒1. Following the notation of Lemma 6.4, if $\boldsymbol{\Omega}^{-1}$ is a tri-diagonal matrix then all the elements of $\boldsymbol{B}$ are 0 except an element at the extreme left and lowest. Then, for any partition the conditional distribution $f(z_1 \mid z_2)$ depends only on the first element of z_2 by virtue of Lemma 6.4.

1.⇒2. Let $z_1 = z_1$, $z_2 = (z_2, \ldots, z_a)'$, then $\boldsymbol{A}$ is a scalar and $\boldsymbol{B}$ is an $(a-1)$-dimensional row vector. For the conditional distribution $f(z_1 \mid z_2)$ to be dependent only on z_2 it is necessary and sufficient that the first element of $\boldsymbol{B}$ is non-zero and all other elements are zero. Then obviously $\boldsymbol{A}^{-1}\boldsymbol{B}z_2$ is a function of only z_2. Next consider the partition $z_1 = (z_1, z_2)'$, $z_2 = (z_3, \ldots, z_a)'$ and the corresponding partitioned covariance matrix. Then $\boldsymbol{A}$ is a 2×2 matrix and $\boldsymbol{B}$ a $2 \times (a-2)$ matrix with first row $\mathbf{0}$. Then we have

$$\boldsymbol{A}^{-1}\boldsymbol{B}z_2 = \begin{bmatrix} \boldsymbol{A}^{12}\boldsymbol{b}_2' \\ \boldsymbol{A}^{22}\boldsymbol{b}_2' \end{bmatrix} z_2$$

where $\boldsymbol{A}^{ij}$ is the (i, j)th element of $\boldsymbol{A}^{-1}$ and $\boldsymbol{b}_2'$ is the second row of $\boldsymbol{B}$. Since $\boldsymbol{A}^{22}$ is non-zero, it is necessary and sufficient that the first element of $\boldsymbol{b}_2'$ is non-zero and all other elements are zero for $\boldsymbol{A}^{-1}\boldsymbol{B}z_2$ to be a function of only z_3. Continuing the same argument, it is proved that $\boldsymbol{\Omega}^{-1}$ should be a tri-diagonal matrix with all $(i, i\pm 1)$ elements non-zero.

2.⇒3. This proof is purely a problem of linear algebra. Define two matrices

$$\boldsymbol{L}(\gamma_1, \ldots, \gamma_n) = \begin{bmatrix} 1 & \gamma_1/\gamma_2 & \gamma_1/\gamma_3 & \cdots & \gamma_1/\gamma_n \\ \gamma_1/\gamma_2 & 1 & \gamma_2/\gamma_3 & \cdots & \gamma_2/\gamma_n \\ \gamma_1/\gamma_3 & \gamma_2/\gamma_3 & 1 & \cdots & \gamma_3/\gamma_n \\ \vdots & \vdots & \vdots & \ddots & \vdots \\ \gamma_1/\gamma_n & \gamma_2/\gamma_n & \gamma_3/\gamma_n & \cdots & 1 \end{bmatrix},$$

$$H(d_1, \ldots, d_n) = \begin{bmatrix} 1 & d_1 & 0 & 0 & \cdots & \cdots & \cdots & 0 \\ d_1 & 1 & d_2 & 0 & \cdots & \cdots & \cdots & 0 \\ 0 & d_2 & 1 & d_3 & \cdots & \cdots & \cdots & 0 \\ 0 & 0 & d_3 & 1 & \cdots & \cdots & \cdots & 0 \\ & & & & \ddots & & & \\ 0 & \cdots & \cdots & 0 & \cdots & 1 & d_{n-2} & 0 \\ 0 & \cdots & \cdots & 0 & \cdots & d_{n-2} & 1 & d_{n-1} \\ 0 & \cdots & \cdots & 0 & \cdots & 0 & d_{n-1} & 1 \end{bmatrix},$$

where $\boldsymbol{L}$ and $\boldsymbol{H}$ are non-singular matrices with $\gamma_i \neq 0$, $d_i \neq 0$ for all i. To show 2.⇒3. it is sufficient to show that $\boldsymbol{H}^{-1}$ takes the form of $\boldsymbol{L}$. Now, the co-factors of $\boldsymbol{H}$ are expressed as

$$H^{jj} = \left|\boldsymbol{H}\left(d_1, \ldots, d_{j-2}\right)\right| \times \left|\boldsymbol{H}\left(d_{j+1}, \ldots, d_{n-1}\right)\right|$$

$$H^{jk} = (-1)^{j+k}\left|\boldsymbol{H}\left(d_1, \ldots, d_{j-2}\right)\right| \times d_j \cdots d_{k-1} \times \left|\boldsymbol{H}(d_{k+1}, \ldots, d_{n-1})\right|, \; k \geq j+1,$$

where $|\boldsymbol{H}(d_1, \ldots, d_k)|$ is the determinant of $\boldsymbol{H}(d_1, \ldots, d_k)$. Since they are all non-zero, by the assumption we have

$$\frac{H^{ij}}{H^{jj}} = \frac{H^{ik}}{H^{jk}} = (-1)^{i+j} d_i \cdots d_{j-1} \frac{D(d_1, \ldots, d_{i-2})}{D(d_1, \ldots, d_{j-2})}, \; 1 \leq i < j \leq k \leq n-1,$$

and this implies that $\boldsymbol{H}^{-1}$ is in the form of $\boldsymbol{L}$.

3.⇒2. Let $|\boldsymbol{L}(\gamma_1, \ldots, \gamma_k)| = \det[\boldsymbol{L}(\gamma_1, \ldots, \gamma_k)]$. Then the co-factors of $\boldsymbol{L}$ are expressed as

$$L^{jj} = \left|\boldsymbol{L}\left(\gamma_1, \ldots, \gamma_{j-1}, \gamma_{j+1}, \cdots, \gamma_n\right)\right|,$$

$$L^{jj+1} = -\left|\boldsymbol{L}\left(\gamma_1, \ldots, \gamma_{j-1}\right)\right| \times \left|\boldsymbol{L}\left(\gamma_{j+2}, \ldots, \gamma_n\right)\right| \times \frac{\left(\gamma_j^2 - \gamma_{j-1}^2\right)\left(\gamma_{j+2}^2 - \gamma_{j+1}^2\right)}{\gamma_j \gamma_{j+1} \gamma_{j+2}^2},$$

$$L^{jk} = 0, \; k > j+1,$$

and this implies that $\boldsymbol{L}^{-1}$ is a tri-diagonal matrix.

To show the Markov property of the components $\boldsymbol{t} = (t_1, \ldots, t_{a-1})'$ of max acc. $t1$ we may show either 2. or 3. Then it is most easy to show that the variance–covariance matrix of $\boldsymbol{t}$ is

$$V(\boldsymbol{t}) = \begin{bmatrix} 1 & \sqrt{\lambda_1/\lambda_2} & \sqrt{\lambda_1/\lambda_3} & \cdots & \sqrt{\lambda_1/\lambda_{a-1}} \\ \sqrt{\lambda_1/\lambda_2} & 1 & \sqrt{\lambda_2/\lambda_3} & \cdots & \sqrt{\lambda_2/\lambda_{a-1}} \\ \sqrt{\lambda_1/\lambda_3} & \sqrt{\lambda_2/\lambda_3} & 1 & \cdots & \sqrt{\lambda_3/\lambda_{a-1}} \\ \vdots & \vdots & \vdots & \ddots & \vdots \\ \sqrt{\lambda_1/\lambda_{a-1}} & \sqrt{\lambda_2/\lambda_{a-1}} & \sqrt{\lambda_3/\lambda_{a-1}} & \cdots & 1 \end{bmatrix} \quad (6.28)$$

where $\lambda_k = N_k/N_k^*$ and it is exactly in the form 3. of Lemma 6.5.

(d) **Algorithm for calculating p-value**. For the algorithm in the normal distribution we define the conditional joint probability of $(t_1, \ldots, t_k)$: $F_k(t_k, t_0|\sigma) = \Pr(t_1 < t_0, \ldots, t_k < t_0 | t_k, \sigma)$, $k = 1, \ldots, a-1$. Then we have

$$F_{k+1}(t_{k+1}, t_0|\sigma) = \Pr(t_1 < t_0, \ldots, t_k < t_0, t_{k+1} < t_0 | t_{k+1}, \sigma)$$

$$= \int_{t_k} \Pr(t_1 < t_0, \ldots, t_k < t_0, t_{k+1} < t_0 | t_k, t_{k+1}, \sigma) \times f_k(t_k|t_{k+1}) \mathrm{d}t_k \quad (6.29)$$

$$= \begin{cases} \int_{t_k} F_k(t_k,\ t_0|\sigma) \times f_k(t_k|t_{k+1}) \mathrm{d}t_k \text{ if } t_{k+1} < t_0 \\ 0, \text{ otherwise,} \end{cases} \quad (6.30)$$

where $f_k(t_k|t_{k+1})$ is the conditional distribution of t_k given t_{k+1}. This is easily obtained as the normal density $N\left\{\sqrt{\lambda_k/\lambda_{k+1}}\, t_{k+1}, (\lambda_{k+1} - \lambda_k)/\lambda_{k+1}\right\}$ under the null hypothesis by virtue of Lemma 6.4. For the non-null distribution the expectations of t_k and t_{k+1} should be taken into consideration appropriately. Equation (6.29) is due to the law of total probability and (6.30) to the Markov property of the sequence $t_k, k=1,\ldots,a-1$. The p-value is obtained at the final step as

$$p = 1 - F_a(t_a, t_0 | \sigma), \tag{6.31}$$

where t_0 is the observed maximum value. To run the formula (6.30) up to $k=a$, we need to define $t_a = -\infty$ so that the inequality $t_a < t_0$ holds always, where the conditional density $f_{a-1}(t_{a-1}|t_a)$ is the unconditional density of the standard normal distribution.

When σ is unknown and replaced by $\hat{\sigma}$ in formula (6.31), the expectation with respect to $\hat{\sigma}/\sigma$ is required to obtain the p-value. Its distribution is a constant times the chi-distribution with df f of $\hat{\sigma}$ and there is no difficulty in executing the calculation.

We denote the rejection region of max acc. $t1$ by

$$R : \max(t_1, \ldots, t_{a-1}) > T_\alpha(\boldsymbol{n}, f),$$

where $T_\alpha(\boldsymbol{n}, f)$ is an upper α point of the null distribution of max acc. $t1$ with $\boldsymbol{n} = (n_1, \ldots, n_a)$ and f the degrees of freedom of $\hat{\sigma}$.

Generally, the multiple integrations are computationally very hard even for a moderate a, but the recursion formula (6.30) converts the multiple integrations to repetitions of the single integration and works very efficiently for any number of levels. The FORTRAN program for p-value is given by Hirotsu *et al.* (1997). It can also be used to calculate the percentiles. For the balanced case, $T_\alpha(\boldsymbol{n}, f)$ is obtained from $t_\alpha(a, f)$ in Table A of the Appendix, where the tabulation is indexed by a and f.

(e) **Characterization of max acc. *t*1**. The max acc. $t1$ is characterized by the following.

1. It comes out directly from a complete class lemma for the test of the monotone hypothesis H_{mon} (Hirotsu, 1982).

2. The components of max acc. $t1$ are essentially the projections of the observation vector $\boldsymbol{y} = (y_{1\cdot}, \ldots, y_{a\cdot})'$ onto the corner vectors of the convex polyhedral cone defined by H_{mon} and consequently all the monotone contrasts can be expressed by a unique and positive linear combination of those components (Hirotsu and Marumo, 2002). For this reason we may call σt_k, $k=1,\ldots,a-1$, the basic contrasts.

3. Accordingly, the simultaneous confidence intervals based on the basic contrasts of max acc. $t1$ can be extended uniquely to all the monotone contrasts, whose significance can therefore be evaluated (Hirotsu and Srivastava, 2000).

4. An exact and very efficient algorithm for calculating the distribution function is available for any a due to the Markov property of the successive components of max acc. $t1$.

5. It has been shown on several occasions that max acc. $t1$ keeps high power against a wide range of ordered alternative hypotheses H_{mon} compared with other well-known tests such as $\bar{\chi}^2$ of Bartholomew (1959a,b) and Williams (1971) (see Hirotsu *et al.*, 1992).

6. max acc. $t1$ is also an efficient score for testing the step change-point hypothesis and therefore gives a unifying approach to the shape and change-point hypotheses.

In spite of these attractive features, there have been several attempts to add some monotone contrasts to the $a-1$ basic contrasts of max acc. $t1$, intending to detect some response patterns more efficiently. However, if we add another monotone contrast to the basic contrasts, the uniqueness is lost and if we replace some of the basic contrasts by other contrasts, then the positivity will be lost in 2. and 3. It is even remarkable that adding a linear trend contrast to the basic contrasts cannot improve the power at all for detecting the linear trend (see Section 6.5.3 (1) (h) and Hirotsu *et al.* (2011) for more details). Also, by these changes of the basic contrasts the Markov property fails and the recursion formula of 4. for an efficient calculation is no longer available.

(f) **Simultaneous lower bounds (SLB) for the isotonic contrasts**. Now, considering the natural ordering, the monotone patterns of interest are given generally by

$$\mu_1 = \cdots = \mu_i < \mu_{i+1} \le \cdots \le \mu_{j-1} < \mu_j = \cdots = \mu_a, \tag{6.32}$$

whose size is defined by

$$\mu(i,j) = \frac{n_j\mu_j + \cdots + n_a\mu_a}{N_{j-1}^*} - \frac{n_1\mu_1 + \cdots + n_i\mu_i}{N_i},\ 1 \le i < j \le a, \tag{6.33}$$

representing the difference between the highest and lowest dose levels under the model (6.32). If the estimated response pattern is (6.32), the optimum level should be j in considering that the lower level is safer with a good cost performance. We represent a strictly monotone pattern by a linear regression model

$$\mu(\text{linear}) : \mu_k = \beta_0 + \beta_1 k.$$

In this case the difference between the highest and lowest dose levels corresponding to (6.33) is given by

$$\Delta\mu(\text{linear}) = (a-1)\beta_1$$

since β_1 represents the difference between the subsequent dose levels. If this pattern is suggested, then the optimum level should be a. It should be noted here that every monotone contrast – including the linear regression type – can be expressed by a unique and positive linear combination of the basic contrasts $\mu(k, k+1), k = 1, \ldots, a-1$.

Now, the simultaneous lower confidence bounds $\mathrm{SLB}(k,k+1)$ of the basic contrasts $\mu(k,k+1)$ can be obtained directly by the inversion of max acc. $t1$ as

$$\mathrm{SLB}(k,k+1)=(\bar{Y}_k^*-\bar{Y}_k)-\hat{\sigma}\left(\frac{1}{N_k}+\frac{1}{N_k^*}\right)^{1/2}T_\alpha(\boldsymbol{n},N-a). \tag{6.34}$$

Then they are uniquely extended to the general monotone contrasts of interest, (6.33) and $\Delta\mu$(linear) by a positive linear combination such as

$$\mathrm{SLB}(i,j+1)=\frac{N_j}{N}\mathrm{SLB}(j,j+1)+\frac{N_i^*}{N}\mathrm{SLB}(i,i+1),\ i\le j, \tag{6.35}$$

$$\begin{aligned}\mathrm{SLB}(\text{linear})&=(a-1)\left[\frac{1}{N}\textstyle\sum_{k=1}^{a-1}\left\{N_k^*\sum_{i=1}^{k-1}(k-i)n_i+N_k\sum_{i=k+1}^{a}(i-k)n_i\right\}\right]^{-1}\\&\quad\times\textstyle\sum_{k=1}^{a-1}\left\{\left(\frac{1}{N_k}+\frac{1}{N_k^*}\right)^{-1}\mathrm{SLB}(k,k+1)\right\}.\end{aligned} \tag{6.36}$$

The proof of formula (6.36) is a little complicated, but given in Hirotsu *et al.* (2011). Further, by the monotone assumption in $\boldsymbol{\mu}$, the bound $\mathrm{SLB}(i,j)$ is improved as

$$\mu(i,j)\ge \max_{i\le l<m\le j}\mathrm{SLB}(l,m). \tag{6.37}$$

Again it should be stressed that it is only max acc. $t1$ that can give such a unique and positive extension of the SLB of the basic contrasts to all the monotone contrasts. Then, an interesting method is to choose a pattern which gives the highest lower confidence bound among the monotone contrasts (6.34)~(6.37).

(g) **Method for estimating the dose–response pattern based on max acc. *t*1**. This idea of estimating the dose–response pattern can be combined with the classical closed test procedure, which is a decision-theoretic approach. We apply max acc. $t1$ downward until it becomes non-significant for the first time. Suppose it stops after exactly k rejections, which we call k-stopping. If $k=0$, we accept the null hypothesis. The suggested model with k-stopping is $\mu_1=\cdots=\mu_{a-k}<\mu_{a-k+1}\le\cdots\le\mu_a$. Then we proceed to compare the lower confidence bounds $\mathrm{SLB}(a-k,a-k+1),\ldots,\mathrm{SLB}(a-k,a)$, corresponding to (6.33) with $i=a-k$ and $j=a-k+1,\ \ldots,\ a$ and choose the one which gives the highest lower confidence bound. For $(a-1)$-stopping we include SLB (linear) for the candidate model μ (linear). This procedure indeed satisfies the requirements to confirm the existence of the dose response by a significance test, giving simultaneously the optimal dose level and the lower confidence bound of the difference between the optimal and base levels.

Example 6.6. Study of red cell counts after dosage of a medicine. The study was a balanced one-way layout with rats at four dose levels (10, 100, 1000, 10,000 ppm) with repetition number five (Furukawa and Tango, 1993). For the data, Furukawa and Tango proved a significant decrease in red cell counts by linear regression analysis. Since the approach by max acc. $t1$ has been given here as a right one-sided test, we subtract the original data from 10 as given in Table 6.7, with box plot as shown in Fig. 6.1.

The SLB for the basic change-point contrasts are obtained by (6.34) as $\mathrm{SLB}(1,2)=0.177$, $\mathrm{SLB}(2,3)=0.131$, $\mathrm{SLB}(3,4)=0.002$. They are extended to $\mathrm{SLB}(i,j+1)$ by (6.35) and SLB(linear) by (6.36), as shown in Table 6.8 (1). The improved lower bounds from formula (6.37) are shown in Table 6.8 (2), where the trace of improvements is shown by ⇛. These alterations suggest that the approach of a rigid linear regression model might be inappropriate. Now, applying the closed test downward by max acc. $t1$, we find it to be significant at all three

Table 6.7 Decrease in red cell counts (10 – original data).

	Repetition				
Dose	1	2	3	4	5
10	1.94	1.73	1.55	1.49	1.86
100	2.03	2.34	1.95	1.70	1.97
1000	2.34	2.29	2.12	1.95	2.20
10,000	2.00	2.11	2.21	2.09	2.60

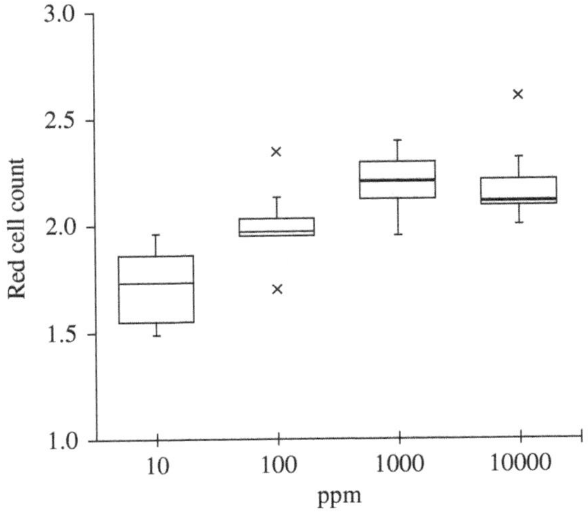

Figure 6.1 Decrease in red cell counts.

Table 6.8 Simultaneous lower bounds SLB $(i, j+1)$ for $p(i, j)$.

(1) Basic formula (6.34)

	$j+1$		
i	2	3	4
1	0.177	0.198	0.135
2		0.131	0.067
3			0.002
		SLB (linear) = 0.159	

(2) Improvement by formula (6.37)

	$j+1$			
i	2	3		4
1	0.177	0.198	⇒	0.198
2		0.131	⇒	0.131
3				0.002
		SLB (linear) = 0.159		

steps (3-stopping), since they are 4.242, 3.519, and 2.118 against the critical values of 2.222, 2.108, and 1.859 at one-sided $\alpha = 0.05$, respectively. Therefore, we compare SLB(1, 2), SLB(1, 3), SLB(1, 4), and SLB(linear) to find $\mathrm{SLB}(1,3) = \mathrm{SLB}(1,4) = 0.198$ as the largest, suggesting a pattern $\mu_1 < \mu_2 < \mu_3 = \mu_4$ as the best-fit model. This coincides very well with a first-glance impression of the box plot in Fig. 6.1.

Example 6.7. Study of the half-life of NFLX after dosage (Hirotsu *et al.*, 2011). The study was originally a balanced one-way layout of five repetitions with rats, comparing the half-life of antibiotic NFLX at five dose levels (5, 10, 25, 50, 200 mg/kg/day). The data of 25 and 200 (mg/kg/day) have been utilized in Example 5.1 to explain the two-sample problem. We noticed there that the largest data point 2.14 of 25 (mg/kg/day) should be an outlier and it is excluded in Table 6.9, thus giving an example of unbalanced data. The box plots are given in Fig. 6.2.

The SLB for basic contrasts are obtained as $\mathrm{SLB}(1,2) = 0.280$, $\mathrm{SLB}(2,3) = 0.375$, $\mathrm{SLB}(3,4) = 0.292$, $\mathrm{SLB}(4,5) = 0.372$ and extended to other monotone contrasts $\mathrm{SLB}(i,j+1)$ as shown in Table 6.10. In this case there is no alteration with formula (6.37), since all the $\mathrm{SLB}(i,j+1)$ satisfy the monotone relationship. Applying

Table 6.9 Half-life of NFLX (antibiotics).

Dosage (mg/kg/day)	Half-life (hr)				
5	1.17	1.12	1.07	0.98	1.04
10	1.00	1.21	1.24	1.14	1.34
25	1.55	1.63	1.49	1.53	
50	1.21	1.63	1.37	1.50	1.81
200	1.78	1.93	1.80	2.07	1.70

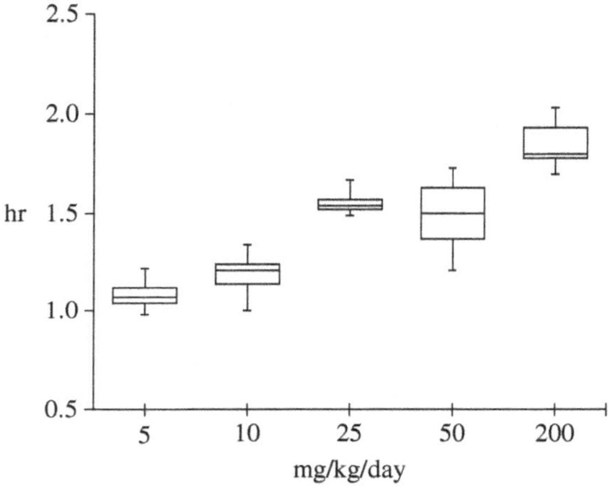

Figure 6.2 NFLX half-life data.

Table 6.10 SLB $(i, j+1)$ for the isotonic contrasts.

	$j+1$			
i	2	3	4	5
1	0.280	0.378	0.393	0.517
2		0.375	0.389	0.514
3			0.292	0.417
4				0.372
	SLB(linear) = 0.529			

max acc. $t1$ downward we stop after three rejections (3-stopping) to accept the hypothesis $H: \mu_1 = \mu_2 (< \mu_3 \leq \mu_4 \leq \mu_5)$. Then, we proceed to compare the SLB(2, 3), SLB(2, 4), and SLB(2, 5) to find $\mathrm{SLB}(2, 5) = 0.514$ as the largest. Therefore, we select the model $\mu_1 = \mu_2 < \mu_3 \leq \mu_4 < \mu_5$, which coincides very well again with the box plots in

Fig. 6.2. It should be noted that the response pattern can be specified so far without assuming any parametric model.

A computer program for calculating the SLB as well as the p-value of max acc. $t1$ is given on the author's website.

(h) **Power comparisons**. The detailed power comparisons of max acc. $t1$ with other monotone tests have been made on several occasions. In Table 6.11 we show the result of comparisons by Hirotsu *et al.* (1992), where the linear trend test (linear score), maximin linear test of Abelson and Tukey (1963) (A–T score), restricted likelihood ratio test of Bartholomew (1959a) (LR), contrast test of Williams (1971), modified Williams by Marcus (1976), and max acc. $t1$ are compared. The underline shows the highest power among the competitors. Then max acc. $t1$, which used to be called max t at that time, shows excellent behavior. Only in the bottom case of very late start-up of the dose–response curve is the modified Williams slightly better.

Hirotsu *et al.* (2011) compared max acc. $t1$ with other maximal contrast tests including Dunnett's test (1964), the dose–response method of Maurer *et al.* (1995) and Hsu and Berger (1999), the maximal standardized subsequent difference of means by Liu *et al.* (2000), and the excellence of max acc. $t1$ was shown again. Further, the effects of adding a linear trend contrast to the basic change-point contrasts of max acc. $t1$ were investigated by introducing the max acc. $t1 + t_l$(linear trend contrast) method. Adding the linear trend contrast destroys the beautiful Markov property of basic contrasts for the probability calculation, so that a time-consuming calculation is made. The closed tests up- and downward with $\alpha = 0.05$ one-sided were employed for each of max acc. $t1$ and max acc. $t1+t_l$, followed by choosing the maximal standardized contrast method among the candidate models:

$$\begin{aligned}
&M_1 : \mu_1 = \mu_2 = \mu_3 < \mu_4,\\
&M_2 : \mu_1 = \mu_2 < \mu_3 = \mu_4,\\
&M_3 : \mu_1 < \mu_2 = \mu_3 = \mu_4,\\
&M_4 : \mu_1 = \mu_2 < \mu_3 < \mu_4,\\
&M_5 : \mu_1 < \mu_2 < \mu_3 = \mu_4,\\
&M_6 : \mu_1 < \mu_2 < \mu_3 < \mu_4.
\end{aligned} \tag{6.38}$$

The SLB method of the previous section cannot be applied to the max acc. $t1+t_l$ method since it doesn't have such a beautiful structure as described in Section 6.5.3 (1) (e) 3. It should be noted that in this case the pattern $\mu_1 < \mu_2 = \mu_3 < \mu_4$ is purposely excluded from the candidate models (6.38), since usually the sigmoid pattern is considered to be more reasonable. The contrasts corresponding to the models of (6.38) are $\mu(3, 4)$ for M_1, $\mu(2, 3)$ for M_2, $\mu(1, 2)$ for M_3, $\mu(2, 4)$ for M_4, $\mu(1, 3)$ for M_5, and $\mu(1, 4)$ for M_6, respectively. However, in producing

Table 6.11 (1) Power comparisons by numerical integration.

μ_1 μ_2 μ_3 μ_4	max acc. $t1$	Linear score	A–T score	LR	Williams	Modified Williams
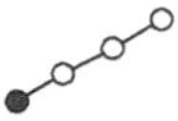	14.0	11.7	11.8	13.7	13.5	13.0
	41.7	41.1	41.1	41.7	41.6	41.3
	66.9	66.9	66.9	66.9	66.9	66.9
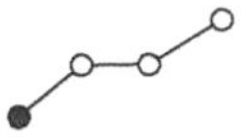	9.4	5.7	5.8	8.9	8.7	7.9
	32.2	25.0	25.1	31.5	31.1	29.3
	61.1	52.6	52.7	60.4	60.0	58.6
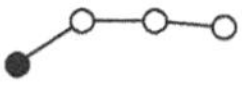	7.9	3.6	3.8	7.3	7.2	6.3
	28.4	16.8	17.6	27.1	26.9	24.8
	58.3	42.1	43.7	57.0	56.8	54.6
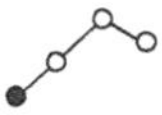	12.3	8.0	8.0	11.8	11.7	10.9
	40.9	36.6	36.3	40.6	40.5	39.9
	66.9	66.1	65.9	66.8	66.8	66.8
	15.8	14.8	13.7	15.0	13.2	14.0
	41.4	40.6	40.3	43.0	37.7	41.7
	73.5	68.3	68.3	71.8	65.9	73.0
	16.8	16.5	17.2	16.7	15.8	17.8
	43.0	38.1	40.2	42.5	37.6	44.5
	74.3	64.3	67.3	73.5	65.0	73.4

Table 6.11 (2) Power comparisons by simulation.

μ_1 μ_2 μ_3 μ_4 μ_5 μ_6 μ_7 μ_8	max acc. $t1$	Linear score	A–T score	LR	Williams	Modified Williams
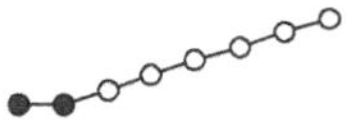	15.0	14.6	13.5	15.0	12.9	13.3
	42.5	40.6	40.3	42.8	37.8	42.1
	72.8	68.4	68.4	73.1	66.0	72.5
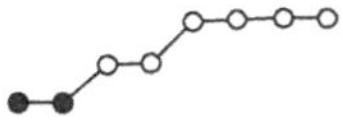	10.7	7.4	6.1	10.2	8.0	7.6
	36.9	32.9	29.1	36.7	29.1	31.7
	69.9	64.2	60.3	69.9	58.9	65.5
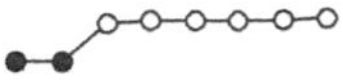	7.5	2.7	2.3	6.6	5.0	4.5
	31.8	18.5	15.1	30.5	21.3	22.3
	68.4	54.0	47.2	67.8	52.5	59.1
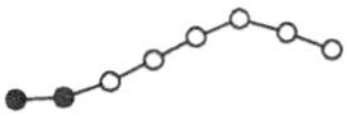	15.2	14.4	13.2	15.1	12.9	13.4
	42.7	40.6	40.3	43.0	37.8	42.2
	72.5	68.2	68.2	72.7	65.7	72.3
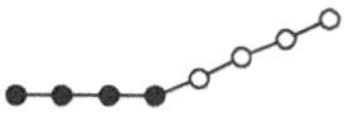	15.1	14.6	14.4	15.3	12.6	13.8
	42.2	35.2	39.4	42.8	36.5	44.0
	73.9	60.0	66.4	74.3	64.6	75.5
	14.5	13.4	14.3	15.0	12.6	14.3
	40.9	30.5	38.0	41.8	36.1	44.9
	73.7	52.5	64.9	74.1	64.2	76.9

Table 6.12 Probability of selecting acceptable models by max acc. $t1$ and max acc. $t1 + t_l$

True pattern	Selection	Downward max acc. $t1$	Downward max acc. $t1+t_l$	Upward max acc. $t1$	Upward max acc. $t1+t_l$
M_1	◎ M_1	85.1	85.1	62.9	61.9
	◎ or ⊙ (M_1, M_4, M_6)	88.4	88.5	80.1	79.8
M_2	◎ M_2	67.8	67.7	67.8	67.7
	◎ or ⊙ (M_2, M_5)	71.2	71.1	72.0	72.0
M_3	◎ M_3	63.3	63.1	86.0	86.0
	◎ or ⊙ (M_3)	63.3	63.1	86.0	86.0
M_4	◎ M_4	9.2	9.4	27.0	27.0
	◎ or ⊙ (M_1, M_4, M_6)	78.4	78.7	57.0	56.8
M_5	◎ M_5	27.9	27.9	9.7	9.9
	◎ or ⊙ (M_2, M_5)	52.1	52.5	20.6	20.8
M_6	◎ M_6	10.1	9.9	10.1	10.0
	◎ or ⊙ (M_1, M_4, M_6)	49.8	50.3	24.6	24.4

the data for simulation, we actually represent strict inequality by a linear trend, since we need to specify definitely the model creating the data. Therefore, model M_6 is actually a linear trend and max acc. $t1+t_l$ is expected to behave better for M_6. We show the simulation result by 10^6 replications in Table 6.12. The correct choice of true model is marked by ◎. Suppose the true model is M_1, for example, then M_4 and M_6 are acceptable as well, since they suggest the correct optimal level four. Therefore, we mark by ⊙ those models which are not correct but acceptable.

Now by Table 6.12 the two methods are similar on the whole and it is even remarkable that adding a linear trend contrast to the basic change-point contrasts cannot contribute at all to improving the probability of a correct decision when the true model is a linear trend. The theoretical basis for this unexpected result is given in the original paper by Hirotsu *et al.* (2011). Then, max acc. $t1$ should be preferred because of the properties 1.~6. in its characterization. Finally, regarding the up- or downward procedure, the latter is generally recommended unless a rapid start of the dose–response curve like M_3 is expected by prior information, where the upward procedure behaves better.

(2) **Cumulative chi-squared statistic for a trend test**

(a) **Cumulative chi-squared χ^{*2}.** As an overall directional test like $\bar{\chi}^2$ of Bartholomew (1959a,b), the cumulative chi-squared χ^{*2} has been repeatedly shown to have excellent power. χ^{*2} is defined as the sum of squares of the $(a-1)$ standardized components of max acc. $t1$ (6.27):

$$\chi^{*2} = \sum_{k=1}^{a-1} t_k^2. \tag{6.39}$$

Therefore, it has an ellipsoidal acceptance region compared with the polygon of max acc. $t1$. A sketch of the rejection regions of these test statistics is given in Hirotsu (1979b). The cumulative chi-squared doesn't belong to the complete class of tests for the simple ordered alternative in the strict sense, but is very robust. As a positive quadratic form of the normal variables, the null and non-null distributions of χ^{*2} are well approximated by a constant times χ^2 distribution $d\chi_f^2$ by adjusting the first two cumulants. More definitely, under the null hypothesis the constants are obtained by solving

$$\kappa_1 = E(\chi^{*2}) = \mathrm{tr}\{V(t)\} = a-1 = df,$$

$$\kappa_2 = V(\chi^{*2}) = 2\mathrm{tr}\{V^2(t)\} = 2\left\{a-1+2\left(\frac{\lambda_1}{\lambda_2}+\cdots+\frac{\lambda_1+\cdots+\lambda_{a-2}}{\lambda_{a-1}}\right)\right\} = 2d^2f,$$

where $V(t)$ and λ_k have been given in (6.28). The approximation can be further improved by adjusting the third cumulant $\kappa_3 = 8\mathrm{tr}\{V(t)\}^3$. The improved upper percentile is given by

$$(1+\Delta)d\chi_f^2(\alpha),$$

$$\Delta = \frac{\delta}{3(f+2)(f+4)}\left\{\left(\chi_f^2(\alpha)\right)^2 - 4\left(\frac{f}{2}+2\right)\chi_f^2(\alpha) + 4\left(\frac{f}{2}+1\right)\left(\frac{f}{2}+2\right)\right\},$$

$$\delta = \frac{\kappa_1\kappa_3}{2\kappa_2{}^2} - 1 = \frac{\mathrm{tr}\{V(t)\}\times\mathrm{tr}\{V^3(t)\}}{[\mathrm{tr}\{V^2(t)\}]^2} - 1.$$

In contrast, the upper tail probability is approximated by

$$\Pr(\chi^{*2} \ge x) = \Pr\left(\chi_f^2 \ge x/d\right) + \frac{2f\delta}{3(f+2)(f+4)} g_2^{p+1}\left(\frac{x}{2d}\right) f_{p+1}\left(\frac{x}{2d}\right),$$

$$g_2^{p+1}(s) = s^2 - 2(p+2)s + (p+1)(p+2), p = f/2,$$

$$f_{p+1}(s) = \frac{1}{\Gamma(p+1)} s^p e^{-s},$$

(see Hirotsu, 1979a for details). When the variance σ^2 is unknown and replaced by $\hat{\sigma}^2$, the statistic $F^* = \chi^{*2}_{\sigma^2=\hat{\sigma}^2}/(df)$ is approximately distributed as an F-distribution with df (f, f_2), where $\chi^{*2}_{\sigma^2=\hat{\sigma}^2}$ implies that σ^2 is replaced by $\hat{\sigma}^2$ in (6.39) and f_2 is the df of the unbiased variance estimator $\hat{\sigma}^2$. In this case the upper percentile is well approximated by

$$(1+\Delta)F_{f,f_2}(\alpha), \tag{6.40}$$

$$\Delta = \frac{\delta}{3(f+2)(f+4)}\left[(f+2)(f+4) - \frac{2(f+f_2)(f+4)}{1+f_2/\{f\times F_{f,f_2}(\alpha)\}} + \frac{(f+f_2)(f+f_2+2)}{[1+f_2/\{f\times F_{f,f_2}(\alpha)\}]^2}\right].$$

Table 6.13 Accuracy of approximation (6.40) at upper five percentile.

n	Exact	First approx. ($\Delta = 0$)	Second approx. (6.40)
2	9.66	9.73	9.69
3	5.33	5.39	5.34
5	4.09	4.14	4.08

Similarly, the upper tail probability is approximated by

$$\Pr(F^* \ge z) = \Pr\{F(f, f_2) \ge z\} + \left\{ \frac{\delta}{3(f+2)(f+4)\mathrm{B}\left(\frac{1}{2}f, \frac{1}{2}f_2\right)} \right\} \left(1 + \frac{fz}{f_2}\right)^{-\frac{(f+f_2)}{2}} \left(\frac{fz}{f_2}\right)^{\frac{f}{2}}$$
$$\times \left\{ (f+2)(f+4) - \frac{2(f+f_2)(f+4)}{1+f_2/(fz)} + \frac{(f+f_2)(f+f_2+2)}{\{1+f_2/(fz)\}^2} \right\}$$

The accuracy of the approximation (6.40) is practically sufficient, as shown in Table 6.13.

(b) **Characterization of χ^{*2}.** The cumulative chi-squared χ^{*2} is most well characterized as a directional goodness-of-fit test of the assumed model. Assuming a balanced one-way layout with repetition number m and known σ^2, the cumulative chi-squared can be expressed as

$$\chi^{*2} = (m/\sigma^2) \|\boldsymbol{P}_a^{*\prime} \bar{\boldsymbol{y}}_{(i)\cdot}\|^2, \qquad (6.41)$$

where $\bar{\boldsymbol{y}}_{(i)\cdot} = (\bar{y}_{1\cdot}, \ldots, \bar{y}_{a\cdot})'$ is a column vector of the averages and

$$\boldsymbol{P}_a^{*\prime} = \boldsymbol{D}\left(\boldsymbol{D}_a'\boldsymbol{D}_a\right)^{-1}\boldsymbol{D}_a'$$

with $\boldsymbol{D}$ a diagonal matrix for standardization so that the squared norm of each row vector of $\boldsymbol{P}_a^{*\prime}$ is unity. $\boldsymbol{P}_a^{*\prime}\boldsymbol{P}_a^*$ coincides with $V(\boldsymbol{t})$ of (6.28) with $n_i \equiv m$ (balanced case). Now, the χ^{*2} can be expanded in the series of independent chi-squared variables as

$$\chi^{*2} = (m/\sigma^2) \sum_{i=1}^{a-1} \left\{ \frac{a}{i(i+1)\|\boldsymbol{p}_{ia}\|^2} \left(\boldsymbol{p}_{ia}' \bar{\boldsymbol{y}}_{(i)\cdot}\right)^2 \right\},$$

where $\boldsymbol{p}_{ia}$ is the ith eigenvector of $\boldsymbol{P}_a^{*\prime}\boldsymbol{P}_a^*$ corresponding to the eigenvalue $a/\{i(i+1)\}$. It is an a-dimensional column vector composed of

$$p_{i,a-1}(x) = \sum_{k=0}^{i} (-1)^k \binom{i}{k} \binom{i+k}{k} \binom{x}{k} \binom{a-1}{k}, \; x = 0, \ldots, a-1,$$

and represents Chebyshev's orthogonal polynomials of order i for a points. This means that the cumulative chi-squared is expanded in the form

$$\chi^{*2} = \left\{ \frac{a}{1\times 2}\chi^2_{(1)} + \frac{a}{2\times 3}\chi^2_{(2)} + \cdots + \frac{a}{(a-1)\times a}\chi^2_{(a-1)} \right\}, \tag{6.42}$$

where $\chi^2_{(1)}, \chi^2_{(2)}, \ldots$ are linear, quadratic, and so on chi-squared components each with df 1 and mutually independent, see Hirotsu (1986) for details. This suggests that the cumulative chi-squared is not only testing the monotone alternative but also an appropriate goodness-of-fit test statistic for a hypothesized model against a systematic departure, mainly but not exclusively linear. It is interesting that the statistic (6.42) has a quadratic component with weight one-third of the linear component, and the coefficients are rapidly decreasing. As a byproduct, it is found that the asymptotic value of f in the chi-squared approximation $\chi^{*2} \sim d\chi^2_f$ is

$$\lim_{a\to\infty} f = \left(\pi^2/3-3\right)^{-1} \doteqdot 3.45.$$

This should be contrasted with df 1 of the t-test for a linear trend and the infinite degrees of freedom of the F-test. The t-test corresponds to weight one for the first component and zero for all others, and the F-test to equal weights of unity for all components in the expansion (6.42). Thus, the cumulative chi-squared is known to be a nicely restricted statistic for the monotone hypothesis, whereas the F-test is omnibus and the t-test is too restricted. Finally, it should be stressed that χ^{*2} is much easier to handle and interpret compared with $\bar{\chi}^2$.

6.5.4 Convexity and slope change-point hypotheses

(1) **General theory**. The idea of Section 6.5.3 is further extended to the convexity hypothesis, which is also an essential shape constraint as the monotone hypothesis in the non-parametric dose–response analysis. Assume a one-way layout with repetition number n_i and consider the convexity hypothesis

$H_{\text{con}}: \mu_i - 2\mu_{i+1} + \mu_{i+2} \ge 0,\ i = 1, \ldots, a-2$ with at least one inequality strong.

This hypothesis can be rewritten in matrix form as

$$H_{\text{con}}: \boldsymbol{L}_a' \boldsymbol{\mu} \ge 0 \tag{6.43}$$

by a second-order differential matrix

$$\boldsymbol{L}_a' = \begin{bmatrix} 1 & -2 & 1 & 0 & 0 & \cdots & 0 & 0 & 0 \\ 0 & 1 & -2 & 1 & 0 & \cdots & 0 & 0 & 0 \\ \vdots & \vdots & \vdots & \vdots & \vdots & \cdots & \vdots & \vdots & \vdots \\ 0 & 0 & 0 & 0 & 0 & \cdots & 1 & 2 & 1 \end{bmatrix} \tag{6.44}$$

instead of the first-order differential matrix $\boldsymbol{D}_a'$ (6.24) in the previous section, where $\boldsymbol{\mu}$ is a column vector of μ_i's. The null model $\boldsymbol{L}_a'\boldsymbol{\mu}=0$ is obviously equivalent to a linear regression model $H_0:\mu_i=\beta_0+\beta_1 i,\ i=1,\ldots,a$, and the sufficient statistics under the null model are

$$Y_a=y_{1.}+\cdots+y_{a.} \text{ and } T_a=y_{1.}+2y_{2.}+\cdots+ay_{a.}. \tag{6.45}$$

In this case an explicit form of the key vector $\boldsymbol{t}=\left(\boldsymbol{L}_a'\boldsymbol{L}_a\right)^{-1}\boldsymbol{L}_a'\boldsymbol{y}_{(i).}$ is also obtained, where $\boldsymbol{y}_{(i).}$ is a column vector of $y_{i.}$ but its expression looks very complicated. However, it can be shown that the rows of $\left(\boldsymbol{L}_a'\boldsymbol{L}_a\right)^{-1}\boldsymbol{L}_a'\boldsymbol{y}_{(i).}$ represent the slope change-point contrasts and are essentially doubly accumulated statistics as follows. First, it is interesting to note that every $\boldsymbol{\mu}$ satisfying (6.43) can be expressed as

$$\boldsymbol{\mu}=(\boldsymbol{\Pi}_B+\boldsymbol{\Pi}_{L_a})\boldsymbol{\mu}=\boldsymbol{B}(\beta_0,\beta_1)'+\boldsymbol{L}_a\left(\boldsymbol{L}_a'\boldsymbol{L}_a\right)^{-1}\boldsymbol{h},\ \boldsymbol{h}\ge 0, \tag{6.46}$$

where β_0 and β_1 are arbitrary regression coefficients with $\boldsymbol{B}=\begin{bmatrix}1 & 1 & 1 & \cdots & 1\\ 1 & 2 & 3 & \cdots & a\end{bmatrix}'$. In equation (6.46), $\boldsymbol{\Pi}_B=\boldsymbol{B}(\boldsymbol{B}'\boldsymbol{B})^{-1}\boldsymbol{B}'$ and $\boldsymbol{\Pi}_{L_a}=\boldsymbol{L}_a\left(\boldsymbol{L}_a'\boldsymbol{L}_a\right)^{-1}\boldsymbol{L}_a'$ are projection matrices with order 2 and $a-2$, and orthogonal to each other so that $\boldsymbol{\Pi}_B+\boldsymbol{\Pi}_{L_a}$ is an identity matrix of order a. Then it can be shown that

$$\boldsymbol{L}_a\left(\boldsymbol{L}_a'\boldsymbol{L}_a\right)^{-1}=(\boldsymbol{b}_1,\boldsymbol{b}_2,\ldots,\boldsymbol{b}_{a-2}),$$
$$\boldsymbol{b}_k=(\boldsymbol{I}-\boldsymbol{\Pi}_B)(0\,0\cdots0\,1\,2\cdots a-k-1)',\ k=1,\ldots,a-2,$$

since we obviously have $\boldsymbol{L}_a'(0\,0\cdots0\,1\,2\cdots a-k-1)'=(0\cdots010\cdots0)'$ with unity as its kth element. Then it is easy to show that $\boldsymbol{B}(\beta_0,\beta_1)'+\boldsymbol{b}_k\Delta_k$ is one expression of the slope change-point model at time point $k+1$:

$$\mu_2-\mu_1=\cdots=\mu_{k+1}-\mu_k=\beta_k;\mu_{k+2}-\mu_{k+1}=\cdots=\mu_a-\mu_{a-1}=\beta_k+\Delta_k.$$

Therefore, the kth column of $\boldsymbol{L}_a\left(\boldsymbol{L}_a'\boldsymbol{L}_a\right)^{-1}$ in (6.46) represents a slope change-point model at time point $k+1$, $k=1,\ldots,a-2$, and orthogonal to the linear regression part $\boldsymbol{B}(\beta_0,\beta_1)'$. Thus, the rows of the key vector $\boldsymbol{t}=\left(\boldsymbol{L}_a'\boldsymbol{L}_a\right)^{-1}\boldsymbol{L}_a'\boldsymbol{y}_{(i).}$ represent the slope change-point contrasts, in contrast to the step change-point contrasts of the previous section. As an example, we give the case of $a=5$ below, where bold type denotes the slope change-point:

$$\boldsymbol{L}_5\left(\boldsymbol{L}_5'\boldsymbol{L}_5\right)^{-1}=\frac{1}{10}\begin{bmatrix}4 & 4 & 2\\ \mathbf{-4} & -1 & 0\\ -2 & \mathbf{-6} & -2\\ 0 & -1 & \mathbf{-4}\\ 2 & 4 & 4\end{bmatrix}_{5\times 3}.$$

Since the components of the key vector are linear functions of $\boldsymbol{y}_{(i)\cdot}$, we can develop a similar probability theory to the previous section, noting the second-order Markov property of the subsequent components in this case. However, this form is not very convenient for dealing with discrete models, since the linear functions can be negative and also take non-integer values. Therefore, a more convenient basic variable is searched for in the following. We obviously have

$$\boldsymbol{y}_{(i)\cdot} = (\boldsymbol{\Pi}_{L_a} + \boldsymbol{\Pi}_B)\boldsymbol{y}_{(i)\cdot} = \boldsymbol{L}_a\boldsymbol{t} + \boldsymbol{\Pi}_B\boldsymbol{y}_{(i)\cdot}.$$

Since $\boldsymbol{\Pi}_B\boldsymbol{y}_{(i)\cdot}$ is a function of the sufficient statistics (6.45), we can base our test on $\boldsymbol{s}$ satisfying $\boldsymbol{y}_{(i)\cdot} = \boldsymbol{L}_a\boldsymbol{s}$, $\boldsymbol{s} = (S_1, \ldots, S_{a-2})'$ instead of $\boldsymbol{t}$ discarding $\boldsymbol{\Pi}_B\boldsymbol{y}_{(i)\cdot}$, where we ignore the last two rows of the equality. The difference between $\boldsymbol{s}$ and $\boldsymbol{t}$ is only a function of the sufficient statistics (6.45). By accumulating both sides of the equation $\boldsymbol{y}_{(i)\cdot} = \boldsymbol{L}_a\boldsymbol{s}$, we have

$$Y_k = y_{1\cdot} + \ldots + y_{k\cdot} = S_k - S_{k-1},\ k = 1, \ldots, a-2,$$

where S_0 is defined to be 0. Therefore we have

$$S_k = \sum_{i=1}^{k} Y_i = ky_{1\cdot} + (k-1)y_{2\cdot} + \ldots + y_{k\cdot},\ k = 1, \ldots, a-2, \tag{6.47}$$

and call them the doubly accumulated statistics. This suggests that the method is a natural extension of max acc. $t1$, which is based on the singly accumulated statistics. This idea also makes it possible to develop a distribution theory for discrete models beyond the normal model in later sections (see also Hirotsu, 2013; Hirotsu *et al.*, 2016). Now, going back to the normal model, we apply the Corollary to Lemma 6.1 to $\bar{\boldsymbol{y}}_{(i)\cdot} \sim N(\boldsymbol{\mu}, \boldsymbol{\Omega})$ for testing H_{con} (6.43), where $\bar{\boldsymbol{y}}_{(i)\cdot} = (\bar{y}_{1\cdot}, \ldots, \bar{y}_{a\cdot})'$ and $\boldsymbol{\Omega} = \text{diag}(\sigma^2/n_i)$. Then, our basic standardized statistics $\boldsymbol{s}_k^* = \left(s_1^*, \ldots, s_{a-2}^*\right)'$ become explicit in the form

$$s_k^* = \text{diag}\left(m_k^{-1/2}\right)\left(\boldsymbol{L}_a'\boldsymbol{\Omega}\boldsymbol{L}_a\right)^{-1}\boldsymbol{L}_a'\bar{\boldsymbol{y}}_{(i)\cdot},\ k = 1, \ldots, a-2,$$

where

$$m_k = \boldsymbol{b}_k'\boldsymbol{\Omega}^{-1}\left\{\boldsymbol{\Omega} - \boldsymbol{B}\left(\boldsymbol{B}'\boldsymbol{\Omega}^{-1}\boldsymbol{B}\right)^{-1}\boldsymbol{B}'\right\}\boldsymbol{\Omega}^{-1}\boldsymbol{b}_k$$

is the kth diagonal element of the variance of $\left(\boldsymbol{L}_a'\boldsymbol{\Omega}\boldsymbol{L}_a\right)^{-1}\boldsymbol{L}_a'\bar{\boldsymbol{y}}_{(i)\cdot}$, namely $\left(\boldsymbol{L}_a'\boldsymbol{\Omega}\boldsymbol{L}_a\right)^{-1}$, which is also the conditional variance of $\left(\boldsymbol{L}_a'\boldsymbol{L}_a\right)^{-1}\boldsymbol{L}_a'\boldsymbol{\Omega}^{-1}\bar{\boldsymbol{y}}_{(i)\cdot} = \left(\boldsymbol{L}_a'\boldsymbol{L}_a\right)^{-1}\boldsymbol{L}_a'\boldsymbol{y}_{(i)\cdot}/\sigma^2$. Thus, s_k^* is the standardized version of S_k with mean 0 and variance 1 under the null model $\boldsymbol{L}_a'\boldsymbol{\mu} = 0$, which corresponds to t_k of (6.27) in Section 6.5.3 (1). Then, two promising statistics are again the maximal contrast statistic max s_k^* and the cumulative chi-squared $\chi^{\dagger 2}$ defined in (5) of this section. We call the former max acc. $t2$ in contrast to max acc. $t1$, which is based on singly accumulated statistics Y_k. It is naturally expected to inherit many of the good properties of max acc. $t1$, which have been proved in various situations.

(2) **Max acc. $t2$ for testing the convexity hypothesis**. The doubly accumulated statistics of independent variables obviously possess the second-order Markov property. After conditioning by the sufficient statistics, the component statistics s_k^* still have that property. Now, the covariance matrix of $\boldsymbol{s}_k^*$ is

$$\mathrm{diag}\left(m_k^{-1/2}\right)\left(\boldsymbol{L}_a'\boldsymbol{\Omega}\boldsymbol{L}_a\right)^{-1}\mathrm{diag}\left(m_k^{-1/2}\right). \tag{6.48}$$

From the form of equation (6.44) and since $\boldsymbol{\Omega}$ is a diagonal matrix, it is obvious that the covariance matrix is an inverse of a penta-diagonal matrix. Then, by similar arguments to Lemma 6.5 that connect the tri-diagonal of an inverse of a covariance matrix to the first-order Markov property, the second-order Markov property of the sequence of $s_1^*, \ldots, s_{a-2}^*$ follows immediately. Then we can give an exact and efficient recursion formula for calculating the distribution function of max s_k^* just as for max acc. $t1$ in Section 6.5.3 (1) (d). Define the conditional probability

$$F_{k+1}\left(s_k^*, s_{k+1}^*, s_0|\sigma\right) = \Pr\left(s_1^* < s_0, \ldots, s_k^* < s_0, s_{k+1}^* < s_0 | s_k^*, s_{k+1}^*, \sigma\right),$$

where s_{a-1}^* and s_a^* are defined to be $-\infty$, although their conditional variances are zero, so that the inequalities related to them hold always. Then we have a recursion formula

$$F_{k+2}\left(s_{k+1}^*, s_{k+2}^*, s_0|\sigma\right) = \Pr\left(s_1^* < s_0, \ldots, s_{k+1}^* < s_0, s_{k+2}^* < s_0 | s_{k+1}^*, s_{k+2}^*, \sigma\right) =$$

$$\int_{s_k^*} \Pr\left(s_1^* < s_0, \ldots, s_k^* < s_0, s_{k+1}^* < s_0, s_{k+2}^* < s_0 | s_k^*, s_{k+1}^*, s_{k+2}^*, \sigma\right) \times f_k\left(s_k^* | s_{k+1}^*, s_{k+2}^*\right) \mathrm{d}s_k^* \tag{6.49}$$

$$= \begin{cases} \displaystyle\int_{s_k^*} F_{k+1}\left(s_k^*, s_{k+1}^*, s_0|\sigma\right) \times f_k\left(s_k^* | s_{k+1}^*, s_{k+2}^*\right) \mathrm{d}s_k^*, & \text{if } s_{k+2}^* < s_0 \\ 0, \text{ otherwise.} \end{cases} \tag{6.50}$$

Equation (6.49) is due to the law of total probability, and equation (6.50) is due to the second-order Markov property. The conditional distribution $f_k\left(s_k^*|s_{k+1}^*, s_{k+2}^*\right)$ is easily obtained from the correlation structure (6.48) by applying Lemma 6.4. However, the last two steps of the recursion formula need some caution, since s_k^* is defined for $1 \le k \le a-2$, and s_{a-1}^* and s_a^* are the constants. We can deal with this simply by extending the correlation $\{\tau_{ij}\}$ among s_k^*'s up to $1 \le i, j \le a$ by defining $\tau_{ij} = 0$ if i and/or j is equal to $a-1$ or a. By this definition, $f_{a-2}\left(s_{a-2}^*|s_{a-1}^*, s_a^*\right)$ is nothing but an unconditional distribution of s_{a-2}^*, for example. Thus we can have

$$p = 1 - F_a\left(s_{a-1}^*, s_a^*, s_0|\sigma\right)$$

for max s_k^* at the final step.

In the case of unknown σ, we need to replace it by the usual estimate $\hat{\sigma}$ and take expectations with respect to $\hat{\sigma}$. This is equivalent to replacing s_0 by $\sqrt{\chi_\nu^2/\nu}\,s_0$ in

(6.50) and taking expectations with respect to χ_ν, where χ_ν is a chi-variable with df ν. Refer to Hirotsu and Marumo (2002) for more details.

(3) **Max acc. *t*2 as an efficient score test of the slope change-point model**. The *k*th component of max acc. *t*2 is nothing but the efficient score for testing the slope change-point model at $i = k+1$, which is defined as

$$M_k: \begin{cases} \mu_i = \beta_0 + \beta_1 i, \ i = 1, \ldots, k+1, \\ \mu_i = \beta_0^* + \beta_1^* i, \ i = k+2, \ldots, a, \end{cases}$$

where $\beta_0 + \beta_1(k+1) = \beta_0^* + \beta_1^*(k+1)$. Define $\Delta = \beta_1^* - \beta_1$, then the model M_k can be rewritten as

$$\begin{cases} \mu_i = \beta_0^* + \Delta(k+1) + (\beta_1^* - \Delta)i, \ i = 1, \ldots, k+1, \\ \mu_i = \beta_0^* + \beta_1^* i, \ i = k+2, \ldots, a. \end{cases}$$

Assuming the independent normal model, we have a log likelihood function

$$\log L = \text{const.} - \frac{1}{(2\sigma^2)}\left[\sum_{i=1}^{k+1}\sum_{j=1}^{n_i}\{y_{ij} - \beta_0^* - \Delta(k+1) - (\beta_1^* - \Delta)i\}^2 + \sum_{i=k+2}^{a}\sum_{j=1}^{n_i}(y_{ij} - \beta_0^* - \beta_1^* i)^2\right].$$

Then it is easy to verify that the efficient score with respect to Δ evaluated at the null hypothesis is

$$\frac{\partial \log L}{\partial \Delta}\Big|_{\Delta=0} = \frac{1}{\sigma^2}\left\{\sum_{i=1}^{k+1}\sum_{j=1}^{n_i}(y_{ij} - \beta_0^* - \beta_1^* i)(k+1-i)\right\}. \tag{6.51}$$

The essential part of (6.51) is nothing but

$$S_k = \sum_{i=1}^{k+1}(k+1-i)y_{i\cdot} = ky_{1\cdot} + (k-1)y_{2\cdot} + \cdots + y_{k\cdot}$$

of (6.47). Therefore, max acc. *t*2 is appropriate also for testing the slope change-point hypothesis M_k at unknown k, $k = 1, \ldots, a-2$.

(4) **Power comparisons with maximin linear and polynomial tests**. Max acc. *t*1 for the monotone hypothesis has been verified to keep high power for the wide range of simple ordered alternatives. However, it has been pointed out that these maximal contrast type tests will not be so useful if the maximal angle of the polyhedral cone representing the restricted alternative is large. In particular, an Abelson and Tukey (1963) type maximin linear test is said to be useful only for the limited case in Robertson *et al.* (1988, Sec. 4.2~4.4), although it is recommended in the book of BANOVA by Miller (1998) to escape from the complicated restricted maximum likelihood approach. Now, it is a matter of simple algebra to show that the cosines of the maximum angle are $1/(a-1)$ and $2/(a-1)$ for the monotone and concavity hypotheses, respectively. This suggests that the max acc. *t*2 introduced in this section is even more appropriate

as a directional test than max acc. *t*1 for the monotone hypothesis. It suggests, however, that the maximin linear test might also do as well for the convexity or concavity hypothesis. Therefore, for $a=6$ and 8 of the balanced case, the maximin linear tests are searched for in Hirotsu and Marumo (2002) on the corners, edges, and faces of the polyhedral cone according to Abelson and Tukey (1963) to obtain the coefficients for the concavity test

$$a=6: -0.5773,\ 0.2829,\ 0.2944,\ 0.2944,\ 0.2829,\ -0.5773,$$

$$a=8: -0.6108, 0.1673, 0.2036, 0.2399, 0.2399, 0.2036, 0.1673, -0.6108.$$

For comparison, we add a simple linear test with coefficients of quadratic pattern

$$a=6: -5,\ 1,\ 4,\ 4,\ 1,\ -5,$$

$$a=8: -7, -1, 3, 5, 5, 3, -1, -7,$$

which we call a polynomial test according to Hirotsu and Marumo (2002).

In Table 6.14 the powers are compared in the directions of the corner vectors and also in a quadratic pattern, where the non-centrality parameter $m\sum(\mu_i-\bar{\mu}_.)^2/\sigma^2$ is fixed at 6 so that the powers are around 0.70. It should be noted that the polynomial test is the most powerful test against the quadratic pattern alternative, giving the upper bound for all available tests. Max acc. *t*2 is seen to keep relatively high power in the wide range of the concavity hypothesis compared with linear tests. The

Table 6.14 Power comparisons of max acc. *t*2, maximin, and polynomial linear tests.

	Alternative hypothesis ($\boldsymbol{\mu}$)								Max acc. *t*2	Maximin linear	Polynomial test
	−10	8	5	2	−1	−4			0.698	0.657	0.615
	−20	2	24	11	−2	−15			0.721	0.657	0.747
$a=6$	−15	−2	11	24	2	−20			0.721	0.657	0.747
	−4	−1	2	5	8	−10			0.698	0.657	0.615
	Quadratic pattern								0.747	0.751	0.790
	−7	4	3	2	1	0	−1	−2	0.674	0.623	0.535
	−70	−5	60	41	22	3	−16	−35	0.702	0.626	0.696
	−35	−10	15	40	23	6	−11	−28	0.710	0.623	0.755
$a=8$	−28	−11	6	23	40	15	−10	−35	0.710	0.623	0.755
	−35	−16	3	22	41	60	−5	−70	0.702	0.626	0.696
	−2	−1	0	1	2	3	4	−7	0.674	0.623	0.535
	Quadratic pattern								0.737	0.726	0.790

polynomial test looks very good when the change-point is located in the middle but not when it is at the end, so it cannot be recommended generally. Another advantage of max acc. *t*2 is that it can suggest a change- point. The maximin linear test looks generally no good.

(5) **Cumulative chi-squared $\chi^{\dagger 2}$ for a goodness-of-fit test.** According to a complete class lemma by Hirotsu (1982), again an appropriate statistic for testing the two-sided version of H_{con} is

$$\chi^{\dagger 2} = m\|\boldsymbol{P}_a^{\dagger\prime}\bar{\boldsymbol{y}}_{(i)\cdot}\|^2,$$

$$\boldsymbol{P}_a^{\dagger\prime} = \boldsymbol{D}\left(\boldsymbol{L}_a'\boldsymbol{L}_a\right)^{-1}\boldsymbol{L}_a',$$

where we assume equal repetition number m and $\boldsymbol{D} = \text{diag}\left(\xi_k^{1/2}\delta_k\right)$ is for standardization with $\xi_k - 1 = 2(k-1)(a-k-2)/(3(a-1))$ and $\delta_k = -[6a(a^2-1)/k(k+1)(a-k)(a-k-1)\{(a-1)(2k+1)-2(k^2-1)\}]^{1/2}$, $k = 1, \ldots, a-2$. The statistic $\chi^{\dagger 2}$ is based on the doubly accumulated statistics (6.47), suggesting that it is a natural extension of χ^{*2} (6.41). It can be expanded just like equation (6.42) for χ^{*2} in the form

$$\chi^{\dagger 2} = \left\{\frac{2a(a+1)}{1\times 2\times 3\times 4}\chi_{(2)}^2 + \frac{2a(a+1)}{2\times 3\times 4\times 5}\chi_{(3)}^2 + \cdots + \frac{2a(a+1)}{(a-2)(a-1)a(a+1)}\chi_{(a-1)}^2\right\}\sigma^2. \tag{6.52}$$

The asymptotic value of f, the degrees of freedom of the approximated chi-squared distribution at $a = \infty$, is 1.70 suggesting that $\chi^{\dagger 2}$ is a more strongly directed statistic than χ^{*2}. Therefore, it is naturally expected to inherit the excellent properties of χ^{*2} as a directional test. Interesting applications of max acc. *t*2 and cumulative chi-squared $\chi^{\dagger 2}$ are given in the examples of Sections 7.3.2, 7.3.3, and 13.2.

6.5.5 Sigmoid and inflection point hypotheses

(1) **General theory**. The idea of previous sections is further extended to the sigmoid hypothesis, which is also an essential shape constraint in the non-parametric dose–response analysis. The sigmoid hypothesis is defined by

$$H_{\text{sig}}: \mu_3 - 2\mu_2 + \mu_1 \geq \cdots \geq \mu_a - 2\mu_{a-1} + \mu_{a-2}, \text{ with at least one inequality strong,} \tag{6.53}$$

extending the convexity hypothesis of previous sections. The null hypothesis is defined by all the equalities in (6.53), which is equivalent to $H_0: \beta_0 + \beta_1 i + \beta_2 i^2$. The sufficient statistics under the null model are $\sum_{i=1}^a y_{i\cdot}$, $\sum_{i=1}^a (iy_{i\cdot}), \sum_{i=1}^a (i^2 y_{i\cdot})$. Equation (6.53) can be expressed in matrix form as $\boldsymbol{Q}_a'\boldsymbol{\mu} \geq 0$, just like equation

(6.43) for the convexity hypothesis H_{con}. $\boldsymbol{Q}_a'$ is a third-order differential matrix and its explicit form and relationship with the inflection-point model are given by Hirotsu and Marumo (2002). Then, by similar arguments to the previous section, triply accumulated statistics are derived as the basic variables. They are expressed in terms of doubly accumulated statistics as

$$W_k = \sum_{i=1}^{k} S_i,\ k = 1, \ldots, a-3. \tag{6.54}$$

Then, a set of sufficient statistics can be expressed as Y_a, S_{a-1}, and W_{a-2}. Let w_k^* be the standardized version of W_k, by the mean and variance under the null model: $\boldsymbol{Q}_a'\boldsymbol{\mu} = 0$. Two promising statistics are again the maximal contrast statistic max w_k^* and the cumulative chi-squared $\chi^{\#2} = \sum_{k=1}^{a-2} w_k^{*2}$. We call the former max acc. *t*3, in contrast to max acc. *t*1 and *t*2, which are based on singly and doubly accumulated statistics Y_k and S_k, respectively.

(2) **Max acc. *t*3 for testing the sigmoid hypothesis.** The triply accumulated statistics w_k^* possess the third-order Markov property under the conditional distribution given the sufficient statistics. Then we can give an exact and efficient recursion formula for probability calculation of max w_k^*, just as for max acc. *t*1 and max acc. *t*2 in previous sections. Define the conditional probability

$$F_{k+2}\left(w_k^*, w_{k+1}^*, w_{k+2}^*, w_0 | \sigma\right) = \Pr\left(w_1^* < w_0, \ldots, w_{k+1}^* < w_0, w_{k+2}^* < w_0 | w_k^*, w_{k+1}^*, w_{k+2}^*, \sigma\right).$$

Then we have a recursion formula

$$F_{k+3}\left(w_{k+1}^*, w_{k+2}^*, w_{k+3}^*, w_0 | \sigma\right) = \Pr\left(w_1^* < w_0, \ldots, w_{k+2}^* < w_0, w_{k+3}^* < w_0 | w_{k+1}^*, w_{k+2}^*, w_{k+3}^*, \sigma\right) =$$
$$\int_{w_k^*} \Pr\left(w_1^* < w_0, \ldots, w_{k+2}^* < w_0, w_{k+3}^* < w_0 | w_k^*, w_{k+1}^*, w_{k+2}^*, w_{k+3}^*, \sigma\right) \times f_k\left(w_k^* | w_{k+1}^*, w_{k+2}^*, w_{k+3}^*\right) dw_k^* \tag{6.55}$$

$$= \begin{cases} \int_{w_k^*} F_{k+2}\left(w_k^*, w_{k+1}^*, w_{k+2}^*, w_0 | \sigma\right) \times f_k\left(w_k^* | w_{k+1}^*, w_{k+2}^*, w_{k+3}^*\right) dw_k^*, & \text{if } w_{k+3}^* < w_0, \\ 0, & \text{otherwise.} \end{cases} \tag{6.56}$$

Equation (6.55) is due to the law of total probability, and equation (6.56) is due to the third-order Markov property of w_k^*, $k = 1, \ldots, a-3$. The *p*-value is obtained as

$$p = 1 - F_a\left(w_{a-2}^*, w_{a-1}^*, w_a^*, w_0 | \sigma\right),$$

where the triplet $(w_{a-2}^*, w_{a-1}^*, w_a^*)$ is defined appropriately so as to satisfy the inequality, always making the distribution f_{a-3} actually unconditional. In the case of unknown σ, an expectation with respect to $\hat{\sigma}$ is required just as in previous sections. One should refer to the original paper of Hirotsu and Marumo (2002) for more detailed calculations.

(3) **Power comparisons with maximin linear and polynomial tests**. Now it is a matter of simple algebra to show that the cosine of the maximum angle is $3/(a-1)$ for the sigmoid hypothesis, which is even larger than those of the monotone and convexity hypotheses. This suggests that max acc. *t*3 is even more appropriate than max acc. *t*2 as a directional test. It suggests again, however, that the maximin linear test might also do as well and need to be compared. The coefficients of the maximin linear test for the sigmoid hypothesis for $a=6$ and 8 of the balanced case are as follows:

$$a=6: 2,\ -3,\ -1,\ 1,\ 3,\ -2,$$

$$a=8: 5,\ -5,\ -3,\ -1,\ 1,\ 3,\ 5,\ -5.$$

A polynomial test with coefficients of cubic pattern is given by

$$a=6: 5,\ -7,\ -4,\ 4,\ 7,\ -5$$

$$a=8: 7,\ -5,\ -7,\ -3,\ 3,\ 7,\ 5,\ -7.$$

In Table 6.15 the powers are compared in the directions of the corner vectors and also in a cubic pattern, where the non-centrality parameter $m\sum(\mu_i-\bar{\mu}.)^2/\sigma^2$ is fixed again at 6 so that the powers are around 0.70. It should be noted that the polynomial test is the most powerful test against the cubic pattern alternative, and its power gives

Table 6.15 Power comparisons of max acc. *t*3, maximin, and polynomial linear tests.

	Alternative hypothesis ($\boldsymbol{\mu}$)								Max acc. *t*3	Maximin linear	Polynomial test
	3	−3	−4	0	9	−5			0.730	0.707	0.697
$a=6$	15	−19	−18	18	19	−15			0.748	0.737	0.774
	5	−9	0	4	3	−3			0.730	0.707	0.697
	Cubic pattern								0.763	0.780	0.790
	7	−9	−3	1	3	3	1	−3	0.705	0.665	0.621
	21	−17	−24	0	13	15	6	−14	0.730	0.689	0.745
$a=8$	7	−4	−8	−5	5	8	4	−7	0.735	0.690	0.776
	14	−6	−15	−13	0	24	17	−21	0.730	0.689	0.745
	3	−1	−3	−3	−1	3	9	−7	0.705	0.665	0.621
	Cubic pattern								0.751	0.748	0.790

the upper bound for all available tests. Max acc. *t*3 is seen to keep relatively high power in the wide range of the sigmoid hypothesis compared with linear tests. The polynomial test looks very good when the change-point is located in the middle but not when it is at the end, so it cannot be recommended generally. The maximin linear test looks generally no good again.

(4) **Cumulative chi-squared statistic $\chi^{\#2}$ for a goodness-of-fit test.** According to a complete class lemma by Hirotsu (1982), again the appropriate statistic for testing the two-sided version of H_{sig} (6.53) is

$$\chi^{\#2} = m\|\boldsymbol{D}(\boldsymbol{Q}_a'\boldsymbol{Q}_a)^{-1}\boldsymbol{Q}_a'\bar{\boldsymbol{y}}_{(i)\cdot}\|^2,$$

where we assume equal repetition number m and $\boldsymbol{D}$ is a matrix for a standardization producing w_k^*. The statistic $\chi^{\#2}$ is based on the triply accumulated statistics (6.54), suggesting that it is a natural extension of χ^{*2} and $\chi^{\dagger 2}$.

6.5.6 Discussion

Since Page (1954, 1961), cumulative sum statistics based approaches have been widely developed in the statistical process control; see also Montgomery (2012) for an explanatory example and related topics. More recently, these have also been extended to the field of environmental statistics (for example, by Manly and Mackenzie, 2003). However, as stated in the review paper of Amiri and Allahyari (2012), most papers assume step, linear trend, and monotonic changes, and it seems that the slope change-point and inflection point models are not popular in these fields. The convex and sigmoid restrictions of this chapter are closely related to the slope change-point and inflection point models, for which the doubly and triply accumulated statistics are newly developed in this book. The basic statistics are so simple that they can be extended to discrete models almost as they are. These include the binomial and Poisson distributions of Chapters 7 and 8, respectively, and interesting applications beyond the statistical process control are given there. They are extended further to two-way interaction problems in Chapters 10, 11, and 13.

References

Abelson, P. R. and Tukey, J. W. (1963) Efficient utilization of non-numerical information in quantitative analysis: General theory and the case of simple order. *Ann. Math. Statist.* **34**, 1347–1369.

Amiri, A. and Allahyari, S. (2012) Change point estimation methods for control chart postsignal diagnostics: A literature review. *Qual. Rel. Eng. Int.* **28**, 673–685.

Armitage, P. and Palmer, M. (1986) Some approaches to the problem of multiplicities in clinical trials. *Proc. 23rd Int. Biometrics Conf., Invited Papers*, 1–15.

Barlow, R. E., Bartholomew, D. J., Bremner, J. M. and Brunk, H. D. (1972) *Statistical inference under order restrictions*. Wiley, New York.

Bartholomew, D. J. (1959a) A test of homogeneity for ordered alternatives. *Biometrika* **46**, 36–48.

Bartholomew, D. J. (1959b) A test of homogeneity for ordered alternatives II. *Biometrika* **46**, 328–335.

Birnbaum, A. (1955) Characterization of complete classes tests of some multi-parametric hypotheses, with applications to likelihood ratio tests. *Ann. Math. Statist.* **26**, 21–36.

Bretz, F., Hothorn, T. and Westfall, P. (2011) *Multiple comparisons using R*. Chapman & Hall/CRC, London.

Cochran, W. G. (1941) The distribution of the largest of a set of estimated variances as a fraction of their total. *Ann. Eugen.* **11**, 47–52.

Cohen, A. and Sackrowitz, H. B. (1991) Tests for independence in contingency tables with ordered categories. *J. Multivar. Anal.* **36**, 56–67.

Dunnett, C. W. (1964) Comparing several treatments with a control. *Biometrics* **20**, 482–491.

Furukawa, T. and Tango, T. (1993) *A new edition of the statistics to medical science*. Asakura Shoten Co., Tokyo (in Japanese).

Hartley, H. O. (1950) The maximum *F* ratio as a short-cut test for heterogeneity of variance. *Biometrika* **37**, 308–312.

Hayter, A. J. (1984) A proof of the conjecture that the Tukey–Kramer multiple comparisons procedure is conservative. *Ann. Statist.* **12**, 61–75.

Hayter, A. J. (1990) A one-sided Studentized range test for testing against a simple ordered alternative. *J. Amer. Statist. Assoc.* **85**, 778–785.

Hirotsu, C. (1976) *Analysis of variance*. Kyoiku-Shuppan, Tokyo (in Japanese).

Hirotsu, C. (1979a) An *F* approximation and its application. *Biometrika* **66**, 577–584.

Hirotsu, C. (1979b) The cumulative chi-squares method and a Studentized maximal contrast method for testing an ordered alternative in a one-way analysis of variance model. *Rep. Statist. Appl. Res. JUSE* **26**, 12–21.

Hirotsu, C. (1982) Use of cumulative efficient scores for testing ordered alternatives in discrete models. *Biometrika* **69**, 567–577.

Hirotsu, C. (1986) Cumulative chi-squared statistic as a tool for testing goodness of fit. *Biometrika* **73**, 165–173.

Hirotsu, C. (1997) Two-way change point model and its application. *Austral. J. Statist.* **39**, 205–218.

Hirotsu, C. (2013) Theory and application of cumulative, two-way cumulative and doubly cumulative statistics. *Jap. J. Appl. Statist.* **42**, 121–143 (in Japanese).

Hirotsu, C. and Marumo, K. (2002) Change point analysis as a method for isotonic inference. *Scand. J. Statist.* **29**, 125–138.

Hirotsu, C. and Srivastava, M. S. (2000) Simultaneous confidence intervals based on one-sided max *t* test. *Statist. Prob. Lett.* **49**, 25–37.

Hirotsu, C., Kuriki, S. and Hayter, A. J. (1992) Multiple comparison procedures based on the maximal component of the cumulative chi-squared statistic. *Biometrika* **79**, 381–392.

Hirotsu, C., Nishihara, K. and Sugihara, M. (1997) Algorithm for p-value, power and sample size calculation for max t test. *Jap. J. Appl. Statist.* **26**, 1–16 (in Japanese).

Hirotsu, C., Yamamoto, S. and Hothorn, L. (2011) Estimating the dose–response pattern by the maximal contrast type test approach. *Statist. Biopharm. Res.* **3**, 40–53.

Hirotsu, C., Yamamoto, S. and Tsuruta, H. (2016) A unifying approach to the shape and change-point hypotheses in the discrete univariate exponential family. *Comput. Statist. Data Anal.* **97**, 33–46.

Hochberg, Y. and Tamhane, A. C. (1987) *Multiple comparison procedures*. Wiley, New York.

Holm, S. A. (1979) A simple sequentially rejective multiple test procedure. *Scand. J. Statist.* **6**, 65–70.

Hsu, J. C. and Berger, R. L. (1999) Stepwise confidence intervals without multiplicity adjustment for dose–response and toxity studies. *J. Amer. Statist. Assoc.* **94**, 468–482.

Liu, W., Miwa, T. and Hayter, A. J. (2000) Simultaneous confidence interval estimation for successive comparisons of ordered treatment effects. *J. Statist. Plan. Inf.* **88**, 75–86.

Manly, B. F. J. and Mackenzie, D. (2003) CUSUM environmental monitoring in time and space. *Environ. Statist.* **10**, 231–247.

Marcus, R. (1976) The powers of some tests of the equality of normal means against an ordered alternative. *Biometrika* **63**, 177–183.

Marcus, R., Peritz, E. and Gabriel, K. R. (1976) On closed testing procedures with special reference to ordered analysis of variance. *Biometrika* **63**, 655–660.

Maurer, W., Hothorn, L. A. and Lehmacher, W. (1995) Multiple comparisons in drug clinical trials and preclinical assays: A-priori ordered hypotheses. In Vollmar, J. (ed.), *Biometrie in der Chemisch-Pharmazeutischen Industrie* **6**. Fischer Verlag, Stuttgart, 3–18.

Miller, R. G. (1998) *BEYOND ANOVA: Basics of applied statistics*. Chapman & Hall/CRC Texts in Statistical Science, New York.

Montgomery, D. C. (2012) *Introduction to statistical quality control*, 6th edn. Wiley, New York.

Moriguti, S. (ed.) (1976) *Statistical methods*, new edn. Japanese Standards Association, Tokyo (in Japanese).

Page, E. S. 1954. Continuous inspection schemes. *Biometrika* **41**, 1–9.

Page, E. S. 1961. Cumulative sum control charts. *Technometrics* **3**, 1–19.

Robertson, T., Wright, F. T. and Dykstra, R. L. (1988) *Order restricted statistical inference*. Wiley, New York.

Schaffer, J. P. (1986) Modified sequentially rejective multiple test procedures. *J. Amer. Statist. Assoc.* **81**, 826–831.

Scheffé, H. (1953) A method for judging all contrasts in the analysis of variance. *Biometrika* **40**, 87–104.

Takeuchi, K. (1979) Test and estimation problems under restricted null and alternative hypotheses. *J. Econ.* **45**, 2–10 (in Japanese).

Wald, A. (1950) *Statistical decision functions*. Wiley, New York.

Williams, D. A. (1971) A test for differences between treatment means when several dose levels are compared with a zero dose control. *Biometrics* **27**, 103–117.

7

One-Way Layout, Binomial Populations

7.1 Introduction

The data of binomial, or more generally multinomial, distributions from a one-way layout are presented as the two-way data of treatment × categorical response, as seen in Tables 3.6 and 11.8 for example. In those tables we assume a multinomial distribution $M(n_i, \boldsymbol{p}_i), \boldsymbol{p}_i = (p_{i1}, \ldots, p_{ib})', p_{i\cdot} = 1$ for the ith row, $i = 1, \ldots, a$, where $a = b = 2$ in Table 3.6. Then, the null hypothesis of homogeneity of treatment effects is expressed as

$$H_0 : \boldsymbol{p}_1 = \cdots = \boldsymbol{p}_a. \tag{7.1}$$

In contrast, in Example 3.5 we have introduced the test of independence hypothesis

$$H_0' : p_{ij} = p_{i\cdot} \times p_{\cdot j},\ i = 1, \ldots, a, j = 1, \ldots, b \tag{7.2}$$

assuming an overall multinomial distribution with $p_{\cdot\cdot} = 1$. We first show that these two formulations are mathematically equivalent.

Equation (7.2) is equivalent to

$$p_{ij} / p_{i\cdot} = p_{\cdot j},\ i = 1, \ldots, a, j = 1, \ldots, b$$

or

$$(p_{i1} / p_{i\cdot}, \ldots, p_{ib} / p_{i\cdot}) \text{ are the same for all } i = 1, \ldots, a. \tag{7.3}$$

Advanced Analysis of Variance, First Edition. Chihiro Hirotsu.

However, equation (7.3) is equivalent to equation (7.1), where the p_{ij} are standardized to satisfy $p_{i\cdot} = 1$. Equation (7.3) implies the response profiles of a rows are the same, which is the same as declaring no treatment effect. Thus, the null hypothesis of homogeneity of treatment effects is equivalent to the independence hypothesis in a two-way contingency table. Therefore, the likelihood ratio test and the goodness-of-fit chi-squared test derived in Example 3.5 are also useful for the overall homogeneity test of treatment effects. It should be noted that these tests are asymptotically equivalent to conditional tests given all the row and column marginal totals, so that the difference in sampling schemes is totally irrelevant. We have already seen in Example 5.13 that the goodness-of-fit chi-squared test is exactly the same with the test of equality of two binomial probabilities by normal approximation in the case $a = b = 2$.

Also, it should be noted that the null hypothesis of independence (7.2) can be rewritten as

$$\log p_{ij} = \log p_{i\cdot} + \log p_{\cdot j}.$$

This is equivalent to the null hypothesis of no interaction in the log linear model

$$\log p_{ij} = \mu + \alpha_i + \beta_j + (\alpha\beta)_{ij}.$$

Therefore, testing the homogeneity of treatment effects is nothing but testing the interaction effects as mentioned in Section 1.7. Thus, the topics of this chapter will be more generally dealt with as an analysis of two-way categorical data. In the meantime, however, the simplest case of the binomial distribution can also be analyzed as one-way data. Therefore, we discuss in this chapter only the multiple comparisons and directional tests of binomial distributions.

7.2 Multiple Comparisons

The multiple comparisons of binomial distributions by the normal approximation are straightforward. We assume a binomial distribution $B(n_i, p_i)$ for the ith observation y_i. Then, the average $\bar{y}_{i\cdot}$ in Section 6.4.2 is y_i/n_i here and the basic statistic for paired comparisons is $y_i/n_i - y_j/n_j$. We use

$$\hat{p}(1-\hat{p}) = \left(\sum_i y_i / \sum_i n_i\right)\left\{1 - \left(\sum_i y_i / \sum_i n_i\right)\right\} \tag{7.4}$$

as a variance estimator under the null hypothesis of homogeneity of p_i, which corresponds to $\hat{\sigma}^2$ (6.6). Then, the procedures 6.4.2 (1) ~ (3) can be applied as they are by replacing $\hat{\sigma}^2$ with (7.4).

7.3 Directional Tests

7.3.1 Monotone and step change-point hypotheses

(1) **Max acc. *t*1 in phase II clinical trial for dose finding**

(a) **Max acc. *t*1 in the binomial case**. We give the result of a typical phase II clinical trial for dose finding in Table 7.1, where the last three columns are for calculation of the test statistic.

For the data, the most well-known approach will be that of the Cochran (1955) and Armitage (1955) test. However, it assumes a linear trend as a principal dose–response and cannot reserve high power against the wide range of the monotone hypothesis, in particular against very early or late start-up. Further, it cannot suggest an optimal dose even if it could detect a significant departure from the null hypothesis. Then, the most appropriate approach to prove the effective dose–response suggesting simultaneously an optimal dose level will again be max acc. $t1$. We assume a binomial distribution $B(n_i, p_i)$ for the observed number of 'yes' (y_i) and apply Lemma 6.2 for the monotone hypothesis H_{mon} (6.23) in the natural parameter $\theta_i = \text{logit}\, p_i = \log\{p_i/(1-p_i)\}$. It should be noted that a monotone hypothesis is unchanged by this transformation from an expectation parameter p_i to a natural parameter θ_i. We have already shown that the key vector in this case is formed by the accumulated statistics in Example 6.5. Then, the standardized accumulated statistics corresponding to (6.27) of normal case is

$$t_k(Y_k) = t_k = \left\{ \left(\frac{N_a - N_k}{N_a N_k} \right) \hat{p}\hat{q} \right\}^{-1/2} \left(\hat{p} - \frac{Y_k}{N_k} \right), \; k = 1, \ldots, a-1, \tag{7.5}$$

Table 7.1 Improvement rate of a drug for heart disease.

	Improved					
Dosage (mg)	No	Yes	Total	N_k	Y_k	$t_k(Y_k)$
100	16	20	36	36	20	
150	18	23	41	77	43	1.319
200	9	27	36	113	70	2.276
225	9	26	35	148	96	1.161
300	5	9	14	162	105	−0.043
Total	57	105	162			

where $Y_k = y_1 + \ldots + y_k, N_k = n_1 + \ldots + n_k, \hat{p} = Y_a/N_a, \hat{q} = 1 - \hat{p}$. It should be noted that Y_a is the sufficient statistic under the null model $\theta_1 = \cdots = \theta_a$, and the conditional null distribution of Y_k given Y_a is a hypergeometric distribution

$$H(Y_k \mid Y_a, N_k, N_a) \sim \binom{Y_a}{Y_k}\binom{N_a - Y_a}{N_k - Y_k} \Big/ \binom{N_a}{N_k} \tag{7.6}$$

with parameters (Y_a, N_k, N_a). The exact variance of Y_k is $\{N_k(N_a - N_k)/\ (N_a - 1)\}\hat{p}\hat{q}$ and therefore the variance employed for standardization in (7.5) is $(N_a - 1)/N_a$ times an under-estimate, but its effect should be practically negligible.

(b) **Markov property**. In this case the Markov property of the sequence $t_1, \ldots, t_{a-1}$ is shown by the form of the joint conditional distribution of $Y_1, \ldots, Y_{a-1}$ given Y_a instead of the covariance structure in the normal case. Now it is well known that the joint null distribution of y_k is factorized in the form

$$\Pi_1^a \binom{n_k}{y_k} p^{\sum y_k}(1-p)^{\sum (n_k - y_k)} = \Pi_1^{a-1} H(Y_k \mid Y_{k+1}, N_k, N_{k+1}) \times \binom{N_a}{Y_a} p^{Y_a}(1-p)^{N_a - Y_a}, \tag{7.7}$$

where $H(Y_k \mid Y_{k+1}, N_k, N_{k+1})$ is the probability function of the hypergeometric distribution (7.6) and the last part is the probability function of the binomial distribution $B(N_a, p)$ for Y_a. Then, the conditional joint distribution of $Y_1, \ldots, Y_{a-1}$ given Y_a is given by the first part of (7.7), which is the products of the hypergeometric distributions. The Markov property of Y_k follows immediately from Lemma 7.1, which is an extension of Lemma 6.5 for the normal distribution to the general non-normal distribution.

Lemma 7.1. Let $f(z_1, \ldots, z_n)$ be the density of $(z_1, \ldots, z_n)$ with respect to the dominating measure $\mu_1 \times \cdots \times \mu_n$. Then the Markov property

$$f(z_1, \ldots, z_l \mid z_{l+1}, \ldots, z_n) = f(z_1, \ldots, z_l \mid z_{l+1}),\ l < n \tag{7.8}$$

holds if and only if there is a factorization of f such that

$$f(z_1, \ldots, z_n) = f_1(z_1, z_2) \times f_2(z_2, z_3) \times \cdots \times f_{n-1}(z_{n-1}, z_n), \tag{7.9}$$

where the functions $f_1, \ldots, f_{n-1}$ are not necessarily densities.

Proof. Suppose that (7.8) holds, then we have

$$\begin{aligned} f(z_1, \ldots, z_n) &= f(z_n) \times f(z_{n-1}|z_n) \times f(z_{n-2}|z_{n-1}, z_n) \times \cdots \times f(z_1|z_2, \ldots, z_n) \\ &= f(z_n) \times f(z_{n-1}|z_n) \times f(z_{n-2}|z_{n-1}) \times \cdots \times f(z_1|z_2) \end{aligned}$$

Therefore, it is necessary for f to be factorized as in (7.9). Next assume (7.9) and define

$$f^*(z_{l+1},\ldots,z_{m-1};z_m)=\frac{f_l(z_l,z_{l+1})\times\cdots\times f_{m-1}(z_{m-1},z_m)}{\int\cdots\int f_l(z_l,z_{l+1})\times\cdots\times f_{m-1}(z_{m-1},z_m)\mathrm{d}\mu_l\cdots\mathrm{d}\mu_{m-1}}$$

if the denominator is non-zero, and zero otherwise. Then it is easy to verify that $f^*(z_{l+1},\ldots,z_{m-1};z_m)$ is a version of both of the conditional densities

$$f\left(z_l,\ldots,z_{m-1}\mid z_m,\ldots,z_n\right)\text{ and }f\left(z_l,\ldots,z_{m-1}\mid z_m\right).$$

(c) **Algorithm for calculating the p-value.** The recursion formula for max acc. $t1$ corresponding to equation (6.30) is as follows. First, define the conditional probability given Y_k as $F_k(Y_k,t_0)=\Pr(t_1<t_0,\ldots,t_k<t_0|Y_k)$. It should be noted that, differently from the normal case, the recursion formula is constructed via Y_k and not directly in terms of t_k. This is because the factorization of the joint distribution into the products of conditional distributions is obtained in terms of Y_k. Then, we have the recursion formula

$$\begin{aligned}F_{k+1}(Y_{k+1},\ t_0)&=\Pr(t_1<t_0,\ldots,t_k<t_0,t_{k+1}<t_0|Y_{k+1})\\&=\textstyle\sum_{Y_k}\Pr(t_1<t_0,\ldots,t_k<t_0,t_{k+1}<t_0|Y_k,Y_{k+1})f_k(Y_k|Y_{k+1})\\&=\begin{cases}\sum_{Y_k}F_k(Y_k,\ t_0)f_k(Y_k|Y_{k+1})\ \textit{if}\ t_{k+1}(Y_{k+1})<t_0,\\ \qquad 0,\ \text{otherwise},\end{cases}\end{aligned}\tag{7.10}$$

where $f_k(Y_k|Y_{k+1})$ is a conditional distribution of Y_k given Y_{k+1}. In this case it is a hypergeometric distribution $H\left(Y_k\mid Y_{k+1},N_k,N_{k+1}\right)$. Defining $t_a(Y_a)=-\infty$, we have the p-value for the observed maximum t_0 at the final step as $p=1-F_a(Y_a,t_0)$. Thus, the algorithm is essentially the same as the normal case except that the integration is now the summation. The algorithm of exact calculation of the p-value for max acc. $t1$, in the binomial case, is supported on the author's website.

Example 7.1. Analysis of Table 7.1. The necessary calculations have been shown in Table 7.1. Noting that $\hat{p}=105/162=0.648$, $\hat{q}=57/162=0.352$, we get $t_k(Y_k)$ as in the last column of Table 7.1. Therefore, we have max acc. $t1=2.276$ and its right one-sided p-value calculated by the recursion formula (7.10) is 0.044. A step change is observed between the dose levels 150 mg and 200 mg.

Table 7.2 Simultaneous lower bounds SLB $(i, j+1)$ for $p(i, j)$.

	$j+1$			
i	2	3	4	5
1	−0.077	−0.056	−0.118	−0.330
	−0.077	(0.008)	(0.008)	(0.008)
2		0.008	−0.054	−0.266
		0.008	(0.008)	(0.008)
3			−0.083	−0.267
			−0.083	(−0.083)
4				−0.295
				−0.295

We can apply the simultaneous lower bounds (6.34) and (6.35) for

$$p(i, j) = \frac{n_j p_j + \ldots + n_a p_a}{N_{j-1}^*} - \frac{n_1 p_1 + \ldots + n_i p_i}{N_i}, 1 \le i < j \le a,$$

by the normal approximation. For the basic contrasts, it becomes

$$\text{SLB}(k, k+1) = \left(\bar{Y}_k^* - \bar{Y}_k\right) - \left\{\hat{p}\hat{q}\left(\frac{1}{N_k} + \frac{1}{N_k^*}\right)\right\}^{1/2} T_\alpha(\boldsymbol{n}, \infty),$$

where $Y_k^* = y_{k+1} + \ldots + y_a, N_k^* = n_{k+1} + \ldots + n_a$ and $\bar{Y}_k$, $\bar{Y}_k^*$ are the averages. The formula (6.37) for improvement is also valid here. In this case, the upper α point $T_\alpha(\boldsymbol{n}, \infty) = 2.173\ (\alpha = 0.05)$ is obtained by the normal theory in Section 6.5.3 (1) (d) and the simultaneous lower bounds SLB$(i, j+1)$ are shown in Table 7.2. The approximate $T_\alpha(\boldsymbol{n}, \infty)$ by the balanced case is 2.151 from Table A of the Appendix at $a = 5$ and df ∞, and there is not a big difference from the exact value.

In Table 7.2 the improved lower bounds by formula (6.37) are shown in parentheses. These results clearly show that there is a change-point between dose levels 2 and 3 at one-sided significance level 0.05.

(2) **Cumulative chi-squared statistic χ^{*2}.** The cumulative chi-squared χ^{*2} has been introduced by Takeuchi and Hirotsu (1982) as the sum of t_k^2 of (7.5), just as equation (6.39) for the normal model. Its distribution is well approximated by a constant times the chi-squared distribution, and the formulae in Section 6.5.3 (2) (a) are valid with $N_i = n_1 + \ldots + n_i$ and $N_i^* = n_{i+1} + \ldots + n_a$ for calculating λ_i. The original idea of this was introduced by Taguchi (1966), as an accumulation analysis. Although some mathematical refinement was required, it was an initial approach to the efficient analysis of ordered alternatives.

Example 7.2. Example 7.1 continued. The cumulative chi-squared statistic is obtained from Table 7.1 as

$$\chi^{*2} = \sum_{k=1}^{a-1} t_k^2 = 1.319^2 + 2.276^2 + 1.161^2 + (-0.043)^2 = 8.268.$$

Then we have $\lambda_1 = 36/126$, $\lambda_2 = 77/85$, $\lambda_3 = 113/49$, $\lambda_4 = 148/14$ from (6.28) and therefore $d^2f = 6.326$ by the formula in Section 6.5.3 (2) (a). Since df is obviously 4, we have $d = 1.58$ and $f = 2.53$. The upper tail probability of $8.268/1.58 = 5.23$ as the chi-squared distribution with df 2.53 is 0.113 by 'keisan.casio.com'. This is a two-sided test and slightly conservative compared with the result of max acc. $t1$, which suggested significance at 0.05 one-sided. Another example of the application of max acc. $t1$ and χ^{*2} is given in Example 7.7.

7.3.2 Maximal contrast test for convexity and slope change-point hypotheses

(1) **Weighted doubly accumulated statistics and max acc.** $t2$. In this section we consider a convexity hypothesis in $\theta_i = \log\{p_i/(1-p_i)\}$. It is useful for a directional goodness-of-fit test of the linear regression model and also for confirming a downturn in the dose–response. It should be noted here that the convexity hypothesis depends on the spacing of events, whereas monotony is a property independent of spacing. Therefore, we consider here a general case of unequal spacing and denote the time or location of the ith event by x_i. Then, the convexity hypothesis is defined by

$$H_{\text{con}}: \boldsymbol{L}_a^{*\prime}\boldsymbol{\theta} \geq 0, \text{ with at least one inequality strong,} \tag{7.11}$$

where $\boldsymbol{L}_a^{*\prime}$ is a second-order differential matrix, defined by

$$\boldsymbol{L}_a^{*\prime} = \begin{bmatrix} \frac{1}{x_2-x_1} & \frac{1}{x_1-x_2}+\frac{1}{x_2-x_3} & \frac{1}{x_3-x_2} & 0 & 0 & \cdots & 0 \\ 0 & \frac{1}{x_3-x_2} & \frac{1}{x_2-x_3}+\frac{1}{x_3-x_4} & \frac{1}{x_4-x_3} & 0 & \cdots & 0 \\ & & & & & \cdots & \\ 0 & 0 & 0 & \cdots & \frac{1}{x_{a-1}-x_{a-2}} & \frac{1}{x_{a-2}-x_{a-1}}+\frac{1}{x_{a-1}-x_a} & \frac{1}{x_a-x_{a-1}} \end{bmatrix}_{(a-2)\times a},$$

similarly as in Section 6.5.4. It has been shown by Hirotsu and Marumo (2002) that

$$\boldsymbol{L}_a^* \left(\boldsymbol{L}_a^{*\prime}\boldsymbol{L}_a^*\right)^{-1} = [\boldsymbol{b}_1, \boldsymbol{b}_2, \ldots, \boldsymbol{b}_{a-2}],$$

$$\boldsymbol{b}_k = (\boldsymbol{I} - \boldsymbol{\Pi}_{\boldsymbol{B}})\ (0, \ldots, 0, x_{k+2} - x_{k+1}, \ldots, x_a - x_{k+1})',$$

where $\boldsymbol{\Pi}_B = \boldsymbol{B}(\boldsymbol{B}'\boldsymbol{B})^{-1}\boldsymbol{B}'$ is defined in Section 6.5.4(1) with

$$\boldsymbol{B} = \begin{pmatrix} 1, \cdots, 1 \\ x_1, \cdots, x_a \end{pmatrix}'$$

here reflecting the unequal spacing. This equation reduces to $\boldsymbol{b}_k$ in Section 6.5.4 (1) in the case of equal spacing. Then,

$$M_k: \ \boldsymbol{\theta} = (\boldsymbol{B}\ \boldsymbol{b}_k)(\beta_0\ \beta_1\ \Delta_k)' = \boldsymbol{B}(\beta_0\ \beta_1)' + \boldsymbol{b}_k\Delta_k$$

is a slope change-point model with change Δ_k at time point x_{k+1} for $k = 1,\ldots,a-2$. Therefore, in this case again each corner vector corresponds to a slope change-point model. The null model H_0: $\boldsymbol{L}_a^{*\prime}\boldsymbol{\theta} = 0$ is obviously equivalent to a linear regression model $H_0 : \theta_i = \beta_0 + \beta_1 x_i$.

We consider here a general exponential family

$$a(\theta_i)b(y_i)\exp(\theta_i y_i), \quad i = 1,\ldots,a \tag{7.12}$$

as the underlying distribution for dealing with the binomial and Poisson distributions simultaneously. The function $b(y_i)$ is

$$b(y_i) = \{y_i!(n_i - y_i)!\}^{-1} \text{ for the binomial model,}$$
$$b(y_i) = (y_i!)^{-1} \text{ for the Poisson model.}$$

In this case also we can consider the essential part $\boldsymbol{t} = \left(\boldsymbol{L}_a^{*\prime}\boldsymbol{L}_a^*\right)^{-1}\boldsymbol{L}_a^{*\prime}\boldsymbol{y}$ of the key vector $\boldsymbol{z} = \left(\boldsymbol{L}_a^{*\prime}\boldsymbol{L}_a^*\right)^{-1}\boldsymbol{L}_a^{*\prime}\boldsymbol{\nu}\left(\hat{\boldsymbol{\theta}}_0\right)$, discarding $\hat{E}_0(\boldsymbol{y})$ from $\boldsymbol{\nu}\left(\hat{\boldsymbol{\theta}}_0\right)$ since it is a function of sufficient statistics

$$Y_a = y_1 + y_2 + \cdots + y_a \quad \text{and} \quad T_a = x_1 y_1 + x_2 y_2 + \cdots + x_a y_a$$

under $H_0 : \theta_i = \beta_0 + \beta_1 x_i$. After standardization under the null distribution given (Y_a, T_a), the maximal elements based on $\boldsymbol{t}$ and $\boldsymbol{z}$ will coincide. However, this form of $\boldsymbol{t}$ is still not very convenient for dealing with the discrete model, since the elements of $\boldsymbol{t}$ can be negative and do not take integer values because of the complicated coefficient matrix. It is so complicated that it looks formidable to develop an exact test. It should be noted that in case of the normal distribution discussed in Section 6.5.4, it was rather easy to deal with the linear contrasts in $\boldsymbol{y}$ as they are, since they are distributed as the normal defined only by mean and variance. Therefore, a more convenient basic variable for dealing with the discrete distribution has been searched for. This approach has already been introduced in Section 6.5.4 (1), deriving the doubly accumulated statistics in case of equal spacing. First, by noting that $\boldsymbol{L}_a^*\left(\boldsymbol{L}_a^{*\prime}\boldsymbol{L}_a^*\right)^{-1}\boldsymbol{L}_a^{*\prime} + \boldsymbol{\Pi}_B$ is an identity matrix of order a, we have

$$y=\left\{L_a^*\left(L_a^{*\prime}L_a^*\right)^{-1}L_a^{*\prime}+\Pi_B\right\}y=L_a^*\,t+\Pi_B y.$$

In this equation $\Pi_B y$ is a function of the sufficient statistics (Y_a, T_a) and can be discarded for developing a similar test. Then, the similar test should be based on $S=(S_1,\ldots,S_{a-2})'$ that satisfies $y=L_a^*S$ instead of t, where we ignore the last two rows of $y=L_a^*S$. By accumulating both sides of $y=L_a^*S$, the equation

$$Y_k=y_1+\cdots+y_k=\frac{S_k-S_{k-1}}{x_{k+1}-x_k},\quad k=1,\ldots,a-2,$$

is obtained, where S_0 is defined to be zero. By accumulating Y_k further after multiplying by $(x_{k+1}-x_k)$, the equation

$$S_k=\sum_{i=1}^{k}(x_{i+1}-x_i)Y_i=(x_{k+1}-x_1)y_1+\cdots+(x_{k+1}-x_k)y_k,\quad k=1,\ldots,a-2,\qquad (7.13)$$

is obtained. In case of equal spacing, the statistics S_k (7.13) reduce to the doubly accumulated statistics of y_k, like (6.47) obtained in Section 6.5.4 (1). The extended equation (7.13) is a weighted doubly accumulated statistic but might be called simply the doubly accumulated statistic hereafter, and the maximal standardized element s_m^* of S_k is employed as the test statistic for the max acc. *t*2 method. The component of max acc. *t*2 is shown to be an efficient score for testing the slope change-point model M_k by the same arguments as in Section 6.5.4 (3) (see Hirotsu *et al.*, 2016).

(2) **Second-order Markov property and a factorization of the conditional distribution**. In case of the normal distribution, the conditional distribution given the complete sufficient statistics and its factorization are very simple, resulting in the respective normal distributions defined by the conditional mean and variance as described in Section 6.5.4 (2). In case of the Poisson and binomial distributions, the conditional distribution given Y_a only and its factorization into the products of serial conditional distributions are also well known and have already been utilized in Section 7.3.1 (1) (b). However, we have an additional conditioning variable T_a here, and several steps will be required to obtain the exact conditional distribution in tractable form (Hirotsu, 2013; Hirotsu *et al.*, 2016). The conditional null distribution given (Y_a, T_a) is obviously in the form

$$G(y|Y_a,T_a)=C_{a-1}^{-1}(Y_a,T_a)\Pi_{k=1}^{a}b(y_k),$$

where the constant $C_{a-1}(Y_a,T_a)$ is determined so that the total probability of y is unity conditionally given Y_a and T_a. Then, by the relationship

$$S_k=S_{k-1}+(x_{k+1}-x_k)Y_k,\quad k=1,\ldots,a-2\qquad (7.14)$$

and $y_k = Y_k - Y_{k-1}$, we can rewrite the probability function in terms of S_k as

$$G(y|Y_a, T_a) = C_{a-1}^{-1}(Y_a, T_a)\Pi_{k=1}^{a} b\left(\frac{S_k - S_{k-1}}{x_{k+1} - x_k} - \frac{S_{k-1} - S_{k-2}}{x_k - x_{k-1}}\right), \tag{7.15}$$

where S_0 and S_{-1} are defined to be zero and $S_{a-1} = x_a Y_a - T_a$, $S_a = S_{a-1} + (x_{a+1} - x_a)Y_a$ with x_{a+1} an arbitrary number. It should be noted that S_{a-1} is an extension of the definition (7.13) to $k = a-1$, but it cannot be done for S_a without introducing a hypothetical value x_{a+1}. For notational convenience anyway, S_{a-1} and S_a are employed as conditioning variables instead of Y_a and T_a. Also, it is sometimes convenient to use S_{a-1} and Y_a as conditioning variables. The one-to-one correspondence among the set of variables (Y_a, T_a), (S_{a-1}, S_a), and (S_{a-1}, Y_a) is obvious. Equation (7.15) implies the second-order Markov property of the sequence $S_1, \ldots, S_{a-2}$. The proof is essentially the same as that of Lemma 7.1. Because of this property, there is available a very efficient and exact algorithm for calculating the p-value of the maximal statistic s_m^*, as well as the normalizing constant $C_{a-1}(S_{a-1}, S_a)$ and the moments for standardization. Now, because of the second-order Markov property, the null distribution $G(y|Y_a, T_a)$ can be factorized in terms of S_k:

$$G(y|Y_a, T_a) = \Pi_{k=1}^{a-2} f_k(S_k|S_{k+1}, S_{k+2}),$$

where $f_k(S_k|S_{k+1}, S_{k+2})$ is the conditional distribution of S_k given S_{k+1} and S_{k+2}. Then, the kth conditional distribution should be in the form

$$f_k(S_k|S_{k+1}, S_{k+2}) = C_{k+1}^{-1}(S_{k+1}, S_{k+2})C_k(S_k, S_{k+1}) \times b\left(\frac{S_{k+2} - S_{k+1}}{x_{k+3} - x_{k+2}} - \frac{S_{k+1} - S_k}{x_{k+2} - x_{k+1}}\right),$$
$$k = 1, \ldots, a-2, \tag{7.16}$$

where C_{k+1} is the normalizing constant and the initial constant is defined as

$$C_1(S_1, S_2) = b\left(\frac{S_1}{x_2 - x_1}\right) \times b\left(\frac{S_2 - S_1}{x_3 - x_2} - \frac{S_1}{x_2 - x_1}\right). \tag{7.17}$$

It should be noted that in equation (7.16) the random variable S_k is included also in the normalizing constant $C_k(S_k, S_{k+1})$ of the previous step as the conditioning variable. Starting from C_1, all the C_k can be calculated recursively by the equation

$$C_{k+1}(S_{k+1}, S_{k+2}) = \sum_{S_k} C_k(S_k, S_{k+1}) \times b\left(\frac{S_{k+2} - S_{k+1}}{x_{k+3} - x_{k+2}} - \frac{S_{k+1} - S_k}{x_{k+2} - x_{k+1}}\right). \tag{7.18}$$

Then, at the final step the overall normalizing constant $C_{a-1}(S_{a-1}, S_a)$ is obtained and the distribution $G(y)$ is determined in terms of $S_k, k = 1, \ldots, a-2$. It should be noted that $C_{a-1}(S_{a-1}, S_a)$ is well defined, since $(S_a - S_{a-1})/(x_{a+1} - x_a)$ is simply Y_a, but in the following the notation $C_{a-1}(S_{a-1}, Y_a)$ is employed instead of $C_{a-1}(S_{a-1}, S_a)$. This factorization of the simultaneous distribution follows the same idea as employed in

Hirotsu *et al.* (2001) to obtain the exact factorization of the null distribution on the three-factor interaction in a $2 \times J \times K$ contingency table (see Section 14.1.1 (1) (c)). In this case, however, the sample space of $S = (S_1, \ldots, S_{a-2})$ is unknown and also explosive when a is large. Therefore, to execute the recursion formula efficiently, several inequalities have been introduced in Hirotsu *et al.* (2016) and we give a summary in the following:

$$\text{Absolute}: \max\{0, S_{a-1} - (x_a - x_{k+1})Y_a\} \le S_k \le \frac{x_{k+1} - x_1}{x_a - x_1} S_{a-1}, \quad k = 1, \ldots, a-2. \quad (7.19)$$

$$\text{Relative}: \frac{x_{k+1} - x_1}{x_k - x_1} S_{k-1} \le S_k \le \frac{1}{x_a - x_k}\{(x_a - x_{k+1})S_{k-1} + (x_{k+1} - x_k)S_{a-1}\}, \quad k = 2, \ldots, a-2. \quad (7.20)$$

The absolute inequality gives restrictions on S_k in terms of S_{a-1} and Y_a, which are constants, and the relative one is useful for the bottom-up procedure to construct conditional probabilities giving the possible range of S_k in terms of S_{k-1}. This inequality can be rewritten as

$$\frac{1}{x_a - x_{k+1}}\{(x_a - x_k)S_k - (x_{k+1} - x_k)S_{a-1}\} \le S_{k-1} \le \frac{x_k - x_1}{x_{k+1} - x_1} S_k, \quad k = 2, \ldots, a-2, \quad (7.21)$$

which is useful for defining the range of S_{k-1} based on S_k. Further, there is an obvious relationship among three S_k's: $(S_k - S_{k-1})/(x_{k+1} - x_k) \le (S_{k+1} - S_k)/\ (x_{k+2} - x_{k+1})$ or equivalently

$$y_{k+1} = \frac{S_{k+1} - S_k}{x_{k+2} - x_{k+1}} - \frac{S_k - S_{k-1}}{x_{k+1} - x_k} \ge 0. \quad (7.22)$$

These inequalities are general for the exponential family (7.12) and sometimes additional inequalities are required. For example, for the independent binomial distribution $B(n_i, p_i)$ in this section the inequality (7.22) is changed to

$$0 \le \frac{S_{k+1} - S_k}{x_{k+2} - x_{k+1}} - \frac{S_k - S_{k-1}}{x_{k+1} - x_k} \le n_{k+1}. \quad (7.23)$$

(3) **Constructing conditional probabilities by a bottom-up procedure**. The construction of the conditional probabilities $f_k(S_k | S_{k+1}, S_{k+2})$ should be bottom-up since $C_k(S_k, S_{k+1})$ is necessary for calculating $C_{k+1}(S_{k+1}, S_{k+2})$. First, the possible combinations of S_1 and S_2 satisfying the inequalities (7.19) and (7.20) are found, and $C_1(S_1, S_2)$ is calculated by (7.17). For those combinations of S_1 and S_2, the range of the

variable S_3 can be determined by the inequalities (7.19), (7.20), and (7.22). Then, the coefficient $C_2(S_2, S_3)$ can be calculated by equation (7.18):

$$C_2(S_2,S_3)=\sum_{S_1} C_1(S_1,S_2)\times b\left(\frac{S_3-S_2}{x_4-x_3}-\frac{S_2-S_1}{x_3-x_2}\right),$$

where the summation is with respect to S_1. In executing the recursion formula, however, a difficulty arises since the S_k satisfying equation (7.13) does not necessarily take successive integers differently from Y_k. It does not occur in the equal spacing where x_k can be defined as k without any loss of generality. Therefore, the inequalities are converted to Y_k, which takes successive integers and generates conformable S_k through equation (7.14). The inequality for Y_k is as follows:

$$\max\left\{\frac{S_{a-1}-S_{k-1}-(x_a-x_{k+1})Y_a}{x_{k+1}-x_k},\frac{S_{k-1}}{x_k-x_1},Y_{k-1}\right\}\le Y_k$$
$$\le \min\left\{\frac{S_{a-1}-S_{k-1}}{x_a-x_k},\frac{1}{x_{k+1}-x_k}\left(\frac{x_{k+1}-x_1}{x_a-x_1}S_{a-1}-S_{k-1}\right)\right\}. \qquad (7.24)$$

A simple example is given below of how to construct the conformable sequence.

Example 7.3. Generating the sequence S conformable to $y=(1,1,1,1,1)$ with $x=(1,2,4,7,8)$. In this example the sufficient statistics are $Y_5=1+1+1+1+1=5$ and $S_{5-1}=7y_1+6y_2+4y_3+y_4=8Y_5-T_5=18$. Then, by the absolute inequality (7.19), the range of $S_1=Y_1$ is obtained as $-12\le S_1=Y_1\le \frac{18}{7}$ and therefore S_1 and Y_1 take 0, 1, and 2. Then, by the inequality (7.24), the conformable Y_2 for $(Y_1,S_1)=(0,0),(1,1)$, and $(2,2)$ are $0\le Y_2\le 3$, $1\le Y_2\le 2$, and $Y_2=2$, respectively. Therefore, the respective conformable S_2 are $(0,2,4,6)$, $(3,5)$, and (6) by $S_2=S_1+2Y_2$. Then, from the pair (Y_2, S_2), the conformable Y_3 can be determined by equation (7.24) and S_3 from S_2 and Y_3 by (7.14). The process is summarized in Table 7.3, where NG implies there is no conformable Y_3 satisfying the inequality. In this example the number of partitions of integer 5 into five integers is 126, among which only five sequences in Table 7.3 are conformable also to $S_4=18$. In contrast, under equal spacing ($x_k=k$), 12 sequences are found to be conformable to the fixed sufficient statistics $Y_5=5$ and $S_4=10$.

Now, given the underlying distribution, the initial constant $C_1(S_1, S_2)$ can be calculated by equation (7.17) at stage 2. Then, at stage 3 the normalizing constant $C_2(S_2, S_3)$ and the conditional probability $f_1(S_1|S_2,S_3)$ can be calculated by equations (7.18) and (7.16), respectively. At stage 4, C_3 and f_2 are calculated; at stage 5, C_4 and f_3 are calculated; and finally, the total probability $G(y|S_4,Y_5)$ is obtained by the product of f_1, f_2, and f_3.

Table 7.3 Process to determine the conformable sequence.

Stage 1	Stage 2		Stage 3		Stage 4		Stage 5
$Y_1 = S_1$	Y_2	$S_2 = S_1 + 2Y_2$	Y_3	$S_3 = S_2 + 3Y_3$	Y_4	$S_4 = S_3 + Y_4$	Y_5
0	0	0	NG	—	—	—	—
0	1	2	4	14	4	18	5
0	2	4	3	13	5	18	5
0	3	6	3	15	3	18	5
1	1	3	NG	—	—	—	—
1	2	5	3	14	4	18	5
2	2	6	3	15	3	18	5

Example 7.4. Example 7.3 continued. As an example, the Poisson model with mean Λ_i can be dealt with by taking $\theta_i = \log \Lambda_i$, $a(\theta_i) = \exp(-\Lambda_i) = \exp\{-\exp(\theta_i)\}$, and

$$b(y_i) = (y_i!)^{-1} = \left\{ \left(\frac{S_i - S_{i-1}}{x_{i+1} - x_i} - \frac{S_{i-1} - S_{i-2}}{x_i - x_{i-1}} \right)! \right\}^{-1}.$$

Starting from the initial constant (7.17), the normalizing constants and conditional probabilities are calculated recursively by (7.18) and (7.16), respectively. To continue the current example at stage 2, we obtain

$$C_1(0,2) = \left\{ \left(\frac{0}{2-1} \right)! \right\}^{-1} \times \left\{ \left(\frac{2-0}{4-2} - \frac{0}{2-1} \right)! \right\}^{-1} = 1$$

and similarly $C_1(0,4) = 1/2$, $C_1(0,6) = 1/6$, $C_1(1,5) = 1, C_1(2,6) = 1/2$. At stage 3, noting that the pair $(S_2, S_3) = (2,14)$ has only one root $S_1 = 0$, $C_2(2, 14)$ is obtained as

$$C_2(2,14) = C_1(0,2) \times \left\{ \left(\frac{14-2}{7-4} - \frac{2-0}{4-2} \right)! \right\}^{-1} = \frac{1}{6}$$

and the conditional probability $f_1(0|2,14)$ is equal to unity. In contrast, the pair $(S_2, S_3) = (6,15)$ has two roots $S_1 = 0$ and 2 so that

$$C_2(6,15) = \frac{1}{6} \times \left\{ \left(\frac{15-6}{3} - \frac{6-0}{2} \right)! \right\}^{-1} + \frac{1}{2} \times \left\{ \left(\frac{15-6}{3} - \frac{6-2}{2} \right)! \right\}^{-1} = \frac{2}{3}.$$

Then, the conditional probabilities are calculated as

$$f_1(0|6,15) = \left(\frac{2}{3} \right)^{-1} \times \frac{1}{6} \times \left\{ \left(\frac{15-6}{3} - \frac{6-0}{2} \right)! \right\}^{-1} = \frac{1}{4},$$

$$f_1(2|6,15) = \left(\frac{2}{3} \right)^{-1} \times \frac{1}{2} \times \left\{ \left(\frac{15-6}{3} - \frac{6-2}{2} \right)! \right\}^{-1} = \frac{3}{4}.$$

It should be noted that the inequalities (7.21) and (7.22) are useful for finding out the roots S_{k-1} for the pair (S_k, S_{k+1}). Other C_2's are $C_2(4,13) = 1/2$, $C_2(5,14) = 1$, and the conditional probabilities $f_1(0|4,13), f_1(1|5,14)$ are both unity since those pairs (S_2, S_3) have only one root $S_1 = 0$ and 1, respectively. The calculations of $C_3(S_3, S_4)$, $C_4(S_4, Y_5)$ and the conditional probabilities $f_2(S_2|S_3, S_4), f_3(S_3|S_4, Y_5)$ go similarly as $C_2(S_2, S_3)$ and $f_1(S_1|S_2, S_3)$, and the result is summarized in Table 7.4. In this small example all the conformable sequences might be written down, but it is generally impossible to figure out all the sequences even for moderate a and Λ_i, and the bottom-up procedure utilizing these inequalities is inevitable.

(4) **Recursion formulae for moments and tail probability**

(a) **Calculating moments**. To standardize the test statistics, it is necessary to calculate the mean and variance of $S_k, k = 1, \ldots, a-2$. They are also calculated recursively by the formula

$$E(S_k^l|S_{a-1}, Y_a) = \sum_{S_{a-2}} \cdots \sum_{S_{k+1}} \sum_{S_k} S_k^l f_k(S_k|S_{k+1}, S_{k+2}) \times f_{k+1}(S_{k+1}|S_{k+2}, S_{k+3}) \times \cdots \times f_{a-2}(S_{a-2}|S_{a-1}, Y_a). \tag{7.25}$$

(b) **Calculating *p*-value of the convexity test** Let s_k^* be the standardized version of S_k, $s_k^* = \dfrac{\{S_k - E(S_k)\}}{V^{1/2}(S_k)}$, $k = 1, \ldots, a-2$. Then the test statistic is

$$s_m^* = \max_{k=1,\cdots,a-2} s_k^*.$$

For the recursion formula, define the conditional probability

$$\begin{aligned} F_k(S_{k-1}, S_k, d) &= \Pr(s_1^* < d, \ldots, s_k^* < d|S_{k-1}, S_k), \\ &= \Pr(S_1 < d_1^*, \ldots, S_k < d_k^*|S_{k-1}, S_k), \ k = 2, \ldots, a, \end{aligned}$$

where $d_k^* = E(S_k) + V^{1/2}(S_k)d, \ \ k = 1, \ldots, a-2$ and s_{a-1}^*, s_a^* are defined as $-\infty$, although their conditional variances are zero so that the inequality always holds. It should be noted that the recursion formulae (7.26) ~ (7.28) are given in terms of S_k, since the distribution theory has been obtained in terms of S_k. This is contrasted with the normal theory, where the recursion formulae are given directly in terms of s_k^*. Correspondingly, d_{a-1}^* and d_a^* are set to $S_{a-1} + \delta$ and $S_a + \delta$, respectively, with δ a positive small number. Then, a recursion formula for F_k is obtained as

$$\begin{aligned} F_{k+1}(S_k, S_{k+1}, d) &= \Pr(S_1 < d_1^*, \ldots, S_k < d_k^*, S_{k+1} < d_{k+1}^*|S_k, S_{k+1}) \\ &= \sum_{S_{k-1}} \Pr(S_1 < d_1^*, \ldots, S_k < d_k^*, S_{k+1} < d_{k+1}^*|S_{k-1}, S_k, S_{k+1}) \\ &\quad \times f_{k-1}(S_{k-1}|S_k, S_{k+1}) \end{aligned} \tag{7.26}$$

Table 7.4 Bottom-up process for constructing the probability distribution.

y_1	y_2	y_3	y_4	y_5	S_1	S_2	S_3	S_4	Y_5	C_1	C_2	C_3	f_1	f_2	f_3	$G(\mathbf{y} \mid S_4, Y_5)$
0	1	3	0	1	0	2	14	18	5	1	1/6	7/6	1	1/7	2/3	2/21
0	2	1	2	0	0	4	13	18	5	1/2	1/2	1/4	1	1	1/7	3/21
0	3	0	0	2	0	6	15	18	5	1/6	2/3	2/3	1/4	1	4/21	1/21
1	1	1	1	1	1	5	14	18	5	1	1	–	1	6/7	2/3	12/21
2	0	1	0	2	2	6	15	18	5	1/2	–	–	3/4	1	4/21	3/21

$$= \begin{cases} \Pr\left(S_1 < d_1^*, \ldots, S_k < d_k^* | S_{k-1}, S_k\right) \times f_{k-1}\left(S_{k-1} | S_k, S_{k+1}\right) & \text{if } S_{k+1} < d_{k+1}^*, \\ 0, \quad \text{otherwise.} \end{cases} \tag{7.27}$$

Equation (7.26) is due to the law of total probability, and equation (7.27) is due to the second-order Markov property of S_k. Thus, essentially the recursion formula is obtained as

$$F_{k+1}(S_k, S_{k+1}, d) = \sum\nolimits_{S_{k-1}} F_k(S_{k-1}, S_k, d) \times f_{k-1}(S_{k-1} | S_k, S_{k+1}). \tag{7.28}$$

There is no difficulty in extending the formula to F_a by taking S_{a-1} and Y_a as fixed values satisfying the inequality always. Then, the p-value of the observed maximum s_m^* is obtained at the final step by

$$p = 1 - F_a(S_{a-1}, Y_a, d) \text{ at } d = s_m^*. \tag{7.29}$$

It should be noted that the procedure converts the multiple summation into the repetition of a single summation, so that the calculation is feasible for large a.

Example 7.5. Example 7.3 continued. Again, the procedure is explained via Example 7.2 with the observed sequence $\boldsymbol{y} = (1,1,1,1,1)'$. First, the moments of S_k are calculated by method (a) of this section, to obtain

$$E(S_1) = 6/7, \quad E(S_2) = 100/21, \quad E(S_3) = 295/21,$$
$$V(S_1) = 20/7^2, \quad V(S_2) = 500/21^2, \quad V(S_3) = 146/21^2.$$

Then, the standardized statistics are obtained as $s_1^* = 0.22361$, $s_2^* = 0.22361$, $s_3^* = -0.08276$, and therefore $s_m^* = 0.22361$, which gives $d_1^* = 1.000$, $d_2^* = 5.000$, $d_3^* = 14.17628$. Of course, d_4^* and d_5^* are set as $18 + \delta$ and $S_a + \delta$ with δ an arbitrary positive constant. Then, the calculation of the p-value by the recursion formula goes as follows.

First, $F_2(S_1, S_2)$ is set to unity for possible combination of (S_1, S_2) with $S_1 < 1.000$ and $S_2 < 5.000$, namely for $F_2(0, 2)$ and $F_2(0, 4)$, while it is set to zero for other combinations. For convenience, we omit here the constant $d\left(= s_m^*\right)$ from the notation $F_2(S_1, S_2, d)$. Then, the possible combinations of (S_2, S_3) with $S_2 < 5.000$ and $S_3 < 14.176$ are found to be (2, 14) and (4, 13). Since these combinations have only one root $(S_1, S_2) = (0,2)$ and (0, 4), respectively, $F_3(2, 14)$ and $F_3(4, 13)$ are found to be unity while F_3 is set to zero for other combinations. Then, the F_4's are calculated as

$$F_4(13,18) = 1 \times 1 = 1 \text{ and } F_4(14,18) = 1 \times 1/7 = 1/7.$$

Finally, $F_a(S_{a-1}, Y_a)$ is calculated as

$$F_5(18,5) = 1 \times 1/7 + 1/7 \times 2/3 = 5/21$$

and therefore the p-value is obtained as

$$p = 1 - 5/21 = 16/21 = 0.762.$$

This is the sum of the probabilities of the last three sequences of Table 7.4.

(c) **Calculating p-value of the concavity test**. To test the concavity hypothesis $H_{\text{con}}^{-} : \boldsymbol{L}_a^{*\prime}\boldsymbol{\theta} \le 0$, with at least one inequality strong, the test should be based on $-S_k$, and the maximal statistic

$$s_m^* = \max_{k=1,\cdots,a-2} s_k^{**}$$

is calculated from the standardized statistics $s_k^{**} = \frac{\{-S_k - E(-S_k)\}}{V^{1/2}(-S_k)}$, $k = 1, \ldots, a-2$. Then, only a slight modification is necessary from convexity test (b). Define the conditional probability

$$F_k(S_{k-1}, S_k) = \Pr\left(s_1^{**} < d, \ldots, s_k^{**} < d | S_{k-1}, S_k\right),$$
$$= \Pr\left(S_1 > d_1^{**}, \ldots, S_k > d_k^{*} | S_{k-1}, S_k\right), \quad k = 2, \ldots, a,$$

where

$$d_k^{**} = E(S_k) - V^{1/2}(S_k)d, \quad k = 1, \ldots, a-2, d_{a-1}^{**} = S_{a-1} - \delta, d_a^{**} = S_a - \delta \quad (\delta > 0).$$

The recursion formula (7.28) and p-value calculation (7.29) are exactly the same.

(d) **Calculating p-value of the two-sided test**. The p-value for the two-sided alternative $H_{\text{con}}^{\pm} : \boldsymbol{L}_a^{*\prime}\boldsymbol{\theta} \ge 0$ or $\boldsymbol{L}_a^{*\prime}\boldsymbol{\theta} \le 0$ with at least one inequality strong, can be obtained by defining $s_m^* = \max |s_k^*| = \max |s_k^{**}|$ and

$$F_k(S_{k-1}, S_k) = \Pr\left(|s_1^*| < d, \ldots, |s_k^*| < d | S_{k-1}, S_k\right),$$
$$= \Pr\left(d_1^{**} < S_1 < d_1^*, \ldots, d_k^{**} < S_k < d_k^* | S_{k-1}, S_k\right), \; k = 2, \ldots, a.$$

Software for these calculations is provided on the author's website.

(e) **Comparison of computing time**. In Table 7.5 we give the number of conformable sequences and the computing time of probabilities by the methods with or without recursion formula to the outcome $y_i \equiv 2$, $i = 1, \ldots, a$, for some a assuming a Poisson model. The asterisk ($*$) implies an estimate by log linear extrapolation. The number is easily explosive at around $a = 15$, and the naïve calculation method soon becomes infeasible.

7.3.3 Cumulative chi-squared test for convexity hypothesis

An important application of the convexity and concavity tests is a directional goodness-of-fit test of a dose–response model. For this purpose, the cumulative

Table 7.5 Number of conformable Poisson sequences and computing time of probabilities.

a	Number of sequences	Computing time	
		Recursion formula	Naïve method
5	55	<1 s	<1 s
6	252	<1 s	<1 s
7	1,242	<1 s	<1 s
8	6,375	<1 s	<1 s
9	33,885	<1 s	<1 s
10	184,717	<1 s	4 s
11	1,028,172	<1 s	35 s
12	5,820,904	<1 s	268 s (4 m 28 s)
13	33,427,622	<1 s	2,062 s (34 m 22 s)
14	194,299,052	<1 s	15,793 s (4 h 23 m 13 s)
15	1,141,190,188	<1 s	1.5* d
20	$4.1\times10^{12*}$	1 s	$1\times10^{2*}$ y
30	$8\times10^{19*}$	8 s	$1\times10^{11*}$ y
40	$1\times10^{27*}$	45 s	$1\times10^{20*}$ y

CPU: Intel i7 4770 3.4 GHz, OS: Windows 7 Professional, Compiler: Delphi (Pascal).
s: seconds, m: minutes, h: hours, d: days, y: years.

chi-squared statistic is proposed. It is denoted by $\chi^{\dagger 2}$ and defined by the sum of squares of the standardized components of the key vector: $\chi^{\dagger 2}=\sum_1^{a-2}s_k^{*2}$. Its null distribution is well approximated by a constant times the chi-squared distribution $d\chi_f^2$, where constants d and f are determined by adjusting the first two cumulants:

$$df=E\left\{\sum_{k=1}^{a-2}s_k^{*2}\right\}=a-2$$

$$2d^2f=E\left\{\sum_{k=1}^{a-2}s_k^{*4}+2\sum\sum_{1\le k<l\le a-2}\left(s_k^{*2}s_l^{*2}\right)\right\}-(a-2)^2.$$

The calculation of the joint moments can be carried out by extending the formula in Section 7.3.2 (4) (a). If necessary, an improved formula has been obtained by Hirotsu (1979) which adjusts the first three cumulants. For the $\chi^{\dagger 2}$ -test of the normal case in Section 6.5.4 (5), a diagonal matrix $\boldsymbol{D}$ was introduced for standardization. In the more complicated case of $\boldsymbol{L}_a^{*\prime}$ here, such a matrix is not available and a simple sum of squares of s_k^* is employed. However, the characteristics will not change greatly.

Table 7.6 Mortality by sex of donor and degree of erythroblastosis.

		Number of:		
Degree of disease	Sex of donor	Deaths	Survivors	Total
None	M	2	21	23
	F	0	10	10
Mild	M	2	40	42
	F	0	18	18
Moderate	M	6	33	39
	F	0	10	10
Severe	M	17	16	33
	F	0	4	4

Example 7.6. Directional goodness-of-fit test for a logit linear model

(1) **max acc. *t*2**. The data in Table 7.6 are taken from Allen *et al.* (1949). Erythroblastosis is a disease in certain newborn infants, and can sometimes be fatal. It is caused by the transmission of anti-Rh antibody from an Rh− mother into the blood of an Rh + baby. One form of treatment is an exchange transfusion, in which as much of the infant's blood as possible is replaced by a donor's blood that is free of anti-Rh antibody. Of a total of 179 cases in which this treatment was used in a Boston hospital, no infant deaths occurred in the 42 cases in which a female donor was used, whereas there were 27 infant deaths out of the 137 cases in which a male donor was used.

Cochran (1954) carried out a detailed analysis of the apparent difference between the male and female donors, and proved the difference to be highly significant. Therefore, a separate analysis is necessary for males and females. In this case, however, the interpretation for female donors is obvious, since there is no infant death and a detailed analysis of the association between deaths and the degree of the disease is required only for male donors. Then, if the interest lies in the dependence of the death rates on the degree of the disease, a common procedure for dealing with this type of data is to assume a logit linear model

$$\log\{p_i/(1-p_i)\}=\beta_0+\beta_1 x_i \tag{7.30}$$

under the binomial distribution

$$\binom{n_i}{y_i} p_i^{y_i}(1-p_i)^{n_i-y_i},$$

since the linear model in p_i easily goes out of the range [0,1]. However, the goodness-of-fit of the model (7.30) should also be confirmed before application. Then, the convexity and/or the concavity test is useful as a directional goodness-of-fit test for the linearity of model (7.30). The problem can be dealt with by taking

$$\theta_i = \log\{p_i/(1-p_i)\} \quad a(\theta_i) = n_i!\{1+\exp(\theta_i)\}^{-n_i} \quad b(y_i) = \{y_i!(n_i-y_i)!\}^{-1}$$

$$= \left[\left(\frac{S_i - S_{i-1}}{x_{i+1}-x_i} - \frac{S_{i-1}-S_{i-2}}{x_i - x_{i-1}}\right)! \times \left\{n_i - \left(\frac{S_i - S_{i-1}}{x_{i+1}-x_i} - \frac{S_{i-1}-S_{i-2}}{x_i - x_{i-1}}\right)\right\}!\right]^{-1}.$$

In this case the initial value of $C_i(S_i, S_{i+1})$ is

$$C_1(S_1, S_2) = \left\{\left(\frac{S_1}{x_2 - x_1}\right)! \times \left(n_1 - \frac{S_1}{x_2 - x_1}\right)!\right\}^{-1}$$

$$\times \left[\left(\frac{S_2 - S_1}{x_3 - x_2} - \frac{S_1}{x_2 - x_1}\right)! \times \left\{n_2 - \left(\frac{S_2 - S_1}{x_3 - x_2} - \frac{S_1}{x_2 - x_1}\right)\right\}!\right]^{-1}.$$

The degree of disease is given by an ordinal qualitative measure in Table 7.6 and tentatively defined as $x_i = i$ here. Since this is a binomial case, the inequality (7.23) is applied instead of (7.22) in executing the recursion formula (7.28). Then, the two-sided p-value 0.076 is obtained by the formula of Section 7.3.2 (4) (d).

(2) **Cumulative chi-squared $\chi^{\dagger 2}$.** The cumulative chi-squared is obtained as $\chi^{\dagger 2} = 8.545$ with $d = 1.311$ and $f = 1.526$. The p-value is 0.023 by the chi-squared approximation. In this size of data, an exact p-value is also available as 0.036 (see Section 7.3.5 (3)). Some characterization of the power of the cumulative chi-squared and the two-sided maximal contrast tests is given in the next section.

The result of the goodness-of-fit test suggests that the logit linear model is inappropriate and the use of some non-parametric test will be required.

Example 7.7. Example 7.6 continued (non-parametric test). As a non-parametric test, the common goodness-of-fit χ^2 for independence is 29.25 with df 3 and highly significant at $p = 1.98 \times 10^{-6}$, but it cannot tell us any relationship between the death rates and the degree of disease. Chassan (1960, 1962) proposed a particular method of test, taking the degree of disease into consideration and extending the one-sided test of the two-sample problem. Bartholomew (1963) claimed a defect in Chassan's extension and suggested the use of a $\bar{\chi}^2$-test as more appropriate for the monotone alternative. It should be noted that monotony is an assumption appropriate for the ordinal qualitative covariate, since it is invariant against spacing. However, as mentioned in Section 6.5.1, an exact analysis by $\bar{\chi}^2$ is possible only in a limited number of cases. Then, max acc. $t1$ for the monotone hypothesis of the binomial data in Section 7.3.1 is simple and more appropriate. For Table 7.6 it gives one-sided p-value 1.2×10^{-6}, suggesting a step change-point between moderate and severe. Thus, it is one of the merits of max acc. $t1$ that it can suggest an outline of the response curve without assuming any rigid parametric model (see also Section 6.5.3 (1)). This implies also that the difference between males and females is not homogeneous, suggesting a kind of interaction. For the two-sided problem of the monotone hypothesis, the cumulative chi-squared χ^{*2} has been proposed in Section 7.3.1 (2), which gives an approximate p-value 3.5×10^{-8}, whereas the exact two-sided p-value of max acc. $t1$ is 2.5×10^{-6}. The approximation is via a constant times the chi-squared variable adjusted for the first two cumulants. An improvement by adjusting the first three cumulants is given in Hirotsu (1979). However, so small a p-value by the chi-squared approximation is not precise anyway, and should be read just as very highly significant, whereas the p-value by max acc. $t1$ is exact by counting up all possible cases – just as for the convexity test (max acc. $t2$) proposed in Section 7.3.2 (4).

7.3.4 Power comparisons

Max acc. $t1$ for the monotone hypothesis has been verified to keep high power for the wide range of the monotone hypothesis. For the normal model, max acc. $t2$ is also shown in Section 6.5.4 (4) to keep relatively high power in the wide range of the convexity hypothesis, compared with linear tests. In the following we compare the power of max acc. $t2$ and the cumulative chi-squared $\chi^{\dagger 2}$ with the test of the quadratic term of the logistic regression model by simulation of 10,000 replications, assuming a binomial distribution with $n_i \equiv 20$. We call the last test a polynomial test according to Hirotsu and Marumo (2002). The comparisons are made as two-sided tests in the direction of corner vectors, and also for the logistic quadratic pattern for each of $a=6$ and 10. The corner vectors and quadratic pattern are as follows:

	Corner vector 1	−10	8	5	2	−1	−4
	Corner vector 2	−20	2	24	11	−2	−15
$a=6$:	Corner vector 3	−15	−2	11	24	2	−20
	Corner vector 4	−4	−1	2	5	8	−10
	Quadratic pattern	−5	1	4	4	1	−5

	Corner vector 1	−36	16	13	10	7	4	1	−2	−5	−8
	Corner vector 2	−168	−28	112	87	62	37	12	−13	−38	−63
	Corner vector 3	−189	−70	49	168	122	76	30	−16	−62	−108
	Corner vector 4	−36	−17	2	21	40	26	12	−2	−16	−30
$a = 10$:	Corner vector 5	−30	−16	−2	12	26	40	21	2	−17	−36
	Corner vector 6	−108	−62	−16	30	76	122	168	49	−70	−189
	Corner vector 7	−63	−38	−13	12	37	62	87	112	−28	−168
	Corner vector 8	−8	−5	−2	1	4	7	10	13	16	−36
	Quadratic pattern	−6	−2	1	3	4	4	3	1	−2	−6

Then obviously the polynomial test is most powerful for the quadratic pattern among all the available tests. The non-centrality parameter $\sum(\theta_i - \bar{\theta}.)^2$ is fixed for the power of the polynomial test to be around 0.80 for the quadratic pattern.

The simulation results are shown in Table 7.7 also for the null model, where Const. means $\theta_i = \text{constant}$ and Slope means the regression model $\theta_i = \beta_0 + \beta_1 i$.

There is observed some conservatism for max acc. *t*2, but this is inevitable for an exact discrete test and in the acceptable range. In contrast, the other tests become sometimes slightly aggressive. The similarity between the corner vectors at the symmetric location suggests the reliability of the simulation. It is verified that max acc. *t*2 keeps relatively high power in the wide range of the convexity hypothesis. The quadratic test looks very good when the change-point is located in the middle, but not when it is at the end. Therefore, it cannot be recommended without prior information on the configuration of the mean vector. The cumulative chi-squared $\chi^{\dagger 2}$ looks to be located between max acc. *t*2 and the polynomial test. It is expected since the leading term of $\chi^{\dagger 2}$ is the chi-squared for the quadratic pattern in the expansion in the independent chi-squared series in the balanced normal model, see (6.52). The comparison between max acc. *t*2 and the polynomial test coincides very well with that of the normal case given in Table 6.14. Another advantage of max acc. *t*2 is that it can suggest a change-point.

The exact non-null distributions of max acc. *t*1 and *t*2 are also easily available. They are obtained by a simple modification of the kernel function $b(y_i)$ and details are given for the case of the Poisson distribution in Chapter 8.

Table 7.7 Power comparisons of max acc. $t2$, $\chi^{\dagger 2}$, and the polynomial test.

	Alternative hypothesis (θ)	Max acc. $t2$	Cumulative chi-squared	Polynomial test
	Const.	0.047	0.053	0.053
	Slope	0.046	0.052	0.046
	Corner vector 1	0.681	0.649	0.590
$a=6$	Corner vector 2	0.712	0.749	0.750
	Corner vector 3	0.715	0.744	0.752
	Corner vector 4	0.678	0.650	0.593
	Quadratic pattern	0.751	0.804	0.802
	Const.	0.048	0.053	0.050
	Slope	0.045	0.047	0.052
	Corner vector 1	0.644	0.519	0.418
	Corner vector 2	0.692	0.682	0.625
	Corner vector 3	0.707	0.738	0.726
$a=10$	Corner vector 4	0.715	0.760	0.763
	Corner vector 5	0.713	0.755	0.772
	Corner vector 6	0.709	0.742	0.734
	Corner vector 7	0.691	0.684	0.626
	Corner vector 8	0.652	0.529	0.422
	Quadratic pattern	0.738	0.799	0.808

7.3.5 Maximal contrast test for sigmoid and inflection point hypotheses

(1) **Weighted triply accumulated statistics and max acc. $t3$.** The sigmoidicity hypothesis for a general unequal spacing is defined by

$$H_{\rm sig}: \frac{1}{x_3-x_1}\left(\frac{\theta_3-\theta_2}{x_3-x_2}-\frac{\theta_2-\theta_1}{x_2-x_1}\right) \geq \cdots \geq \frac{1}{x_a-x_{a-2}}\left(\frac{\theta_a-\theta_{a-1}}{x_a-x_{a-1}}-\frac{\theta_{a-1}-\theta_{a-2}}{x_{a-1}-x_{a-2}}\right), \quad (7.31)$$

extending the convexity hypothesis of the previous section. The null hypothesis is defined by all the equalities in (7.31), which is equivalent to $H_0: \theta_i=\beta_0+\beta_1 x_i+\beta_2 x_i^2$. Equation (7.31) can be expressed in matrix form as $\boldsymbol{Q}_a^{*\prime}\boldsymbol{\theta} \geq 0$, just like equation (7.11) for the convexity hypothesis in Section 7.3.2 (1). The explicit form of $\boldsymbol{Q}_a^{*\prime}$ and its relationship with the inflection point model are given by Hirotsu and Marumo (2002). As the basic variables from the key vector, the weighted triply accumulated statistics are derived after similar algebra to deriving the weighted doubly accumulated statistics for the convexity hypothesis. They are denoted by W_k and expressed in terms of S_i as

$$W_k = \sum_{i=1}^{k} (x_{i+2} - x_i) S_i, \; k = 1, \ldots, a-2.$$

Then the maximal contrast test is defined by

$$w_m^* = \max_k w_k^*,$$

where w_k^* is the standardized version of $W_k, k = 1, \ldots, a-3$. The sufficient statistics are Y_a, S_{a-1}, and W_{a-2} as described in Section 6.5.5, and we can derive the factorization of the conditional probability $f(W_1, \ldots, W_{a-3} \mid Y_a, S_{a-1}, W_{a-2})$ into the products of conditional probabilities $\Pi_{k=1}^{a-3} f_k(W_k \mid W_{k+1}, W_{k+2}, W_{k+3})$. Here we employed the notation W_{a-2}, W_{a-1}, W_a, but actually they are expressed by the sufficient statistics Y_a, S_{a-1}, W_{a-2}. The calculation is essentially the same as given in Section 7.3.2 (3), except the inequalities for W_k for efficient execution of the bottom-up procedure.

Absolute: $$W_{a-2} - \{(x_{k+3} - x_{k+1}) + (x_{k+4} - x_{k+2}) + \cdots + (x_a - x_{a-2})\} \times \frac{x_{a-1} - x_1}{x_a - x_1} S_{a-1}$$

$$\leq W_k \leq \frac{x_{k+1} + x_{k+2} - x_1 - x_2}{x_{a-1} + x_a - x_1 - x_2} W_{a-2}, \; k = 1, \ldots, a-3.$$

Relative: $$\frac{x_{k+1} + x_{k+2} - x_1 - x_2}{x_k + x_{k+1} - x_1 - x_2} W_{k-1} \leq W_k$$

$$\leq \frac{(x_{k+2} - x_k)}{(x_{k+2} - x_k) + \cdots + (x_a - x_{a-2})} W_{a-2} + \frac{(x_{k+3} - x_{k+1}) + \cdots + (x_a - x_{a-2})}{(x_{k+2} - x_k) + \cdots + (x_a - x_{a-2})} W_{k-1},$$

$k = 1, \ldots, a-3,$

$$0 \leq \frac{1}{x_{k+1} - x_k} \left(\frac{W_k - W_{k-1}}{x_{k+2} - x_k} - \frac{W_{k-1} - W_{k-2}}{x_{k+1} - x_{k-1}} \right) - \frac{1}{x_k - x_{k-1}} \left(\frac{W_{k-1} - W_{k-2}}{x_{k+1} - x_{k-1}} - \frac{W_{k-2} - W_{k-3}}{x_k - x_{k-2}} \right)$$

$$\leq n_k, \; k = 1, \ldots, a; \; W_{-2} = W_{-1} = W_0 = 0.$$

The calculation of moments and the tail probability for the sigmoid test are essentially the same as given in the previous section for convexity, although the conditioning variables become three, making the computation a little harder for a large sequence. The maximal contrast test might be called max acc. $t3$.

(2) **Cumulative chi-squared statistic $\chi^{\#2}$**. For a directional goodness-of-fit test, the cumulative chi-squared statistic

$$\chi^{\#2} = \sum_{k=1}^{a-3} w_k^{*2}$$

is proposed again as a promising test. The chi-squared approximation of its distribution is obtained in exactly the same way as given in Section 7.3.3, just by replacing s_k^* by w_k^* and $(a-2)$ by $(a-3)$ in calculating the cumulants.

(3) **More direct method for calculating p-value in the non-explosive sequence.** If it is possible to write down all the conformable sequences in the case of a non-explosive sample space, a simpler method for calculating p-value is available. First, search for the conformable sequences by the bottom-up procedure, utilizing efficiently the inequalities, and attach them to the probabilities obtained by the products of conditional probabilities. Then, sum up those probabilities for the sequences whose test statistics are equal to or larger than the observed test statistic. The procedure is applicable to both the maximal contrast and cumulative chi-squared tests, and is particularly useful for a directional goodness-of-fit test of the dose–response curve, where the sequence is usually not large. This method is employed in Examples 7.6 and 7.8.

Example 7.8. Estimating the upper confidence bound at the lowest dose of an inverted sigmoidally constrained dose–response curve. The data in Table 7.8 are from Schmoyer (1984), who summarized and analyzed the experiment of Dalbey and Lock (1982). Schmoyer gave a smoothed estimate of the dose–response curve under the sigmoid assumption and an upper confidence bound 0.057 of the risk at the lowest dose under an additional assumption that the response rate at zero dose is zero, which improves a naïve estimate 0.095 $\left(=1-0.05^{1/30}\right)$ at significance level 0.05. Hirotsu and Srivastava (2000) obtained an upper confidence bound 0.056 under the monotone assumption of the dose–response curve and 0.035 under the same assumption with Schmoyer. The proof of sigmoidicity has been, however, based on the normal approximation of the binomial distribution, which might be difficult to assume for the extremely low and high doses of the experiment. Therefore, we apply the exact max acc. $t3$ test developed in this section, (1) ~ (3), to obtain the p-values 0.025 for inverted sigmoidicity, 0.814 for sigmoidicity, and 0.051 for two-sided. In contrast, the two-sided p-value is 0.056 by the cumulative chi-squared $\chi^{\#2}$. This suggests an inverted sigmoid departure of the response from a logit linear model. Then, the logit

Table 7.8 Results of diesel fuel aerosol experiment.

Dose x_i	n_i	y_i	y_i/n_i
8	30	0	0
16	40	1	0.025
24	40	2	0.05
28	10	5	0.5
32	30	12	0.4
48	20	16	0.8
64	10	6	0.6
72	10	10	1.0

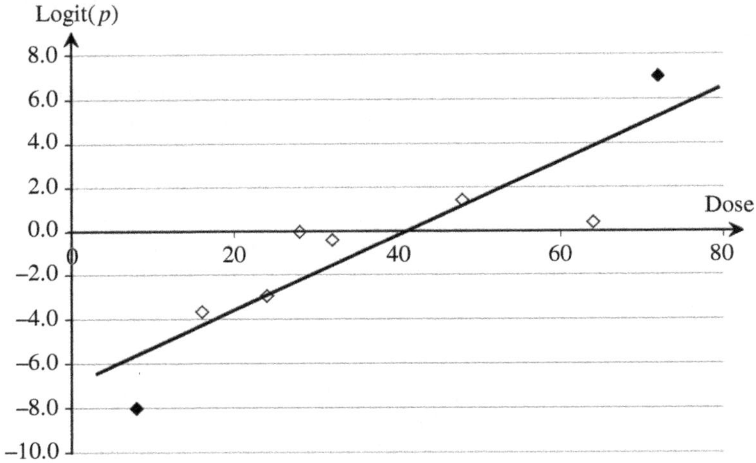

Figure 7.1 Fitting a logit linear regression curve.

linear regression will give a higher response rate than the true response rate at the lowest dose, giving a conservative estimate of the risk (see Fig. 7.1). Thus we have a conservative upper confidence bound 0.055 of the risk at the lowest dose at significance level 0.05 by a standard software for logit regression analysis.

It is interesting to see that several different methods suggest a rather similar estimate. Then, the present method should be appealing by an exact test of the sigmoidicity and the maximum likelihood estimation of only two parameters, utilizing the whole data without any additional assumption like zero response rate at zero dose. Other methods are utilizing only a part of data.

Finally, while accumulated statistics are very popular in statistical applications, the use of doubly and triply accumulated statistics is rather novel and expected to have new applications. Some of the software is provided on the author's website.

References

Allen, F. H., Diamond, L. K. and Watrous, J. B. (1949) Erythroblastosis fetalis: The value of blood from female donors for exchange transfusion. *New Engl. J. Med.* **241**, 799–806.

Armitage, P. (1955) Tests for linear trends in proportions and frequencies. *Biometrics* **11**, 375–386.

Bartholomew, D. J. (1963) On Chassan's test for order. *Biometrics* **19**, 188–191.

Chassan, J. B. (1960) On a test for order. *Biometrics* **16**, 119–121.

Chassan, J. B. (1962) An extension of a test for order. *Biometrics* **18**, 245–247.

Cochran, W. G. (1954) Some methods of strengthening the common chi-square tests. *Biometrics* **10**, 417–451.

Cochran, W. G. (1955) A test of a linear function of the deviations between observed and expected numbers. *J. Amer. Statist. Assoc.* **50**, 377–397.

Dalbey, W. and Lock, S. (1982) Inhalation toxicology of diesel fuel obsculant aerosol in Sprague-Dawley rats. ORNL/TM-8867, Oak Ridge National Laboratory, Oak Ridge, TN.

Hirotsu, C. (1979) An *F* approximation and its application. *Biometrika* **66**, 577–584.

Hirotsu, C. (2013) Theory and its application of the cumulative, two-way cumulative and doubly cumulative sum statistics. *Jap. J. Appl. Statist.* **42**, 121–143 (in Japanese).

Hirotsu, C. and Marumo, K. (2002) Change point analysis as a method for isotonic inference. *Scand. J. Statist.* **29**, 125–138.

Hirotsu, C. and Srivastava, M. S. (2000) Simultaneous confidence intervals based on one-sided max *t* test. *Statist. Prob. Lett.* **49**, 25–37.

Hirotsu, C., Aoki, S., Inada, T. and Kitao, Y. (2001) An exact test for the association between the disease and alleles at highly polymorphic loci with particular interest in the haplotype analysis. *Biometrics* **57**, 769–778.

Hirotsu, C., Yamamoto, S. and Tsuruta, H. (2016) A unifying approach to the shape and change-point hypotheses in the discrete univariate exponential family. *Comput. Statist. Data Anal.* **97**, 33–46.

Schmoyer, R. L. (1984) Sigmoidally constrained maximal likelihood estimation in quantal bioassay. *J. Amer. Statist. Assoc.* **79**, 448–453.

Taguchi, G. (1966) *Statistical analysis*. Maruzen, Tokyo (in Japanese).

Takeuchi, K. and Hirotsu, C. (1982) The cumulative chi-squares method against ordered alternatives in two-way contingency tables. *Rep. Statist. Appl. Res., JUSE* **29**, 1–13.

8

Poisson Process

Max acc. $t1$ and max acc. $t2$ for the binomial distribution can be applied almost as they are for the change-point hypotheses in a Poisson sequence. As a real example, a sequence of spontaneous reporting of drug adverse events caused by the administration of some compound drug for interstitial pneumonia is given in Table 8.1. The data are the number of reportings per month from November 2003 to May 2010 at PMDA (Pharmaceutical and Medical Device Agency, Japan) and an independent Poisson sequence is assumed. It is a very serious problem at PMDA how to detect a significant change of time series in as short a time as possible. Again, we are interested in detecting a change-point as well as a change of general tendency. Then the first choice will be max acc. $t1$.

8.1 Max acc. $t1$ for the Monotone and Step Change-Point Hypotheses

8.1.1 Max acc. $t1$ statistic in the Poisson sequence

We assume an independent Poisson distribution $Po(\Lambda_i)$ for the observed sequence y_i and apply Lemma 6.2 for $\theta_i = \log \Lambda_i$. Then the standardized accumulated statistics corresponding to (7.5) of the binomial case are

$$t_k(Y_k) = t_k = \left\{ \left(\frac{a-k}{ak} \right) \hat{\Lambda} \right\}^{-1/2} \left(\hat{\Lambda} - \frac{Y_k}{k} \right),\ k = 1,\ \ldots,\ a-1, \tag{8.1}$$

Advanced Analysis of Variance, First Edition. Chihiro Hirotsu.

Table 8.1 Spontaneous reporting of adverse events per month at PMDA.

k	1	2	3	4	5	6	7	8	9	10	11	12	13	14	15
y_k	1	4	1	1	1	1	3	0	4	1	3	0	2	4	3
k	16	17	18	19	20	21	22	23	24	25	26	27	28	29	30
y_k	3	2	4	1	4	1	4	2	1	2	2	1	0	1	5
k	31	32	33	34	35	36	37	38	39	40	41	42	43	44	45
y_k	1	4	1	4	2	3	7	3	3	4	1	5	4	5	6
k	46	47	48	49	50	51	52	53	54	55	56	57	58	59	60
y_k	2	4	9	3	4	1	1	6	3	5	8	1	1	6	3
k	61	62	63	64	65	66	67	68	69	70	71	72	73	74	75
y_k	3	1	2	3	1	3	4	3	3	5	2	2	0	4	4
k	76	77	78	79											
y_k	4	2	2	4											

where $Y_k = y_1 + \ldots + y_k$, $\hat{\Lambda} = Y_a/a$. It should be noted that Y_a is the sufficient statistic under the null model $\theta_1 = \cdots = \theta_a$, and the conditional null distribution of Y_k given Y_a is a binomial distribution

$$B\left(Y_k \mid Y_a, \frac{k}{a}\right) \sim \binom{Y_a}{Y_k} \left(\frac{k}{a}\right)^{Y_k} \left(1 - \frac{k}{a}\right)^{Y_a - Y_k},$$

so that the conditional variance given Y_a is $\{k(a-k)/\mathrm{a}\}\hat{\Lambda}$.

8.1.2 Distribution function of max acc. $t1$ under the null model

In this case the Markov property of $t_k(Y_k)$, $k = 1, \ldots, a-1$ (8.1) is shown by the form of the joint conditional distribution of $Y_1, \ldots, Y_{a-1}$ given Y_a. Now it is well known that the joint null distribution of y_k at $\Lambda_i \equiv \Lambda$ is factorized in the form

$$\Pi_{k=1}^{a}\{e^{-\Lambda}\Lambda^{y_k}/y_k!\} = \frac{Y_a!}{\Pi y_i!}\left(\frac{1}{a}\right)^{Y_a} \times \frac{e^{-a\Lambda}(a\Lambda)^{Y_a}}{Y_a!},$$

where the last part is the Poisson distribution $Po(a\Lambda)$ for Y_a and the first part is a multinomial distribution for $y_1, \ldots, y_a$ given Y_a. This part is further factorized into the products of binomial distributions as

$$\Pi_{k=1}^{a-1} \binom{Y_{k+1}}{Y_k} \left(\frac{k}{k+1}\right)^{Y_k} \left(\frac{1}{k+1}\right)^{Y_{k+1} - Y_k}$$

to give the conditional joint distribution of $Y_1, \ldots, Y_{a-1}$ given Y_a. Then, the Markov property of Y_k follows immediately from Lemma 7.1. The recursion formula (7.10) for max acc. $t1$ is valid as it is, except that the conditional distribution is now a binomial distribution

$$f_k(Y_k|Y_{k+1}) = \binom{Y_{k+1}}{Y_k}\left(\frac{k}{k+1}\right)^{Y_k}\left(\frac{1}{k+1}\right)^{Y_{k+1}-Y_k}.$$

Example 8.1. Change-point analysis of Table 8.1. Max acc. $t1$ is applied by the estimate of $\hat{\Lambda}=2.835$. The observed maximum is max $t_k(Y_k)=3.497$ at April 2006 ($k=29$) with p-value 0.0096 by the recursion formula (7.10). It should be noted that large $t_k(Y_k)$ suggests a shift of mean between k and $k+1$, and we call this a change at $k+1$. Therefore, max acc. $t1$ suggests a shift of means at $k+1=30$. Then it is interesting to confirm whether it changed to a decreasing tendency at some point after that time ($k+1=30$) by an appropriate action. For this purpose, max acc. $t2$ for the concavity hypothesis can be applied (see Section 8.2.1).

8.1.3 Max acc. $t1$ under step change-point model

In the change-point analysis, it is also of interest to make a statistical inference on the step change-point k. For this purpose, the non-null distribution is obtained first.

(1) **Non-null distribution.** For the normal model, the maximal accumulated statistics have been derived as efficient scores for the step change-point and slope change-point models in Sections 6.5.3 (1) (b) and 6.5.4 (3), respectively. In this section we develop the non-null distribution for the step change-point model in the mean of a Poisson sequence:,

$$M_k \begin{cases} \theta_i = \log \Lambda_i = \theta, \ i = 1, \cdots, k, \\ \theta_i = \log \Lambda_i = \theta + \Delta, \ i = k+1, \ldots, a. \end{cases} \tag{8.2}$$

Now, it is an easy calculation to obtain the factorization of the joint conditional distribution given Y_a under model (8.2) as

$$G(y \mid Y_a, \Delta) = \Pi_{i=1}^{a-1} f_i^*(Y_i|Y_{i+1}), \tag{8.3}$$

where

$$\begin{cases} f_i^*(Y_i|Y_{i+1}) = \binom{Y_{i+1}}{Y_i}\left(\frac{i}{i+1}\right)^{Y_i}\left(\frac{1}{i+1}\right)^{Y_{i+1}-Y_i}, i=1, \dots, k-1, \\ f_i^*(Y_i|Y_{i+1}) = C_{i+1}^{-1}(Y_{i+1},\Delta)\binom{Y_{i+1}}{Y_i}\left(\frac{i}{i+1}\right)^{Y_i}\left(\frac{1}{i+1}\right)^{Y_{i+1}-Y_i} e^{-Y_i\Delta}, i=k, \\ f_i^*(Y_i|Y_{i+1}) = C_{i+1}^{-1}(Y_{i+1},\Delta)C_i(Y_i,\Delta)\binom{Y_{i+1}}{Y_i}\left(\frac{i}{i+1}\right)^{Y_i}\left(\frac{1}{i+1}\right)^{Y_{i+1}-Y_i}, i=k+1,\dots,a-1 \end{cases}$$

with $C_i(Y_i,\Delta)$, $k+1 \le i \le a$, normalizing constants. It should be noted that in Section 7.3.1 (1) under the null hypothesis of the step change-point model, the coefficients C_i are known and need not be calculated numerically. For the non-null distribution where such a coefficient is unknown, we can apply a similar recursion formula to calculate these constants as employed in Section 7.3.2 (2) for the doubly accumulated statistics. The method is particularly useful for obtaining a distribution further conditioned on Y_k to obtain a confidence region for a change-point in the next section. The method is also systematically extended to the slope change-point model in Section 8.2. Now we express $f_i^*(Y_i|Y_{i+1})$ as

$$\begin{cases} f_i^*(Y_i|Y_{i+1}) = C_{i+1}^{-1}(Y_{i+1})C_i(Y_i)\{(Y_{i+1}-Y_i)!\}^{-1}, i=1, \dots, k-1, \\ f_k^*(Y_k|Y_{k+1}) = C_{k+1}^{-1}(Y_{k+1}, \Delta)C_k(Y_k)e^{-Y_k\Delta}\{(Y_{k+1}-Y_k)!\}^{-1}, i=k, \\ f_i^*(Y_i|Y_{i+1}) = C_{i+1}^{-1}(Y_{i+1}, \Delta)C_i(Y_i, \Delta)\{(Y_{i+1}-Y_i)!\}^{-1}, i=k+1,\dots,a-1, \end{cases} \tag{8.4}$$

by changing the definition of the coefficients C_i slightly, and calculate the constants recursively starting from $C_1 = 1/Y_i!$ by

$$\begin{cases} C_{i+1}(Y_{i+1}) = \sum_{Y_i} C_i(Y_i)\{(Y_{i+1}-Y_i)!\}^{-1}, i=1, \dots, k-1, \\ C_{k+1}(Y_{k+1}, \Delta) = \sum_{Y_k} C_k(Y_k)e^{-Y_k\Delta}\{(Y_{k+1}-Y_k)!\}^{-1}, i=k, \\ C_{i+1}(Y_{i+1}, \Delta) = \sum_{Y_i} C_i(Y_i, \Delta)\{(Y_{i+1}-Y_i)!\}^{-1}, i=k+1,\dots,a-1. \end{cases} \tag{8.5}$$

The recursion formula (7.10) is valid as it is again, by substituting the conditional distribution (8.4), and gives a useful method for calculating the power at given Δ. Equation (8.4) shows also that the accumulated statistic Y_k is the efficient score with respect to Δ for the step change-point model (8.2).

(2) **Confidence region for a step change-point.** We have given the confidence interval on the amount of change Δ in the sections of Chapter 6. However, in Example 8.1

we are also interested in the confidence region on the change-point $k+1$. As usual, it is obtained as the set of $K+1$ that are not rejected at level α by the test of the null hypothesis on the change-point,

$$H_0^{K+1}: k+1 = K+1, \tag{8.6}$$

against the alternative hypothesis

$$H_1^{K+1}: k+1 \neq K+1$$

asserting $K+1$ not to be a change-point. An appropriate test statistic is again $\max_k t_k(Y_k)$ of (8.1), but the maximization is with respect to $k=1,\ldots,a-1$, $k \neq K$ and its null distribution should be defined under the null hypothesis H_0^{K+1}. This statistic is asymptotically equivalent to the likelihood ratio statistic by Worsley (1986). In this case the null distribution contains a nuisance parameter Δ. However, according to Worsley (1986), we can make the inference free from Δ by conditioning on the sufficient statistic Y_K under H_0^{K+1}. The conditional null distribution is most easily obtained by running the recursion formulae (8.5) and (7.10), fixing Y_K at the observed value. This is easily done by altering the inequality for restricting Y_K to the one point of the observed value in running the recursion formula for calculating power in Section 8.1.3 (1). Then, the confidence region eventually collects those $K+1$ for which $t_K(Y_K)$ is sufficiently close to the observed maximum $\max_k t_k(Y_k)$. Worsley (1986) proposed a different recursion formula considering independent Markov processes for both sides of the assumed change-point. However, our method is most easy to extend to the second order Markov sequence in Section 8.2. An all in one program for p-value, power, and confidence region is given in Hirotsu and Tsuruta (2017).

Example 8.2. Example 8.1 continued. We test the null hypothesis (8.6) for $K=1, \ldots, a-1$, applying the recursion formulae (8.5) and (7.10), and collect those $K+1$ with two-sided p-value larger than or equal to 0.10. Then the confidence region at confidence coefficient 0.90 is obtained as an interval $27 \leq K+1 \leq 43$.

8.2 Max acc. *t*2 for the Convex and Slope Change-Point Hypotheses

8.2.1 Max acc. *t*2 statistic in the Poisson sequence

The max acc. *t*2 method based on the statistic S_k has been developed in Section 7.3.2, assuming a discrete univariate exponential family and that all the procedures for a binomial distribution are valid also for a Poisson distribution with only a slight modification. The function $b(y_k)$ is changed to $(y_k!)^{-1}$, so that we use the function $b(y_k) = \{(S_k - 2S_{k-1} + S_{k-2})!\}^{-1}$ in the probability calculation of Section 7.3.2 (2), since the data are taken every one-month interval. The formulae in 7.3.2 (2) ~ (4)

are all valid, including the conditional distribution of (7.16), except that n_{k+1} is not necessary in equation (7.23).

Example 8.3. Example 8.1 continued. In this example we are interested in testing a downturn tendancy and therfore perform a concavity test based on $-s_m^*$. A downturn is detected at October 2007 $(K+1=48)$ with observed maximum $-s_m^*=2.858$ and p-value 0.0093 by the formula given in Section 7.3.2 (4) (c).

8.2.2 Max acc. *t*2 under slope change-point model

(1) **Non-null distribution.** In this section we consider a slope change-point model at $x=x_{k+1}$ assuming general unequal intervals,

$$M_k\begin{cases}\theta_i=\log\Lambda_i=\beta_0+\beta_k x_i,\ i=1,\ \ldots,k+1,\\ \theta_i=\log\Lambda_i=\beta_0^*+\beta_k^* x_i,\ i=k+2,\ \ldots,\ a,\end{cases}$$

where $\beta_0+\beta_k x_{k+1}=\beta_0^*+\beta_k^* x_{k+1}$. Defining $\Delta_k=\beta_k^*-\beta_k$, model M_k can be rewritten as

$$M_k\begin{cases}\theta_i=\log\Lambda_i=\beta_0^*+\Delta_k x_{k+1}+(\beta_k^*-\Delta_k)x_i,\ i=1,\ \ldots,k+1\\ \theta_i=\log\Lambda_i=\beta_0^*+\beta_k^* x_i,\ i=k+2,\ \ldots,\ a.\end{cases}\tag{8.7}$$

Then the conditional distribution $G(\mathbf{y}|Y_a,T_a,\Delta_k)$ is obtained as

$$G(\mathbf{y}|Y_a,T_a,\Delta_k)=C^{-1}(Y_a,T_a,\Delta_k)\Pi_{i=1}^{a-2}\left\{\left(\frac{S_{i+2}-S_{i+1}}{x_{i+3}-x_{i+2}}-\frac{S_{i+1}-S_i}{x_{i+2}-x_{i+1}}\right)!\right\}^{-1}e^{S_k\Delta_k},\tag{8.8}$$

where the definitions of Y_a, T_a, and S_i are the same as in Section 7.3.2. The factorization of the conditional distribution $G(\mathbf{y}|Y_a,T_a,\Delta_k)$ is obtained by a slight modification of Section 7.3.2. The idea is very similar to the previous section, and we give only the result in the following:

$$\begin{cases}f_i^*(S_i|S_{i+1},S_{i+2})=C_{i+1}^{-1}(S_{i+1},S_{i+2})C_i(S_i,S_{i+1})\left\{\left(\frac{S_{i+2}-S_{i+1}}{x_{i+3}-x_{i+2}}-\frac{S_{i+1}-S_i}{x_{i+2}-x_{i+1}}\right)!\right\}^{-1},\ 1\le i\le k-1\\ f_k^*(S_k|S_{k+1},S_{k+2})=C_{k+1}^{-1}(S_{k+1},S_{k+2},\Delta_k)C_k(S_k,S_{k+1})\left\{\left(\frac{S_{k+2}-S_{k+1}}{x_{k+3}-x_{k+2}}-\frac{S_{k+1}-S_k}{x_{k+2}-x_{i+1}}\right)!\right\}^{-1}e^{S_k\Delta_k},\ i=k,\\ f_i^*(S_i|S_{i+1},S_{i+2})=C_{i+1}^{-1}(S_{i+1},S_{i+2},\Delta_k)C_i(S_i,S_{i+1},\Delta_k)\left\{\left(\frac{S_{i+2}-S_{i+1}}{x_{i+3}-x_{i+2}}-\frac{S_{i+1}-S_i}{x_{i+2}-x_{i+1}}\right)!\right\}^{-1},\ k+1\le i\le a-2.\end{cases}$$

We can calculate the normalizing constants recursively, noting the change of kernel for calculating $C_{k+1}^{-1}(S_{k+1},S_{k+2},\Delta_k)$ at $i=k$. The overall constant $C(Y_a,T_a,\Delta_k)$ is obtained at the final step. It should be noted that the initial value of the normalizing constant is

$$C_1(S_1,S_2)=\left\{\left(\frac{S_1}{x_2-x_1}\right)!\right\}^{-1}\left\{\left(\frac{S_2-S_1}{x_3-x_2}-\frac{S_1}{x_2-x_1}\right)!\right\}^{-1}.$$

If the change-point is at $x = x_2$, then the recursion formula starts from the second equation. Equation (8.8) shows again that the accumulated statistic S_k is the efficient score with respect to Δ_k for the slope change-point model (8.7).

(2) **Confidence region for a slope change-point.** The confidence set of the change-point x_{k+1} is obtained as the set of x_{K+1} that are not rejected at level α by the test of null hypothesis on the change-point,

$$H_0^{x_{K+1}} : x_{k+1} = x_{K+1},$$

against the alternative hypothesis,

$$H_1^{x_{K+1}} : x_{k+1} \neq x_{K+1}$$

asserting for x_{K+1} not to be a change-point. An appropriate test statistic is again the maximal standardized statistic of S_k in Section 7.3.2, but the maximization is with respect to $k = 1, \ldots, a-2$, $k \neq K$ and its null distribution should be defined under the null hypothesis $H_0^{x_{K+1}}$. Again we make a conditional inference given the sufficient statistic S_K under $H_0^{x_{K+1}}$ to be free from the nuisance parameter Δ_K. The conditional null distribution is most easily obtained by running the recursion formula fixing S_K at the observed value. This is just as in Section 8.1.3 (2).

Example 8.4. Example 8.1 continued. Applying the recursion formula for $K = 1, \ldots., a-2$, the confidence region at confidence coefficient 0.90 is obtained as an interval $35 \leq x_{K+1} \leq 58$.

The non-null distributions of max acc. $t1$ and $t2$ in this chapter can be applied to the binomial distribution by a simple alteration of the kernel function as given in Example 7.6.

References

Hirotsu, C. and Tsuruta, H. (in press) An algorithm for a new method of change-point analysis in the independent Poisson sequence. *Biometrical Letters* **54**.

Worsley, K. J. (1986) Confidence regions and tests for a change-point in a sequence of exponential family of random variables. *Biometrika* **73**, 91–104.

9

Block Experiments

The role of the block factor has already been mentioned in Sections 1.6 and 5.1.3. In this chapter we discuss a general a-sample problem and non-parametric tests in block experiments.

9.1 Complete Randomized Blocks

In comparing a treatments, we prepare b blocks and randomize a treatments in each block. We assume no-interaction between treatment and block, so that the statistical model is expressed as

$$y_{ij} = \alpha_i + \beta_j + e_{ij},\ i = 1,\ \dots,\ a,\ j = 1, \dots,\ b, \tag{9.1}$$

where α_i denotes the treatment effect, β_j the block effect, and the error e_{ij} are distributed independently and identically as normal $N(0,\ \sigma^2)$. Let $\boldsymbol{y}$ be a column vector of y_{ij} arranged in dictionary order, and model (9.1) be expressed in matrix form as

$$\boldsymbol{y} = \boldsymbol{X}_\alpha \boldsymbol{\alpha} + \boldsymbol{X}_\beta \boldsymbol{\beta} + \boldsymbol{e},$$

where $\boldsymbol{\alpha} = (\alpha_1, \dots, \alpha_a)'$, $\boldsymbol{\beta} = (\beta_1, \dots, \beta_b)'$, $\boldsymbol{X}_\alpha = \begin{bmatrix} \boldsymbol{j}_b & \boldsymbol{0} & \cdots & \boldsymbol{0} \\ \boldsymbol{0} & \boldsymbol{j}_b & \cdots & \boldsymbol{0} \\ & & \ddots & \\ \boldsymbol{0} & \boldsymbol{0} & \cdots & \boldsymbol{j}_b \end{bmatrix}_{n \times a}$, and $\boldsymbol{X}_\beta = \begin{bmatrix} \boldsymbol{I}_b \\ \boldsymbol{I}_b \\ \vdots \\ \boldsymbol{I}_b \end{bmatrix}_{n \times b}$

with $n = ab$. This is a simple linear model, and we can derive an analysis of treatment effects $\boldsymbol{\alpha}$ with a nuisance parameter $\boldsymbol{\beta}$. However, we can also derive a standard form by an orthonormal transformation

$$\boldsymbol{M}' = \begin{bmatrix} n^{-1/2}\boldsymbol{j}'_n \\ b^{-1/2}\boldsymbol{P}'_a\boldsymbol{X}'_\alpha \\ a^{-1/2}\boldsymbol{P}'_b\boldsymbol{X}'_\beta \\ \boldsymbol{Q}' \end{bmatrix}.$$

Then it is a matter of simple algebra to derive

$$\boldsymbol{M}'\boldsymbol{y} = \begin{bmatrix} z_\mu \\ z_\alpha \\ z_\beta \\ z_e \end{bmatrix} = \begin{bmatrix} n^{1/2}(\bar{\alpha}. + \bar{\beta}.) \\ b^{1/2}\boldsymbol{P}'_a\alpha \\ a^{1/2}\boldsymbol{P}'_b\beta \\ \boldsymbol{0} \end{bmatrix} + \boldsymbol{M}'\boldsymbol{e}. \tag{9.2}$$

In equation (9.2) the estimable functions are nicely separated, which leads to the partition of sums of squares:

$$\begin{aligned} \|\boldsymbol{M}'\boldsymbol{y}\|^2 &= \boldsymbol{y}'\boldsymbol{y} = \textstyle\sum_i\sum_j y_{ij}^2 \\ &= n^{-1}(\boldsymbol{j}'_n\boldsymbol{y})^2 + b^{-1}\|\boldsymbol{P}'_a\boldsymbol{X}'_\alpha\boldsymbol{y}\|^2 + a^{-1}\|\boldsymbol{P}'_b\boldsymbol{X}'_\beta\boldsymbol{y}\|^2 + \|\boldsymbol{Q}'\boldsymbol{y}\|^2 \\ &= n^{-1}y_{..}^2 + \left(b^{-1}\textstyle\sum_i y_{i.}^2 - n^{-1}y_{..}^2\right) + \left(a^{-1}\textstyle\sum_j y_{.j}^2 - n^{-1}y_{..}^2\right) + \|\boldsymbol{Q}'\boldsymbol{y}\|^2 \\ &= S_\mu \quad + \quad S_\alpha \quad\quad + \quad S_\beta \quad\quad + S_e \end{aligned} \tag{9.3}$$

where S_μ is for a general mean and sometimes called a correction term (CT). In contrast, the total sum of squares is obtained as

$$S_T = \textstyle\sum_i\sum_j y_{ij}^2 - y_{..}^2/n \tag{9.4}$$

and S_e is calculated by

$$S_e = S_T - S_\alpha - S_\beta \tag{9.5}$$

with df $ab - a - b + 1$. The unbiased variance is obtained as

$$\hat{\sigma}^2 = S_e/(ab - a - b + 1). \tag{9.6}$$

The components S_β and S_e in equation (9.3) partition the error sum of squares of a one-way layout into two parts. In other words, S_e (9.5) is obtained by eliminating block effects S_β from S_e in Section 6.1. S_T is nothing but S_0 in (6.3), and S_α is S_H in (6.4) with $n_{i.} \equiv b$. The non-centrality parameter of

Table 9.1 ANOVA table for block experiment.

Factor	Sum of squares	df	Mean squares	F	Non-centrality
Main effects A	S_α (9.3)	$a-1$	$S_\alpha/(a-1)$	F (9.7)	γ_α (9.8)
Block effects B	S_β (9.3)	$b-1$	$S_\beta/(b-1)$		
Error	S_e (9.5)	$ab-a-b+1$	$S_e/(ab-a-b+1)=\hat{\sigma}^2$ (9.6)		
Total	S_T (9.4)	$ab-1$			

$$F=\frac{S_\alpha/(a-1)}{S_e/(ab-a-b+1)} \tag{9.7}$$

is

$$\gamma_\alpha=b\sum_i(\alpha_i-\bar{\alpha}.)^2/\sigma^2, \tag{9.8}$$

which is the same as γ (6.6) of a one-way layout when the repetition number is b. These results are summarized in ANOVA Table 9.1.

Thus, the structure is the same as a one-way layout, except that the error sum of squares has the block effects S_β subtracted. Therefore an overall homogeneity test, Scheffé, Tukey, and Dunnett-type multiple comparisons, and directional tests are performed in the same way as for a one-way layout using the unbiased variance $\hat{\sigma}^2$ (9.6). However, if there is a missing value, then the structure becomes unbalanced and we have to go back to a general linear statistical inference.

Example 9.1. Measurements of enzymatic activation. The data of Table 9.2 are for evaluating the methods A_1 and A_2 to measure enzymatic activation (Hirotsu, 1992). Two samples from each of eight lots are evaluated by A_1 and A_2, respectively. From the totals given in Table 9.2, we easily obtain

Table 9.2 Measurements of enzymatic activation.

	Lot j								
Method i	1	2	3	4	5	6	7	8	Total ($y_{i\cdot}$)
1	434	431	454	441	428	420	448	432	3488
2	430	436	441	434	423	410	442	432	3448
Total ($y_{\cdot j}$)	864	867	895	875	851	830	890	864	6936 ($=y_{\cdot\cdot}$)
Difference	4	-5	13	7	5	10	6	0	40 ($=y_{1\cdot}-y_{2\cdot}$)

Table 9.3 ANOVA table for measurements of enzymatic activation.

Factor	Sum of squares	df	Mean squares	F	Non-centrality
Main effects A	100	1	100	6.36*	γ_α (9.8)
Block effects B	1510	7	215.7		
Error	110	7	$15.7 = \hat{\sigma}^2$		
Total	1720	15			

$$S_T = 1720,$$
$$S_\alpha = 100,$$
$$S_\beta = 1510,$$
$$S_e = S_T - S_\alpha - S_\beta = 110.$$

These results are summarized in ANOVA Table 9.3. The result is significant at level 0.05 compared with $F_{1,\,7}(0.05) = 5.59$.

If we ignore the variation of the lots and perform the usual ANOVA of a one-way layout, we should have the error sum of squares

$$S'_e = S_e + S_\beta = 1620$$

with df 14 $(=7+7)$, which leads to the Fratio

$$F' = \frac{S_\alpha/1}{S'_e/14} = 0.86$$

and no evidence is obtained for the difference between two measuring methods because of the large variation among the lots.

On the other hand, we may apply the method for a paired sample in Section 5.1.3 since $a = 2$. The error sum of squares for the differences is

$$S''_e = 4^2 + (-5)^2 + \cdots + 0^2 - 40^2/8 = 220,$$

with df 7. This is exactly twice the error sum of squares S_e here. It corresponds to the fact that the variance of the differences $y_{i1} - y_{i2}$ is $2\sigma^2$. Anyway, we have the t-statistic

$$t = \frac{40/8}{\sqrt{220/(8-1)}} = 2.523,$$

whose square exactly coincides with F in Table 9.3. Thus it has been verified that the method of this section coincides with the method for a paired sample when $a = 2$.

9.2 Balanced Incomplete Blocks

When an experiment requires a long time or a large space, it will be impossible to carry out a complete set of a experiments in a block because of its capacity. The possible number of experiments in a block is called the block size, denoted by k in this section. The design of experiments in which k is smaller than the number of treatments a is called an incomplete block design, where a complete set of treatments cannot be performed in a block. Among various incomplete block designs we consider here only the balanced incomplete block design (BIBD), which satisfies three conditions:

1. a treatments are replicated an equal number r times;
2. b blocks are of equal block size k;
3. every pair of treatments occurs simultaneously in the same number λ of blocks $-\lambda$ is called an association number.

Under these conditions, the number of experiments satisfies

$$n = ar = bk.$$

A treatment appears in r blocks and in each block it meets with $k-1$ other treatments. In contrast, it meets λ times with each of the other treatments, which leads to an equation

$$\lambda(a-1) = r(k-1). \tag{9.9}$$

It should be noted that equation (9.9) is a necessary condition for a BIBD and not a sufficient condition. The BIBD is expressed by an incidence matrix $N(a \times b)$ composed of 0, 1 elements, where every row represents treatment, every column represents block, and the (i, j) element is unity if the ith treatment occurs in the jth block and 0 otherwise. An example of the incidence matrix for BIBD $(a=8, k=4, b=14, r=7, \lambda=3)$ is given below:

$$N = \begin{bmatrix} 1\,0\,1\,0\,1\,0\,1\,0\,1\,0\,1\,0\,1\,0 \\ 1\,0\,1\,0\,0\,1\,0\,1\,1\,0\,0\,1\,0\,1 \\ 1\,0\,0\,1\,1\,0\,0\,1\,0\,1\,1\,0\,0\,1 \\ 1\,0\,0\,1\,0\,1\,1\,0\,0\,1\,0\,1\,1\,0 \\ 0\,1\,0\,1\,0\,1\,0\,1\,1\,0\,1\,0\,1\,0 \\ 0\,1\,0\,1\,1\,0\,1\,0\,1\,0\,0\,1\,0\,1 \\ 0\,1\,1\,0\,0\,1\,1\,0\,0\,1\,1\,0\,0\,1 \\ 0\,1\,1\,0\,1\,0\,0\,1\,0\,1\,0\,1\,1\,0 \end{bmatrix}_{8\times 14}. \tag{9.10}$$

It is observed in the incidence matrix (9.10) that every row contains seven unities, every column four unities, and every pair of treatments appears in three blocks simultaneously. It is easy to understand the equation

$$NN' = \begin{bmatrix} r\ \lambda \cdots\cdots \lambda \\ \lambda\ r\ \lambda \cdots \lambda \\ \ddots \\ \lambda\ \lambda \cdots\cdots r \end{bmatrix}_{a\times a}.$$

The structure of the datum y_{ij} of the ith treatment in the jth block is

$$y_{ij} = \alpha_i + \beta_j + e_{ij},\ i = 1, \ldots, a, j = 1, \ldots, b, \tag{9.11}$$

where the error e_{ij} are distributed independently as normal $N(0, \sigma^2)$. Let $\boldsymbol{y}$ be a column vector of y_{ij} arranged in dictionary order, and the model (9.11) expressed in matrix form as

$$\boldsymbol{y} = \boldsymbol{X}_\alpha \boldsymbol{\alpha} + \boldsymbol{X}_\beta \boldsymbol{\beta} + \boldsymbol{e} \tag{9.12}$$

where $\boldsymbol{\alpha} = (\alpha_1, \ldots, \alpha_a)'$, $\boldsymbol{\beta} = (\beta_1, \ldots, \beta_b)'$ as before. Noting $\boldsymbol{X}_\alpha'\boldsymbol{X}_\alpha = r\boldsymbol{I}_a, \boldsymbol{X}_\beta'\boldsymbol{X}_\beta = k\boldsymbol{I}_b$ and $\boldsymbol{N} = \boldsymbol{X}_\alpha'\boldsymbol{X}_\beta$, we can derive a standard form for BIBD. It should be noted that in the complete randomized blocks of Section 9.1, $\boldsymbol{X}_\alpha'\boldsymbol{X}_\beta$ is an $a \times b$ matrix with all the elements unity.

First we derive a normal equation (2.27) for the linear model (9.12) and eliminate the block effects $\boldsymbol{\beta}$ to obtain the adjusted equation (9.13) for treatment effects $\boldsymbol{\alpha}$:

$$\boldsymbol{A}'\boldsymbol{X}_\alpha \hat{\boldsymbol{\alpha}} = \boldsymbol{A}'\boldsymbol{y}, \tag{9.13}$$

$$\boldsymbol{A} = \boldsymbol{X}_\alpha - k^{-1}\boldsymbol{X}_\beta \boldsymbol{X}_\beta' \boldsymbol{X}_\alpha.$$

It should be noted that $\boldsymbol{A}$ is expressed as

$$\boldsymbol{A} = \left(\boldsymbol{I} - \boldsymbol{\Pi}_{X_\beta}\right)\boldsymbol{X}_\alpha,\ \boldsymbol{\Pi}_{X_\beta} = \boldsymbol{X}_\beta\left(\boldsymbol{X}_\beta'\boldsymbol{X}_\beta\right)^{-1}\boldsymbol{X}_\beta',$$

that is, $\boldsymbol{A}$ is an orthogonal projection of $\boldsymbol{X}_\alpha$ onto the orthogonal subspace of $\boldsymbol{X}_\beta$, thus eliminating the block effects $\boldsymbol{\beta}$. The equation $\boldsymbol{A}'\boldsymbol{X}_\beta = \boldsymbol{0}$ is obvious and equation (9.13) is called an adjusted equation.

An $a \times a$ matrix

$$\boldsymbol{A}'\boldsymbol{A} = r\boldsymbol{I}_a - k^{-1}\boldsymbol{N}\boldsymbol{N}'$$

has an obvious eigenvalue zero with related eigenvector $\boldsymbol{j}$, and the eigenvalue $\lambda a/k$ of multiplicity $(a-1)$. The $(a-1)$ eigenvectors of $\lambda a/k$ can be expressed by the columns of $\boldsymbol{P}_a$. Then we can define $(a-1)$ orthonormal vectors

$$(\lambda a/k)^{-1/2}\boldsymbol{P}_a'\boldsymbol{A}'.$$

The matrix $N'N$ has the same non-zero eigenvalues with NN', and in particular the eigenvalue $(r-\lambda)$ of multiplicity $(a-1)$ with related eigenvectors $P_a'N$. It has further a zero eigenvalue of multiplicity $(b-a)$ with related orthonormal eigenvectors Q_b satisfying $NQ_b=\mathbf{0}$, where Q_b is a $b\times(b-a)$ matrix. Finally, we can define an $n\times(n-b-a+1)$ orthonormal matrix Q, which is orthogonal to X_α and X_β and yields the error part. In summary, we have an $n\times n$ orthonormal transformation

$$M' = \begin{bmatrix} n^{-1/2}j_n' \\ \{k(r-\lambda)\}^{-1/2}P_a'NX_\beta' \\ k^{-1/2}Q_b'X_\beta' \\ (\lambda a/k)^{-1/2}P_a'A' \\ Q' \end{bmatrix},$$

which leads to a standard form

$$z=M'y=\begin{bmatrix} z_\mu \\ z_{\beta(\alpha)} \\ z_\beta \\ z_\alpha \\ z_e \end{bmatrix} = \begin{bmatrix} (r\alpha_. + k\beta_.)/\sqrt{n} \\ \{(r-\lambda)/k\}^{1/2}P_a'\alpha + \{(r-\lambda)/k\}^{-1/2}P_a'N\beta \\ k^{1/2}Q_b'\beta \\ (\lambda a/k)^{1/2}P_a'\alpha \\ \mathbf{0} \end{bmatrix} + M'e. \quad (9.14)$$

The sum of squares is expressed as

$$\|M'y\|^2 = \|y\|^2 = \mathrm{CT} + S_T,$$

where $\mathrm{CT}=y_{..}^2/n$ and

$$S_T = \sum_i\sum_j y_{ij}^2 - y_{..}^2/n. \quad (9.15)$$

The partition of the total sum of squares S_T is obtained as

$$S_T = S_\beta + S_\alpha + S_e,$$

where

$$S_\beta = \|z_{\beta(\alpha)}\|^2 + \|z_\beta\|^2 = \sum_j y_{.j}^2/k - y_{..}^2/n, \quad (9.16)$$

$$S_\alpha = \|z_\alpha\|^2 = (\lambda a/k)^{-1}\|A'y\|^2, \quad (9.17)$$

$$S_e = \|z_e\|^2 = \|Q'y\|^2 = S_T - S_\alpha - S_\beta. \quad (9.18)$$

Equation (9.16) is derived since $X'_\beta y$ is a vector of $y_{\cdot j}$ and the rows of

$$\begin{bmatrix} (r-\lambda)^{-1/2}P'_aN \\ Q'_b \end{bmatrix}_{(b-1)\times b}$$

are orthonormal to each other and also orthogonal to j_b.

The unbiased variance is

$$\hat{\sigma}^2 = S_e/(n-a-b+1) \tag{9.19}$$

with df $(n-a-b+1)$. The F-test for the homogeneity of treatment effects is constructed from S_α and S_e as

$$F = \frac{S_\alpha/(a-1)}{S_e/(n-a-b+1)} \tag{9.20}$$

with non-centrality parameter

$$\begin{aligned}\gamma_\alpha &= (\lambda a/k)\|P'_a\alpha\|^2/\sigma^2 = (\lambda a/k)\alpha'\left(I_a - a^{-1}jj'\right)\alpha/\sigma^2 \\ &= (\lambda a/k)\textstyle\sum_i(\alpha_i-\bar{\alpha}_\cdot)^2/\sigma^2.\end{aligned} \tag{9.21}$$

These results are summarized in the ANOVA Table 9.4.

The term 'unadjusted' for block effects implies that the treatment effects are involved in the non-centrality, whereas the block effects have been eliminated from S_α. The S_α here is just like S_α in the one-way layout of Section 6.1, or (9.3) of the previous section with repetition number $\lambda a/k$. The ratio of $\lambda a/k$ to the actual repetition number r is therefore called the efficiency of BIBD. Further, the estimators of the contrasts

$$P'_a\hat{\alpha} = (\lambda a/k)^{-1}P'_aA'y$$

Table 9.4 ANOVA table for BIBD.

Factor	Sum of squares	df	Mean squares	F	Non-centrality
Block effects B	S_β (9.16)	$b-1$	$S_\beta/(b-1)$		Unadjusted
Treatment effects A	S_α (9.17)	$a-1$	$S_\alpha/(a-1)$	F (9.20)	γ_α (9.21)
Error	S_e (9.18)	$n-a-b+1$	$S_e/(n-a-b+1)=\hat{\sigma}^2$ (9.19)		
Total	S_T (9.15)	$n-1$			

are mutually independent with equal variance $(\lambda a/k)^{-1}\sigma^2$, so it is very easy to make multiple comparisons or isotonic inference.

Example 9.2. An experiment on the efficacy of eight antibiotics in blocks of size four. Since a petri dish can contain only four treatments, we employed a design of BIBD $(a=8, k=4, b=14, r=7, \lambda=3)$ whose incidence matrix has been given in (9.10), and obtained the data of Table 9.5 (Hirotsu, 1976). In Table 9.5 some totals necessary for calculating sums of squares are shown. Among them, $\boldsymbol{X}_\alpha'\boldsymbol{y}$ and $\boldsymbol{X}_\beta'\boldsymbol{y}$ would be obvious. The term $k^{-1}\boldsymbol{N}\boldsymbol{X}_\beta'\boldsymbol{y}$ is the total of the block average $k^{-1}\boldsymbol{X}_\beta'\boldsymbol{y}$ in which each treatment is assigned. Then, $\boldsymbol{A}'\boldsymbol{y}$ adjusts each of the treatment totals by $k^{-1}\boldsymbol{X}_\beta'\boldsymbol{y}$ to eliminate the block effects. For example, for treatment 1 the treatment total is $80+79+73+80+80+79+78=549$ and the block average for adjustment is $(172+177+174+178+278+224+179)/4=345.5$. Therefore, the adjusted treatment total is $549-345.5=203.5$. By this operation the block effects are eliminated from the treatment effects. Now the sums of squares are obtained as follows:

$$S_T=80.25^2+79^2+\cdots+12^2-2493^2/56=35795.98,$$

$$S_\beta=\left(172^2+\cdots+183^2\right)/4-2493^2/56=6466.23,$$

$$S_\alpha=\left\{203.5^2+\cdots+(-198.75)^2\right\}/6=29056.29,$$

$$S_e=S_T-S_\alpha-S_\beta=273.46.$$

The unbiased variance is

$$\hat{\sigma}^2=S_e/(n-a-b+1)=273.46/35=7.81.$$

These results are summarized in the ANOVA Table 9.6. The result is highly significant.

There is no difficulty in performing the multiple comparisons of treatment effects. Note again that

$$\boldsymbol{P}_a'\hat{\boldsymbol{\alpha}}=(\lambda a/k)^{-1}\boldsymbol{P}_a'\boldsymbol{A}'\boldsymbol{y}=\boldsymbol{P}_a'\boldsymbol{\alpha}+(\lambda a/k)^{-1}\boldsymbol{P}_a'\boldsymbol{A}'\boldsymbol{e},\quad \boldsymbol{P}_a'\boldsymbol{A}'\boldsymbol{e}\sim N\left\{\boldsymbol{0},(\lambda a/k)\sigma^2\boldsymbol{I}_{a-1}\right\}.$$

Hence, simply consider $\boldsymbol{A}'\boldsymbol{y}$ as a treatment total with an effective repetition number $n_e=\lambda a/k$, which is six instead of seven in this example. In applying Scheffé type multiple comparisons to a contrast $\boldsymbol{L}'\boldsymbol{\alpha}$, we simply employ a point estimate $n_e^{-1}\boldsymbol{L}'\boldsymbol{A}'\boldsymbol{y}$ with variance $n_e^{-1}\boldsymbol{L}'\boldsymbol{L}\sigma^2$, which leads to a simultaneous confidence interval

$$\boldsymbol{L}'\boldsymbol{\alpha}\sim n_e^{-1}\boldsymbol{L}'\boldsymbol{A}'\boldsymbol{y}\pm\sqrt{n_e^{-1}\hat{\sigma}^2(a-1)(\boldsymbol{L}'\boldsymbol{L})F_{a-1,\,n-a-b+1}(\alpha)}.$$

Table 9.5 Efficacy of antibiotics.

	Petri dish														Total		
Antibiotic	1	2	3	4	5	6	7	8	9	10	11	12	13	14	$X'_{\alpha}y$	$k^{-1}NX'_{\beta}y$	$A'y$
1	80		79		73		80		80		79		78		549	345.5	203.5
2	53		52			52		56	58			57		54	382	325.5	56.5
3	31			36	33			34		32	35			34	235	296.75	−61.7
4	8			16		13	10			1		10	12		70	271.75	201.75
5		75		75		81		81	80		78		77		547	351.5	195.5
6		56		52	59		53		60			55		57	392	327.5	64.5
7		40	36			30	35			30	32			38	241	298.75	−57.75
8		15	10		9			13		8		10	12		77	275.75	−198.75
$X'_{\beta}y$	172	186	177	179	174	176	178	184	278	71	224	132	179	183	2493	2493	0.00

Table 9.6 ANOVA table for the antibiotic data.

Factor	Sum of squares	df	Mean squares	F	Non-centrality
Block effects B	$S_\beta = 6466.23$	13	$S_\beta/(b-1)$		Unadjusted
Treatment effects A	$S_\alpha = 29056.29$	7	4150.9	531.49**	γ_α (9.21)
Error	$S_e = 273.46$	35	$S_e/(n-a-b+1) = \hat{\sigma}^2 = 7.81$		
Total	$S_T = 35795.98$	55			

If $\boldsymbol{L} = (1\ 0\ 0\ 1\ 1\ 0\ 0\ 0\ -3)'$, for example, we have 199.5 ± 18.73 at confidence coefficient 0.95.

For pairwise comparisons by Tukey's method, we use

$$\widehat{\alpha_i - \alpha_{i'}} \pm \sqrt{n_e^{-1}\hat{\sigma}^2} q_{a,\, n-a}(\alpha).$$

The analysis explained above is based on the treatment differences within a block and called an intra-block analysis. In contrast, if random effects are assumed for the blocks with expectation zero, the component $z_{\beta(\alpha)}$ also has information on $\boldsymbol{P}_a'\boldsymbol{\alpha}$ with larger variance than the intra-block analysis caused by $\{(r-\lambda)/k\}^{-1/2}\boldsymbol{P}_a'\boldsymbol{N}\boldsymbol{\beta}$ in equation (9.14). There is still a possibility of improving the intra-block estimator of $\boldsymbol{P}_a'\boldsymbol{\alpha}$, called a recovery of inter-block information.

9.3 Non-parametric Method in Block Experiments

9.3.1 Complete randomized blocks

(1) **Two-sample problem of the binary data.** We consider a paired two-sample problem of the binary data. We compare two drugs, where the outcome is success (1) or failure (0). If the outcome is sensitive to various conditions of subjects such as sex, age, body weight, and so on, an overall randomized experiment will suffer from those variations. Then, to reduce the noise in comparisons, the two drugs are assigned to the matched pair with respect to those noise factors. The matched pair can also be the two places of skin of a subject in the test of pasting drugs, or the two data of the same subject in a cross-over clinical trial. In this case the matched pair is considered to be a block of size two. In the case of the normal distribution in Section 5.1.3, the treatment difference within a block has been analyzed. Similarly, we define the difference within a block based on a logit model in this section. As an example we consider the data of Maxwell (1961), where the effects of depression

Table 9.7 Data structure of the jth matched pair.

	Data			Model		
Subject	Success 1	Failure 0	Total	Success 1	Failure 0	Total
Not depersonalized 1	y_{1j}	$1-y_{1j}$	1	p_{1j}	$1-p_{1j}$	1
Depersonalized 2	y_{2j}	$1-y_{2j}$	1	p_{2j}	$1-p_{2j}$	1
Total	$y_{\cdot j}$	$2-y_{\cdot j}$	2			

treatment change between subjects, depersonalized or not. The treatment effects are compared on 23 matched pairs. The data structure of the jth matched pair is given in Table 9.7.

We assume a binomial distribution $B(1, p_{ij})$ for y_{ij}, $i=1, 2$, in Table 9.7. Then, the probability of the jth table is given by the products of two binomial distributions and the probability function for the whole data is

$$\Pr\left(Y_{ij}=y_{ij},\ i=1,2, j=1,\ \ldots,\ 23\right)=\Pi_i\Pi_j\left\{p_{ij}^{y_{ij}}\left(1-p_{ij}\right)^{1-y_{ij}}\right\}.$$

We model the treatment effect by a difference Δ in the logit model as

$$\log\left\{p_{1j}/\left(1-p_{1j}\right)\right\}=\beta_j+\Delta, \tag{9.22}$$

$$\log\left\{p_{2j}/\left(1-p_{2j}\right)\right\}=\beta_j. \tag{9.23}$$

As usual, we assume here that there is no interaction between the treatment and the block. Under the null hypothesis of interest

$$H_0: \Delta=0 \tag{9.24}$$

the sufficient statistics are obviously $y_{\cdot j}, j=1, \ldots, 23$. Therefore we can consider the conditional distribution of y_{1j} given $y_{\cdot j}$. By substituting (9.22) and (9.23) into the probability function, the jth conditional distribution is known to be proportional to $e^{\Delta y_{1j}}, y_{1j}=0, 1$. However, when $y_{\cdot j}=0$ or 2, y_{1j} is uniquely 0 or 1 and we can consider only the case $y_{\cdot j}=1$, which yields the two types of unlike pair as in Table 9.8. For the unlike pair, the conditional distribution of y_{1j} given $y_{\cdot j}$ is

$$\frac{e^{\Delta y_{1j}}}{\sum_{y_{1j}} e^{\Delta y_{1j}}}=\frac{e^{\Delta y_{1j}}}{1+e^{\Delta}},\ y_{1j}=0,\ 1, \tag{9.25}$$

where the summation is for $y_{1j}=0$ and 1. Finally, the probability function of the whole data is obtained by the products of (9.25) as

Table 9.8 Unlike pair.

	Type 1			Type 2		
Subject	Success 1	Failure 0	Total	Success 1	Failure 0	Total
Not depersonalized 1	1	0	1	0	1	1
Depersonalized 2	0	1	1	1	0	1

Table 9.9 Maxwell's data.

	Depersonalized		
Not depersonalized	Success	Failure	Total
Success	x_{11}	x_{12}	$x_{1\cdot}$
Failure	x_{21}	x_{22}	$x_{2\cdot}$
Total	$x_{\cdot 1}$	$x_{\cdot 2}$	$x_{\cdot\cdot}$

$$\Pr\{Y_{1j}=y_{1j}|y_{\cdot j}=1, j=1, \ldots, 23\} = \left(\frac{e^{\Delta}}{1+e^{\Delta}}\right)^{t}\left(\frac{1}{1+e^{\Delta}}\right)^{r-t},$$

where r is the number of unlike pairs and t is the number of $y_{1j}=1$ among unlike pairs. Then, the sufficient statistic for the parameter Δ of interest is t and its distribution given r is a binomial distribution with probability $p=e^{\Delta}/(1+e^{\Delta})$:

$$\Pr(T=t|r) = \binom{r}{t}\frac{e^{\Delta t}}{(1+e^{\Delta})^{r}}.$$

Under H_0 (9.24), $p=1/2$ and therefore the test reduces to testing the null hypothesis $H_0: p=1/2$ based on t, which is distributed as $B(r, p)$. This test is called McNemar's test. It should be noted that the data are summarized as in Table 9.9, where the number of unlike pairs is $r=x_{12}+x_{21}$, among which the number of $(y_{1j}=1)$ is $t=x_{12}$. Then, the p-value is evaluated as a binomial distribution $B(r, 1/2)$. If r and t are moderately large, the likelihood ratio test and the goodness-of-fit test are easily derived.

$$\begin{aligned}&\text{Likelihood ratio test statistic}: 2[t\log(2t/r)+(r-t)\log\{2(r-t)/r\}]\\&\text{Goodness-of-fit test statistic}: (2t-r)^2/r.\end{aligned} \tag{9.26}$$

If the null hypothesis is rejected, then the confidence interval for e^{Δ} or Δ is obtained from the confidence interval for $p=e^{\Delta}/(1+e^{\Delta})$.

Example 9.3. Quantitative analysis of IgE antibody. Two methods of quantitative analysis of IgE antibody – RAST and scratch testing – are evaluated within 166 subjects and the result is summarized in Table 9.10 (Furukawa and Tango, 1993). Table 9.10 is a realization of Table 9.9, where $r = 2 + 18 = 20$ and $t = 2$. The two-sided p-value is calculated as

$$p = 2\left\{\binom{20}{0} + \binom{20}{1} + \binom{20}{2}\right\} / 2^{20} = 0.00040$$

and highly significant, suggesting a high sensitivity of the scratch method. The goodness-of-fit chi-squared test (9.26) gives a similar p-value 0.00035.

(2) ***a*-Sample problem of the binary data.** The response is binary as before, but the number of treatments is a and a complete set can be evaluated in a block – that is, the block size is a. Now we have data in the form of Table 9.11. Here, all the

Table 9.10 Quantitative analysis of IgE antibody.

	Scratch		
RAST	+	−	Total
+	85	2	87
−	18	51	69
Total	103	53	166

Table 9.11 Data structure of the jth matched pair.

	Data		
Treatment	Success 1	Failure 0	Total
1	y_{1j}	$1-y_{1j}$	1
⋮	⋮	⋮	⋮
i	y_{ij}	$1-y_{ij}$	1
⋮	⋮	⋮	⋮
a	y_{aj}	$1-y_{aj}$	1
Total	$y_{\cdot j}$	$a-y_{\cdot j}$	a

y_{ij} are uniquely determined if $y_{\cdot j} = 0$ or a and the other cases correspond to an unlike pair. Under the null hypothesis

$$H_0: \text{All the treatments are equivalent,}$$

the distribution of y_{ij}, $i = 1, \ldots, a-1$ given all the marginal totals in Table 9.11 is a multivariate hypergeometric distribution. Therefore, the expectation and covariance are obtained as

$$E(y_{ij}) = y_{\cdot j}/a, \; i = 1, \ldots, a, \tag{9.27}$$

$$V(y_{ij}) = (y_{\cdot j}/a)(1 - y_{\cdot j}/a), \; i = 1, \ldots, a, \tag{9.28}$$

$$Cov(y_{ij}, y_{i'j}) = -V(y_{ij})/(a-1), \; 1 \le i < i' \le a. \tag{9.29}$$

As a test statistic it is reasonable to compare the number of successes of each treatment in the unlike pair, that is, to test the homogeneity of

$$t_i = \sum_{j^*} y_{ij}, \; i = 1, \ldots, a, \tag{9.30}$$

where the asterisk implies the summation with respect to the unlike pair. Since the data from different blocks are independent of each other, the expectation and covariance of t_i are obtained by the summation of (9.27) ~ (9.29) as

$$E(t_i) = \sum_{j^*} y_{\cdot j}/a,$$

$$V(t_i) = \sum_{j^*} \{(y_{\cdot j}/a)(1 - y_{\cdot j}/a)\},$$

$$Cov(t_i, t_{i'}) = -V(t_i)/(a-1) \text{ for } i \ne i'.$$

The variance–covariance matrix of $\boldsymbol{t} = (t_1, \ldots, t_a)'$ obviously satisfies

$$V(\boldsymbol{P}_a'\boldsymbol{t}) = \left(1 + \frac{1}{a-1}\right) V(t_i)\boldsymbol{I}_{a-1} = \left\{\left(\frac{a}{a-1}\right) V(t_i)\right\}\boldsymbol{I}_{a-1}. \tag{9.31}$$

Therefore, for the contrast of $\boldsymbol{t} = (t_1, \ldots, t_a)'$ the variance and covariance can be calculated as if the t_i were independent with variance $\{a/(a-1)\}V(t_i)$, and this makes it very easy to make inference based on t_i (9.30).

(a) **Cochran's homogeneity test.** The chi-squared statistic for the homogeneity test is given by

$$\chi^2 = \sum_{i=1}^{a} (t_i - \bar{t}_\cdot)^2 / \left\{\left(\frac{a}{a-1}\right) V(t_i)\right\} = \frac{(a-1)\left(\sum t_i^2 - t_\cdot^2/a\right)}{\sum_{j^*} y_{\cdot j} - \sum_{j^*} y_{\cdot j}^2/a}. \tag{9.32}$$

It should be noted that we dealt with the t_i as if it were independent with constant variance $\{a/(a-1)\}V(t_i)$, although actually they are correlated as the multivariate hypergeometric distribution. The statistic (9.32) coincides with the statistic (9.26) when $a=2$, where $t_1=t$, $t_2=r-t$, and $\sum_{j^*} y_{\cdot j} = \sum_{j^*} y_{\cdot j}^2 = r$.

Example 9.4. Cochran's data (Cochran, 1950). The data of Table 9.12 are a summary of the culture test of diphtheria bacilli for comparing four culture mediums $A \sim D$. The number of treatments and the block size are four, and there are 69 blocks. There are five patterns of success and failure as shown in Table 9.12, among which the middle three patterns give an unlike pair – eliminating the cases of four mediums with all success or all failure.

From this table we easily get

$$\sum_{j^*} y_{\cdot j} = 3\times 2 + 3\times 3 + 2\times 1 = 17\ (=t.),$$
$$\sum_{j^*} y_{\cdot j}^2 = 3^2\times 2 + 3^2\times 3 + 2^2\times 1 = 49,$$

and

$$\chi^2 = \frac{3\left(2^2+6^2+3^2+6^2-17^2/4\right)}{17-49/4} = 8.053^*. \tag{9.33}$$

This value is significant compared with $\chi_3^2(0.05) = 7.815$.

Because of a simple covariance structure (9.31), the multiple comparison procedures are also easily derived.

Table 9.12 Summary of the culture test $(a=4)$.

Medium i	Pattern of success and failure					t_i
1 (A)	1	1	0	0	0	$t_1=1\times 2+0\times 3+0\times 1=2$
2 (B)	1	1	1	1	0	$t_2=1\times 2+1\times 3+1\times 1=6$
3 (C)	1	0	1	0	0	$t_3=0\times 2+1\times 3+0\times 1=3$
4 (D)	1	1	1	1	0	$t_4=1\times 2+1\times 3+1\times 1=6$
$y_{\cdot j}$	4	3	3	2	0	
Number of blocks	4	2	3	1	59	(Total 69)

(b) **Turkey's method.** The pairwise difference

$$t(i;i') = \left(\frac{a-1}{\sum_{j^*} y_{\cdot j} - \sum_{j^*} y_{\cdot j}^2/a} \right)^{1/2} |t_i - t_{i'}|$$

is compared with $q_{a,\,\infty}(\alpha)$.

(c) **Scheffé's method.** For any contrast $\sum_i L_i t_i$, $L_\cdot = 0$,

$$\chi^2 = \frac{(a-1)\left(\sum_i L_i t_i\right)^2}{\left(\sum_i L_i^2\right)\left(\sum_{j^*} y_{\cdot j} - \sum_{j^*} y_{\cdot j}^2/a\right)}$$

is compared with $\chi_{a-1}^2(\alpha)$.

(d) **Dunnett's method.** The difference from the standard

$$t(i;1) = \left(\frac{a-1}{\sum_{j^*} y_{\cdot j} - \sum_{j^*} y_{\cdot j}^2/a} \right)^{1/2} |t_i - t_1|/\sqrt{2}$$

is evaluated by the method of Dunnett (1964).

Example 9.5. Example 9.4 continued. We apply Scheffé's method to the contrast $t_2 + t_4 - t_1 - t_3 = 7$. The statistic is

$$3 \times (12-5)^2 / \left\{ 4\left(17 - \frac{49}{4}\right) \right\} = 7.737.$$

This value is very close to $\chi_3^2(0.05) = 7.815$, suggesting a difference between the groups (*B*, *D*) and (*A*, *C*). Actually, this component explains 96% of the total chi-squared (9.33).

(3) ***a*-Sample problem of the ordered categorical data.** Extending the approach to binary data of the previous section, we can also develop methods for ordered categorical data. Extending Table 9.11, the data of the *k*th block, $k = 1, \ldots, n$, is presented in Table 9.13, where the elements y_{ijk} take 1 if the *i*th treatment responds to the *j*th category and 0 otherwise. Thus, all the row totals are unity. When the number of categories *b* is equal to two, this table reduces to Table 9.11 for binary data.

Under the null hypothesis of homogeneity

$$H_0: \text{All the treatments are equivalent,}$$

Table 9.13 Data structure of the kth block, $k = 1, \ldots, n$.

	Ordered categorical response					
Treatment	1	$\cdots$	j	$\cdots$	b	Total
1	y_{11k}	$\cdots$	y_{1jk}	$\cdots$	y_{1bk}	1
$\vdots$	$\vdots$		$\vdots$		$\vdots$	$\vdots$
i	y_{i1k}	$\cdots$	y_{ijk}	$\cdots$	y_{ibk}	1
$\vdots$	$\vdots$		$\vdots$		$\vdots$	$\vdots$
a	y_{a1k}	$\cdots$	y_{ajk}	$\cdots$	y_{abk}	1
Total	$y_{\cdot 1k}$	$\cdots$	$y_{\cdot jk}$	$\cdots$	$y_{\cdot bk}$	a

the distribution of y_{ijk}, $i = 1, \ldots, a-1, j = 1, \ldots, b-1$ given all the marginal totals $y_{\cdot jk}$ and $y_{i\cdot k}$ is again a multivariate hypergeometric distribution independent of each other for different k. Therefore, the expectation and covariance are obtained as

$$E\left(Y_{ijk}\right) = y_{i\cdot k} \times y_{\cdot jk} / y_{\cdot\cdot k} = y_{\cdot jk}/a, \; i = 1, \ldots, a, \tag{9.34}$$

$$Cov\left(Y_{ijk}, Y_{i'j'k}\right) = (a\delta_{ii'} - 1) y_{\cdot jk}\left(a\delta_{jj'} - y_{\cdot j'k}\right) / \left\{a^2(a-1)\right\}. \tag{9.35}$$

In particular, we note that

$$V\left(Y_{ijk}\right) = y_{\cdot jk}\left(a - y_{\cdot jk}\right)/a^2,$$
$$V\left(Y_{\cdot jk}\right) = 0.$$

By the independence of the blocks, we obtain the expectation and covariance of $Y_{ij\cdot}$ by the summation of (9.34) and (9.35) with respect to k as follows:

$$E\left(Y_{ij\cdot}\right) = y_{\cdot j\cdot}/a, \tag{9.36}$$

$$Cov\left(Y_{ij\cdot}, Y_{i'j'\cdot}\right) = \sum_{k=1}^{n}\left\{(a\delta_{ii'} - 1) y_{\cdot jk}\left(a\delta_{jj'} - y_{\cdot j'k}\right)\right\} / \left\{a^2(a-1)\right\}. \tag{9.37}$$

For the test of ordinal effects, the score statistics

$$u_i = \sum_k \sum_j \left(w_j y_{ijk}\right) = \sum_j \left(w_j y_{ij\cdot}\right), \; i = 1, \ldots, a \tag{9.38}$$

with weights $w_1 < \cdots < w_b$ are introduced. From u_i, a homogeneity test is easily obtained. Also, by considering $b-1$ step change-point contrasts, the cumulative chi-squared statistic is constructed.

(a) **Linear score test.** It is easy to derive the expectation of U_i:

$$E(U_i) = a^{-1}\sum_{j=1}^{b}(w_j y_{\cdot j \cdot}) = u_{\cdot}/a.$$

The variance of U_i is

$$V(U_i) = \boldsymbol{w}'\left[a^{-2}\sum_{k=1}^{n}\{y_{\cdot jk}(a\delta_{jj'} - y_{\cdot j'k})\}\right]_{b\times b}\boldsymbol{w}$$

$$= a^{-1}\left[\sum_{j=1}^{b}\left(w_j^2 y_{\cdot j \cdot}\right) - a^{-1}\sum_{k=1}^{n}\left\{\sum_{j=1}^{b}(w_j y_{\cdot jk})\right\}^2\right],$$

where $\boldsymbol{w} = (w_1, \ldots, w_b)'$. Similarly, the covariance is obtained as

$$Cov(U_i, U_{i'}) = -(a-1)^{-1}V(U_i),$$

giving a variance–covariance matrix of $\boldsymbol{U} = (U_1, \ldots, U_a)'$,

$$V(\boldsymbol{U}) = \{a/(a-1)\}\,V(U_i)\left(\boldsymbol{I} - a^{-1}\boldsymbol{jj}'\right).$$

Thus, in considering the contrast in $\boldsymbol{U}$ we can deal with the variables U_i as if they were uncorrelated with equal variance

$$\sigma^2 = \{a/(a-1)\}\,V(U_i) = (a-1)^{-1}\left[\sum_{j=1}^{b}\left(w_j^2 y_{\cdot j \cdot}\right) - a^{-1}\sum_{k=1}^{n}\left\{\sum_{j=1}^{b}(w_j y_{\cdot jk})\right\}^2\right] \tag{9.39}$$

This structure is similar to (9.31) of binary data in the previous section. Then, by the normal approximation we obtain the chi-squared statistic

$$\chi^2 = \frac{1}{\sigma^2}\sum_{i=1}^{a}(u_i - \bar{u}_{\cdot})^2 = \frac{1}{\sigma^2}\left(\sum_{i=1}^{a}u_i^2 - \frac{1}{a}u_{\cdot}^2\right) \tag{9.40}$$

with df $(a-1)$.

As an important special case of $b = a$, if we take a rank within a block as score and if further there is no tie – namely $\boldsymbol{w} = (1, \ldots, a)'$ – then all the column totals $y_{\cdot jk}$ are unity. In this case we have simply

$$\sigma^2 = na(a+1)/12 \tag{9.41}$$

and the chi-squared statistic becomes very simple, which is called Friedman's (1937) test. It is very easy to partition χ^2 (9.40) to obtain the statistics for multiple comparisons.

Turkey's method: The pairwise difference

$$|u_i - u_{i'}|/\sigma$$

is compared with $q_{a,\,\infty}(\alpha)$.

Scheffé's method: For any contrast $\sum_i L_i u_i$, $L_. = 0$,

$$\chi^2 = \frac{\{\sum_i (L_i u_i)\}^2}{\sigma^2 \left(\sum_i L_i^2\right)}$$

is compared with $\chi^2_{a-1}(\alpha)$.

Dunnett's method: The difference from the standard

$$u(i;1) = |u_i - u_1| / \left(\sqrt{2}\sigma\right)$$

is evaluated by the method of Dunnett (1964).

(b) **Cumulative chi-squared test.** We apply step change-point contrast, which implies constructing accumulated statistics as in Table 9.14. This is equivalent to taking a score like

$$w_1 = \cdots = w_J = 1,\ w_{J+1} = \cdots = w_b = 0 \tag{9.42}$$

and we have the variance

$$\sigma_J^2 = \frac{1}{a-1}\left\{\sum_{j=1}^J y_{\cdot j \cdot} - \frac{1}{a}\sum_{k=1}^n \left(\sum_{j=1}^J y_{\cdot jk}\right)^2\right\} \tag{9.43}$$

Table 9.14 Accumulated data according to step change at J.

Treatment	Data: $1 \sim J$	Data: $J+1 \sim b$	Total
1	$\sum_{j=1}^J y_{1j\cdot}$	$n - \sum_{j=1}^J y_{1j\cdot}$	n
⋮	⋮	⋮	⋮
i	$\sum_{j=1}^J y_{ij\cdot}$	$n - \sum_{j=1}^J y_{ij\cdot}$	n
⋮	⋮	⋮	⋮
a	$\sum_{j=1}^J y_{aj\cdot}$	$n - \sum_{j=1}^J y_{aj\cdot}$	n
Total	$\sum_{j=1}^J y_{\cdot j\cdot}$	$an - \sum_{j=1}^J y_{\cdot j\cdot}$	an

from (9.39) and the Jth chi-squared component

$$\chi_J^2 = \frac{1}{\sigma_J^2}\sum_{i=1}^{a}\left(\sum_{j=1}^{J} y_{ij\cdot} - \frac{1}{a}\sum_{j=1}^{J} y_{\cdot j\cdot}\right)^2. \tag{9.44}$$

Then, the cumulative chi-squared is defined by

$$\chi^{*2} = \sum_{J=1}^{b-1}\chi_J^2. \tag{9.45}$$

To derive the distribution of χ^{*2} (9.45), we give a matrix expression

$$\chi^{*2} = \left\| \left\{ \boldsymbol{P}_a' \otimes \begin{bmatrix} \sigma_1^{-1} & 0 & 0 & \cdots 0 & 0 \\ \sigma_2^{-1} & \sigma_2^{-1} & 0 & \cdots 0 & 0 \\ & & \cdots\cdots\cdots & & \\ \sigma_{b-1}^{-1} & \sigma_{b-1}^{-1} & \cdots & \sigma_{b-1}^{-1} & 0 \end{bmatrix} \right\} \boldsymbol{y} \right\|^2,$$

where $\boldsymbol{y}$ is a vector of $y_{ij\cdot}$ arranged in dictionary order. Then, a covariance structure of $U_{iJ} = \sum_{j=1}^{J} y_{ij\cdot}$ and $U_{i'J'} = \sum_{j=1}^{J'} y_{i'j\cdot}$ is required. It is obtained for $J \le J'$ as

$$Cov\,(U_{iJ}, U_{iJ'}) = \frac{1}{a}\left[\sum_{j=1}^{J} y_{\cdot j\cdot} - \frac{1}{a}\sum_{k=1}^{n}\left\{\left(\sum_{j=1}^{J} y_{\cdot jk}\right)\left(\sum_{j=1}^{J'} y_{\cdot jk}\right)\right\}\right],$$

$$Cov\,(U_{iJ}, U_{i'J'}) = -\frac{1}{a-1} Cov\,(U_{iJ}, U_{iJ'}),\ i \ne i'.$$

Similarly, as in the previous section, in considering the contrasts with respect to i we can regard U_{iJ} and $U_{i'J'}$ as uncorrelated, that is,

$$Cov\,(U_{iJ}, U_{i'J'}) = \delta_{i,i'}\frac{1}{a-1}\left[\sum_{j=1}^{J} y_{\cdot j\cdot} - \frac{1}{a}\sum_{k=1}^{n}\left\{\left(\sum_{j=1}^{J} y_{\cdot jk}\right)\left(\sum_{j=1}^{J'} y_{\cdot jk}\right)\right\}\right],\ J \le J'.$$

Let us define a matrix $\boldsymbol{V}$ whose J, J' element is

$$V(J, J') = \frac{1}{a-1}\left[\sum_{j=1}^{J} y_{\cdot j\cdot} - \frac{1}{a}\sum_{k=1}^{n}\left\{\left(\sum_{j=1}^{J} y_{\cdot jk}\right)\left(\sum_{j=1}^{J'} y_{\cdot jk}\right)\right\}\right] / (\sigma_J\sigma_{J'}),\ 1 \le J \le J' \le b-1. \tag{9.46}$$

Then the constants for the chi-squared approximation of χ^{*2} by $d\chi_f^2$ are obtained from

$$df = \mathrm{tr}(\boldsymbol{I}_{a-1}\otimes \boldsymbol{V}) = (a-1)(b-1),$$
$$2d^2f = 2\mathrm{tr}(\boldsymbol{I}_{a-1}\otimes \boldsymbol{V})^2 = 2(a-1)\mathrm{tr}(\boldsymbol{V}^2),$$

namely

$$d = \mathrm{tr}\{\boldsymbol{V}^2/(b-1)\},$$
$$f = (a-1)(b-1)/d.$$

As stated in the linear score test statistic, if the response category is rank without tie, all the formulae become very simple. In this case b is equal to a and all the column totals $y_{\cdot jk}$ are unity. Equations (9.36) and (9.37) become

$$E(Y_{ij\cdot}) = n/a,$$
$$Cov\,(Y_{ij\cdot}, Y_{i'j'\cdot}) = \frac{n}{a-1}(\delta_{ii'} - a^{-1})(\delta_{jj'} - a^{-1}).$$

Equation (9.39) becomes simply

$$\sigma^2 = \frac{n}{a-1}\left\{\textstyle\sum_{j=1}^a w_j^2 - a^{-1}\left(\sum_{j=1}^a w_j\right)^2\right\}. \tag{9.47}$$

Substituting (9.42) into (9.47), we have

$$\sigma_J^2 = \frac{n}{a(a-1)}J(a-J) \tag{9.48}$$

and $V(J, J') = \sqrt{\dfrac{J}{a-J}}\Big/\sqrt{\dfrac{J'}{a-J'}}$ instead of (9.43) and (9.46), respectively. Therefore,

$$\mathrm{tr}(\boldsymbol{V}^2) = a^2\left\{\left(\frac{1}{1\times 2}\right)^2 + \left(\frac{1}{2\times 3}\right)^2 + \cdots + \left(\frac{1}{(a-1)\times a}\right)^2\right\}.$$

Example 9.6. Coating experiment in the electrolytic bath. Four metal plates are coated in one electrolytic bath. The purpose of the experiment is to know whether there is any difference in the thickness of coating by the attached place of metal in the bath – 25 places are selected, and for each point a set of four data measurements is taken. These four data measurements are ranked in order of thickness of the coating (Hirotsu, 1992). The data of the first block in Table 9.15 are in the form of Table 9.16 and coincide exactly with Table 9.13, where $a = b$ and without tie. Since there is no tie, the simple formulae can be applied.

Table 9.15 Ranks of four metals at 25 points.

Points	Metal 1	2	3	4	Total
1	2	3	1	4	10
2	3	4	2	1	10
3	1	4	2	3	10
4	2	1	3	4	10
5	1	3	4	2	10
6	3	4	1	2	10
7	4	1	3	2	10
8	3	4	2	1	10
9	1	4	2	3	10
10	3	2	4	1	10
11	1	3	4	2	10
12	3	1	4	2	10
13	2	3	1	4	10
14	3	4	2	1	10
15	1	4	3	2	10
16	1	4	3	2	10
17	1	4	2	3	10
18	1	3	2	4	10
19	2	1	3	4	10
20	2	1	3	4	10
21	3	4	2	1	10
22	3	1	2	4	10
23	3	1	2	4	10
24	1	3	2	4	10
25	2	4	3	1	10
Total	52	71	62	65	250

Table 9.16 Data structure of the first block.

Metal	Rank 1	2	3	4	Total $y_{i\cdot 1}$
1	0	1	0	0	1
2	0	0	1	0	1
3	1	0	0	0	1
4	0	0	0	1	1
Total $y_{\cdot j1}$	1	1	1	1	$y_{\cdot\cdot 1} = 4$

(a) **Friedman's test.** We put $w_j = j$ and then u_i (9.38) is nothing but the total score of each metal, given in the last row of Table 9.15. We have

$$u_1 = 52,\ u_2 = 71,\ u_3 = 62,\ u_4 = 65.$$

Since $\sum_j \left(w_j^2\right) = 1^2 + 2^2 + 3^2 + 4^2 = 30$, $\sum_j w_j = 10$, we have $\sigma^2 = 125/3$ from (9.39). It is also obtained from (9.41) simply as

$$\sigma^2 = \frac{25 \times 4 \times 5}{12} = 41.6667.$$

Putting these values into (9.40), we have finally

$$\chi^2 = \frac{1}{41.6667}\left(52^2 + 71^2 + 62^2 + 65^2 - \frac{1}{4} \times 250^2\right) = 4.45.$$

This value is non-significant as a chi-squared distribution with df 3.

(b) **Cumulative chi-squared test.** We first prepare Table 9.17 of $y_{ij\cdot}$, counting how many times each metal obtained the ranks 1 ~ 4. Then, we prepare Table 9.18 of $\sum_{j=1}^{J}\left(y_{ij\cdot} - 25/4\right)$, $J = 1, 2, 3$, which are the components of (9.44).

From this table we obtain

$$\chi^{*2} = 3.00 + 3.48 + 10.04 = 16.52.$$

Noting that $a = b$ and there is no tie, the constants for the chi-squared approximation are obtained from the simplified formulae as

Table 9.17 Table of $y_{ij\cdot}$.

	Rank j				
Metal i	1	2	3	4	Total
1	9	6	9	1	25
2	7	1	6	11	25
3	3	11	7	4	25
4	6	7	3	9	25
Total $y_{\cdot j\cdot}$	25	25	25	25	100

Table 9.18 Table of $\sum_{j=1}^{J}\left(y_{ij\cdot}-25/4\right)$.

Metal i	Rank J		
	1	2	3
1	$9-25/4=11/4$	$9+6-2\times25/4=10/4$	$9+6+9-3\times25/4=21/4$
2	$7-25/4=3/4$	$7+1-2\times25/4=-18/4$	$7+1+6-3\times25/4=-19/4$
3	$3-25/4=-13/4$	$3+11-2\times25/4=6/4$	$3+11+7-3\times25/4=9/4$
4	$6-25/4=-1/4$	$6+7-2\times25/4=2/4$	$6+7+3-3\times25/4=-11/4$
Sum of squares	300/16	464/16	1004/16
$\sigma_J^2(9.48)$	$\frac{25\times1\times3}{4\times3}=\frac{25}{4}$	$\frac{25\times2\times2}{4\times3}=\frac{25}{3}$	$\frac{25\times3\times1}{4\times3}=\frac{25}{4}$
Sum ofsquares$/\sigma_J^2$	3.00	3.48	10.04

$$d = = \frac{\operatorname{tr}(V)^2}{b-1} = \frac{4^2}{4-1}\left(\frac{1}{2^2} + \frac{1}{6^2} + \frac{1}{12^2}\right) = 1.519,$$

$$f = (a-1)(b-1)/d = (4-1)^2/d = 5.927.$$

The p-value of $\chi^{*2}/d = 10.876$ is 0.089, as the chi-squared distribution with df 5.927.

9.3.2 Incomplete randomized blocks with block size two

(1) **Homogeneity test.** In comparing a drugs which are sensitive to subject conditions, every combination of two drugs is compared by matched pair. Let the drugs i_1 and i_2 be compared $n_{i_1 i_2}$ $(= n_{i_2 i_1})$ times, among which the number of unlike pairs is $r_{i_1 i_2}(= r_{i_2 i_1})$. Let $y_{i_1 i_2}$ be the number of subjects in unlike pairs who were successful with drug i_1 and unsuccessful with drug i_2, and $y_{i_2 i_1}$ be the number of reverse cases:

$$y_{i_1 i_2} + y_{i_2 i_1} = r_{i_1 i_2}(= r_{i_2 i_1}).$$

Similarly to equations (9.22) and (9.23), we assume the model for $i_1 < i_2$

$$\log\{p_{i_1 j}/(1 - p_{i_1 j})\} = \beta_j + \Delta_{i_1 i_2},$$
$$\log\{p_{i_2 j}/(1 - p_{i_2 j})\} = \beta_j.$$

To eliminate β_j, we follow the formulation of 9.3.1 (1) and obtain the binomial distribution for $y_{i_1 i_2}$ given $r_{i_1 i_2}$ as

$$B\{r_{i_1 i_2}, e^{\Delta_{i_1 i_2}}/(1 + e^{\Delta_{i_1 i_2}})\}.$$

In this experiment we have such a distribution for all the $\binom{a}{2}$ combinations of drugs. Therefore, the data are summarized just like Table 9.19 of professional baseball league results in Japan in 1981, where the draws are eliminated (Hirotsu, 1983b). The data (y_{ij}, y_{ji}) at symmetric positions imply y_{ij} victories and y_{ji} defeats of i against j. The probability function is the product of the binomial distribution

$$\Pi_{1 \le i < j \le a} \binom{r_{ij}}{q_{ij}} q_{ij}^{\,y_{ij}} q_{ji}^{\,y_{ji}}, r_{ij} = y_{ij} + y_{ji}, \tag{9.49}$$

where

$$q_{ij} = e^{\Delta_{ij}}/(1 + e^{\Delta_{ij}}), q_{ji} = 1 - q_{ij}, i < j,$$

Table 9.19 Score sheet of the Pacific League in Japan in 1981.

	Team					
Team	1. Fighters	2. Braves	3. Lions	4. Orions	5. Hawks	6. Buffalos
1. Fighters	*	14	7	13	16	18
2. Braves	12	*	15	8	16	17
3. Lions	16	9	*	12	14	12
4. Orions	12	17	12	*	13	7
5. Hawks	10	10	11	12	*	11
6. Buffalos	4	8	12	16	13	*

which is the percentage of victories of drug i against j. The null hypothesis of the homogeneity of the effects of a drugs is expressed as

$$H_0: \Delta_{ij} = 0 \Leftrightarrow q_{ij} = q_{ji} = \frac{1}{2},\ 1 \le i < j \le a \tag{9.50}$$

and called a hypothesis of symmetry in the two-way square table. Since the expectation of y_{ij} under H_0 (9.50) is

$$E(y_{ij}) = (y_{ij} + y_{ji})/2,$$

we obtain a goodness-of-fit chi-squared

$$\chi^2 = \sum_{i=1}^{a}\sum_{j=1}^{a} \frac{\{y_{ij} - (y_{ij} + y_{ji})/2\}^2}{(y_{ij} + y_{ji})/2} = \sum\sum_{1 \le i < j \le a} \frac{(y_{ij} - y_{ji})^2}{y_{ij} + y_{ji}}, \tag{9.51}$$

which is equivalent to (9.26) when $a = 2$. The degrees of freedom are obtained by a general formula as

$$f = a^2 - a - \binom{a}{2} = \frac{a(a-1)}{2} = \binom{a}{2}.$$

The likelihood ratio test is obtained as

$$-2\log\lambda = 2\sum\sum_{i \ne j} y_{ij} \log \frac{2y_{ij}}{(y_{ij} + y_{ji})},$$

which is asymptotically equivalent to χ^2 (9.51).

(2) **Bradley–Terry model.** A model for probability q_{ij},

$$q_{ij} = \gamma_i / (\gamma_i + \gamma_j) \tag{9.52}$$

is often assumed and called the Bradley–Terry model. It is equivalent to defining the difference Δ_{ij} in log scale as

$$\Delta_{ij} = \log\gamma_i - \log\gamma_j.$$

This model assumes a one-dimensional scale for measuring the strength of the drug, where each drug is scaled by $\log\gamma_i$. Another interpretation is that each drug has a latent variable y, which is distributed according to an exponential distribution $\gamma_i^{-1}e^{-y_i/\gamma_i}$, and drug i defeats j with victory when $y_i > y_j$. In this case the probability that drug i defeats j is

$$\begin{aligned}\Pr(Y_i > Y_j) &= \int\int_{y_i > y_j} \gamma_i^{-1}e^{-y_i/\gamma_i} \times \gamma_j^{-1}e^{-y_j/\gamma_j}\mathrm{d}y_i\mathrm{d}y_j \\ &= \int \gamma_j^{-1}e^{-(\gamma_i^{-1}+\gamma_j^{-1})y_j}\mathrm{d}y_j = \gamma_i/(\gamma_i+\gamma_j)\end{aligned}$$

and coincides with (9.52). On the one hand, the Bradley–Terry model is equivalent to a quasi-symmetry model in the next section.

(3) **Marginal symmetry and quasi-symmetry model.** The symmetry hypothesis (9.50) is partitioned into two hypotheses: marginal symmetry H_m and quasi-symmetry H_q.

$$\begin{aligned}&H_m : q_{i\cdot} = q_{\cdot i},\ i = 1, \ldots, a, \\ &H_q : q_{ij}q_{jk}q_{ki} = q_{ji}q_{ik}q_{kj},\ 1 \le i < j < k \le \mathrm{a}.\end{aligned} \tag{9.53}$$

Marginal symmetry is obviously the symmetry of marginal probabilities. However, if marginal symmetry holds there is still a possibility of asymmetry on the whole. When there are equivalent transitions from i to j, j to k, and k to i, for example, marginal symmetry holds but symmetry on the whole fails. The hypothesis to deny such asymmetry is the quasi-symmetry H_q. The degrees of freedom for H_m are

$$f_m = a - 1$$

and for H_q,

$$f_q = \binom{a}{2} - f_m = \binom{a-1}{2}. \tag{9.54}$$

Very interestingly, the quasi-symmetry model is equivalent to the Bradley–Terry model. First it is obvious that if equation (9.52) holds, then equation (9.53) holds. Next, if equation (9.53) holds then a subset K of it,

$$K : q_{ij}q_{ja}q_{ai} = q_{ji}q_{ia}q_{aj},\ 1 \le i < j < a \tag{9.55}$$

also holds. Since the degrees of freedom of K is also $\binom{a-1}{2}$, K and H_q are equivalent. From (9.55) we easily have

$$\frac{q_{ij}}{q_{ij}+q_{ji}}=\frac{q_{ia}/q_{ai}}{q_{ia}/q_{ai}+q_{ja}/q_{aj}},$$

but this is the same form as (9.52), putting $\gamma_i=q_{ia}/q_{ai}$. A so-called three-way deadlock that i is stronger than j, j is stronger than k, and k is stronger than i is expressed as

$$q_{ij}q_{jk}q_{ki}>q_{ji}q_{ik}q_{kj}.$$

Therefore, the Bradley–Terry model is a model where there is no three-way deadlock.

The likelihood function under the Bradley–Terry model is obtained by putting (9.52) into (9.49) as

$$L=\Pi\Pi_{1\le i<j\le a}\left\{\frac{r_{ij}!}{y_{ij}!y_{ji}!}\times\frac{1}{(\gamma_i+\gamma_j)^{r_{ij}}}\right\}\cdot\Pi_{i=1}^{a}\gamma_i^{t_i},$$

where

$$t_i=\sum\nolimits_{j(\neq i)}y_{ij},\ i=1,\ \ldots,\ a$$

is the total number of victories of the ith drug or team. To obtain the MLE for γ_i, we impose a restriction

$$\sum\nolimits_i\gamma_i=M,$$

since the γ_i are determined essentially up to the ratio. Introducing the Lagrange multiplier λ, we solve $\partial\{\log L-\lambda(\sum_i\gamma_i-M)\}/\partial\gamma_i=0$ to obtain

$$\frac{t_i}{\hat{\gamma}_i}-\sum\nolimits_{j(\neq i)}\frac{r_{ij}}{\hat{\gamma}_i+\hat{\gamma}_j}-\lambda=0. \tag{9.56}$$

By summing up equation (9.56) with respect to i after multiplying $\hat{\gamma}_i$, we obtain

$$\begin{aligned}&\sum\nolimits_i t_i-\sum\sum\nolimits_{i\neq j}r_{ij}\hat{q}_{ij}-\lambda M\\&=\sum\nolimits_i t_i-\sum\sum\nolimits_{1\le i<j\le a}r_{ij}\left(\hat{q}_{ij}+\hat{q}_{ji}\right)-\lambda M\\&=\sum\nolimits_i t_i-\sum\sum\nolimits_{1\le i<j\le a}r_{ij}-\lambda M=-\lambda M=0\end{aligned}$$

That is, we have $\lambda = 0$ and the equations to solve become

$$\hat{\gamma}_i \sum_{j(\neq i)} \frac{r_{ij}}{\hat{\gamma}_i + \hat{\gamma}_j} = t_i, \tag{9.57}$$

$$\sum_i \hat{\gamma}_i = M. \tag{9.58}$$

Equations (9.57) and (9.58) can be solved recursively. Let $\hat{\gamma}_i^0$ be the value at some step and put it into (9.57) to obtain the estimate at the next step,

$$\hat{\gamma}_i = t_i / \left\{ \sum_{j(\neq i)} \frac{r_{ij}}{\hat{\gamma}_i^0 + \hat{\gamma}_j^0} \right\}. \tag{9.59}$$

Then renew it proportionally by

$$\hat{\gamma}_i = \frac{M\hat{\gamma}_i}{\sum \hat{\gamma}_i}$$

so as to satisfy equation (9.58), then go back to the procedure (9.59) until convergence is obtained. The cell frequencies are estimated by

$$\hat{y}_{ij} = r_{ij}\hat{\gamma}_i / (\hat{\gamma}_i + \hat{\gamma}_j).$$

Another iterative method of obtaining the MLE $\hat{y}_{ij}$ more directly is as follows. Let the initial value be

$$\hat{y}_{ij} = \hat{y}_{ji} = (y_{ij} + y_{ji})/2. \tag{9.60}$$

Then apply the iterative scaling procedure to

$$\hat{y}_{i\cdot} = y_{i\cdot}, \tag{9.61}$$

$$\hat{y}_{\cdot j} = y_{\cdot j}, \tag{9.62}$$

$$\hat{y}_{ij} + \hat{y}_{ji} = y_{ij} + y_{ji}, \tag{9.63}$$

until convergence is obtained (see Fienberg, 1980 for example). Finally, we obtain test statistics

$$\text{Goodness-of-fit}: \chi^2 = \sum\sum_{i\neq j} (y_{ij} - \hat{y}_{ij})^2 / \hat{y}_{ij}, \tag{9.64}$$

$$\text{Likelihood ratio}: \lambda = 2\sum\sum_{i\neq j} y_{ij} \log (y_{ij}/\hat{y}_{ij}).$$

These tests are valid when some combinations are not compared by setting the cell frequency zero and subtracting the number of those combinations from the degrees of freedom f_q (9.54).

Table 9.20 Fitted values to the score sheet of the Pacific League in Japan in 1981.

	Team					
Team	1. Fighters	2. Braves	3. Lions	4. Orions	5. Hawks	6. Buffalos
1. Fighters	*	13.398	12.149	13.697	15.763	12.992
2. Braves	12.602	*	12.310	13.317	15.38	14.391
3. Lions	10.851	11.690	*	12.474	14.475	13.511
4. Orions	11.303	11.683	11.526	*	13.990	12.498
5. Hawks	10.237	10.620	10.525	11.010	*	11.608
6. Buffalos	9.008	10.609	10.489	10.502	12.392	*

Example 9.7. Fitting Bradley–Terry model to the data of Table 9.19. By the iterative scaling of (9.60) ~ (9.63), we obtain the fitted values of Table 9.20.

The goodness-of-fit χ^2 (9.64) is 22.39* with $f_q = 10$, which is significant at approximately level 0.01, since $\chi^2_{10}(0.01)$ is 23.21. The result presents evidence that the Bradley–Terry model would not fit well, inviting a more detailed analysis on the three-way deadlocks. We discuss multiple comparison procedures on the three-way deadlocks in the next section.

(4) **Multiple comparisons in the quasi-symmetry model.** The multiple comparisons on treatments are the sort of row-wise multiple comparisons of a two-way table, which will be discussed in detail in Sections 10.3 and 11.3. To define a contrast for specifying the three-way deadlock, we re-parameterize the probability q_{ij} in the log linear model. We define

$$\alpha_{i'} = \log(q_{i'a}/q_{aa}),\ \beta_{j'} = \log(q_{aj'}/q_{aa}),\ \theta_{i'j'} = \log\{(q_{i'j'}q_{aa})/(q_{i'a}q_{aj'})\}$$

for $i', j' = 1, \ldots, a-1$, following Hirotsu (1983a). We further introduce

$$\mu_{i'} = \alpha_{i'} - \beta_{i'} = \log(q_{i'a}/q_{ai'}),\ i' = 1, \ldots, a-1,$$
$$\mu_{i'j'} = \theta_{i'j'} - \theta_{j'i'} = \log\{(q_{i'j'}q_{j'a}q_{ai'})/(q_{j'i'}q_{i'a}q_{aj'})\},\ 1 \le i' < j' \le a-1$$

The conditional inference of the $\mu_{i'}$ and $\mu_{i'j'}$ is based on the conditional distribution

$$\Pr(Y_{ij} = y_{ij} \mid r_{ij}, y_{ii}) = \frac{1}{C(\mu_{i'}, \mu_{i'j'})}\left(\Pi\Pi_{i<j}\frac{r_{ij}!}{y_{ij}!\,y_{ji}!}\right)$$
$$\exp\left(\sum_i y^*_{i'.}\,\mu_{i'} + \sum\sum_{i'<j'} y_{i'j'}\,\mu_{i'j'}\right) \quad (9.65)$$

where $C(\mu_{i'}, \mu_{i'j'})$ is a normalizing constant and $y^*_{i'.} = y_{i'.} - y_{i'i'}$. In the case of data type as in Table 9.19, we can simply take $y_{ii} = 0$. As stated in Section 9.3.2 (3), the Bradley–Terry mode is expressed as the null hypothesis

$$K_0 : \mu_{i'j'} = 0,\ 1 \le i' < j' \le a-1 \tag{9.66}$$

in model (9.65). To define a contrast for the three-way deadlock, let $\boldsymbol{d}(i, j, h)$ be an a^2-dimensional vector defined for a triplet (i, j, h) such that the $a(i-1)+j$th, the $a(j-1)+h$th, and the $a(h-1)+i$th elements are 1, the $a(i-1)+h$th, the $a(j-1)+i$th, and the $a(h-1)+j$th elements are -1, and all other elements 0 if $1 \le i < j < h \le a$ and similarly for other permutations. Let $\boldsymbol{q}$ be an a^2-dimensional vector with $\log q_{ij}$ as its $a(i-1)+j$th element. Then, the row-wise hypothesis for the i, jth row is defined by

$$K_0(i;j) : \boldsymbol{D}'(i;j)\boldsymbol{q} = 0,$$

where $\boldsymbol{D}'(i;j)$ is an $(a-2) \times a^2$ matrix with $\boldsymbol{d}'(i, j, h)$, $h = 1, \ldots, a$, $h \ne i, j$ as its $(a-2)$ rows. The null hypothesis $K_0(i; j)$ implies that the combination (i, j) does not form a three-way deadlock with other $(a-2)$ treatments. Since

$$\mu_{ij} = \boldsymbol{d}'(i, j, a)\boldsymbol{q}, \boldsymbol{d}'(i, j, h) = \boldsymbol{d}'(i, j, a) + \boldsymbol{d}'(j, h, a) - \boldsymbol{d}'(i, h, a)$$

for $1 \le i < j < h < a$, it is also obvious that the overall null hypothesis K_0 (9.66) is equivalent to the hypotheses $K_0(i; j)$ holding for all $i \ne j$.

To test hypotheses $K_0(i; j)$, the distribution of y_{ij} further conditioned on $y^*_{i'.}$ can easily be derived from (9.65). Its use is, however, not practical unless a is typically small. By a general theory for multinomials, the asymptotic conditional distribution of $\boldsymbol{y} = (y_{11}, y_{12}, \ldots, y_{aa})'$ is normal with mean $(m_{11}, m_{12}, \ldots, m_{aa})'$ and covariance matrix $\boldsymbol{V}_q = \boldsymbol{V} - \boldsymbol{VJ}(\boldsymbol{J}'\boldsymbol{VJ})^{-1}\boldsymbol{J}'\boldsymbol{V}$, where $\boldsymbol{V} = \mathrm{diag}(m_{ij})$, a diagonal matrix with m_{ij} as its $a(i-1)+j$th diagonal element and $\boldsymbol{J}$ an $a^2 \times \{a^2 - 2^{-1}(a-1)(a-2)\}$ full rank matrix orthogonal to every $\boldsymbol{d}'(i, j, h)$. In particular, under the null hypothesis K_0 (9.66) the m_{ij} are given by $\hat{y}_{ij}$ of (9.61) ~ (9.63) and m_{ii} are set to zero.

Let $\boldsymbol{l}$ be a vector of $\log y_{ij}$ arranged in dictionary order. Under $K_0(i; j)$ we can expand $\boldsymbol{D}'(i; j)\boldsymbol{l}$ in

$$\boldsymbol{D}'(i;j)\boldsymbol{l} = \boldsymbol{D}'(i;j)\boldsymbol{V}^{-1}\boldsymbol{y} + O(y_{..}^{-1}).$$

Since

$$V\{\boldsymbol{D}'(i;j)\boldsymbol{V}^{-1}\boldsymbol{y}\} = \boldsymbol{D}'(i;j)\boldsymbol{V}^{-1}\boldsymbol{D}(i;j)$$

we have an asymptotic chi-squared statistic

$$\chi^2(i;j) = \{\boldsymbol{D}'(i;j)\boldsymbol{V}^{-1}\boldsymbol{y}\}'\{\boldsymbol{D}'(i;j)\boldsymbol{V}^{-1}\boldsymbol{D}(i;j)\}^{-1}\boldsymbol{D}'(i;j)\boldsymbol{V}^{-1}\boldsymbol{y} \tag{9.67}$$

with df $a-2$. In this calculation, m_{ii} should be set to an arbitrary number except zero for defining V^{-1}. It is obvious that it vanishes after multiplying $D'(i; j)$ and has no effect on the result. The statistic (9.67) is obviously a component of the goodness-of-fit χ^2 (9.64). Therefore, we can conservatively compare $\chi^2(i;j)$ with $\chi^2_{f_q}(\alpha)$.

For the between-groups chi-squared, let an a^2-dimensional vector $d'(S_1, S_2, S_3)$ defined for disjoint subgroups of rows satisfy both orthogonality to every column of J and orthogonality to every $d'(i, j, h)$ with at least two of i, j, h taken from the same set S_1, S_2, or S_3. Such a vector can be obtained as the sum of all different $d'(i, j, h)$, $i \in S_1, j \in S_2, h \in S_3$. Then, an appropriate statistic for evaluating the association among three subgroups is

$$\chi^2(S_1, S_2, S_3) = \{d'(S_1, S_2, S_3)y\}^2 / \{d'(S_1, S_2, S_3)V_q d(S_1, S_2, S_3)\}. \quad (9.68)$$

This definition of chi-squared is not consistent with a single degree of freedom component chi-squared

$$\chi^2(i, j, h) = \{d'(i, j, h)V^{-1}y\}^2 / \{d'(i, j, h)V^{-1}d(i, j, h)\}.$$

The definition (9.68) is, however, to be preferred because it is orthogonal to the between-rows chi-squared $\chi^2(i; j)$ defined for i, j from the same set S_1, S_2, or S_3. If the between-groups chi-squared is not large enough, we can try a two degrees of freedom component chi-squared in $d'(S_1, S_2, S_3)y$ and $d'(S_4, S_5, S_6)y$, where two of three subgroups S_4, S_5, S_6 may coincide with S_1, S_2, or S_3.

Example 9.8. Example 9.7 continued for multiple comparisons. Chi-squared distances between rows are calculated according to (9.67) and given in Table 9.21. The distance $\chi^2(1;4) = 14.83$ elucidates 66% of the overall chi-squared 22.39 with df 15 by a single component of four degrees of freedom. Since its upper tail probability as a chi-squared distribution with df 4 is nearly 0.005, a conservative significance level is 0.075 by Bonferroni's inequality. Table 9.21 also shows that teams 5 and

Table 9.21 Squared distances between teams of the Pacific League in Japan in 1981.

	Team					
Team	1. Fighters	2. Braves	3. Lions	4. Orions	5. Hawks	6. Buffalos
1. Fighters	0	7.84	9.59	14.83	6.24	12.89
2. Braves		0	7.53	11.00	4.52	13.17
3. Lions			0	8.53	3.27	13.00
4. Orions				0	5.24	13.84
5. Hawks					0	4.74
6. Buffalos						0

6 are somewhat distinguishable in that team 5 will not form a three-way deadlock with any other team and team 6 in the opposite extreme.

These considerations suggest homogeneous subgroups $S_1 = (1, 2)$, $S_2 = (3, 4)$, and $S_3 = (6)$, with team 5 being merged or not into any of these subgroups. The observed values of (9.68) for the suggested four kinds of grouping are $\chi^2(S_1, S_2, S_3) = 12.25$, $\chi^2\{(1, 2, 5), (3, 4), (6)\} = 8.40$, $\chi^2\{(1, 2), (3, 4, 5), (6)\} = 9.96$, and $\chi^2\{(1, 2), (3, 4), (5, 6)\} = 9.96$. The between-groups chi-squared $\chi^2(S_1, S_2, S_3)$ elucidates nearly 55% of the overall chi-squared with only 1 df and can explain three conspicuously large components $\chi^2\{(2), (4), (6)\} = 10.09$, $\chi^2\{(1), (3), (6)\} = 8.26$, and $\chi^2\{(1), (4), (6)\} = 7.76$. Further, if it is added by an orthogonal component $\chi^2(3; 4)$ or $\chi^2(1; 2)$, it can elucidate nearly 90% of the overall chi-squared. These considerations may suggest a three-way deadlock among S_1, S_2, and S_3. An interpretation is that team 6 has been too weak against teams 1 and 2, but strong enough against teams 3 and 4. The ranking would therefore have been changed if team 6 were absent (see Hirotsu, 1983a for details).

Kuriki (1991) developed Scheffé type multiple comparison procedures concerning the paired comparisons in a normal model. In particular, he derived a distribution of the largest eigenvalue of a skew-symmetric matrix based on the distribution of the largest eigenvalue of a complex Wishart matrix by Khatri (1964). He applied this to the data of Table 9.19 after an arcsine transformation of the binomial probability, and obtained a conclusion conformable with those obtained in Example 9.8.

References

Cochran, W. G. (1950) The comparison of percentages in matched samples. *Biometrika* **37**, 256–266.

Dunnett, C. W. (1964) Comparing several treatments with a control. *Biometrics* **20**, 482–491.

Fienberg, S. E. (1980) *The analysis of cross classified categorical data.* MIT Press, Cambridge, MA.

Friedman, M. (1937) Use of ranks to avoid the assumption of normality in analysis of variance. *J. Amer. Statist. Assoc.* **32**, 675–701.

Furukawa, T. and Tango, T. (1993) *A new edition of the statistics to medical science.* Asakura Shoten Company, Tokyo (in Japanese).

Hirotsu, C. (1976) *Analysis of variance*. Kyoiku-shuppan, Tokyo (in Japanese).

Hirotsu C. (1983a) Defining the pattern of association in two-way contingency tables. *Biometrika* **70**, 579–590.

Hirotsu, C. (1983b). A goodness-of-fit test for Bradley–Terry model by multiple comparisons approach. *Quality* **13**, 141–149 (in Japanese).

Hirotsu, C. (1992) *Analysis of experimental data: Beyond the analysis of variance*. Kyoritsu-shuppan, Tokyo (in Japanese).

Khatri, C. G. (1964) Distribution of the largest or smallest characteristic root under null hypothesis concerning complex multivariate normal populations. *Ann. Math. Statist.* **35**, 1807–1810.

Kuriki, S. (1991) A construction of simultaneous confidence regions and multiple comparisons procedures concerning the paired comparison model. *Jap. J. Appl. Statist.* **20**, 127–137 (in Japanese).

Maxwell, A. E. (1961) *Analysing qualitative data*. Methuen, London.

10

Two-Way Layout, Normal Model

10.1 Introduction

The analysis of interaction effects is one of the central topics in ANOVA. In previous books, however, mainly an overall F- or χ^2- test has been described. Now, there are several immanent problems in the analysis of two-way data which are not described everywhere.

(1) The characteristics of the rows and columns – such as controllable, indicative, variational, and response – should be taken into consideration.

(2) The degrees of freedom are often so large that an overall analysis can tell almost nothing about the details of the data. In contrast, the multiple comparison procedures based on 1 df contrasts as taken in BANOVA (Miller, 1998) are too lacking in power, and the test result is also usually unclear.

(3) There is often a natural ordering in the rows and/or columns, which should be taken into account in the analysis.

In the usual two-way ANOVA with the controllable factors in the row and column, the purpose of the experiment will be to determine the best combination of the two factors that gives the highest productivity. However, let us consider Example 10.3 in Section 10.3.5, on the international adaptability test of rice varieties. There the rows represent the 44 regions – such as Niigata (Japan), Seoul, Nepal, Egypt, and Mexico – and the columns represent the 18 varieties of rice – such as Rafaelo, Koshihikari, Belle

Advanced Analysis of Variance, First Edition. Chihiro Hirotsu.

Patna, and Hybrid. Then, the columns are controllable but the rows are indicative, and the problem is by no means choosing the best combination of the row and column as in the usual ANOVA. Instead, the purpose should be to assign an optimal variety to each region. Then the row-wise multiple comparison procedures for grouping rows with a similar response profile to columns are interesting. Assigning a common optimal variety to the regions in the same subgroup should be an attractive approach. As another example, let us consider a dose–response analysis based on the ordered categorical data in a phase II clinical trial (see Table 11.8 of Section 11.5.2, for example). Then the rows represent dose levels and are controllable. The columns are the response variables and the data are characterized by the ordinal rows and columns. Of course, the purpose of the trial is to choose an optimal dose level based on the ordered categorical responses. Then, applying the change-point contrast to rows should be an interesting approach for detecting the effective dose. There are several ideas for dealing with the ordered response variables. These examples show that all two-way data requires its own analysis. Indeed, the analysis of two-way data is a rich source of interesting theories and applications (Hirotsu, 1978, 1983a,b, 1991, 1993, 1997, 2009; Hirotsu *et al.*, 2001, 2003).

10.2 Overall ANOVA of Two-Way Data

Randomized experiments are performed in all ab combinations of treatment A (a levels) and treatment B (b levels). The purposes of the experiment are various, as stated in the Introduction. In this section we introduce the basic concept and standard form of interaction commonly applied to all these cases. The observation in the kth experiment of the ith level of A and the jth level of B is denoted by y_{ijk}, and we assume a model

$$y_{ijk} = \mu_{ij} + e_{ijk},\ i = 1, \ldots, a; j = 1,\ \ldots,\ b; k = 1,\ \ldots,\ m, \tag{10.1}$$

where μ_{ij} is the expectation of y_{ijk} and the error e_{ijk} are assumed to be distributed as $N(0, \sigma^2)$ independently of each other. The number of repetitions is not necessarily equal, but we assume an equal number m in this section and give procedures for the unbalanced data in Section 10.5. Then the total number of experiments is $n = abm$. We re-parameterize μ_{ij} of (10.1) as

$$\mu_{ij} = \mu + \alpha_i + \beta_j + (\alpha\beta)_{ij}, \tag{10.2}$$

where the $(\alpha\beta)_{ij}$ express the departure from an additive model and are one form of interaction effect. However, the number of parameters in (10.2) is $ab + a + b + 1$ and therefore these parameters are not one-to-one with μ_{ij}. We give a more exact definition of interaction effects in the following.

Let the weighted average of μ_{ij} with respect to weight w_j be defined by

$$A_i = \sum_{j=1}^{b} w_j \mu_{ij} = \bar{\mu}_{i\cdot},\ \ \sum_{j=1}^{b} w_j = 1,\ i = 1, \ldots, a.$$

If factor B is the supplier of a material and the possible amount of supplied material by supplier j is proportional to w_j, such an average will make sense. Similarly, the weighted average

$$B_j = \sum_{i=1}^{a} v_i \mu_{ij} = \bar{\mu}_{\cdot j},\ \sum_{i=1}^{a} v_i = 1,\ j = 1, \ldots, b$$

is defined. Further, we define the weighted average of A_i or B_j as

$$\mu = \sum_{i=1}^{a} v_i A_i = \sum_{j=1}^{b} w_j B_j \tag{10.3}$$

and we call it a general mean. We define the differences

$$\alpha_i = A_i - \mu, \tag{10.4}$$

$$\beta_j = B_j - \mu, \tag{10.5}$$

and call them the main effects of A_i and B_j, respectively. Now, $\mu_{ij} - \bar{\mu}_{\cdot j}$ is the ith treatment effect of A specific to the jth level of B, and its average with respect to w_j, $\sum_j w_j(\mu_{ij} - \bar{\mu}_{\cdot j})$, is nothing but the main effect α_i. We define the interaction $(\alpha\beta)_{ij}$ as the difference of these two terms,

$$(\alpha\beta)_{ij} = (\mu_{ij} - \bar{\mu}_{\cdot j}) - (\bar{\mu}_{i\cdot} - \mu) = \mu_{ij} - \bar{\mu}_{i\cdot} - \bar{\mu}_{\cdot j} + \mu, \tag{10.6}$$

which is the ith treatment effect of A specific to the jth level of B, or the jth treatment effect of B specific to the ith level of A by symmetry in (10.6). It should be noted that all these effects depend on the weight system $\{v_i\}$ and $\{w_j\}$. Further, defining the treatment effects (10.3) ~ (10.6) is equivalent to imposing constraints

$$\begin{cases} \sum_i v_i \alpha_i = 0, \\ \sum_j w_j \beta_j = 0, \\ \sum_i v_i (\alpha\beta)_{ij} = 0, \\ \sum_j w_j (\alpha\beta)_{ij} = 0 \end{cases} \tag{10.7}$$

in model (10.2). Therefore, it looks as if the test of interaction effects generally depends on how to define the constraints as the identification condition on parameters. In this regard the following theorem is very important.

Theorem 10.1. Invariance of null hypothesis of interaction (Scheffé, 1959). The null hypothesis of interaction $H_{\alpha\beta} : (\alpha\beta)_{ij} = 0, i = 1, \ldots, a, j = 1, \ldots, b$, is invariant for any choice of the weight system.

Proof. We prove that if all the $(\alpha\beta)_{ij}$ are equal to zero by some weight system $\{v_i^0\}$ and $\{w_j^0\}$, then all the $(\alpha\beta)_{ij}$ are equal to zero also by any other weight system $\{v_i\}$ and

$\{w_j\}$. The following formulation is due to Scheffé (1959). Let the effects defined by the weight system $\{v_i^0\}$ and $\{w_j^0\}$ be μ^0, α_i^0, β_j^0, $(\alpha\beta)_{ij}^0$. If $(\alpha\beta)_{ij}^0 \equiv 0$ holds, we have

$$\mu_{ij} = \mu^0 + \alpha_i^0 + \beta_j^0.$$

For such μ_{ij} we redefine the interaction $(\alpha\beta)_{ij}$ by any other weight system $\{v_i\}$ and $\{w_j\}$. Then we have

$$A_i = \sum_{j=1}^b w_j\mu_{ij} = \mu^0 + \alpha_i^0 + \sum_j w_j\beta_j^0,\ i = 1, \ldots, a,$$
$$B_j = \sum_{i=1}^a v_i\mu_{ij} = \mu^0 + \sum_i v_i\alpha_i^0 + \beta_j^0,\ j = 1, \ldots, b,$$
$$\mu = \mu^0 + \sum_i v_i\alpha_i^0 + \sum_j w_j\beta_j^0$$

and therefore we have surely

$$(\alpha\beta)_{ij} = \mu_{ij} - (A_i + B_j - \mu) = 0.$$

Although the definition of interaction $(\alpha\beta)_{ij}$ depends on the employed weight system, it has been verified by Theorem 10.1 that the null hypothesis

$$H_{\alpha\beta}: (\alpha\beta)_{ij} = 0 \Leftrightarrow \mu_{ij} = \bar{\mu}_{i.} + \bar{\mu}_{.j} - \mu \tag{10.8}$$

is a concept that does not depend on the weight system.

Theorem 10.1 is also concerned with the fact that in the original model (10.2), all the interaction contrasts $(\alpha\beta)_{ij} - (\alpha\beta)_{i'j} - (\alpha\beta)_{ij'} + (\alpha\beta)_{i'j'}$ are estimable while any contrast in α_i or β_j is not estimable. After imposing the constraints (10.7), all the parameters in model (10.2) become estimable but those parameters depend on the weight system except for interaction contrasts. It should be stressed that only for the interaction contrasts are the constraints (10.7) regarded just as the identification conditions. Similarly, when $H_{\alpha\beta}$ holds the contrasts in the main effects α_i and β_j do not depend on the weight system. If the situation is like Fig. 1.2 of no interaction, it is obvious that the difference of effects between F_2 and F_1 is the same by any weight of the average with respect to the levels of G. In contrast, if the situation is like Fig. 1.1, it is obvious that a large weight to G_2 acts in favor of F_1 and the reverse if a large weight is given to G_1, thus the inference depends on the constraints imposed.

As stated above, an inference on the interaction contrasts does not depend on the particular weight system. This implies that a statistical analysis may be possible without any constraint on the parameters. Actually, we can take an orthonormal transformation as shown in Example 2.3 or Section 9.1.

Let $\boldsymbol{y}$ be a column vector of y_{ijk} arranged in dictionary order and express model (10.2) as

$$\boldsymbol{y} = \mu\boldsymbol{j} + \boldsymbol{X}_\alpha\boldsymbol{\alpha} + \boldsymbol{X}_\beta\boldsymbol{\beta} + \boldsymbol{X}_{\alpha\beta}(\boldsymbol{\alpha\beta}) + \boldsymbol{e},$$

where $\boldsymbol{X}_\alpha = \boldsymbol{I}_a \otimes \boldsymbol{j}_{bm}$, $\boldsymbol{X}_\beta = \boldsymbol{j}_a \otimes \boldsymbol{I}_b \otimes \boldsymbol{j}_m$, $\boldsymbol{X}_{\alpha\beta} = \boldsymbol{I}_{ab} \otimes \boldsymbol{j}_m$ with $\otimes$ Kronecker's product. When $m = 1$, the matrices $\boldsymbol{X}_\alpha$ and $\boldsymbol{X}_\beta$ coincide with those in Section 9.1. The orthonormal transformation for the standard form is now

$$\boldsymbol{M}' = \begin{bmatrix} n^{-1/2}\boldsymbol{j}_n' \\ (bm)^{-1/2}\boldsymbol{P}_a'\boldsymbol{X}_\alpha' \\ (am)^{-1/2}\boldsymbol{P}_b'\boldsymbol{X}_\beta' \\ (m)^{-1/2}(\boldsymbol{P}_a' \otimes \boldsymbol{P}_b')\boldsymbol{X}_{\alpha\beta}' \\ \boldsymbol{Q}' \end{bmatrix},$$

where $\boldsymbol{P}_a$ is defined in (2.36) and $\boldsymbol{Q}'$ is an $(n - ab) \times n$ $(n = abm)$ orthonormal matrix satisfying

$$\boldsymbol{Q}'[\boldsymbol{j}\ \boldsymbol{X}_\alpha\ \boldsymbol{X}_\beta\ \boldsymbol{X}_{\alpha\beta}] = \boldsymbol{0}.$$

Finally we have a standard form

$$\boldsymbol{z} = \boldsymbol{M}'\boldsymbol{y} = \begin{bmatrix} z_\mu \\ \boldsymbol{z}_\alpha \\ \boldsymbol{z}_\beta \\ \boldsymbol{z}_{\alpha\beta} \\ \boldsymbol{z}_e \end{bmatrix} = \begin{bmatrix} n^{1/2}\left(\mu + \bar{\alpha}_. + \bar{\beta}_. + \overline{(\alpha\beta)}_{..}\right) \\ (bm)^{1/2}\boldsymbol{P}_a'\left\{\boldsymbol{\alpha} + \overline{(\boldsymbol{\alpha\beta})}_{(i)\cdot}\right\} \\ (am)^{1/2}\boldsymbol{P}_b'\left\{\boldsymbol{\beta} + \overline{(\boldsymbol{\alpha\beta})}_{\cdot(j)}\right\} \\ (m)^{1/2}(\boldsymbol{P}_a' \otimes \boldsymbol{P}_b')(\boldsymbol{\alpha\beta}) \\ \boldsymbol{0} \end{bmatrix} + \boldsymbol{M}'\boldsymbol{e}. \qquad (10.9)$$

From (10.9), $\boldsymbol{z}_e$ is an unbiased estimator of zero and $\boldsymbol{z}_{\alpha\beta}$ is the unbiased estimator of the interaction contrast $(\boldsymbol{P}_a' \otimes \boldsymbol{P}_b')(\boldsymbol{\alpha\beta})$. Since no constraint is imposed, $\bar{\alpha}_.$, $\bar{\beta}_.$, and so on remain in the equation. The sum of squares is expressed as

$$\|\boldsymbol{M}'\boldsymbol{y}\|^2 = \|\boldsymbol{y}\|^2 = \mathrm{CT} + S_T$$

and the partition of total sum of squares

$$S_T = \sum_{i=1}^{a}\sum_{j=1}^{b}\sum_{k=1}^{m} y_{ijk}^2 - y_{...}^2/n \qquad (10.10)$$

is obtained as

$$S_T = S_\alpha + S_\beta + S_{\alpha\beta} + S_e,$$

where

$$S_\alpha = \|z_\alpha\|^2 = (bm)^{-1}\sum_{i=1}^{a} y_{i\cdot\cdot}^2 - y_{\cdots}^2/n, \tag{10.11}$$

$$S_\beta = \|z_\beta\|^2 = (am)^{-1}\sum_{j=1}^{b} y_{\cdot j\cdot}^2 - y_{\cdots}^2/n, \tag{10.12}$$

$$S_{\alpha\beta} = \|z_{\alpha\beta}\|^2 = (m)^{-1}\sum_{i=1}^{a}\sum_{j=1}^{b} y_{ij\cdot}^2 - (bm)^{-1}\sum_{i=1}^{a} y_{i\cdot\cdot}^2 - (am)^{-1}\sum_{j=1}^{b} y_{\cdot j\cdot}^2 + y_{\cdots}^2/n, \tag{10.13}$$

$$S_e = \|Q'y\|^2 = S_T - S_\alpha - S_\beta - S_{\alpha\beta}. \tag{10.14}$$

The unbiased variance is

$$\hat{\sigma}^2 = S_e/(n-ab) \tag{10.15}$$

with df $n-ab$. The df of $S_{\alpha\beta}$ is

$$ab-(a-1)-(b-1)-1 = ab-a-b+1 = (a-1)(b-1)$$

and the F-statistic is given by

$$F = \frac{S_{\alpha\beta}/\{(a-1)(b-1)\}}{S_e/(n-ab)} \tag{10.16}$$

with non-centrality parameter

$$\begin{aligned}\gamma_{\alpha\beta} &= m\|(\boldsymbol{P}_a'\otimes\boldsymbol{P}_b')(\boldsymbol{\alpha\beta})\|^2/\sigma^2 \\ &= m\boldsymbol{\mu}'\{(\boldsymbol{I}_a - a^{-1}\boldsymbol{jj}')\otimes(\boldsymbol{I}_b - b^{-1}\boldsymbol{jj}')\}\boldsymbol{\mu}/\sigma^2 \\ &= m\sum_{i=1}^{a}\sum_{j=1}^{b}(\mu_{ij}-\bar{\mu}_{i\cdot}-\bar{\mu}_{\cdot j}+\mu)^2/\sigma^2.\end{aligned} \tag{10.17}$$

When the interaction exists, it is involved in the non-centrality parameter of S_α and S_β. If the interaction does not exist, S_α and S_β are regarded as the sum of squares from a one-way layout with number of repetitions *bm* and *am*, respectively. These results are summarized in Table 10.1.

Example 10.1. Corrosion resistance of aluminum alloy from Davies (1954). The data of Table 10.2 are the corrosion resistance of nine aluminum alloys at four sites of a factory with different chemical atmosphere (Davies, 1954). At each of the sites a plate made from each alloy was exposed for a year. The plates were then submitted to four observers, who assessed their condition visually and awarded marks to each from 0 to 10 according to the degree of resistance to attack. Since, by Davies, there is no evidence of the effect of observers, we deal with them as mere repetitions. Thus, the data are regarded as from 4×9 experiments with four repetitions. The sums of squares are obtained as follows, based on the totals given in Table 10.2.

Table 10.1 ANOVA table for two-way layout.

Factor	Sum of squares	df	Mean squares	F	Non-centrality
Main effects of A	S_α (10.11)	$a-1$	$S_\alpha/(a-1)$		
Main effects of B	S_β (10.12)	$b-1$	$S_\beta/(b-1)$		
Interaction	$S_{\alpha\beta}$ (10.13)	$(a-1)(b-1)$	$S_{\alpha\beta}/(a-1)(b-1)$	F (10.16)	$\gamma_{\alpha\beta}$ (10.17)
Error	S_e (10.14)	$n-ab$	$S_e/(n-ab)=\hat{\sigma}^2$ (10.15)		
Total	S_T (10.10)	$n-1$			

Table 10.2 Corrosion resistance of aluminum alloys.

		Alloy j									
Site i	Observer k	1	2	3	4	5	6	7	8	9	$y_{i\cdot k}$
	1	5	5	5	4	6	6	1	6	7	45
	2	4	5	5	4	5	3	1	5	7	39
1	3	7	7	7	7	8	5	4	7	7	59
	4	6	5	4	5	7	6	3	6	7	49
	$y_{1j\cdot}$	22	22	21	20	26	20	9	24	28	192 ($y_{1\cdot\cdot}$)
	1	8	7	7	7	5	4	5	4	5	52
	2	7	8	6	7	6	5	3	7	8	57
2	3	9	9	9	9	8	6	7	8	8	73
	4	8	8	7	7	5	5	7	4	5	56
	$y_{2j\cdot}$	32	32	29	30	24	20	22	23	26	238 ($y_{2\cdot\cdot}$)
	1	4	4	5	3	4	3	0	5	5	33
	2	1	3	3	2	5	2	0	4	5	25
3	3	5	5	5	6	6	4	3	7	9	50
	4	3	3	7	2	3	3	1	6	6	34
	$y_{3j\cdot}$	13	15	20	13	18	12	4	22	25	142 ($y_{3\cdot\cdot}$)
	1	6	5	6	5	6	4	4	7	5	48
	2	1	3	6	5	5	4	3	6	5	38
4	3	5	5	7	6	8	7	5	8	8	59
	4	5	3	5	3	5	3	3	7	6	40
	$y_{4j\cdot}$	17	16	24	19	24	18	15	28	24	185 ($y_{4\cdot\cdot}$)
$y_{\cdot j\cdot}$		84	85	94	82	92	70	50	97	103	757 ($y_{\cdot\cdot\cdot}$)

Table 10.3 ANOVA table for Example 10.1.

Factor	Sum of squares	df	Mean squares	F	Non-centrality
Main effects of A	128.74	3	42.91		
Main effects of B	128.18	8	16.02		
Interaction	89.32	24	3.72	2.00**	$\gamma_{\alpha\beta}$ (10.17)
Error	201.25	108	1.86 $\left(=\hat{\sigma}^2\right)$		
Total	547.49	143			

$$S_T=\sum_i\sum_j\sum_k y_{ijk}^2-y_{\cdots}^2/n=4527-3979.51=547.49,$$

$$S_\alpha=(bm)^{-1}\sum_i y_{i\cdot\cdot}^2-y_{\cdots}^2/n=4108.25-3979.51=128.74,$$

$$S_\beta=(am)^{-1}\sum_j y_{\cdot j\cdot}^2-y_{\cdots}^2/n=4107.69-3979.51=128.18,$$

$$S_{\alpha\beta}=(m)^{-1}\sum_i\sum_j y_{ij\cdot}^2-(bm)^{-1}\sum_i y_{i\cdot\cdot}^2-(am)^{-1}\sum_i y_{\cdot j\cdot}^2+y_{\cdots}^2/n$$

$$=4325.75-4108.25-4107.69+3979.51=89.32,$$

$$S_e=S_T-S_\alpha-S_\beta-S_{\alpha\beta}=547.49-128.74-128.18-89.32=201.25.$$

These results are summarized in ANOVA Table 10.3. The upper tail probability of $F=2.00$ is 0.0086 and the result is highly significant. We therefore do not proceed to testing the main effects. However, only this result does not tell us any detail of the interaction structure and cannot lead to any action. A more useful approach will be row-wise multiple comparisons given in the next section (see Example 10.2 of Section 10.3.5 (1)).

10.3 Row-wise Multiple Comparisons

10.3.1 Introduction

In this section a method is developed to classify rows and/or columns into subgroups so that additivity holds within each of the sub-tables made up of the grouped rows or columns. The method can also be applied to the case where only one observation is available for each cell, and yields an estimate of the variance in spite of no replicated observation.

Suppose that we are given two-way observations and assume a model (10.1), where the e_{ijk} are independently and normally distributed with mean 0 and

Table 10.4 Yields of corn in bushels per acre (JASA, Johnson and Graybill, 1972).

Liming treatment	Pounds per acre	Minor elements added	Soil type: 1. Very fine sandy loam	2. Sandy clay loam	3. Loamy sand
1. No lime	0	None	11.1	32.6	63.3
2. Course slag	4000	None	15.3	40.8	65.0
3. Medium slag	4000	None	22.7	52.1	58.8
4. Agricultural slag	4000	None	23.8	52.8	61.4
5. Agricultural limestone	4000	None	25.6	63.1	41.1
6. Agricultural slag	4000	B, Z_n, M_n	31.2	59.5	78.1
7. Agricultural limestone	4000	B, Z_n, M_n	25.8	55.3	60.2

variance σ^2. In testing the null hypothesis of interaction it is common to apply the F-test as mentioned in the previous section. If the null hypothesis of interaction is accepted then the μ_{ij} can be modeled by $\mu_{ij}=\bar{\mu}_{i.}+\bar{\mu}_{.j}-\bar{\mu}_{..}$, satisfying $\mu_{ij}-\mu_{i'j}-\mu_{ij'}+\mu_{i'j'}=0$ for all $i\neq i'$ and $j\neq j'$. When the null hypothesis is rejected, however, we are faced with a more complicated model since the degrees of freedom for interaction are usually large and it is desirable to simplify the structure of interaction. In practice, it would be helpful if we can classify the levels of each factor into homogeneous subgroups so that in each of them the interaction vanishes and the interaction exists only among the subgroups; that is, $\mu_{ij}-\mu_{i'j}-\mu_{ij'}+\mu_{i'j'}=0$ unless both i and i', j and j' belong to different subgroups. Therefore, we give a method for finding such a partition so that the probability of judging any partition to be significant when actually the additive model holds is at most equal to a pre-assigned significance level. We call such a model of μ_{ij} a block interaction model. The method is useful for both types of factor, controllable and indicative. To illustrate some of the ideas, an example is taken from Johnson and Graybill (1972), which is the yields of corn in bushels per acre $(m=1)$. The data are shown in Table 10.4 and plotted in two ways in Figs 10.1 and 10.2, where the rows represent the fertilizer and the columns are soil type. They suggest the existence of some interaction. Figure 10.1 suggests that rows 1, 2, and 6 do not interact with the columns and rows 3, 4, and 7 are homogeneous in the same sense. Similarly, Fig. 10.2 suggests that columns 1 and 2 are homogeneous and column 3 interacts with the rows very differently. On the whole, it is expected that the interaction exists among those subgroups of rows and columns, suggesting the necessity of a row and/or column-wise multiple comparison procedure for obtaining such a partition, since those subgroups are not given in advance.

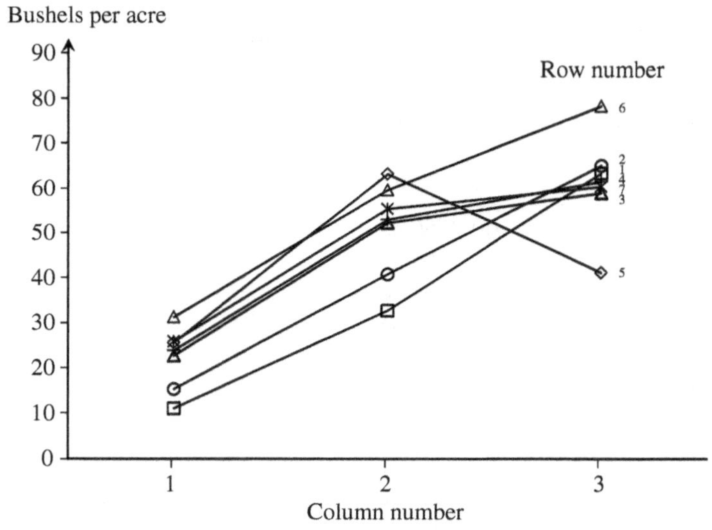

Figure 10.1 Yields of corn in bushels per acre.

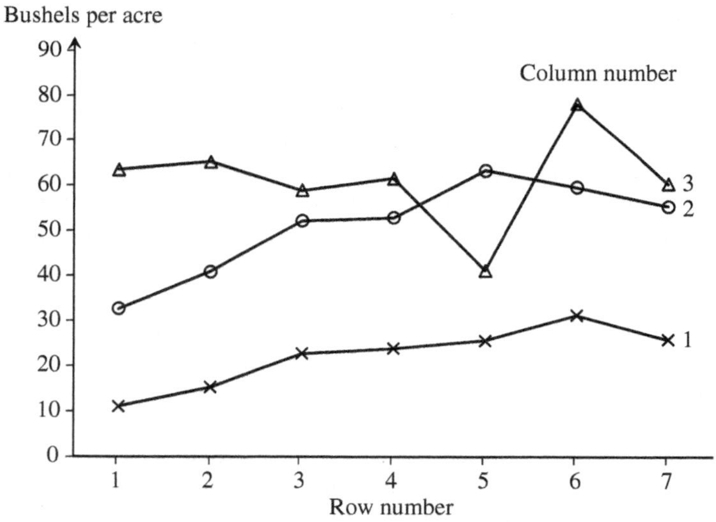

Figure 10.2 Yields of corn in bushels per acre.

10.3.2 Interaction elements

(1) **Pair-wise interaction element.** Re-parameterize the μ_{ij} of (10.1) as

$$\boldsymbol{\mu} = \boldsymbol{\mu}_+ + (\boldsymbol{P}_a \otimes \boldsymbol{P}_b)\boldsymbol{\eta}, \tag{10.18}$$

where $\boldsymbol{\mu}$ is a vector of μ_{ij}'s arranged in dictionary order, $\boldsymbol{\mu}_+$ an additive part of $\boldsymbol{\mu}$ with $b(i-1)+j$th element $\bar{\mu}_{i.} + \bar{\mu}_{.j} - \bar{\mu}_{..}$, and $\boldsymbol{\eta} = (\boldsymbol{P}'_a \otimes \boldsymbol{P}'_b)\boldsymbol{\mu}$ is an interaction part expressed by $(a-1)(b-1)$ parameters orthogonal to each other. It is obvious that $H_{\alpha\beta}$ (10.8) is equivalent to setting $\boldsymbol{\eta} = \boldsymbol{0}$. The contribution of two particular rows, l_1 and l_2, to $\boldsymbol{\eta}$ is given by

$$\boldsymbol{L}(l_1; l_2) = \left\{ \left(0, \cdots, 0, \frac{1}{\sqrt{2}}, 0, \cdots, 0, -\frac{1}{\sqrt{2}}, 0 \cdots, 0\right) \otimes \boldsymbol{P}'_b \right\} = \frac{1}{\sqrt{2}} \boldsymbol{P}'_b (\boldsymbol{\mu}_{l_1} - \boldsymbol{\mu}_{l_2}),$$

where $\boldsymbol{\mu}_i = (\mu_{i1}, \ldots, \mu_{ib})'$. This is called a pair-wise interaction element between two rows. If it is known to be zero, one can pool the rows l_1 and l_2 in further searching for a significant interaction structure.

An estimate of $\boldsymbol{L}(l_1; l_2)$ is obtained simply by replacing $\boldsymbol{\mu}_l$ by $\bar{\boldsymbol{y}}_{l.} = (\bar{y}_{l1.}, \ldots, \bar{y}_{lb.})'$ in its defining equation: $\hat{\boldsymbol{L}}(l_1; l_2) = 2^{-1/2} \boldsymbol{P}'_b (\bar{\boldsymbol{y}}_{l_1.} - \bar{\boldsymbol{y}}_{l_2.})$. Then, its size is evaluated by

$$\chi^2(l_1; l_2) = m \left\| \hat{\boldsymbol{L}}(l_1; l_2) \right\|^2,$$

which may be called the chi-squared distance between the rows l_1 and l_2, since it is distributed as $\sigma^2\chi^2$ with df $b-1$ and the non-centrality parameter $m\|\boldsymbol{L}(l_1; l_2)\|^2$. $\chi^2(l_1; l_2)$ is nothing but the interaction sum of squares for the $2 \times b$ data composed of the l_1th and l_2th rows.

(2) **Interaction element between two subgroups.** Without any loss of generality, let the first subgroup be composed of the first p_1 rows and the second subgroup be the subsequent p_2 rows $(p_1 + p_2 \le a)$. Then, the interaction element between the two subgroups is defined by

$$\begin{aligned} \boldsymbol{L}(1, \ldots, p_1; p_1 + 1, \ldots, p_1 + p_2) = \{p_1 p_2 (p_1 + p_2)\}^{-1/2} \\ \left\{ (p_2, \ldots, p_2, -p_1, \ldots, -p_1, 0, \ldots, 0) \otimes \boldsymbol{P}'_b \right\} \boldsymbol{\mu}. \end{aligned} \tag{10.19}$$

An estimate $\hat{\boldsymbol{L}}(1, \ldots, p_1; p_1 + 1, \ldots, p_1 + p_2)$ is obtained simply by replacing $\boldsymbol{\mu}$ in (10.19) by $\bar{\boldsymbol{y}}_{(i).} = (\bar{\boldsymbol{y}}'_{1.}, \ldots, \bar{\boldsymbol{y}}'_{a.})'$, a vector of $\bar{y}_{ij.}$ arranged in dictionary order. Then its size is evaluated by

$$\chi^2(1, \ldots, p_1; p_1 + 1, \ldots, p_1 + p_2) = m \left\| \hat{\boldsymbol{L}}(1, \ldots, p_1; p_1 + 1, \ldots, p_1 + p_2) \right\|^2.$$

This should be called the chi-squared distance between two subgroups. It should be noted that a pair-wise interaction element is a special case of the interaction element between two subgroups where $p_1 = p_2 = 1$.

(3) **Generalized interaction element among any number of subgroups.** Without any loss of generality, we assume a partition of rows into K subgroups:

$$H_1\{1, \ldots, p_1\}, H_2\{p_1+1,\ldots,p_1+p_2\}, \ldots, H_K\{p_1+\cdots+p_{K-1}+1,\ldots,p_1+\cdots+p_K(=a)\}.$$

Then the generalized interaction element is defined by

$$(\boldsymbol{\rho}'\otimes \boldsymbol{P}_b')\boldsymbol{\mu},\ \boldsymbol{\rho}'\boldsymbol{j}=0,\ \|\boldsymbol{\rho}\|^2=1,\ \rho_i\equiv\zeta_k\{n(H_k)\}^{-1/2}\ \text{for}\ i\in H_k,\ k=1,\ \ldots,\ K, \qquad (10.20)$$

with $n(H_k)$ the number of rows in H_k. It should be noted that the ρ_i are bounded to a constant for the rows in the same subgroup k. Now, unlike the previous interaction elements the definition of the generalized interaction element is not unique and there are several choices of $\boldsymbol{\rho}$. Therefore, we choose $\boldsymbol{\rho}$ so that the chi-squared distance becomes largest subject to restrictions (10.20):

$$\chi^2(H_1,\ \ldots,\ H_K)=\max_{\boldsymbol{\rho}'\boldsymbol{j}=0,\ \|\boldsymbol{\rho}\|^2=1} m\left\|(\boldsymbol{\rho}'\otimes \boldsymbol{P}_b')\bar{\boldsymbol{y}}_{(i)\cdot}\right\|^2,\ \rho_i=\zeta_k\{n(H_k)\}^{-1/2}\ \text{for}\ i\in H_k. \qquad (10.21)$$

Let $\bar{\boldsymbol{Y}}_{k\cdot}=\sum_{i\in H_k}\bar{\boldsymbol{y}}_{i\cdot}$ and $\boldsymbol{w}_k=\{n(H_k)\}^{-1/2}\boldsymbol{P}_b'\bar{\boldsymbol{Y}}_{k\cdot}$. It should be noted that $\{n(H_k)\}^{-1/2}$ has been introduced in (10.20) for normalizing the variance of $\boldsymbol{w}_k$. Then we get an equation

$$\chi^2(H_1,\ \ldots,\ H_K)=\max_{\sum\{n(H_k)\}^{1/2}\zeta_k=0,\ \|\boldsymbol{\zeta}\|^2=1} m\left\|\sum_{k=1}^K(\zeta_k\boldsymbol{w}_k)\right\|^2.$$

Therefore, the maximization of (10.21) is actually with respect to $\boldsymbol{\zeta}=(\zeta_1,\ldots,\zeta_K)$. Then, change $\bar{\boldsymbol{Y}}_{k\cdot}$ into $\sum_{i\in H_k}\left(\bar{\boldsymbol{y}}_{i\cdot}-a^{-1}\sum_{i=1}^a\bar{\boldsymbol{y}}_{i\cdot}\right)$ in the definition of $\boldsymbol{w}_k$, so that $\sum_k\{n(H_k)\}^{1/2}\boldsymbol{w}_k$ becomes zero. By this change of $\bar{\boldsymbol{Y}}_{k\cdot}$, the maximal value remains unchanged by virtue of the relation $\sum_{k=1}^K\{n(H_k)\}^{1/2}\zeta_k=\sum_{i=1}^a\rho_i=0$. Define $\boldsymbol{W}=(\boldsymbol{w}_1,\ldots,\boldsymbol{w}_K)$ by the renewed $\boldsymbol{w}_k$. Then $\chi^2(H_1,\ldots,\ H_K)$ is obtained as the largest eigenvalue of the matrix $m\ \boldsymbol{W}'\boldsymbol{W}$ with $\boldsymbol{\zeta}$ an associated eigenvector. Since $\{\{n(H_1)\}^{1/2},\ \ldots,\ \{n(H_K)\}^{1/2}\}'$ is an eigenvector of $m\ \boldsymbol{W}'\boldsymbol{W}$ corresponding to the zero root, the orthogonality relation $\sum\{n(H_k)\}^{1/2}\zeta_k=0$ is automatically satisfied as the two orthogonal eigenvectors of $\boldsymbol{W}'\boldsymbol{W}$.

10.3.3 Simultaneous test procedure for obtaining a block interaction model

The simultaneous test procedure for obtaining a block interaction model is achieved by the following distribution theory for a maximal statistic.

Lemma 10.1. Lemma 1 of Hirotsu (1983a) and Lemma 2.1 of Hirotsu (1983b). The maximal value of $m\|\{(a_1, \ldots, a_a)\otimes \boldsymbol{P}_b'\}\bar{y}_{(i)\cdot}\|^2$ with respect to a_i subject to restrictions $\sum a_i = 0$, $\sum a_i^2 = 1$ is the largest eigenvalue W_1 of a Wishart matrix, which is distributed as $W\{\sigma^2 \boldsymbol{I}_{\min(a-1,\, b-1)}, \max(a-1, b-1)\}$ under the null hypothesis of interaction (10.8).

Proof. See Lemma 1 of Hirotsu (1983a) and also Lemma 2.1 of Hirotsu (1983b).

It should be noted that the maximal statistic W_1 in Lemma 10.1 is nothing but a generalized chi-squared distance among a subgroups, where each subgroup is composed of just one row. It is also the largest eigenvalue of $m\,\boldsymbol{Z}'\boldsymbol{Z}$ when $\boldsymbol{Z}'$ is an $a\times b$ matrix with $\bar{y}_{ij\cdot}-\bar{y}_{i\cdot\cdot}-\bar{y}_{\cdot\cdot j}+\bar{y}_{\cdots}$ as its (i, j) element, where $\mathrm{tr}\,(m\boldsymbol{Z}'\boldsymbol{Z}) = \mathrm{tr}\,(m\boldsymbol{Z}\boldsymbol{Z}') = S_{\alpha\beta}$. It is therefore obvious that all the chi-squared distances of the previous section are bounded above by the largest eigenvalue of the Wishart matrix in Lemma 10.1. Therefore, the distribution of W_1 can be used as the reference distribution conservatively for all the chi-squared distances. However, the distribution includes an unknown nuisance parameter σ^2 and there are two cases for dealing with it.

(1) **There is repetition.** In this case, there is available the usual unbiased estimate of variance $\hat{\sigma}^2$ (10.15). Then, the distribution of $W_1/\hat{\sigma}^2 = (W_1/\sigma^2)/(\hat{\sigma}^2/\sigma^2)$ is that of the largest eigenvalue W_2 of the Wishart matrix $W\{I_{\min(a-1,\, b-1)}, \max(a-1, b-1)\}$ divided by χ_f^2/f, where χ_f^2 is a chi-squared variable with df $f = ab(n-1)$ and independent of W_2. The upper tail probability of W_2 is obtained by equation (3.13) and Corollary 3.2 of Kuriki and Takemura (2001) as

$$\Pr(W_2 \ge w_0) \cong \sum_{i=0}^{\min(q,\nu)-1} \delta_{q+\nu-1-2i}\bar{G}_{q+\nu-1-2i}(w_0),$$

$$\delta_{q+\nu-1-2i} = (-1)^i 2^{q+\nu-2-i}\frac{\Gamma\left\{\frac{1}{2}(q+1)\right\}\Gamma\left\{\frac{1}{2}(\nu+1)\right\}\Gamma\left\{\frac{1}{2}(q+\nu-1)-i\right\}}{\sqrt{\pi}\,\Gamma(q-i)\Gamma(\nu-i)(i!)} \qquad (10.22)$$

where $q = \min(a-1, b-1)$, $\nu = \max(a-1, b-1)$, and $\bar{G}_l(w_0)$ is the upper tail probability of the chi-squared distribution with df l. Therefore, the upper tail probability of $W_3 = W_1/\hat{\sigma}^2$ is given by

$$\Pr(W_3 \ge w_0) = E\left\{\Pr\left(W_2 \ge w_0\chi_f^2/f\right)\right\},$$

where the expectation is taken with respect to χ_f^2, which is $f \times (\hat{\sigma}^2/\sigma^2)$. That is, replace w_0 by $w_0\chi_f^2/f$ in (10.22) and take expectations of the resulting equation with respect to χ_f^2. It should be noted that the formula (10.22) is an approximation due to the tube method, and recently an exact method has been proposed by Marco (2014). The upper percentiles of the largest eigenvalue of the Wishart matrix are given for $\alpha = 0.01$ and 0.05 in Table C of the Appendix.

(2) **There is no repetition.** In this case the statistic $W_1/S_{\alpha\beta}$ is free from a nuisance parameter σ^2, where $S_{\alpha\beta} = \|\{\boldsymbol{P}_a' \otimes \boldsymbol{P}_b'\}\boldsymbol{y}\|^2 = \sum_i \sum_j (y_{ij} - \bar{y}_{i.} - \bar{y}_{.j} + \bar{y}_{..})^2$ is the total sum of squares for interaction, which is also equal to $\mathrm{tr}(\boldsymbol{W}\boldsymbol{W}') = \mathrm{tr}(\boldsymbol{W}'\boldsymbol{W})$ when $K = a$. The upper percentiles h_α of the null distribution of $W_1/S_{\alpha\beta}$ have been obtained by Johnson and Graybill (1972): $\Pr(W_1/S_{\alpha\beta} \geq h_\alpha) = \alpha$ for $\alpha = 0.01$ and 0.05. The entries of the table are exact for $b = 3$, a all numbers, and $b = 4$, $a = 5, 7$, but the others are by approximation (see Johnson and Graybill, 1972 for details). We can apply the distribution conservatively to any chi-squared distance divided by $S_{\alpha\beta}$, since it is bounded above by $W_1/S_{\alpha\beta}$. Also we have

$$\Pr\left[S \geq \{h_\alpha/(1-h_\alpha)\}(S_{\alpha\beta} - W_1) \mid H_0\right] \leq \alpha,$$

where S stands for any chi-squared distance.

The accurate tail probability $\Pr(W_1/S_{\alpha\beta} \geq w_0)$ under the null hypothesis for $w_0 \geq \frac{1}{2}$ is easily calculated by the formula (10.22) just by replacing $\bar{G}_{q+\nu-1-2i}(w_0)$ by $\bar{B}_{\frac{1}{2}(q+\nu-1)-i, \frac{1}{2}(\nu q - q - \nu + 1)+i}(w_0)$, where $\bar{B}_{l,m}(w_0)$ is the upper tail probability of beta random variable with parameter (l, m). It is the result of Corollary 3.2 and equation (3.9) of Kuriki and Takemura (2001).

10.3.4 Constructing a block interaction model

By summing up the significance information of interaction elements, we can derive a useful block interaction model. We explain the procedure via an example of the yields of corn (Table 10.4), which is an example without repetition ($m = 1$).

The chi-squared distances among rows defined in 10.3.2 (1) are calculated as in Table 10.5 (1). We rearrange the rows to obtain Table 10.5 (2), so that the rows with a small chi-squared distance are placed close. The clustering algorithm of rows in Section 11.3.3 can also be applied here based on the chi-squared distances.

It is obvious that row 5 behaves very differently from the other rows. We therefore try the chi-squared distance between two subgroups $H_1 = (1, 2, 3, 4, 6, 7)$ and $H_2 = (5)$ to find $\chi^2(H_1; H_2) = 647.80$. For the data it is easy to calculate $S_{\alpha\beta} = 947.43$ and the largest eigenvalue $W_1 = 943.04$. Therefore, the reference value for any chi-squared distance is obtained as $\{h_\alpha/(1-h_\alpha)\}(S_{\alpha\beta} - W_1) = 48.37$ at $h_{0.05} = 0.9168$ and 101.91 at $h_{0.01} = 0.9587$, respectively, from Table 1 of Johnson and Graybill (1972). The chi-squared distance $\chi^2(H_1; H_2)$ is seen to be highly significant. Then, any pair chosen one from each of the subgroups $H_{11}(1, 2, 6)$ and $H_{12}(4, 3, 7)$

Table 10.5 Matrix of squared distance.

(1) Original

Row number	1	2	3	4	5	6	7
1	0	13.4	149.6**	126.3**	730.0**	37.4	174.4**
2		0	84.4*	67.0*	574.7**	8.1	103.8**
3			0	1.0	218.9**	42.8	1.0
4				0	249.5**	30.8	4.0
5					0	451.5**	190.8**
6						0	57.6*
7							0

(2) Rearranged

Row number	1	2	6	4	3	7	5
1	0	13.4	37.4	126.3**	149.6**	174.4**	730.0**
2		0	8.1	67.0*	84.4*	103.8**	574.7**
6			0	30.8	42.8	57.6*	451.5**
4				0	1.0	4.0	249.5**
3					0	1.0	218.9**
7						0	190.8**
5							0

of H_1 shows somewhat large chi-squared distance. Therefore, we try the chi-squared distance between the two subgroups H_{11} and H_{12} to find $\chi^2(H_{11};H_{12})=258.67$, which is orthogonal to $\chi^2(H_1;H_2)$ and highly significant also at $\alpha=0.01$. The sum $\chi^2(H_1;H_2)+\chi^2(H_{11};H_{12})=906.48$ is equivalent to the interaction sum of squares of the proportionally unbalanced 3×3 table obtained by collapsing the sub-tables H_{11} and H_{12}, respectively. The proportionally unbalanced two-way data mean that the repetition number m_{ij} of each cell satisfies $m_{ij}=m_{i\cdot}\cdot m_{\cdot j}/m_{\cdot\cdot}$. This explains 95.7% of the total interaction sum of squares $S_{\alpha\beta}$ = 947.43, reflecting the homogeneity in each subgroup.

In this case the rows and columns are both nominal and dealt with symmetrically. The chi-squared distances among columns are calculated as $\chi^2(1;2)=70.44$, $\chi^2(1;3)=467.43$, and $\chi^2(2;3)=883.27$. The chi-squared distance between two subgroups $K_1=(1,2)$ and $K_2=(3)$ is $\chi^2(K_1;K_2)$ = 876.99, which is even larger than $\chi^2(H_1;H_2)$ and highly significant. $\chi^2(K_1;K_2)$ is equal to $S_{\alpha\beta}-\chi^2(1;2)$ and also to the interaction sum of squares of the 7×2 table obtained by collapsing the subgroup K_1, and explains 93.3% of the total interaction sum of squares. Although $\chi^2(1;2)$ is also significant at level 0.05, its percentage contribution to the total sum of squares $S_{\alpha\beta}$

is small. It is of course easy to continue the analysis without collapsing columns 1 and 2, but we employ here the model

$$\boldsymbol{\mu}=\boldsymbol{\mu}_{+}+\left\{(\boldsymbol{q}_1, \boldsymbol{q}_2)\otimes 6^{-1/2}(1,\ 1,\ -2)'\right\}(\eta_1, \eta_2)', \tag{10.23}$$

where $\boldsymbol{q}_1=42^{-1/2}(1,\ 1,\ 1,\ 1,\ -6,\ 1,\ 1)', \boldsymbol{q}_2=6^{-1/2}(1,\ 1,\ -1,\ -1, 0,\ 1,\ -1)'$. This model is a simplification of model (10.18) and agrees well with what is suggested by a first glance at Figs 10.1 and 10.2. The contrast $\boldsymbol{q}_1$ reflects a comparison of H_1 and H_2, $\boldsymbol{q}_2$ a comparison of H_{11} and H_{12}. Similarly, the contrast $(1,\ 1,\ -2)$ represents a comparison of K_1 and K_2.

The LS estimates of η_1 and η_2 are obtained by replacing $\boldsymbol{\mu}$ in equation (10.23) by the observation vector $\boldsymbol{y}$ and multiplying $\left\{(\boldsymbol{q}_1, \boldsymbol{q}_2)'\otimes 6^{-1/2}(1,\ 1,\ -2)\right\}$ to both sides of it, that is,

$$(\hat{\eta}_1, \hat{\eta}_2)'=\left\{(\boldsymbol{q}_1, \boldsymbol{q}_2)'\otimes 6^{-1/2}(1,\ 1,\ -2)\right\}\boldsymbol{y}.$$

They are $\hat{\eta}_1=-24.542$ and $\hat{\eta}_2=-15.667$ in this case, respectively. The departures of the $\hat{\mu}_{ij}$ from $\hat{\mu}_{+}=\bar{y}_{i\cdot}+\bar{y}_{\cdot j}-\bar{y}_{\cdot\cdot}$ are given by the last term of (10.23) with η_i replaced by $\hat{\eta}_i$, $i=1,2$. The same and an easier expression of the model (10.23) will be

$$\mu_{ij}=\bar{\mu}_{i.}+\bar{\mu}_{.j}-\bar{\mu}_{..}+\eta_{ij}\ \text{with}\ \eta_{ij}=\eta_{i'j'}=\eta_{uv}\ \text{if}\ i,\ i'\in H_u\ \text{and}\ j, j'\in K_v, \tag{10.24}$$

where H_u, $u=1,2,3$ and $K_v, v=1,2$, denote homogeneous subgroups of rows and columns, respectively. It should be noted that we renumbered H_{11}, H_{12}, and H_2 as $H_1(1,\ 2,\ 6)$, $H_2(4,\ 3,\ 7)$, and $H_3(5)$ for convenience. The model (10.24) deserves the name of a block interaction model. The LS estimator of parameters in (10.24) is given by

$$\hat{\mu}_{ij}=\hat{\mu}_{+}+\left[\sum\nolimits_{i\in H_u}\sum\nolimits_{j\in K_v}y_{ij}/\{n(H_u)n(K_v)\}-\sum\nolimits_{i\in H_u}\bar{y}_{i.}/n(H_u)-\sum\nolimits_{j\in K_v}\bar{y}_{.j}/n(K_v)+\bar{y}_{..}\right], \tag{10.25}$$

where $\hat{\mu}_{+}=\bar{y}_{i\cdot}+\bar{y}_{\cdot j}-\bar{y}_{\cdot\cdot}$ is an additive part and the last term represents the departure of the $\hat{\mu}_{ij}$ from $\hat{\mu}_{+}$. These are given in Table 10.6 (1) and show nicely the interaction pattern, where rows are rearranged. These departures coincide with the estimate $\left\{(\boldsymbol{q}_1, \boldsymbol{q}_2)\otimes 6^{-1/2}(1,\ 1,\ -2)'\right\}(\hat{\eta}_1, \hat{\eta}_2)'$ of the last term of (10.23). For example, the $(1,\ 1)$ element is calculated as $(42\times 6)^{-1/2}\hat{\eta}_1+(6\times 6)^{-1/2}\hat{\eta}_2=-4.157$ and coincides with the estimate by the last term of (10.25). In Table 10.6 (2) we give $\hat{\mu}_{ij}$ by recovering $\hat{\mu}_{+}$ as well as the coefficients of variance calculated by

$$V(\hat{\mu}_{ij})=[(a+b-1)/ab+\{a-n(H_u)\}\{b-n(K_v)\}/\{abn(H_u)n(K_v)\}](\sigma^2/m) \tag{10.26}$$

Table 10.6 Estimation of μ_{ij}.

	(1) Departure of the $\hat{\mu}_{ij}$ from $\bar{y}_{i.}+\bar{y}_{.j}-\bar{y}_{..}$			(2) Estimate $\hat{\mu}_{ij}$ and its variance		
	Column j			Column j		
Row i	1	2	3	1	2	3
1	−4.16	−4.16	8.32	9.0 $\left(\frac{29}{72}\right)$	37.7 $\left(\frac{29}{72}\right)$	60.4 $\left(\frac{5}{9}\right)$
2	−4.16	−4.16	8.32	13.7 $\left(\frac{29}{72}\right)$	45.4 $\left(\frac{29}{72}\right)$	65.1 $\left(\frac{5}{9}\right)$
6	−4.16	−4.16	8.32	**29.6** $\left(\frac{29}{72}\right)$	**58.3** $\left(\frac{29}{72}\right)$	**81.0** $\left(\frac{5}{9}\right)$
4	1.07	1.07	−2.13	24.5 $\left(\frac{29}{72}\right)$	53.2 $\left(\frac{29}{72}\right)$	60.3 $\left(\frac{5}{9}\right)$
3	1.07	1.07	−2.13	23.1 $\left(\frac{29}{72}\right)$	51.7 $\left(\frac{29}{72}\right)$	58.8 $\left(\frac{5}{9}\right)$
7	1.07	1.07	−2.13	25.6 $\left(\frac{29}{72}\right)$	54.3 $\left(\frac{29}{72}\right)$	61.4 $\left(\frac{5}{9}\right)$
5	9.28	9.28	−18.55	**30.0** $\left(\frac{4}{7}\right)$	**58.7** $\left(\frac{4}{7}\right)$	41.1 (1)

in parentheses. They should be read as $V(\hat{\mu}_{11})=\frac{29}{72}\sigma^2$, $V(\hat{\mu}_{12})=\frac{29}{72}\sigma^2$, $V(\hat{\mu}_{13})=\frac{5}{9}\sigma^2$, and so on. On the whole, it is observed that those variances of $\hat{\mu}_{ij}$ of the block interaction model reduce to one-half of the variance σ^2 of the naïve estimator y_{ij}. It should be noted that formula (10.26) is for a general case and here it is applied for $m=1$.

In this case, the usual unbiased estimate $\hat{\sigma}^2$ is not available. However, an estimate of σ^2 is obtained by subtracting the estimated interaction effects from the total sum of squares for the interaction as follows:

$$\tilde{\sigma}^2=\sum_i\sum_j\left(y_{ij}-\hat{\mu}_{ij}\right)^2/f=\left(S_{\alpha\beta}-\|\hat{\boldsymbol{\eta}}\|^2\right)/f,$$

where $f=(a-1)(b-1)-(A-1)(B-1)$ with $(A-1)(B-1)$ the number of orthogonal interaction contrasts in model (10.23) or (10.24). It should be noted that $\hat{\mu}_{ij}$ and $\tilde{\sigma}^2$ are the better estimators of the μ_{ij} and σ^2 than those which might be obtained assuming the additive model in separate sub-tables. For the current example, it is obtained as

$$\tilde{\sigma}^2=\left(S_{\alpha\beta}-847.79\right)/(12-2)=9.97 \text{ with df } 10.$$

Johnson and Graybill (1972) give three estimates of σ^2 as 1.43 (df 3.06), 5 (df 5), and 11.75 (df 6) without referring to any preference among them. Our estimate is within their range and seems more reliable because of the larger degrees of freedom. Therefore, the standardized error (SE) of the $\hat{\mu}_{ij}$ is around 2.4. Finally, if the column is considered as an indicative factor, the best choice of treatment will be 6 for loamy sand and 5 or 6 for soil types 1 and 2. Treatment 6 is characterized to be generally good for all soil types.

10.3.5 Applications

We give two further such examples.

Example 10.2. Example 10.1 continued. The purpose of the experiment of Example 10.1 was to choose an appropriate alloy for each of four sites, considered to be an uncontrollable factor. Then, it is preferable if an alloy is suitable for as many sites as possible, since it would be inconvenient to have to use different alloys at different sites of a factory. We therefore apply the row-wise multiple comparison procedure of Section 10.3.3 to the data of Example10.1. The squared distances are obtained as in Table 10.7.

In this case we already have $\hat{\sigma}^2 = 1.86$ with df 108. The degree of freedom is so large that we can deal with it as if it were a constant, and we evaluate the squared distances approximately by 1.86 times the largest eigenvalue of the Wishart matrix $W\{\boldsymbol{I}_3, 8\}$. Then, the upper percentiles are 42.07 ($\alpha = 0.05$) and 51.19 ($\alpha = 0.01$). Therefore, $S(2; 4)$ is significant at level 0.05 and $S(2; 3)$ is highly significant as shown in Table 10.7. The squared distance between two subgroups (1, 3, 4) and (2) is $S(1, 3, 4; 2) = 64.04$ and elucidates 71.70% of the total sum of squares for interaction $S_{\alpha\beta} = 89.32$. The averaged responses are as given in Table 10.8. Now the interpretation is clear, and we conclude that for sites 1, 3, and 4 alloy 9 or 8 would be appropriate and alloy 1 or 2 for site 2. If the sum of squares among observers is subtracted from the error sum of squares, the unbiased variance $\hat{\sigma}^2$ becomes 0.90 with df 105 and the evidence becomes even stronger. The same conclusion has also been obtained by another approach in Hirotsu (1983a).

Table 10.7 Matrix of squared distance.

Row number	1	2	3	4
1	0	38.36	8.53	17.69
2		0	52.75**	49.61*
3			0	11.69
4				0

Table 10.8 Averaged response.

	Alloy j								
Site i	1	2	3	4	5	6	7	8	9
1, 3, 4	4.33	4.42	5.42	4.33	5.67	4.17	2.33	6.17	6.42
2	8.00	8.00	7.25	7.50	6.00	5.00	5.50	5.75	6.50

Example 10.3. International adaptability test of rice varieties. This example, given in Hirotsu (1976), is a two-way layout without repetition examining the adaptability of 18 rice varieties to 44 combinations of regions and years. The row-wise multiple comparison procedure nicely classifies the varieties into four types: Formosan, Indian, Japanese and Korean, and Hybrid (specific to Mexico). The regions are also classified properly into six groups: Korea and the northern part of Japan, the southern part of Japan, tropical regions, Nepal, Egypt, and Mexico (see Hirotsu, 1976 for more details). For somewhat large data like this we need some automatic procedure for clustering rows. We therefore give an algorithm for clustering rows in Section 11.3.3, so that the generalized chi-squared distance among clusters is large, achieving simultaneously homogeneity within each cluster.

10.3.6 Discussion on testing the interaction effects under no replicated observation

As seen in the above examples, there are often cases where no replicated observation is available in the two-way data. Then, the usual unbiased estimate of variance $\hat{\sigma}^2$ is not available and therefore the textbook usually assumes replicated observations as in Section 10.2. Under these circumstances, Tukey (1949) proposed a test of a 1 df non-additivity model without repetition. He assumes a particular interaction model

$$\Psi: y_{ij} = \mu + \alpha_i + \beta_j + g\alpha_i\beta_j + e_{ij},\ i = 1, \ldots, a; j = 1, \ldots, b, \tag{10.27}$$

with $\alpha. = \beta. = 0$, e_{ij} independently distributed as $N(0, \sigma^2)$ and considers testing the null hypothesis

$$H_g : g = 0.$$

This is a non-linear model and out of the range of linear statistical inference. However, there is available an F-test based on

$$S_g = \frac{\left(\sum\sum \hat{\alpha}_i \hat{\beta}_j y_{ij}\right)^2}{\sum \hat{\alpha}_i^2 \sum \hat{\beta}_j^2},\ \hat{\alpha}_i = \bar{y}_{i\cdot} - \bar{y}_{\cdot\cdot},\ \hat{\beta}_j = \bar{y}_{\cdot j} - \bar{y}_{\cdot\cdot}$$

$$S_e^* = S_{\alpha\beta} - S_g = \sum\sum \left(y_{ij} - \bar{y}_{i\cdot} - \bar{y}_{\cdot j} + \bar{y}_{\cdot\cdot}\right)^2 - S_g.$$

It is shown that S_e^* is distributed as $\sigma^2\chi^2_{(a-1)(b-1)-1}$ independently of S_g under the null hypothesis H_g. Therefore, the test statistic

$$F = \frac{S_g/1}{S_e^*/\{(a-1)(b-1)-1\}}$$

is distributed as an F-distribution with df$\{1; (a-1)(b-1)-1\}$ under H_g. This is theoretically a very interesting and sophisticated 1 df test for interaction, but in practice the modeling by linear-by-linear interaction based on main effects α_i and β_j seems too restricted without any prior information on the interaction. Johnson and Graybill (1972) extended the model (10.27) to

$$\Psi' : y_{ij} = \mu + \alpha_i + \beta_j + g u_i v_j + \varepsilon_{ij}$$

and Mandel (1971) also proposed the model

$$\Psi'' : y_{ij} = \mu + \alpha_i + \beta_j + \sum_{k=1}^{K} g_k u_{ki} v_{kj} + \varepsilon_{ij}.$$

These models are more flexible than Tukey's model (10.27), but still based on linear-by-linear interaction and might be unclear even if $K = 1$ or 2. In contrast, our block interaction model is very easy to interpret and seems more robust without assuming any restricted model in advance. As an exception, for the quantitative factors a simple response surface model like linear-by-linear can be successful (see Cox, 1958). A removal interaction by transformation is also described there.

10.4 Directional Inference

In this section we consider the cases where the row and/or column categories follow a natural ordering such as time, temperature, and dose levels. Then we are interested in testing some systematic effects such as monotone and convex along the ordered categories. In case of the normal model with natural ordering in both rows and columns, a multivariate (two-way) one-sided ordered alternative

$$T_1 : \mu_{i+1j+1} - \mu_{ij+1} - \mu_{i+1j} + \mu_{ij} \geq 0,\ i = 1, \ldots, a-1; j = 1, \ldots, b-1 \tag{10.28}$$

and the two-sided version T_2 of T_1 have been proposed in the literature. In (10.28) the differences $(\mu_{ij+1} - \mu_{ij})$ and $(\mu_{i+1j} - \mu_{ij})$ are increasing in i and j, respectively. Further, a useful two-sided ordered alternative

$$T_3 : \mu_{ij+1} - \mu_{i'j+1} - \mu_{ij} + \mu_{i'j} \geq 0, j = 1, \ldots, b-1,\ i \neq i' \tag{10.29}$$

has been introduced by Hirotsu (1978) when there is a natural ordering only in columns. Hypothesis T_3 implies that the differences $\mu_{ij} - \mu_{i'j}$ are tending upwards in j and are essentially two-sided, since the rows i and i' are permutable. It should be noted that T_3 (10.29) is different from the two-sided ordered alternative T_2, since there is assumed no ordering in rows. As stated before, the restricted maximum likelihood approach is too complicated for these interaction analyses. However, the methods based on the accumulated statistics can be very naturally extended just by replacing the orthogonal contrasts by the change-point contrasts as follows. Of course, in applying these sums of squares they should be divided by the unbiased variance $\hat{\sigma}^2$ (10.15) to cancel out the unknown σ^2.

10.4.1 Ordered rows or columns

(1) **Overall analysis of ordered columns by the cumulative chi-squared statistic χ^{*2}.** When testing the null hypothesis of interaction against the two-sided alternative T_3, simply replace the orthogonal contrasts $\boldsymbol{P}_b'$ by the change-point contrasts $\boldsymbol{P}_b^{*\prime}$ in the usual sum of squares for interaction to obtain

$$\begin{aligned} S_{\alpha\beta^*} &= m\|(\boldsymbol{P}_a' \otimes \boldsymbol{P}_b^{*\prime})\bar{\boldsymbol{y}}_{(i)(j)\cdot}\|^2 \\ &= \left\{\frac{b}{1\times 2}\chi^2_{(1)} + \frac{b}{2\times 3}\chi^2_{(2)} + \cdots + \frac{b}{(b-1)\times b}\chi^2_{(b-1)}\right\}\sigma^2, \end{aligned} \tag{10.30}$$

where $\bar{\boldsymbol{y}}_{(i)(j)\cdot}$ is another expression of $\bar{\boldsymbol{y}}_{(i)\cdot}$ for the mean observation vector of $\bar{y}_{ij\cdot}$ arranged in dictionary order with a two-way suffix, m the number of repetitions and $\chi^2_{(j)}$ the chi-squared for Chebyshev's jth-order orthogonal polynomial with df $a-1$. This is obviously a very natural extension of the usual sum of squares $S_{\alpha\beta} = m\|(\boldsymbol{P}_a' \otimes \boldsymbol{P}_b')\bar{\boldsymbol{y}}_{(i)(j)\cdot}\|^2$, which is expanded in the form $\{\chi^2_{(1)} + \chi^2_{(2)} + \cdots + \chi^2_{(b-1)}\}\sigma^2$. For the statistic $S_{\alpha\beta^*}$, a very good chi-squared approximation is available extending the method of one-way layout in Section 6.5.3 (2) (a) as

$$\begin{aligned} &\kappa_1 = E(\chi^{*2}) = (a-1)(b-1) = df \\ &\kappa_2 = V(\chi^{*2}) = 2(a-1)\left\{b-1+2\left(\frac{\gamma_1}{\gamma_2}+\ldots+\frac{\gamma_1+\ldots+\gamma_{b-1}}{\gamma_{b-1}}\right)\right\} = 2d^2f, \\ &\gamma_j = j/(b-j) \end{aligned}$$

An improvement by using $\kappa_3 = 8(a-1)\mathrm{tr}\left(\boldsymbol{P}_b^{*\prime}\boldsymbol{P}_b^*\right)^3$ is also applicable. It is divided by the unbiased variance $\hat{\sigma}^2$ to cancel out σ^2, for which a very good F-approximation is available in Hirotsu (1979). A real example of this statistic is given by Hirotsu (1978). When $m=1$ we introduce a particular sum of squares for cancelling out σ^2 which has an inverse characteristic to the systematic statistic like $S_{\alpha\beta^*}$ (10.30), see Chapter 13.

(2) **Scheffé type multiple comparisons of rows with natural ordering only in columns.** The method of Section 10.3 is also very naturally extended to accommodate the ordered effects along columns. For the monotone hypothesis, the interaction elements and their chi-squared distances are defined simply by replacing $\boldsymbol{P}_b'$ by $\boldsymbol{P}_b^{*\prime}$ in their defining equations of 10.3.2 (1), (2), and (3). Then a reference distribution is given by the following lemma.

Lemma 10.2. Lemma 3.1 of Hirotsu (1983a). When $a \ge b$, the maximum value of $m\|\{(a_1, \ldots, a_a)\otimes \boldsymbol{P}_b^{*\prime}\}\bar{\boldsymbol{y}}_{(i)(j)\cdot}\|^2$ with respect to a_i subject to restrictions $\sum a_i = 0$, $\sum a_i^2 = 1$ is the largest eigenvalue W_1^* of a Wishart matrix distributed as $W\left(\sigma^2 \boldsymbol{P}_b^{*\prime}\boldsymbol{P}_b^*, a-1\right)$ under H_0.

Proof. The proof is similar to Lemma 10.1 and omitted.

The convexity hypothesis is also very easily dealt with by replacing $\boldsymbol{P}_b'$ by $\boldsymbol{P}_b^{\dagger\prime}$ in the defining equations of interaction elements. The distribution concerned is the largest eigenvalue of a Wishart matrix for variance not proportional to an identity matrix, which is very difficult to handle compared with the case given in (10.22). However, a very nice chi-squared approximation has been obtained by Hirotsu (2009), which is described in Section 11.4.2 (2). In its applications to χ^{*2} and $\chi^{\dagger 2}$ in this section, the largest eigenvalues of $\boldsymbol{P}_b^{*\prime}\boldsymbol{P}_b^*$ and $\boldsymbol{P}_b^{\dagger\prime}\boldsymbol{P}_b^{\dagger}$ dominate the other eigenvalues and the chi-squared approximation using the dominant term only is good enough. Actually, the largest eigenvalue is larger than one-half of $\mathrm{tr}\left(\boldsymbol{P}_b^{*\prime}\boldsymbol{P}_b^*\right)$ and three-quarters of $\mathrm{tr}\left(\boldsymbol{P}_b^{\dagger\prime}\boldsymbol{P}_b^{\dagger}\right)$, for the respective cases. Thus, our approach to the shape and change-point hypotheses is very easy to extend, even to two-way problems, and really unifying. Interesting applications of these methods to the profile analysis of repeated measurements are given in Chapter 13 according to Hirotsu (1991) and Hirotsu *et al.* (2003), where there is no repetition in the cell $(m = 1)$. In Section 10.3.3 (2) the largest eigenvalue was divided by the total sum of squares $S_{\alpha\beta}$ for interaction. In Chapter 13 we further introduce the sum of squares which has an inverse characteristic to the systematic statistic such as χ^{*2} or $\chi^{\dagger 2}$ to cancel out σ^2 in the case where there is natural ordering in columns.

(3) **Multiple comparisons of ordered rows.** If there is natural ordering in the rows, all the permutations of rows make no sense and the multiple comparisons are restricted to $a-1$ step change-point contrasts of the cumulative chi-squared statistic $S_{\alpha^*\beta} = m\|\left(\boldsymbol{P}_a^{*\prime} \otimes \boldsymbol{P}_b'\right)\bar{\boldsymbol{y}}_{(i)(j)\cdot}\|^2$,

$$\max \mathrm{acc.}\chi^2\left(\boldsymbol{P}_a^{*\prime}\right) = \max_k m\|\{\boldsymbol{r}^{*\prime}(1,\ldots,k;k+1,\ldots,a)\otimes \boldsymbol{P}_b'\}\bar{\boldsymbol{y}}_{(i)(j)\cdot}\|^2$$

where $\boldsymbol{r}^{*\prime}(1,\ldots,k;k+1,\ldots,a)$ is the kth row of $\boldsymbol{P}_a^{*\prime}$ denoting the kth step change-point contrast and the maximization is with respect to $k = 1,\ldots,a-1$. Max acc.χ^2 is a natural extension of max acc. $t1$, so that we call this method max acc. $\chi^2(\boldsymbol{P}_a^{*\prime})$, making clear the range of maximization. However, in its introductory time, the p-value was evaluated by Bonferroni inequality based on the bivariate chi-squared distribution. Later, in Hirotsu *et al.* (1992), the exact algorithm for probability calculation was obtained and formulae in a closed form were given up to $a=5$, as well as for max acc. $t1$ based on the Markov property of the component statistics. We give an extended algorithm for the p-value in Section 11.4.3 and an interesting application in Section 11.4.4 (3). The upper percentiles for $\chi^2(\boldsymbol{P}_a^{*\prime})$ divided by $\hat{\sigma}^2$ are given for some a, b, and m in Table B of the Appendix.

10.4.2 Ordered rows and columns

(1) **Two-way cumulative chi-squared method.** For the two-way two-sided ordered alternative

$$T_2: \quad \mu_{ij+1}-\mu_{i'j+1}-\mu_{ij}+\mu_{i'j}\geq 0,\ i=1,\ldots,a-1; j=1,\ldots,b-1,$$
$$\text{or}\ \ \mu_{ij+1}-\mu_{i'j+1}-\mu_{ij}+\mu_{i'j}\leq 0, i=1,\ldots,a-1; j=1,\ldots,b-1,$$

the two-way cumulative chi-squared statistic

$$S_{\alpha^*\beta^*}=m\|(\boldsymbol{P}_a^{*\prime}\otimes\boldsymbol{P}_b^{*\prime})\bar{\boldsymbol{y}}_{(i)(j)\cdot}\|^2 \tag{10.31}$$

is obtained by replacing both of $\boldsymbol{P}_a'$ and $\boldsymbol{P}_b'$ by $\boldsymbol{P}_a^{*\prime}$ and $\boldsymbol{P}_b^{*\prime}$ in $S_{\alpha\beta}=m\|(\boldsymbol{P}_a'\otimes\boldsymbol{P}_b')\bar{\boldsymbol{y}}_{(i)(j)\cdot}\|^2$, respectively. For this statistic also a very good chi-squared approximation is available, extending the method of one-way layout (Section 6.5.3 (2)). The details are given in Section 11.5.1, and we can apply the formula here simply as the case of $R_i\equiv 1$ and $C_j\equiv 1$.

(2) **Two-way maximal contrast method.** In the case with natural ordering in both the rows and columns, max acc. χ^2 has been extended by Hirotsu (1993, 1997) to

$$\max\max\ \chi^2=\max_k\max_l m\|\{\boldsymbol{r}^{*\prime}(1,\ldots,k;k+1,\ldots,a)\otimes\boldsymbol{c}^{*\prime}(1,\ldots,l;l+1,\ldots,a)\}\bar{\boldsymbol{y}}_{(i)(j)\cdot}\|^2, \tag{10.32}$$

where $\boldsymbol{c}^{*\prime}(1,\ldots,l;l+1,\ldots,a)$ is the lth row of $\boldsymbol{P}_b^{*\prime}$. It is extended also to

$$\max\ \chi^{*2}=\max_k m\|\{\boldsymbol{r}^{*\prime}(1,\ldots,k;k+1,\ldots,a)\otimes\boldsymbol{P}_b^{*\prime}\}\bar{\boldsymbol{y}}_{(i)(j)\cdot}\|^2, \tag{10.33}$$

$$\max\ \mathrm{Wil}\ \chi^2=\max_k m\|\{\boldsymbol{r}^{*\prime}(1,\ldots,k;k+1,\ldots,a)\otimes\boldsymbol{s}\}\bar{\boldsymbol{y}}_{(i)\cdot(j)\cdot}\|^2, \tag{10.34}$$

where $\boldsymbol{s}$ is an appropriate score vector such as Wilcoxon's rank score. It is obvious that statistics (10.32) ~ (10.34) are the components of (10.31). An exact analysis is possible for max max χ^2 (10.32) when a and b are not large, while approximate distributions are available for (10.33) and (10.34). Thus, the two-way data analysis is indeed a rich source of interesting statistical theories and applications. The analytical methods introduced in this chapter are extended to discrete two-way data in Chapter 11, and interesting examples are given there.

10.5 Easy Method for Unbalanced Data

10.5.1 Introduction

When the repetition number is unequal for each cell, the two-way data are said to be unbalanced. The analysis of unbalanced two-way data is not so popular compared with unbalanced one-way data. The reason lies not only in the complexity of calculation, but also in the non-uniqueness of the hypotheses tested. The meaning of the null hypothesis of interaction is clear. However, there is confusion in testing the null hypotheses of the main effects. There are choices to define the main effects by the simple average or the weighted average proportional to the repetition number. Also, the problem arises since the main effects of the two factors are not orthogonal to each other, caused by the unbalance. Further, it is controversial whether the main effects should be tested under the assumption of no interaction or not. As stated in Section 10.2, we basically recommend the inference of main effects under the assumption of no interaction unless there is a particular reason for the inference of the main effects under the existence of interaction effects. One should refer to Kutner (1974) for various approaches in this situation. Further, the problem is more serious for multiple comparison procedures and it seems that there is no established approach to multiple comparisons. If there is no empty cell, then an easy method of this section may give an appropriate procedure both theoretically and computationally.

10.5.2 Sum of squares based on cell means

We assume model (10.1) with repetition number m_{ij}. For the unbalanced two-way data, a simple standard form is no longer available and we have to go back to the general linear model as shown in Section 10.5.7 (2). However, if there is no empty cell $\left(m_{ij} \geq 1, i = 1, \ldots, a, j = 1, \ldots, b\right)$ we can analyze $x_{ij} = \bar{y}_{ij\cdot}$ as if it were from a balanced model with equal repetition $m = \left(\frac{1}{ab}\sum_i \sum_j \frac{1}{m_{ij}}\right)^{-1}$. By considering x_{ij} we can reduce the problem of unbalanced data to unequal variance σ^2/m_{ij}. The F-test is known to be robust against unequal variance and we can also make adjustments easily since its structure is known.

Now, the model is the same as in Section 10.2 but we assume an unequal number m_{ij} of repetitions. First,

$$S_e = \sum_{i=1}^{a} \sum_{j=1}^{b} \sum_{k=1}^{m_{ij}} \left(y_{ijk} - \bar{y}_{ij\cdot}\right)^2 \tag{10.35}$$

is distributed as $\sigma^2 \chi^2_{n-ab}$ with df $n-ab$, $n = m_{\cdots}$. Therefore, $S_e/(n-ab)$ is the best unbiased estimator of σ^2 as usual. Then, we form the usual sum of squares (10.11) ~ (10.13) of x_{ij} with repetition number 1 as

$$S_\alpha = b^{-1} \sum_i x_{i\cdot}^2 - (ab)^{-1} x_{\cdot\cdot}^2, \tag{10.36}$$

$$S_\beta = a^{-1} \sum_j x_{\cdot j}^2 - (ab)^{-1} x_{\cdot\cdot}^2, \tag{10.37}$$

$$S_{\alpha\beta} = \sum_i \sum_j x_{ij}^2 - b^{-1} \sum_i x_{i\cdot}^2 - a^{-1} \sum_j x_{\cdot j}^2 + (ab)^{-1} x_{\cdot\cdot}^2. \tag{10.38}$$

Next, noting that the variance of x_{ij} is

$$V(\boldsymbol{x}) = \text{diag}\left(\sigma^2/m_{ij}\right), \boldsymbol{x} = (x_{11}, x_{12}, \ldots, x_{ab})', \tag{10.39}$$

the expectations of the sum of squares (10.36) ~ (10.38) are easily obtained as

$$\begin{aligned} E(S_\alpha) &= b\sum_i (\bar{\mu}_{i\cdot} - \bar{\mu}_{\cdot\cdot})^2 + \{(a-1)/m\}\sigma^2, \\ E(S_\beta) &= a\sum_j (\bar{\mu}_{\cdot j} - \bar{\mu}_{\cdot\cdot})^2 + \{(b-1)/m\}\sigma^2, \\ E(S_{\alpha\beta}) &= \sum_i \sum_j (\mu_{ij} - \bar{\mu}_{i\cdot} - \bar{\mu}_{\cdot j} + \bar{\mu}_{\cdot\cdot})^2 + \{(a-1)(b-1)/m\}\sigma^2, \end{aligned} \tag{10.40}$$

suggesting non-centrality parameters similar to the balanced case and easy to interpret.

10.5.3 Testing the null hypothesis of interaction

To test the null hypothesis

$$H_{\alpha\beta}: \mu_{ij} = \bar{\mu}_{i\cdot} + \bar{\mu}_{\cdot j} - \bar{\mu}_{\cdot\cdot},$$

a statistic

$$F_{\alpha\beta} = \frac{mS_{\alpha\beta}/\{(a-1)(b-1)\}}{S_e/(n-ab)} \tag{10.41}$$

is suggested by equation (10.40). If the unbalance in m_{ij} is small, this is approximately distributed as an F-statistic with df $\{(a-1)(b-1), n-ab\}$. If the unbalance is moderate, we can employ the chi-squared approximation $d\chi_f^2$ for $S_{\alpha\beta}$, adjusting the first two cumulants under $H_{\alpha\beta}$ by

$$E\left(S_{\alpha\beta}/\sigma^2\right) = \{(a-1)(b-1)/m\} = dE\left(\chi_f^2\right) = df,$$

$$V\left(S_{\alpha\beta}/\sigma^2\right) = 2\left\{f_{\alpha\beta}\left(m_{ij}^{-1}\right) + m^{-2}\right\} = d^2V\left(\chi_f^2\right) = 2d^2f,$$

where

$$f_{\alpha\beta}\left(u_{ij}\right) = \left\{\frac{(a-2)(b-2)}{(ab)}\right\}\sum_i\sum_j u_{ij}^2 + \left(\frac{a-2}{a}\right)\sum_i\left(b^{-1}\sum_j u_{ij}\right)^2 + \left(\frac{b-2}{b}\right)\sum_j\left(a^{-1}\sum_i u_{ij}\right)^2.$$

Then we have constants for approximation,

$$f = \frac{\{(a-1)(b-1)/m\}^2}{f_{\alpha\beta}\left(m_{ij}^{-1}\right) + m^{-2}},\ d = \frac{(a-1)(b-1)}{mf}. \tag{10.42}$$

Since $mS_{\alpha\beta}/(a-1)(b-1) = \left(S_{\alpha\beta}/\sigma^2\right)/(df) \sim \chi_f^2/f$, we can simply consider $F_{\alpha\beta}$ (10.41) as an F-statistic with df $\{f,\ (n-ab)\}$. Therefore, the adjusted method is very easy using the same statistic (10.41), just by altering the degrees of freedom to (10.42). If necessary, we can employ the F-approximation of Hirotsu (1979) using the third term of Laguerre's orthogonal polynomial expansion. However, this seems unnecessary by the simulation results in Table 10.10. By a simple calculation, f (10.42) can be expressed as

$$f = (a-1)(b-1)\left\{1 - \frac{f_{\alpha\beta}\left(m_{ij}^{-1} - m^{-1}\right)}{f_{\alpha\beta}\left(m_{ij}^{-1}\right) + m^{-2}}\right\}. \tag{10.43}$$

The last term of (10.43) can be used as a measure of unbalance. Further, since

$$f_{\alpha\beta}\left(m_{ij}^{-1}\right) + m^{-2} = (a-1)(b-1)/m^2$$

for $m_{ij} \equiv m$, a more simple function

$$\lambda_{\alpha\beta} = m^2 f_{\alpha\beta}\left(m_{ij}^{-1} - m^{-1}\right)/\{(a-1)(b-1)\} \tag{10.44}$$

is a good measure of unbalance.

Since $S_{\alpha\beta} = \|\left(\boldsymbol{P}_a' \otimes \boldsymbol{P}_b'\right)\boldsymbol{x}\|^2$, or $F_{\alpha\beta}$ (10.41), is of so simple a structure, the row and/or column-wise multiple comparisons in Section 10.3 and directional inference in Section 10.4 can be applied almost as they are. In this sense, the easy method here is very useful.

10.5.4 Testing the null hypothesis of main effects under $H_{\alpha\beta}$

When the null hypothesis $H_{\alpha\beta}$ is not rejected, we can test the null hypothesis

$$H_\alpha : \mu_{ij} = \mu + \beta_j$$

by

$$F_\alpha = \frac{mS_\alpha/(a-1)}{S_e/(n-ab)}, \tag{10.45}$$

which is approximately distributed as an F-statistic with df $\{a-1, (n-ab)\}$ if the unbalance is small. If the unbalance is moderate, it is well approximated by the F-statistic with df $\{f, (n-ab)\}$, where

$$f = (a-1)\left\{1 - \frac{f_\alpha\left(b^{-1}\sum_j m_{ij}^{-1} - m^{-1}\right)}{f_\alpha\left(b^{-1}\sum_j m_{ij}^{-1}\right) + m^{-2}}\right\},$$

$$f_\alpha(u_i) = \left(\frac{a-2}{a}\right)\sum_i u_i^2$$

is obtained by adjusting the first two cumulants. A simple measure of unbalance in this case is

$$\lambda_\alpha = m^2 f_\alpha\left(b^{-1}\sum_j m_{ij}^{-1} - m^{-1}\right)/(a-1). \tag{10.46}$$

By symmetry, the null hypothesis

$$H_\beta : \mu_{ij} = \mu + \alpha_i$$

can be tested by

$$F_\beta = \frac{mS_\beta/(b-1)}{S_e/(n-ab)}, \tag{10.47}$$

with

$$f = (b-1)\left\{1 - \frac{f_\beta\left(a^{-1}\sum_i m_{ij}^{-1} - m^{-1}\right)}{f_\beta\left(a^{-1}\sum_i m_{ij}^{-1}\right) + m^{-2}}\right\},$$

$$f_\beta(u_j) = \left(\frac{b-2}{b}\right)\sum_j u_j^2.$$

A simple measure of unbalance in this case is

$$\lambda_\beta = m^2 f_\beta\left(a^{-1}\sum_i m_{ij}^{-1} - m^{-1}\right)/(b-1). \tag{10.48}$$

The multiple comparison procedures of Chapter 6 are also easily applied, which is a merit of this easy method. Thus, the analysis of $x_{ij} = \bar{y}_{ij.}$ gives very simple and useful procedures only if there is no empty cell and the unbalance is not too heavy.

10.5.5 Accuracy of approximation by easy method

In previous sections the simple measure of unbalance (10.44), (10.46), and (10.48) have been introduced. Hirotsu (1969) examined the size of the naïve approximate test without adjustment by the asymptotic expansion of the characteristic function of the test statistic. The results are given in the fourth column of Table 10.10, where the designs examined are rather heavily unbalanced (as given in Table 10.9). It is found that the unbalance measure λ's are at most 0.1+ for those designs with maximin ratio of m_{ij} up to 4. The increase in type I error is roughly 0.02λ ($2\lambda\%$), so that the naïve approximate method will be acceptable for λ up to 0.05. For λ beyond 0.05 the adjustment by the first two cumulants will work well. We therefore make a simulation for the performance of the adjusted method as well as the naïve method, and the results are shown in Table 10.10.

10.5.6 Simulation

We performed a simulation for the performance of the naïve and adjusted approximate methods. For each design of $D1 \sim D5$, the necessary number of independent normal variables was generated. We took the average of the 12 uniform numbers to generate a normal random variable. The replication number of simulation is 5×10^5 for each of $D1 \sim D5$. At each time, the p-value of each test is calculated, which should be distributed as a uniform distribution if the test is exact. We show only the percentage of the p-value less than 0.05 by each method in the fifth and sixth columns of Table 10.10, but the results are similar for other points. The standard deviation (SD) of the estimated test size by simulation is 0.03(%) as a binomial distribution. Now, the simulation supports the evaluation of the naïve method by the asymptotic expansion of the characteristic function of the test statistic in the previous section. Then, the adjustments by the first two cumulants work remarkably well, keeping a precise and slightly conservative test size for all cases.

10.5.7 Comparison with the LS method on real data

Example 10.4. Analysis of Table 10.11. The data of Table 10.11 were originally from Afifi and Azen (1972); Kutner (1974) dropped 14 data points marked by $*$ randomly to create the unbalanced data for his explanation of various approaches. We also employ the data for comparisons of the easy method with a standard linear model approach. The purpose of the experiment was to evaluate the effect of four drugs

Table 10.9 Repetition number.

		D1			D2			D3			D4					D5				
		Column			Column			Column			Column					Column				
		1	2	3	1	2	3	1	2	3	1	2	3	4	5	1	2	3	4	5
Row	1	1	2	2	1	2	2	1	1	4	1	1	1	1	1	2	3	1	1	3
	2	2	2	3	2	3	2	1	3	1	2	2	2	2	1	2	1	2	4	1
	3	2	2	1	1	3	1	4	1	1	3	3	3	3	3	3	2	1	1	2

Table 10.10 Size of the naïve approximate test and related measure of unbalance.

Design	Hypothesis	λ	Size of naïve test (%)	Simulation: Naive	Simulation: Adjusted
D1	H_α	0.016	5.03	5.03	4.99
	H_β	0.007	5.01	5.01	4.99
	$H_{\alpha\beta}$	0.047	5.09	5.06	4.95
D2	H_α	0.024	5.04	4.98	4.92
	H_β	0.042	5.07	5.05	4.95
	$H_{\alpha\beta}$	0.079	5.14	5.09	4.92
D3	H_α	0.0001	5.00	4.91	4.91
	H_β	0.0001	5.00	4.92	4.92
	$H_{\alpha\beta}$	0.051	5.10	4.89	4.78
D4	H_α	0.090	5.22	5.15	4.85
	H_β	0.008	5.02	4.98	4.95
	$H_{\alpha\beta}$	0.106	5.29	5.18	4.89
D5	H_α	0.0002	5.00	4.90	4.90
	H_β	0.032	5.10	5.05	4.95
	$H_{\alpha\beta}$	0.093	5.25	5.18	4.92

crossed with three experimentally induced diseases. Each drug–disease combination was applied to six randomly selected dogs. The measurement to be analyzed was the increase in systolic pressure (mmHg) due to the treatment. First we analyze the data by our naïve and adjusted methods. Then it is analyzed by a standard linear model.

(1) **Analysis by the easy method.** The results of analysis by the easy method are shown in Table 10.12. It is seen that there is only a very small difference between the naïve and adjusted methods for the unbalance of m_{ij} from 3 to 6 in Table 10.11.

Table 10.11 Increase in systolic pressure due to the treatment.

	Disease j		
Drug i	1	2	3
1	42, 44, 36, 13, 19, 22	33, 40*, 26, 34*, 33, 21	31, −3, 19*, 25, 25, 24
2	28, 40*, 23, 24, 42, 13	31*, 34, 33, 31, 33*, 36	3, 26, 28, 32, 4, 16
3	28*, 21*, 1, 29, 6*, 19	−4*, 11, 9, 7, 1, −6	21, 1, 2*, 9, 3, 9*
4	24, 19*, 9, 22, −2,15	27, 12, 12, −5, 16, 15	22, 7, 25, 5, 12, 7*

(2) **Analysis by a linear model.** Next we compare the results with a linear statistical inference. There have been proposed various approaches to the analysis of the unbalanced data as described in Kutner (1974). Here we take an approach to testing the main effects only under the assumption of no interaction, which can avoid the effects of the identification conditions described by Kutner. First, the sum of squares for the error is $S_e = 5040.82$ by (10.35) and the unbiased variance is

$$\hat{\sigma}^2 = S_e/(n-ab) = 5040.817/(58-12) = 109.583,$$

which is used commonly for tests of interaction and main effects.

(a) **Test of the null hypothesis $H_{\alpha\beta}$ (10.8).** According to the procedure explained in Section 3.4.1, we need to calculate the residual sum of squares under $H_{\alpha\beta}$: $\mu_{ij} = \mu + \alpha_i + \beta_j$. For this purpose, we express $\boldsymbol{\mu} = E(\boldsymbol{y})$ in matrix form as in Section 10.2. However, to solve the normal equation for the LS estimates it is convenient to take the form

$$\boldsymbol{y} = \begin{bmatrix} \boldsymbol{j} \ \boldsymbol{X}_\alpha^* \ \boldsymbol{X}_\beta^* \end{bmatrix} \begin{bmatrix} \mu \\ \boldsymbol{\alpha}^* \\ \boldsymbol{\beta}^* \end{bmatrix} + \boldsymbol{e}, \tag{10.49}$$

where $\boldsymbol{\alpha}^* = (\alpha_1, \alpha_2, \alpha_3)'$, $\boldsymbol{\beta}^* = (\beta_1, \beta_2)'$ dropping α_4 and β_3. This is equivalent to imposing the constraints $\alpha_4 = 0$ and $\beta_3 = 0$, but it is obvious that the choice of constraints does not affect the inference. It is an easy task to write down explicitly the design matrix of (10.49) but it is $58 \times (1+3+2)$ and too large to present here. Then, the normal equation can be solved easily since the coefficient matrix is non-singular,

Table 10.12 Analysis of variance by easy method.

Factor	F-statistics	Naïve df	Naïve p-Value	Adjusted df df	Adjusted df p-Value
Main effects of A	8.9711 (10.45)	3	0.000087**	2.9681	0.000083**
Main effects of B	1.6415 (10.47)	2	0.2048	1.9984	0.2048
Interaction	1.1454 (10.41)	6	0.3518	5.876	0.3517

which gives $\hat{\mu} = 10.0729$, $\hat{\boldsymbol{\alpha}} = (12.5115,\ 11.7035,\ -4.5592,\ 0)'$, and $\hat{\boldsymbol{\beta}} = (5.8792,\ 4.2495,\ 0)'$. The residual sum of squares for model (10.49) is obtained as

$$S_{0\alpha\beta} = \sum_i \sum_j \sum_k \left\{ y_{ijk} - \left(\hat{\mu} + \hat{\alpha}_i + \hat{\beta}_j \right) \right\}^2 = 5771.42.$$

Finally, the sum of squares for interaction is obtained as the increase in residual sum of squares from S_e as

$$S_{\alpha\beta} = S_{0\alpha\beta} - S_e = 5771.42 - 5040.82 = 730.60$$

with df $(a-1)(b-1) = 6$. Then, the F-statistic is obtained as

$$F = \frac{S_{\alpha\beta}/\{(a-1)(b-1)\}}{\hat{\sigma}^2} = \frac{730.60/6}{109.583} = 1.1112,$$

whose p-value is 0.3706 and non-significant. Therefore, we proceed to test the main effects.

(b) **Test of the null hypothesis $H_\alpha : \mu_{ij} = \mu + \beta_j$ assuming $H_{\alpha\beta}$.** The estimate of μ_{ij} under $H_\alpha \cap H_{\alpha\beta}$ is obviously

$$\hat{\mu}_{ij} = \left(\widehat{\mu + \beta_j} \right) = y_{\cdot j \cdot} / m_{\cdot j \cdot}.$$

The residual sum of squares is then

$$S_{0(\alpha \cap \alpha\beta)} = \sum_i \sum_j \sum_k \left\{ y_{ijk} - \left(\widehat{\mu + \beta_j} \right) \right\}^2 = \sum_i \sum_j \sum_k y_{ijk}^2 - \sum_j \left(y_{\cdot j \cdot}^2 / m_{\cdot j \cdot} \right) = 8722.05.$$

The increase in residual sum of squares is therefore

$$S_\alpha = S_{0(\alpha \cap \alpha\beta)} - S_{0\alpha\beta} = 8722.05 - 5771.42 = 2950.63$$

with df $(a-1)=3$. The F-statistic is obtained as

$$F=\frac{S_\alpha/(a-1)}{\hat{\sigma}^2}=\frac{2950.63/3}{109.583}=8.9753,$$

whose p-value is 8.6×10^{-5} and highly significant.

(c) **Test of the null hypothesis $H_\beta: \mu_{ij}=\mu+\alpha_i$ assuming $H_{\alpha\beta}$.** The estimate of μ_{ij} under $H_\beta \cap H_{\alpha\beta}$ is obviously

$$\hat{\mu}_{ij}=(\widehat{\mu+\alpha_i})=y_{i\cdot\cdot}/m_{i\cdot\cdot}.$$

The residual sum of squares is then

$$S_{0(\beta\cap\alpha\beta)}=\sum_i\sum_j\sum_k\left\{y_{ijk}-(\widehat{\mu+\alpha_i})\right\}^2=\sum_i\sum_j\sum_k y_{ijk}^2-\sum_i\left(y_{i\cdot\cdot}^2/m_{i\cdot\cdot}\right)=6130.92.$$

The increase in residual sum of squares is therefore

$$S_\beta=S_{0(\beta\cap\alpha\beta)}-S_{0\alpha\beta}=6130.92-5771.42=359.50$$

with df $(b-1)=2$. The F-statistic is obtained as

$$F=\frac{S_\beta/(b-1)}{\hat{\sigma}^2}=\frac{359.50/2}{109.583}=1.640, \tag{10.50}$$

whose p-value is 0.2051 and non-significant.

These results are summarized in ANOVA Table 10.13.
This result and the easy methods are compared in Table 10.14. It is observed again that there is very little difference among the procedures compared. Then, our methods are

Table 10.13 ANOVA by a linear model.

Factor	Sum of Squares	F-statistic	df	p-Value
Main effects of A	2950.629	8.9753	3	0.000086**
Main effects of B	359.503	1.6403	2	0.2051
Interaction $A\times B$	730.597	1.1112	6	0.3706

Table 10.14 Comparison of p-values of the easy methods with a standard inference.

Factor	Linear model	Naïve	Adjusted df
Main effects of A	0.000086**	0.000087**	0.000083**
Main effects of B	0.2051	0.2048	0.2048
Interaction $A\times B$	0.3706	0.3518	0.3517

very simple to apply and easy to interpret. Further, it is straightforward to apply our method to multiple comparisons and directional inference.

Next we try to test the null hypothesis H_β assuming $H_\alpha \cap H_{\alpha\beta}$; that is, ignoring the effects of $\boldsymbol{\alpha}$ instead of adjusting it, which should give the same result as (c) for the balanced case because of the orthogonality of the two main effects. Then, what shall happen for the unbalanced data? The residual sum of squares under $H_\alpha \cap H_\beta \cap H_{\alpha\beta}$ is obviously

$$S_{0(\alpha\cap\beta\cap\alpha\beta)} = \sum_i \sum_j \sum_k y_{ijk}^2 - y_{\cdots}^2 / m_{\cdots} = 9136.017 \tag{10.51}$$

and its increase from $S_{0(\alpha\cap\alpha\beta)}$ is

$$S_\beta^* = S_{0(\alpha\cap\beta\cap\alpha\beta)} - S_{0(\alpha\cap\alpha\beta)} = \sum_j \left(y_{\cdot j \cdot}^2 / m_{\cdot j \cdot} \right) - y_{\cdots}^2 / m_{\cdots} = 413.975,$$

with df $(b-1) = 2$. The F-statistic is then

$$F^* = \frac{S_\beta^*/(b-1)}{\hat{\sigma}^2} = \frac{413.975/2}{5040.82/46} = 1.8887 \tag{10.52}$$

and the associated p-value is 0.1628. This result suggests a relatively large difference from other methods because of the influence of the main effect $\boldsymbol{\alpha}$. The analysis by F (10.50) escapes from this spurious effect and is called adjusted for $\boldsymbol{\alpha}$. The S_α in Table 10.13 are also adjusted for factor B. It should be noted here that the sum of squares (10.51) is the so-called total sum of squares S_T of the ANOVA table. However, for the unbalanced data the sum of squares S_e and S_α, S_β, $S_{\alpha\beta}$ in Table 10.13 does not add up to S_T. Instead, the sum of S_e, S_α, S_β^* and $S_{\alpha\beta}$ is equal to S_T. Also, S_e, S_α^*, S_β, and $S_{\alpha\beta}$ add up to S_T, where

$$S_\alpha^* = S_{0(\alpha\cap\beta\cap\alpha\beta)} - S_{0(\beta\cap\alpha\beta)} = \sum_i \left(y_{i\cdot\cdot}^2 / m_{i\cdot\cdot} \right) - y_{\cdots}^2 / m_{\cdots}.$$

Therefore, sometimes two ANOVA tables are presented where either one of the main effects is adjusted and the other unadjusted, thus making the sum of squares for $A, B, A \times B$ and error equal to the total sum of squares S_T. Including this phenomenon, the analysis of unbalanced two-way data is a little controversial. Our method is essentially the type II analysis of Yates (1934), but recommends testing the main effects only after verifying no interaction. The analysis by F (10.52) is type I and generally not recommended. The test of the main effects under the existence of interaction effects in type III analysis is also not recommended, since it is an analysis of non-estimable parameters and suffers from the effects of identification conditions if the interaction actually exists.

10.5.8 Estimation of the mean μ_{ij}

When an interaction exists, $\hat{\mu}_{ij} = x_{ij} \left(= \bar{y}_{ij\cdot} \right)$ is BLUE. Therefore, we consider here the estimation of $\mu_{ij} = \mu + \alpha_i + \beta_j$ by

$$\hat{\mu}_{ij} = \bar{x}_{i\cdot} + \bar{x}_{\cdot j} - \bar{x}_{\cdot\cdot}.$$

It is obvious that this estimator is unbiased. To evaluate the variance of $\hat{\mu}_{ij}$, we introduce a vector $\boldsymbol{v}_{ij} = (v_{11}, v_{12}, \ldots, v_{ab})'$, where

$$v_{kl} = \frac{1}{b}\delta_{ik} + \frac{1}{a}\delta_{jl} - \frac{1}{ab},\ k = 1, \ldots, a,\ l = 1, \ldots, b.$$

Since $\hat{\mu}_{ij}$ is expressed as $\boldsymbol{v}_{ij}'\boldsymbol{x}$, we have from (10.39) the variance

$$\begin{aligned} \mathrm{V}(\hat{\mu}_{ij}) &= \boldsymbol{v}_{ij}'\mathrm{diag}\left(\sigma^2/m_{ij}\right)\boldsymbol{v}_{ij} \\ &= \frac{1}{ab}\left\{\frac{1}{m} + (b-2)\frac{1}{a}\sum\nolimits_i \frac{1}{m_{ij}} + (a-2)\frac{1}{b}\sum\nolimits_j \frac{1}{m_{ij}} + \frac{2}{m_{ij}}\right\}\sigma^2. \end{aligned} \tag{10.53}$$

Equation (10.53) coincides of course with

$$V(\hat{\mu}_{ij}) = \frac{a+b-1}{abm}\sigma^2$$

of the BLUE in the balanced case. The variance $V(\hat{\mu}_{ij})$ (10.53) has been compared with the BLUE in an unbalanced case by Hirotsu (1969). The difference is negligible for $\hat{\mu}_{11}, \hat{\mu}_{12}, \hat{\mu}_{22}$ in $D1$, whereas BLUE is recommended for the extreme case of $\hat{\mu}_{13}, \hat{\mu}_{31}$ in $D3$ and $\hat{\mu}_{24}$ in $D5$. In other cases, the increase of SD is approximately 5%, widening the confidence interval at this rate. An easy method for estimating the mean is not so important compared with multiple comparisons and directional inferences, since the LS method is also easily applicable. But still it is useful, since the difference from BLUE is negligible for a slight unbalance.

References

Afifi, A. A. and Azen, S. P. (1972) *Statistical analysis: A computer oriented approach.* Academic Press, New York.

Cox, D. R. (1958) *Planning of experiments.* Wiley, New York.

Davies, O. L. (1954) *The design and analysis of industrial experiments.* Oliver & Boyd, London.

Hirotsu, C. (1969) Properties of the analysis of variance tests for the fixed effects model in the unbalanced case. *Rep. Statist. Appl. Res., JUSE* **16**, 44–59.

Hirotsu, C. (1976) *Analysis of variance.* Kyoiku Shuppan, Tokyo (in Japanese).

Hirotsu, C. (1978) Ordered alternatives for interaction effects. *Biometrika* **65**, 561–570.

Hirotsu, C. (1979) An *F* approximation and its application. *Biometrika* **66**, 577–584.

Hirotsu, C. (1983a) An approach to defining the pattern of interaction effects in a two-way layout. *Ann. Inst. Statist. Math.* **A35**, 77–90.

Hirotsu, C. (1983b) Defining the pattern of association in two-way contingency tables. *Biometrika* **70**, 579–590.

Hirotsu, C. (1991) An approach to comparing treatments based on repeated measures. *Biometrika* **78**, 583–594.

Hirotsu, C. (1993) Beyond analysis of variance techniques: Some applications in clinical trials. *Int. Statist. Rev.* **61**, 183–201.

Hirotsu, C. (1997) Two-way change point model and its application. *Austral. J. Statist.* **39**, 205–218.

Hirotsu, C. (2009) Clustering rows and/or columns of a two-way contingency table and a related distribution theory. *Comput. Statist. Data Anal.* **53**, 4508–4515.

Hirotsu, C., Kuriki, S. and Hayter, A. J. (1992) Multiple comparison procedures based on the maximal component of the cumulative chi-squared statistic. *Biometrika* **79**, 381–392.

Hirotsu, C., Aoki, S., Inada, T. and Kitao, Y. (2001) An exact test for the association between the disease and alleles at highly polymorphic loci with particular interest in the haplotype analysis. *Biometrics* **57**, 769–778.

Hirotsu, C., Ohta, E., Hirose, N. and Shimizu, K. (2003) Profile analysis of 24-hours measurements of blood pressure. *Biometrics* **59**, 907–915.

Johnson, D. E. and Graybill, F. A. (1972) An analysis of a two-way model with interaction and no replication. *J. Amer. Statist. Assoc.* **67**, 862–868.

Kuriki, S. and Takemura, A. (2001) Tail probabilities of the maxima of multilinear forms and their applications. *Ann. Statist.* **29**, 328–371.

Kutner, M. H. (1974) Hypothesis testing in linear model. *Am. Statist.* **28**, 98–100.

Mandel, J. (1971) A new analysis of variance model for non additive data. *Technometrics* **13**, 1–18.

Marco, C. (2014) Distribution of largest eigenvalue for real Wishart and Gaussian random matrices and a simple approximation for the Tracy–Widom distribution. *J. Multivar. Anal.* **129**, 69–81.

Miller, R. G. (1998) *BEYOND ANOVA: Basics of applied statistics*. Chapman & Hall/CRC Texts in Statistical Science, New York.

Scheffé, H. (1959) *The analysis of variance*. Wiley, New York.

Tukey, J. W. (1949) One degree of freedom for non-additivity. *Biometrics* **5**, 232–242.

Yates, F. (1934) The analysis of multiple classifications with unequal numbers in the different classes. *J. Amer. Statist. Assoc.* **29**, 51–56.

11

Analysis of Two-Way Categorical Data

11.1 Introduction

An overall goodness-of-fit chi-square (3.57) for independence is a well-known approach to a contingency table, just like an F-test in the two-way ANOVA. It cannot, however, give any detailed information on the association between the rows and columns. As an example, we give Table 11.1 which was first analyzed by Hirotsu (1977). It gives the number of patients cross-classified by their occupation and severity of illness at their first visit at the National Cancer Institute of Japan. The goodness-of-fit chi-square is 95.75, with df 18 by (3.57), which is extremely highly significant. It suggests that there is a strong association between occupation and severity of illness, probably reflecting the differences of the system at that time for detecting cancer at an early stage at each enterprise. The result, however, cannot tell us any details of the association. We may evaluate the departure of each cell from independence by the formula

$$e_{ij} = \left(y_{ij} - \frac{y_{i\cdot} y_{\cdot j}}{y_{\cdot\cdot}} \right) \Big/ \sqrt{\frac{y_{i\cdot} y_{\cdot j} (y_{\cdot\cdot} - y_{i\cdot})(y_{\cdot\cdot} - y_{\cdot j})}{y_{\cdot\cdot}^3}}, \tag{11.1}$$

which is asymptotically distributed as a standard normal variable under the null hypothesis of independence. These are given in parentheses in Table 11.1. There are observed several large deviations – such as the cells (3, 1), (3, 2), and (10, 1) – but again they cannot give any clear-cut explanation of the data. Therefore, some multiple-comparison approaches have been proposed, among which the row- and/or column-wise multiple comparisons proposed by Hirotsu (1977, 1983) have been

Advanced Analysis of Variance, First Edition. Chihiro Hirotsu.

Table 11.1 Number of cancer patients by occupation and severity of illness at their first visit at the National Cancer Institute of Japan.

	Severity			
Occupation	1. Slight	2. Medium	3. Serious	Total
1. Professional and technical workers	148 (2.53) ((148.5))	444 (−2.58) ((452.4))	86 (0.63) ((77.1))	678
2. Managers and officials	111 (2.10) ((112.1))	352 (−0.58) ((341.6))	49 (−1.53) ((58.2))	512
3. Clerical and related workers	645 (6.68) ((631.4))	1911 (−4.88) ((1924.4))	328 (1.04) ((328.1))	2884
4. Sales workers	165 (−2.25) ((160.7))	771 (2.36) ((764.0))	119 (−0.67) ((130.3))	1055
5. Farmers, lumbermen, fishermen, quarrymen	383 (−4.41) ((384.3))	1829 (3.21) ((1827.2))	311 (0.72) ((311.5))	2523
6. Workers in transport and communication systems	96 (2.11) ((95.5))	293 (−1.24) ((290.9))	47 (−0.75) ((49.6))	436
7. Craftsmen	98 (1.15) ((106.4))	330 (−0.98) ((324.3))	58 (0.01) ((55.3))	486
8. Production process workers	199 (−1.91) ((187.1))	874 (1.03) ((889.3))	155 (0.34) ((151.6))	1228
9. Service workers	59 (1.02) ((63.1))	199 (−0.30) ((192.2))	30 (0.79) ((32.8))	288
10. Persons without regular occupations	262 (−4.54) ((276.9))	1320 (2.74) ((1316.6))	236 (1.53) ((224.5))	1818
Total	2166	8323	1419	11908

verified to be useful on several occasions compared with other multiple-comparison approaches – see Greenacre (1988), Hirotsu (2009), and the examples in Sections 11.3.3 and 11.4.4 (re-analysis of Table 11.1). The row-wise multiple comparisons are essential if the data are taken as the one-way layout with categorical responses instead of the usual normal response variables. Then, it is nothing but an analysis of treatment effects. The multiple comparison procedure proposed first by Hirotsu (1977, 1983) is essentially of Scheffé type, but the actual procedure was limited to comparisons of the squared distances between every two rows or two clusters of rows, which are uniquely defined by the normalization and orthogonality conditions. Then there was an inevitable loss of power in the procedure. It has been extended to the generalized squared distance among any number of rows or clusters of rows, as explained in Section 10.3.2 (3), and the loss of power was dissolved in the process of Scheffé type multiple comparisons.

An interesting extension of the method is to the one-way layout with ordered categorical responses. In this case the procedure is essentially unchanged, except for the definition of the order-sensitive chi-squared distance and the related asymptotic distribution as explained in Section 10.4.1(2). Then, the reference distribution becomes that of the largest eigenvalue of a Wishart matrix, which is very difficult to handle. The normal approximation given by Anderson (2003) is quite unsatisfactory, especially when the first- and second-largest eigenvalues are close to each other. We therefore propose a chi-squared approximation as a more reasonable one in Section 11.4.2 (2). It nicely improves the normal approximation of Anderson and also the first-order chi-squared approximation introduced by Hirotsu (1991).

The row-wise multiple comparisons can be applied to moderately large data with a hundred rows. Then we need some automatic procedure for clustering. We therefore give an algorithm for clustering rows in Section 11.3.3, so that the generalized chi-squared distance among clusters is large, achieving simultaneously homogeneity within each cluster.

11.2 Overall Goodness-of-Fit Chi-Square

Let a two-way contingency table be denoted by $\{y_{ij}\}_{a\times b}$ and the row, column, and grand totals by $R_i = y_{i\cdot}$, $C_j = y_{\cdot j}$, and $N = y_{\cdot\cdot}$, respectively. We assume a multinomial distribution with cell probabilities $\{p_{ij} \mid p_{\cdot\cdot} = 1\}$. The null hypothesis of independence is

$$H_0 : p_{ij} = p_{i\cdot}p_{\cdot j} \text{ for all } i \text{ and } j,$$

and the statistical inference is based on the conditional distribution given the sufficient statistics R_i and C_j under H_0, which is the multivariate hypergeometric distribution.

The well-known goodness-of-fit chi-square,

$$\chi^2 = \sum_{i=}^{a} \sum_{j=1}^{b} \frac{\left(y_{ij} - R_iC_j/N\right)^2}{R_iC_j/N}, \tag{11.2}$$

has already been derived as equation (3.57) in Example 3.5, which is asymptotically distributed as a chi-squared distribution with df $(a-1)(b-1)$ under H_0.

To partition χ^2, we introduce two vectors

$$\boldsymbol{r} = N^{-1/2}\left(R_1^{1/2}, \ldots, R_a^{1/2}\right)', \boldsymbol{c} = N^{-1/2}\left(C_1^{1/2}, \ldots, C_b^{1/2}\right)'$$

and define $\boldsymbol{R}'_{(a-1)\times a}$ and $\boldsymbol{C}'_{(b-1)\times b}$ so that $\begin{pmatrix} \boldsymbol{r}' \\ \boldsymbol{R}' \end{pmatrix}$ and $\begin{pmatrix} \boldsymbol{c}' \\ \boldsymbol{C}' \end{pmatrix}$ are the a- and b-dimensional orthonormal matrices, respectively. They correspond to $\boldsymbol{P}'_a$ and $\boldsymbol{P}'_b$ in the ANOVA model of previous sections. Define a column vector z with elements

$z_{ij} = y_{ij}/\left(R_iC_j/N\right)^{1/2}$ arranged in dictionary order. Then, under the null hypothesis H_0, the conditional expectation and variance of $(\boldsymbol{R}'\otimes\boldsymbol{C}')\boldsymbol{z}$ as the multivariate hypergeometric distribution given R_i, C_j are

$$E\{(\boldsymbol{R}'\otimes\boldsymbol{C}')\boldsymbol{z}\} = \boldsymbol{0}_{(a-1)(b-1)}, \tag{11.3}$$

$$V\{(\boldsymbol{R}'\otimes\boldsymbol{C}')\boldsymbol{z}\} = \{N/(N-1)\}\boldsymbol{I}_{(a-1)(b-1)}, \tag{11.4}$$

with $\boldsymbol{0}_n$ and $\boldsymbol{I}_n$ the n-dimensional zero vector and the identity matrix, respectively. It should be noted that in $(\boldsymbol{R}'\otimes\boldsymbol{C}')\boldsymbol{z}$, every row of $\boldsymbol{R}'$ makes the orthogonal contrast in rows and every row of $\boldsymbol{C}'$ makes the orthogonal contrast in columns, thus together making the interaction contrast. In the following we ignore the coefficient $\{N/(N-1)\}$ in the variance, since our contingency table example is usually large. Then, every element of $(\boldsymbol{R}'\otimes\boldsymbol{C}')\boldsymbol{z}$ is standardized as expectation 0 and variance 1, orthogonal to each other. The sum of squares

$$\chi^2 = \|(\boldsymbol{R}'\otimes\boldsymbol{C}')\boldsymbol{z}\|^2 \tag{11.5}$$

is nothing but the goodness-of- fit chi-squared (11.2) and every element of $(\boldsymbol{R}'\otimes\boldsymbol{C}')\boldsymbol{z}$ gives the partition of χ^2 into one degree of freedom. The overall goodness-of-fit χ^2 test cannot give any details on the two-way data, even if it were significant (as mentioned in Section 11.1). It is just like the F-test in the ANOVA model. In contrast, multiple comparisons based on the 1 df component statistics of (11.5) cannot have reasonable power if the two-way table is moderately large. Also, the interpretation of the test results is usually unclear, as stated in Section 11.1 regarding Table 11.1. Then, the row-wise multiple comparison procedure proposed in Section 10.3 for the ANOVA model is again an attractive approach. Incidentally, e_{ij} of (11.1) is a component of (11.5) defined by $\{\boldsymbol{r}'(i;\bar{i})\otimes\boldsymbol{c}'(j;\bar{j})\}\boldsymbol{z}$, where

$$\boldsymbol{r}'(i;\bar{i}) = \left(R_i^{-1}+N_{\bar{i}}^{-1}\right)^{-1/2}\left(R_1^{1/2}/N_{\bar{i}},\cdots,R_{i-1}^{1/2}/N_{\bar{i}},-R_i^{1/2}/R_i,R_{i+1}^{1/2}/N_{\bar{i}},\cdots,R_a^{1/2}/N_{\bar{i}}\right),$$

with $N_{\bar{i}} = \sum_{i\in\bar{i}} R_i$, $\bar{i} = (1, \ldots, i-1, i+1, \ldots, a)$ and $\boldsymbol{c}'(j;\bar{j})$ similarly defined.

11.3 Row-wise Multiple Comparisons

11.3.1 Chi-squared distances among rows

(1) **Pair-wise chi-squared distance.** Those chi-squared distances introduced in this section are a generalization of those introduced in Section 10.3.2 for discrete data. The chi-squared distance between the lth and mth rows is defined as a component of (11.5) like

$$\chi^2(l_1;l_2) = \|\{\boldsymbol{r}'(l_1;l_2)\otimes\boldsymbol{C}'\}\boldsymbol{z}\|^2, \tag{11.6}$$

$$r'(l_1;l_2) = \left(R_{l_1}^{-1} + R_{l_2}^{-1}\right)^{-1/2}\left(0\cdots 0 R_{l_1}^{-1/2} 0\cdots 0 - R_{l_2}^{-12} 0\cdots 0\right),\ l_1, l_2 = 1, \ldots, a; l_1 \neq l_2.$$

(2) **Chi-squared distance between two subgroups of rows.** Without any loss of generality, let the first subgroup $H_1\{1, \ldots, p_1\}$ be composed of the first p_1 rows and the second subgroup $H_2\{p_1+1, \ldots, p_1+p_2\}$ be composed of the subsequent p_2 rows $(p_1+p_2 \le a)$. Then, the chi-squared distance between the two subgroups is defined by

$$\chi^2(H_1;H_2) = \|\{r'(H_1;H_2)\otimes C'\}z\|^2, \tag{11.7}$$

$$r'(H_1;H_2) = \left(T_1^{-1} + T_2^{-1}\right)^{-1/2}\left(\frac{R_1^{1/2}}{T_1}\cdots\frac{R_{p_1}^{1/2}}{T_1} - \frac{R_{p_1+1}^{1/2}}{T_2}\cdots - \frac{R_{p_1+p_2}^{1/2}}{T_2} 0\cdots 0\right),$$

$$T_1 = \sum\nolimits_{i\in H_1} R_i,\ T_2 = \sum\nolimits_{i\in H2} R_i.$$

(3) **Generalized chi-squared distance among any number of subgroups.** Without any loss of generality, we assume a partition of rows into m subgroups:

$$H_1\{1, \ldots, p_1\}, H_2\{p_1+1, \ldots, p_1+p_2\}, \ldots, H_K\{p_1+\cdots+p_{K-1}+1, \ldots, p_1+\cdots+p_K(=a)\}.$$

Then, the generalized chi-squared distance is defined by

$$\chi^2(H_1;\ldots;H_K) = \max_{\rho' r=0,\ \|\rho\|^2=1} \|(\rho'\otimes C')z\|^2,\ \rho_i = \zeta_k(R_i/T_k)^{1/2} \text{for } i\in H_k, \tag{11.8}$$

where $T_k = \sum_{i\in H_k} R_i$.

The maximization is actually with respect to $\boldsymbol{\zeta} = (\zeta_1, \ldots, \zeta_K)$ under the condition

$$\sum T_k^{1/2}\zeta_k = 0,\ \sum\nolimits_{k=1}^{K}\zeta_k^2 = 1. \tag{11.9}$$

Let $Y_{kj} = \sum_{i\in H_k} y_{ij}$, $k = 1, \ldots, K$, denote the frequency of the kth cluster at the jth column, so that $\{Y_{kj}\}$ gives a $K\times b$ table with the row total T_k collapsing these pooled rows. Then, equation (11.8) becomes

$$\chi^2(H_1;\ldots;H_K) = \max_{\zeta}\boldsymbol{\zeta}'\begin{pmatrix} w_1' \\ \vdots \\ w_K' \end{pmatrix}(w_1\cdots w_K)\boldsymbol{\zeta}$$

$$\text{with } w_k = (T_k/N)^{-1/2} C'\left(C_1^{-1/2}Y_{k1}, \ldots, C_b^{-1/2}Y_{kb}\right)'. \tag{11.10}$$

In particular, we have

$$(w_1,\dots,w_K)\left(T_1^{1/2},\ \dots,\ T_K^{1/2}\right)' = \sum_{k=1}^{K}\left(T_k^{1/2}w_k\right) = NC'c = 0$$

suggesting $\left(T_1^{1/2},\ \dots,\ T_K^{1/2}\right)'$ to be an eigenvector of the matrix $(w_1\cdots w_K)'(w_1\cdots w_K)$ corresponding to a zero eigenvalue. Then, the maximization with respect to $\boldsymbol{\zeta}$ reduces to the problem of the largest eigenvalue of $(w_1\cdots w_K)'(w_1\cdots w_K)$ and the condition (11.9) is automatically satisfied.

11.3.2 Reference distribution for simultaneous inference in clustering rows

All the chi-squared distances (11.6) ~ (11.8) introduced in the previous section are bounded above by $\chi^2(\{1\};\dots;\{a\}) = \max_{\rho' r=0,\ \|\rho\|^2=1} \|\{\rho' \otimes C'\}z\|^2$ which is the largest eigenvalue of $W'W$, where $W = (w_1\cdots w_a), w_i = (R_i/N)^{-1/2}C'\left(C_1^{-1/2}y_{i1},\ \dots,\ C_b^{-1/2}y_{ib}\right)'$. Its asymptotic null distribution is that of the largest eigenvalue of $W\{I_{\min(a-1,\ b-1)},\ \max(a-1,\ b-1)\}$, as given in Lemma 10.1 for the ANOVA model.

11.3.3 Clustering algorithm and a stopping rule

To execute the row-wise multiple comparison procedure for a moderately large two-way table, it is convenient to have software working automatically. We therefore introduce the clustering algorithm proposed by Hirotsu (2009) to obtain a classification of rows such that the generalized squared distance among clusters is significantly large, achieving simultaneously homogeneity within each cluster. We have already introduced a chi-squared distance among clusters, and therefore we can propose the following algorithm based on this.

(1) Specify K, the number of clusters.

(2) Start from a clusters, each of which is composed of one row.

(3) Let $G_1,\ \dots,G_{a-k+1}$ be the clusters at the kth stage. Find two clusters G_i and $G_{i'}$ that give the smallest squared distance $\chi^2(G_i;G_{i'})$ among all possible combinations of two clusters from $G_1,\ \dots,G_{a-k+1}$. Then, combine these two clusters to form $(a-k)$ clusters for the next $(k+1)$ th stage.

(4) Continue procedure 3 until the number of clusters becomes the pre-specified number K. The resulting partition is denoted by $G_1,\ \dots,G_K$. Then, make an adjustment by the next algorithm.

(5) First calculate the squared distance between row 1 and clusters $G_1(\bar{1}), \ldots, G_K(\bar{1})$, where $G_k(\bar{1})$ denotes the fact that row 1 is eliminated from G_k. Then classify row 1 into the cluster that gives the smallest squared distance $\chi^2(1;G_k(\bar{1}))$ among $k = 1, \ldots, K$. Do the same thing between row 2 and the renewed clusters with row 2 eliminated. Continue the process repeatedly until no reduction in the generalized squared distance $\chi^2(G_1; \ldots; G_K)$ is obtained.

(6) *Stopping rule:* We begin with $K = 2$ and continue the process until the generalized squared distance among the clusters $(G_1, \ldots, G_K)$ becomes significant for the first time at the pre-specified level α_1. Then we evaluate the variation within each cluster by the maximum eigenvalue at level α_2. If all K clusters show non-significant within variation we stop here, concluding that there are K clusters and giving their interpretation. Example of section 11.4.4 at $K = 2$ is typical for this case. If all K clusters show significant within variation we proceed to the $K+1$ th cluster and continue the process. As an intermediate case, let the within variation be significant for K_1 clusters and non-significant for K_2 $(= K - K_1)$ clusters. Then we fix those K_2 clusters and apply the clustering procedure anew to the rows in K_1 clusters.

As an example, we obtain a significant classification at $K = 2$ in Example 11.1 below with two clusters $G_1(1)$ and $G_2(2, \ldots, 8)$. The within variation in G_2 is significant, so $K_1 = K_2 = 1$ in this case. We fix and eliminate G_1 and apply the clustering procedure anew to G_2. In this particular case we need not adjust the significance level α_1, since it follows the closed testing procedure of Marcus *et al.* (1976). This suggests that it is reasonable to take $\alpha_1 = \alpha_2$, since otherwise we apply different α's to G_2 in testing within variation and the clustering procedure, respectively. We apply this generally in the following, since there is no reason to choose any particular value for α_2. The subgroup G_2 was not analyzed separately, but as part of the original table in Hirotsu (2009). This leads to putting the coefficient ρ_i to zero for the eliminated rows in calculating the generalized squared distance among the clusters from G_2. Then, the maximization does not reduce to the maximum eigenvalue problem and requires a very complicated optimization procedure. The difference lies only in the treatment of the column totals of the two-way table, and there is only a slight difference in the outcome. Therefore, we deal with G_2 independently from the eliminated rows in this book.

Example 11.1. Israeli adults cross-classified by worries and country of origin. The data in Table 11.2 were reported by Guttman (1971) and have been analyzed by Greenacre (1988) by the method of Hirotsu (1983). 1554 Israeli adults are cross-tabulated according to the row categories of principal worries and the column categories depending on their country of origin, with abbreviations as follows: 1. ASAF, Asia-Africa; 2. IFAA, Israel, father Asia-Africa; 3. IFI, Israel, father Israel; 4. IFEA, Israel, father Europe-America; 5. EUAM, Europe-America.

Table 11.2 Israeli adults cross-classified by principal worries and residence area.

	Residence area				
Worries	1. ASAF	2. IFAA	3. IFI	4. IFEA	5. EUAM
1. Personal economics	104	14	9	16	48
2. Other worries	81	14	12	52	128
3. Sabotage	70	9	7	24	117
4. Enlisted relative	61	8	5	22	104
5. Military situation	97	12	14	28	218
6. More than one worry	20	2	0	6	42
7. Economic situation	4	1	1	2	11
8. Political situation	32	6	7	28	118

First applying the clustering algorithm to rows, we obtain a highly significant squared distance 77.90 between clusters $G_1(1)$ and $G_2(2, \ldots, 8)$ at $K=2$ for the reference value 23.55 at $\alpha=0.05$ from the Wishart distribution $W(\boldsymbol{I}_4, 7)$. The within variation 24.66 of the cluster G_2 evaluated as the largest root of $W(\boldsymbol{I}_4, 6)$ is significant with p-value 0.018. Therefore, we separate G_1 (Personal economics) and apply the clustering procedure anew to G_2. The generalized chi-squared distances $\chi^2(2;3, 4, 5, 6, 7, 8)=20.77$ at $K=2$ and $\chi^2(2;8;3, 4, 5, 6, 7) = 20.99$ at $K=3$ are non-significant at significance level 0.05 as the largest eigenvalue of $W(\boldsymbol{I}_4, 6)$, and we obtain a significant result first at $K=4$ with chi-squared distance $\chi^2(2;3, 4, 6;5, 7;8)=23.195$ and related p-value 0.030. The generalized chi-squared distance among five clusters $G_1(1)$, $G_2(2)$, $G_3(3,4, 6)$, $G_5(5, 7)$, and $G_8(8)$ is 91.81, and its relative contribution to the largest root 92.73 of the original table is 0.99.

Now the software first presents the original data and calculates the Wishart matrix $\boldsymbol{W}'\boldsymbol{W}$ via the vector $\boldsymbol{w}_k$(11.10) with $K=a$ and its largest eigenvalue as 92.73, which is evaluated as highly significant by the Wishart distribution $W(\boldsymbol{I}_4, 7)$. Then the clustering algorithm starts for the pre-specified number of clusters $K=(2,\ldots,8)$. The search for significant clustering at pre-specified $\alpha=0.05$ goes like this. First try $K=2$ to find clustering in $G_1(1)$ and $G_2(2, 3, 4, 5, 6, 7, 8)$ is highly significant with p-value 0.13×10^{-10}. Then check the within variation of sub-clusters to find G_2 is inhomogeneous with $p=0.018$ by the Wishart $W(\boldsymbol{I}_4, 6)$. Therefore, after separating $G_1(1)$, re-clustering of G_2 starts and obtains a significant clustering first at $K=4$ for $G_2(2)$, $G_3(3,4, 6)$, $G_5(5, 7)$, and $G_8(8)$ with generalized squared distribution distance 23.20 and related p-value 0.030. The within variation of sub-clusters $G_3(3,4, 6)$ and $G_5(5, 7)$ is evaluated as non-significant with p-values 0.718 and 0.949 by the Wishart distribution $W(\boldsymbol{I}_2, 4)$ and $W(\boldsymbol{I}_1, 4)$, respectively. Therefore, the algorithm stops here and gives a summary of the classification, the generalized squared distance 91.84 among five sub-clusters, and its contribution to the original largest root 92.73.

In this case, however, the number of clusters 5 is too large for $a=8$, and the relative contribution looks excessively high. Therefore we may try another significance level, $\alpha=0.10$ say. Then we can separate row 2 (Other worries) from G_2 with p-value 0.081.

Table 11.3 Collapsed data.

	Collapsed table			Simple departure measure		
Cluster	$F_1(1, 2, 3)$	$F_4(4)$	$F_5(5)$	$F_1(1, 2, 3)$	$F_4(4)$	$F_5(5)$
$G_1(1)$	127	16	48	1.75	0.73	0.5
$G_2(2)$	107	52	128	0.98	1.58	0.88
$G_3(3, \ldots, 8)$	356	110	610	0.87	0.89	1.12

The within variation in the counterpart sub-cluster $G_3(3, \ldots, 8)$ is 15.17 with p-value 0.19 by Wishart distribution $W(\boldsymbol{I}_4, 5)$, and the algorithm stops here. Thus the algorithm first separates 1 (Personal economics) and then 2 (Other worries) from subgroup G_3 related to political matters. The relative contribution of $G_1(1)$, $G_2(2)$, and $G_3(3, \ldots, 8)$ is still 0.88 and reasonably high.

Since the column categories are also nominal, we can apply the same procedures to them as rows and obtain a significant clustering $F_1(1, 2, 3)$ and $F_4(4, 5)$ at $K=2$. This separates the subgroup related to Europe-America from the others. Applying the largest eigenvalue test to each of F_1 and F_4, the former is found to be homogeneous. In contrast, the largest eigenvalue 24.55 of F_4 is significant at $\alpha=0.05$ and immediately suggests that F_4 should be separated into $F_4(4)$ and $F_5(5)$, since this is only one possible classification. The generalized chi-squared distance 90.54 among F_1, F_4, F_5 explains 98% of the largest root 92.73 of the original table. If we employ $\alpha=0.10$, we have the same result.

The collapsed data and the simple departure measure from the independence model $y_{ij}/(R_iC_j/N)$ of Cox and Snell (1981) are given in Table 11.3. It is seen that G_1 is strongly associated with the cluster $F_1(1, 2, 3)$ of columns. G_2 is strongly associated with F_4 and the subgroup $G_3(3, \ldots, 8)$ is associated with F_5. In short, those from Asia-Africa tend toward personal economics while those from Europe-America tend more toward the political situation and other worries. The reader is recommended to compare this analysis with that of Guttman (1971).

11.4 Directional Inference in the Case of Natural Ordering Only in Columns

11.4.1 Overall analysis

In the case where there is natural ordering in columns, we are interested in distinguishing some systematic departure such as the upward or downward tendency along with columns. The hypothesis of interest is

$$T_3' : \frac{p_{i'1}}{p_{i1}} \le \frac{p_{i'2}}{p_{i2}} \le \cdots \le \frac{p_{i'b}}{p_{ib}},\ i \ne i',$$

which is a probability version of (10.29) and also an extension of (5.19) in Section 5.2.3. This is a sort of profile analysis of rows, and order-sensitive squared distances have been proposed by Hirotsu (1983, 1993, 2009). For the upward and downward profiles, the cumulative chi-squared has been proposed, which simply replaces the matrix C' by $C^{*\prime}$ in the definition of goodness-of-fit χ^2 (11.5), where the jth row of $C^{*\prime}$ is

$$\boldsymbol{c}^{*\prime}(1,\ldots,j;j+1,\ldots,b)=\left(\frac{1}{V_j}+\frac{1}{V_j^*}\right)^{-1/2}\left(\frac{\sqrt{C_1}}{V_j}\cdots\frac{\sqrt{C_j}}{V_j}-\frac{\sqrt{C_{j+1}}}{V_j^*}\cdots-\frac{\sqrt{C_b}}{V_j^*}\right), \tag{11.11}$$

$$V_j=\textstyle\sum_{k=1}^{j}C_k,\ V_j^*=\sum_{k=j+1}^{b}C_k,\ \text{for } j=1,\ \ldots,\ b-1. \tag{11.12}$$

We call $\boldsymbol{c}^{*\prime}(1,\ldots,j;j+1,\ldots,b)$ a step change-point contrast between the jth and $j+1$th columns. Thus, an overall cumulative chi-squared statistic is defined by

$$\chi^{*2}=\|(\boldsymbol{R}'\otimes\boldsymbol{C}^{*\prime})\boldsymbol{z}\|^2=\chi_1^{*2}+\cdots+\chi_{b-1}^{*2}, \tag{11.13}$$

where χ_j^{*2} is the goodness-of-fit chi-square for the $a\times 2$ sub-table obtained by collapsing columns $1,\ \ldots,\ j$ and $j+1,\ \ldots,b$, respectively. This obviously corresponds to $S_{\alpha\beta^*}$ (10.30) of the normal model in Section 10.4.1(1). The chi-squared approximation of the null distribution is obtained by extending the method in 6.5.3 (2) (a) to the two-way contingency table:

$$\kappa_1=E(\chi^{*2})=(a-1)(b-1)=df,$$

$$\kappa_2=V(\chi^{*2})=2(a-1)\left\{b-1+2\left(\frac{\gamma_1}{\gamma_2}+\ldots+\frac{\gamma_1+\ldots+\gamma_{b-2}}{\gamma_{b-1}}\right)\right\}=2d^2f,$$

$$\gamma_j=V_j/V_j^*.$$

This formula is the same as obtained in 10.4.1 (1), except for the definition of γ_j. The improved upper percentile and the upper tail probability are obtained by the method in Section 6.5.3 (2) (a), just by changing the definition of kappa to κ_1, κ_2, and $\kappa_3=8(a-1)\mathrm{tr}\{\boldsymbol{C}^{*\prime}\boldsymbol{C}^*\}^3$ based on

$$\boldsymbol{C}^{*\prime}\boldsymbol{C}^*=\begin{bmatrix}1 & \sqrt{\gamma_1/\gamma_2} & \sqrt{\gamma_1/\gamma_3} & \cdots & \sqrt{\gamma_1/\gamma_{b-1}}\\ \sqrt{\gamma_1/\gamma_2} & 1 & \sqrt{\gamma_2/\gamma_3} & \cdots & \sqrt{\gamma_2/\gamma_{b-1}}\\ \sqrt{\gamma_1/\gamma_3} & \sqrt{\gamma_2/\gamma_3} & 1 & \cdots & \sqrt{\gamma_3/\gamma_{b-1}}\\ \vdots & \vdots & \vdots & \vdots & \vdots\\ \sqrt{\gamma_1/\gamma_{b-1}} & \sqrt{\gamma_2/\gamma_{b-1}} & \sqrt{\gamma_3/\gamma_{b-1}} & \cdots & 1\end{bmatrix}.$$

That is, replacing λ_i by γ_j. These formulae coincide with the formulae for the normal model in Section 10.4.1(1) by altering (11.12) to $V_j = j, V_j^* = b-j$, or $\gamma_j = j/(b-j)$.

11.4.2 Row-wise multiple comparisons

(1) **Chi-squared distances.** The chi-squared distance between the lth and mth rows, the chi-squared distance between two subgroups of rows, and the generalized chi-squared distance are defined parallel to Section 11.3, simply replacing C' by $C^{*\prime}$ in equations (11.6) ~ (11.8),

$$\chi^{*2}(l_1;l_2) = \|(r'(l_1;l_2) \otimes C^{*\prime})z\|^2, \tag{11.14}$$

$$\chi^{*2}(H_1;H_2) = \|\{r'(H_1;H_2) \otimes C^{*\prime}\}z\|^2, \tag{11.15}$$

$$\chi^{*2}(H_1;\ldots;H_K) = \max_{\rho' r = 0,\ \|\rho\|^2=1} \|\{\rho' \otimes C^{*\prime}\}z\|^2,\ \rho_i = \zeta_k \{R_i/T_k\}^{1/2} \text{ for } i \in H_k. \tag{11.16}$$

Then all the chi-squared distances (11.14) ~ (11.16) are bounded above by $\chi^{*2}(\{1\};\ldots;\{a\}) = \max_{\rho' r = 0,\ \|\rho\|^2=1} \|\{\rho' \otimes C^{*\prime}\}z\|^2$, which is the largest eigenvalue W_1^* of $W^{*\prime}W^*$, where $W^* = (w_1^*, \ldots, w_a^*), w_i^* = (R_i/N)^{-1/2}$ $C^{*\prime}\left(C_1^{-1/2}y_{i1}, \ldots, C_b^{-1/2}y_{ib}\right)'$. Its null distribution is that of the largest eigenvalue of the Wishart matrix $W(C^{*\prime}C^*, a-1)$ when $a \ge b$. The distribution is very hard to handle, but we can give a very nice chi-squared approximation in the following.

(2) **Chi-squared approximation of the distribution of the largest eigenvalue of the Wishart matrix.** The proposed method adjusts the first two cumulants to $d\chi_f^2$ as before. These cumulants are obtained by asymptotic expansion, applying the method of Sugiura (1973) as

$$E(W_1^*) = \kappa_1 = q\tau_1 + \left(1 - \frac{2}{q}\right)\sum_{2\le j\le p} \frac{\tau_1\tau_j}{\tau_1 - \tau_j} + \frac{2}{q}\sum\sum_{2\le j<k\le p} \frac{\tau_1\tau_j\tau_k}{(\tau_1-\tau_j)(\tau_1-\tau_k)}, \tag{11.17}$$

$$\begin{aligned} V(W_1^*) = \kappa_2 = 2q\tau_1^2 &+ \frac{-2q^3+32q+144}{q^3}\sum_{2\le j\le p}\left(\frac{\tau_1\tau_j}{\tau_1-\tau_j}\right)^2 \\ &+ \frac{2(q^2+6q-4)}{q^2}\sum\sum_{2\le j<k\le p}\frac{\tau_1^2\tau_j\tau_k}{(\tau_1-\tau_j)(\tau_1-\tau_k)}, \end{aligned} \tag{11.18}$$

where $q = a-1, p = b-1$, and τ_j is the jth largest eigenvalue of $C^{*\prime}C^*$ satisfying $\sum \tau_j = \operatorname{tr}(C^{*\prime}C^*) = b-1$. Then $E(W_1^*)$ and $V(W_1^*)$ are set equal to df and $2d^2f$, respectively. In particular, in case of $b=3$ $(p=2)$ they reduce to

$$E(W_1^*) = \kappa_1 = q\tau_1 + \left(1 - \frac{2}{q}\right)\frac{\tau_1\tau_2}{\tau_1 - \tau_2} = df, \tag{11.19}$$

$$V(W_1^*) = \kappa_2 = 2q\tau_1^2 + \frac{-2q^3+32q+144}{q^3}\left(\frac{\tau_1\tau_2}{\tau_1-\tau_2}\right)^2 = 2d^2 f. \tag{11.20}$$

In the balanced case, where all the column totals are equal, the τ_j are explicitly

$$\tau_j = \frac{b}{\{j(j+1)\}}, j=1,\ldots,b-1,$$

so that $\tau_1 = 3/2, \tau_2 = 1/2$ for $b=3$ and $\tau_1 = 2, \tau_2 = 2/3$, $\tau_3 = 1/3$ for $b=4$, for example. Then, for the balanced case the first terms in $E(W_1^*)$ and $V(W_1^*)$ are dominant when q is moderately large, since then the largest eigenvalue is large enough compared with the other eigenvalues. The two-way ANOVA model fits to this situation, and in this case the approximation by $\tau_1\chi_q^2$ is suggested by the form of (11.19) and (11.20). It has actually been shown to behave better than the naïve normal approximation by Anderson (2003). However, if the largest and second-largest eigenvalues are close, making $\tau_1 - \tau_2$ small, then the rest of the terms in (11.17) and (11.18) become non-negligible and we need full terms in these equations. In Table 11.4 we give upper five and one percentiles calculated by several methods, where 0-approx. is the normal approximation by Anderson (2003), 1-approx. is the chi-squared approximation by Hirotsu (1991) using only the dominant terms, and 2-approx. is the chi-squared approximation using full terms (Hirotsu, 2009). Zonal is the value obtained by Aida and Hirotsu (1983) through the Zonal polynomial expansion of the cumulative distribution function of the largest eigenvalue of the Wishart matrix given by James (1964, 1968), and adopted here as reference value. They employed 120 terms to ensure accuracy up to two decimal places. It gives a very good reference value, but obtained only in a very limited case. The vacancies in Table 11.4 mean, for example, that the required accuracy has not been achieved even with 120 terms in the Zonal polynomial expansion. It is seen from this table that the accuracy of the 2-approx. is excellent, whereas the normal approx. is very poor when the largest and second-largest eigenvalues are close to each other. The proposed method behaves much better in those least favorable cases.

11.4.3 Multiple comparisons of ordered columns

Because of the natural ordering, not all the permutations of columns make sense and we are interested in the $(b-1)$ step change-point contrasts $\boldsymbol{c}^{*\prime}(1,\ldots j; j+1,\ldots,b)$ defined in equation (11.11). We have already discussed a mathematically equivalent model in Section 10.4.1 (3) for the normal distribution, where a method is proposed for comparing ordered rows. Here we apply the idea to the ordered columns and propose a maximal statistic

$$\max \text{acc.}\ \chi^2(\boldsymbol{C}^{*\prime}) = \max_j \|\{\boldsymbol{R}' \otimes \boldsymbol{c}^{*\prime}(1,\ldots,j;j+1,\cdots,b)\}\boldsymbol{z}\|^2, \tag{11.21}$$

where the maximization is with respect to $j = 1, \ldots, b-1$. This is actually the maximal component of χ^{*2} defined in (11.13), and detects a cut point of the largest contribution

Table 11.4 Comparing upper percentiles by several approximation methods.

(1) $b=3$

			τ_1					
%	$a-1$	Method	1.2	1.3	1.4	Balanced	1.6	1.7
	5	Zonal	15.01	15.62	16.40	17.27	18.21	19.17
		0-approx.	12.24	13.26	14.28	15.30	16.32	17.34
		1-approx.	13.28	14.39	15.50	16.61	17.71	18.82
		2-approx.	15.23	15.45	16.17	17.06	18.02	19.02
	10	Zonal	23.78	25.05	26.54	28.14	29.79	31.47
		0-approx	20.83	22.56	24.30	26.03	27.77	29.51
5		1-approx.	21.97	23.80	25.63	27.46	29.29	31.12
		2-approx.	21.99	24.30	26.13	27.89	29.62	31.36
	20	Zonal	39.58	42.12	44.91	47.81	50.76	53.75
		0-approx.	36.48	39.52	42.56	45.60	48.64	51.69
		1-approx.	37.69	40.83	43.93	47.12	50.26	53.40
		2-approx.	38.31	41.62	44.65	47.65	50.66	53.69
	5	Zonal	19.72	20.76	21.98	23.28	24.62	25.99
		0-approx.	18.03	18.31	19.17	20.23	21.38	22.58
		1-approx.	18.10	19.61	21.12	22.63	24.14	25.65
		2-approx.	20.14	20.59	21.67	22.97	24.35	25.78
	10	Zonal	29.52	31.35	33.37	35.48	37.62	39.80
		0-approx.	25.43	28.38	30.62	32.71	34.78	36.84
1		1-approx.	27.85	30.17	32.47	34.81	37.13	39.46
		2-approx.	26.36	30.02	32.65	35.04	37.34	39.62
	20	Zonal	46.82	50.06	53.50	57.02	60.59	
		0-approx.	43.75	37.85	51.45	54.97	58.48	61.99
		1-approx.	45.08	48.87	52.59	56.35	60.11	63.86
		2-approx.	44.61	49.17	53.03	56.75	60.44	64.11

(continued overleaf)

Table 11.4 (*Continued*)

(2) $b=4$

%	$a-1$	Method	Case 1	Balanced	3	4	5
	5	Zonal	22.75	20.92	19.48	23.41	24.06
		0-approx.	21.53	19.20	17.14	22.40	23.17
		1-approx.	21.29	18.98	16.94	22.14	22.90
		2-approx.	22.45	21.40	21.18	23.31	23.90
	10	Zonal	36.69	33.35	30.62	37.91	39.28
5		0-approx.	36.09	32.17	28.72	37.53	38.83
		1-approx.	35.21	31.38	28.02	36.61	37.88
		2-approx.	36.33	32.93	30.17	37.63	38.80
	20	Zonal					
		0-approx.	62.30	55.54	49.59	64.79	67.03
		1-approx.	60.40	53.85	48.08	62.82	64.99
		2-approx.	61.68	55.54	50.29	63.96	66.01
	5	Zonal	30.43	27.72	25.49	31.42	32.35
		0-approx.	25.28	22.54	20.12	26.29	27.20
		1-approx.	29.01	25.86	23.09	30.17	31.21
		2-approx.	30.45	28.60	28.31	31.32	32.16
	10	Zonal			37.99		
1		0-approx.	41.38	36.89	32.94	43.04	44.52
		1-approx.	44.63	39.79	35.52	46.42	48.02
		2-approx.	45.41	40.83	36.92	47.18	48.72
	20	Zonal					
		0-approx.	69.79	62.21	55.55	72.53	75.08
		1-approx.	72.24	64.40	57.50	75.13	77.72
		2-approx.	73.27	65.70	59.06	76.09	78.60
			Case 1	Balanced	Case 3	Case 4	Case 5
		τ_2/τ_1	0.3	1/3	0.4	0.5	0.6
		τ_3/τ_2	0.5	1/2	0.4	0.5	0.6
		$\tau_1-\tau_2$	1.45	4/3	1.15	0.86	0.61

to χ^{*2}. It is asymptotically distributed as the maximum of the correlated chi-squared variables under the null hypothesis of independence. We call this method max acc. $\chi^2(C^{*\prime})$. A very efficient algorithm for the p-value calculation is given by Hirotsu *et al.* (1992) based on the Markov property of the successive chi-squared components. The idea is as follows.

Let $\boldsymbol{u}_j = \{\boldsymbol{R}' \otimes \boldsymbol{c}^{*\prime}(1,\ldots,j;j+1,\ldots,b)\}\boldsymbol{z}$. Then the asymptotic null distribution of $\boldsymbol{u}_j$ is $N(\boldsymbol{0}, \boldsymbol{I}_{a-1})$ with covariance $Cov(\boldsymbol{u}_j, \boldsymbol{u}_{j'}) = \tau_{jj'}\boldsymbol{I}_{a-1}$, where $\tau_{jj'} = \sqrt{\gamma_j/\gamma_{j'}}$ is the (j, j') element of $\boldsymbol{C}^{*\prime}\boldsymbol{C}^*$, $1 \le j \le j' \le b-1$. Therefore, $\chi_j^{*2} = \boldsymbol{u}_j^2$ is distributed as the chi-squared distribution with df $a-1$ and the joint distribution of χ_j^{*2} and $\chi_{j'}^{*2}$ is a bivariate chi-squared distribution with correlation $\tau_{jj'}$. The Markov property exists in $\boldsymbol{u}_j$ and we have the factorization of the joint distribution as

$$f(\boldsymbol{u}_1, \ldots, \boldsymbol{u}_{b-1}) = f(\boldsymbol{u}_1 \mid \boldsymbol{u}_2) \times \cdots \times f(\boldsymbol{u}_{b-2} \mid \boldsymbol{u}_{b-1}) \times f(\boldsymbol{u}_{b-1}),$$

where $f(\boldsymbol{u}_j \mid \boldsymbol{u}_{j+1})$ is the conditional normal density of $N\left\{\tau_{jj+1}\boldsymbol{u}_{j+1}, \left(1-\tau_{jj+1}^2\right)\boldsymbol{I}_{a-1}\right\}$ given $\boldsymbol{u}_{j+1}$. Then the conditional distribution of $\|\boldsymbol{u}_j\|^2/\left(1-\tau_{jj+1}^2\right)$ given $\boldsymbol{u}_{j+1}$ is the non-central chi-squared distribution with df $a-1$ and the non-centrality parameter $\tau_{jj+1}^2\|\boldsymbol{u}_{j+1}\|^2/\left(1-\tau_{jj+1}^2\right)$ thus depending only on $\|\boldsymbol{u}_{j+1}\|^2$. It suggests a Markov property of the sequence of χ_j^{*2}, $j = 1, \ldots, b-1$, and leads to the recursion formula for the distribution of max acc. $\chi^2(C^{*\prime})$. The idea is similar to that of Section 6.5.3 (1) (d), which led to the recursion formula for max acc. $t1$.

We define the conditional joint probability of $(\|\boldsymbol{u}_1\|^2, \ldots, \|\boldsymbol{u}_k\|^2)$ given $\|\boldsymbol{u}_k\|^2$ as

$$F_k\left(\|\boldsymbol{u}_k\|^2, c\right) = \Pr\left(\|\boldsymbol{u}_1\|^2 < c, \ldots, \|\boldsymbol{u}_k\|^2 < c \,\|\, \|\boldsymbol{u}_k\|^2\right), k = 1, \ldots, b-1.$$

Then we have

$$\begin{aligned} F_{k+1}\left(\|\boldsymbol{u}_{k+1}\|^2, c\right) &= \Pr\left(\|\boldsymbol{u}_1\|^2 < c, \ldots, \|\boldsymbol{u}_k\|^2 < c, \|\boldsymbol{u}_{k+1}\|^2 < c \,\|\, \|\boldsymbol{u}_{k+1}\|^2\right) \\ &= \int_{\|\boldsymbol{u}_k\|^2} \Pr\left(\|\boldsymbol{u}_1\|^2 < c, \ldots, \|\boldsymbol{u}_k\|^2 < c, \|\boldsymbol{u}_{k+1}\|^2 < c \,\|\, \|\boldsymbol{u}_k\|^2, \|\boldsymbol{u}_{k+1}\|^2\right) \\ &\quad \times f_k\left(\|\boldsymbol{u}_k\|^2 \,|\, \|\boldsymbol{u}_{k+1}\|^2\right) \mathrm{d}\|\boldsymbol{u}_k\|^2. \end{aligned} \tag{11.22}$$

$$= \begin{cases} \displaystyle\int_{\|\boldsymbol{u}_k\|^2} F_k\left(\|\boldsymbol{u}_k\|^2, c\right) \times f_k\left(\|\boldsymbol{u}_k\|^2 \,|\, \|\boldsymbol{u}_{k+1}\|^2\right) \mathrm{d}\|\boldsymbol{u}_k\|^2 \text{if } \|\boldsymbol{u}_{k+1}\|^2 < c, \\ 0, \text{ otherwise,} \end{cases} \tag{11.23}$$

where $f_k\left(\|\boldsymbol{u}_k\|^2 \,|\, \|\boldsymbol{u}_{k+1}\|^2\right)$ is a conditional distribution of $\|\boldsymbol{u}_k\|^2$ given $\|\boldsymbol{u}_{k+1}\|^2$. Equation (11.22) is from the total probability theory and (11.23) is due to the Markov property of the sequence of $\|\boldsymbol{u}_k\|^2$, corresponding to (6.29) and (6.30) of max acc. $t1$, respectively. The conditional distribution is easily obtained by applying the asymptotic expansion of the non-central chi-squared distribution with respect to its non-centrality parameter as

$$f_k\left(\|\boldsymbol{u}_k\|^2 \,|\, \|\boldsymbol{u}_{k+1}\|^2\right) = \sum_{l=0}^{\infty} e^{-\delta/2} \frac{(\delta/2)^l}{l!\left(1-\tau_{k,\,k+1}^2\right)} g_{a-1+2l}\left(\frac{\|\boldsymbol{u}_k\|^2}{1-\tau_{k,\,k+1}^2}\right),$$

$$\delta = \frac{\tau_{k,\,k+1}^2 \|\boldsymbol{u}_{k+1}\|^2}{1-\tau_{k,\,k+1}^2},$$

where $g_\nu(x)$ is a density function of the chi-squared distribution with df ν. The p-value is obtained at the final step by $1-F_b$, where the integration by $\|\boldsymbol{u}_{b-1}\|^2$ is unconditional with respect to the chi-squared distribution $f_{b-1}\left(\|\boldsymbol{u}_{b-1}\|^2 \,|\, \|\boldsymbol{u}_b\|^2\right) = g_{a-1}\left(\|\boldsymbol{u}_{b-1}\|^2\right)$.

In contrast, a closed form is given by Hirotsu *et al.* (1992) up to $k = 3 \sim 5$, expressed in the summation of the chi-squared distribution function $G_{a-1+2k}\left\{c/\left(1-\tau_{k,\,k+1}^2\right)\right\}$. Since they exhibit little complication, we give here only a formula for $b = 3$. This formula for $b = 3$ coincides with the formula obtained by Siotani (1959) through direct inversion of the joint characteristic function of the bivariate chi-squared distribution. These formulae can be applied to Section 10.4.1 (3) with $\boldsymbol{R}^{*\prime}$ instead of $\boldsymbol{C}^{*\prime}$:

$$b = 3 : \Pr\left\{\max \chi^2(\boldsymbol{C}^{*\prime}) \le c\right\} = \left(1-\tau_{12}^2\right)^{f/2} \sum_{k=0}^{\infty} \tau_{12}^{2k} \frac{\Gamma(f/2+k)}{\Gamma(f/2)k!} G_{f+2k}^2\left(\frac{c}{1-\tau_{12}^2}\right), f = a-1.$$

11.4.4 Re-analysis of Table 11.1 taking natural ordering into consideration

(1) **Overall analysis by the cumulative chi-squared statistic.** As a typical example, we analyze Table 11.1. The components of the cumulative chi-squared (11.13) are obtained as the goodness-of-fit chi-square of the two collapsed sub-tables given in Table 11.5. Then we get

$$\chi^{*2} = \chi_1^{*2} + \chi_2^{*2} = 91.251 + 8.390 = 99.641.$$

The constants for the chi-squared approximation are

$$d = 1 + \frac{2}{b-1}\left(\frac{\gamma_1}{\gamma_2}\right) = 1 + \frac{2}{3-1}\left(\frac{2166}{8323+1419} \Big/ \frac{2166+8323}{1419}\right) = 1.030,$$
$$f = (10-1) \times (3-1)/d = 17.374,$$

Table 11.5 Calculation of the components of cumulative chi-squared.

	Severity		Severity		
Occupation	Slight	Medium-Serious	Slight-Medium	Serious	Total
1	148	530	592	86	678
2	111	401	463	49	512
3	645	2239	2556	328	2884
4	165	890	936	119	1055
5	383	2140	2212	311	2523
6	96	340	389	47	436
7	98	388	428	58	486
8	199	1029	1073	155	1228
9	59	229	258	30	288
10	262	1556	1582	236	1818
Total	2166	9742	10489	1419	11908

⇓ $\chi_1^{*2} = 91.251$ ⇓ $\chi_2^{*2} = 8.390$

Table 11.6 Chi-squared distances between two rows (rows rearranged).

Row	10	5	4	8	7	9	2	1	6	3
10	0	0.85	2.52	1.67	8.93	7.72	18.6	18.3	15.3	50.1**
5		0	0.88	0.65	6.86	5.79	15.2	15.9	12.5	47.8**
4			0	1.10	4.71	3.73	9.41	11.4	8.51	23.5*
8				0	3.83	3.95	10.5	9.29	8.35	23.5*
7					0	0.41	1.71	0.68	0.82	1.48
9						0	0.30	1.24	0.30	0.85
2							0	2.7	0.35	1.48
1								0	0.92	1.01
6									0	0.16
3										0

giving $d\chi_f^2 = 29.08$ ($\alpha = 0.05$), 35.14 ($\alpha = 0.01$). In this case the column totals are highly unbalanced in size, concentrating on medium severity, so the gain in power is not so much compared with the usual goodness-of-fit chi-square. It suggests anyway a strong association between occupation and severity of illness, but again cannot tell us any details of the data, because it is an overall test.

(2) **Row-wise multiple comparisons.** For reference, we first give the chi-squared distances (11.14) between two rows in Table 11.6, where the rows are rearranged so that two rows which give a smaller distance come closer to each other. Applying the clustering algorithm of Section 11.3.3 based on the cumulative chi-squared statistics, we have a highly significant chi-squared distance $\chi^{*2}(G_1;G_4)=90.96$ between two clusters $G_1(1, 2, 3, 6, 7, 9)$ and $G_4(4, 5, 8, 10)$ at $K=2$. For reference value we have $\tau_{12}=\sqrt{\gamma_1/\gamma_2}=0.1734$, which gives $\tau_1=1.1734$, $\tau_2=0.8266$. These two eigenvalues are very close, suggesting that this might be a worst case for the chi-squared approximation beyond the range of Table 11.4 (1). However, it can still give an approximate critical value $0.541\chi^2_{23.541}(0.05)=19.39$ by (11.19) and (11.20), which will be enough to evaluate $\chi^{*2}(G_1;G_4)=90.96$. Since the more accurate critical value obtained by Aida and Hirotsu (1983) by Zonal polynomial expansion is 21.85 at $\alpha=0.05$, the approximation is not so bad. However, we still recommend using the formula of Section 11.4.2 (2) within the range of Table 11.4 (1) and (2).

To evaluate the within variation of G_1 and G_4, the largest eigenvalues $W_1^*(G_1)=3.37$ and $W_1^*(G_4)=2.83$ are calculated. Their respective p-values are obtained as 0.86 and 0.34 by the chi-squared approximation for the largest eigenvalues of $W\left(C_1^{*\prime}C_1^*,5\right)$ and $W\left(C_4^{*\prime}C_4^*,3\right)$, where the matrices $C_i^{*\prime}C_i^*$, $i=1,4$ are calculated by equation (11.11) for the respective partitioned sub-tables. Thus the algorithm stops here, declaring that there are two different subgroups of rows.

(3) **Comparing ordered columns.** We apply the max acc. $\chi^2(C^{*\prime})$ (11.21) method of Section 11.4.3. The components of χ^{*2} (11.13) have already been calculated in Table 11.5, and max acc. $\chi^2(C^{*\prime})$ is $\chi_1^{*2}=91.251$. This is evaluated as highly significant by the formula ($b=3$) of Section 11.4.3, suggesting separation between columns 1 and 2. The within variation of the sub-table composed of columns 2 and 3 in the second cluster is evaluated as the usual goodness-of-fit chi-square, since it is composed of just two columns. It is 5.23 and non-significant as chi-squared with df 9. The collapsed sub-tables in two ways and a simple departure measure from independence are given in Table 11.7. Now the interpretation is clear. The cluster $G_1(1, 2, 3, 6, 7, 9)$ is characterized by the high proportion in the 'slight' category of illness relative to the cluster $G_4(4, 5, 8, 10)$. This result suggests also a simple block interaction model

$$p_{ij}=p_{i.}p_{.j}\theta_{lm}\ (l=1\text{ for }i\in G_1,\ l=2\text{ for }i\in G_4;\ m=1\text{ for }j\in F_1,\ m=2\text{ for }j\in F_2). \tag{11.24}$$

The fitted frequencies by model (11.24) are obtained by an iterative scaling procedure, keeping the entries of Table 11.7 in addition to all the marginal totals, and given in Table 11.1 just below the original data in double parentheses. A very nice fit is seen by adding just one parameter for interaction, where the df for interaction θ_{lm} is obviously 1. The goodness-of-fit chi-square of model (11.24) reduces to 8.04 with df 17, from 95.75 for the independence model with df 18. Indeed, the fit is excellent.

Table 11.7 Collapsed data and simple departure measure.

Cluster	Collapsed table		Simple departure measure	
	Slight (F_1)	Medium-Serious (F_2)	Slight (F_1)	Medium-Serious (F_2)
G_1(1, 2, 3, 6, 7, 9)	1157	4127	1.20	0.95
G_4(4, 5, 8, 10)	1009	5615	0.84	1.04

11.5 Analysis of Ordered Rows and Columns

11.5.1 Overall analysis

The idea of the previous section is naturally extended to the case of natural ordering, both in rows and columns. Now the doubly accumulated chi-square is defined by

$$\chi^{**2} = \| (\boldsymbol{R}^{*\prime} \otimes \boldsymbol{C}^{*\prime}) z \|^2, \tag{11.25}$$

where $\boldsymbol{R}^{*\prime}$ is defined similarly to $\boldsymbol{C}^{*\prime}$ by $\boldsymbol{r}^{*\prime}(1,\ldots,i;i+1,\ldots,a)$ instead of $\boldsymbol{c}^{*\prime}(1,\ldots,j;j+1,\ldots,b)$:

$$\boldsymbol{r}^{*\prime}(1,\ldots,i;i+1,\ldots,a) = \left(\frac{1}{U_i} + \frac{1}{U_i^*}\right)^{-1/2} \left(\frac{\sqrt{R_1}}{U_i} \cdots \frac{\sqrt{R_i}}{U_i} - \frac{\sqrt{R_{i+1}}}{U_i^*} \cdots - \frac{\sqrt{R_a}}{U_i^*}\right),$$

$$\text{with } U_i = \textstyle\sum_{k=1}^{i} R_k,\ U_i^* = \sum_{k=i+1}^{a} R_k \text{ for } i = 1, \ldots, a-1. \tag{11.26}$$

We call $\boldsymbol{r}^{*\prime}(1,\ldots,i;i+1,\ldots,a)$ also a step change-point contrast between the ith and $i+1$th rows. χ^{**2} is equivalent to the sum of squares of the goodness-of-fit chi-square $\left(\chi_{ij}^{**2}\right)$ of all the $(a-1)(b-1)$ partitions of the original table into 2×2 sub-tables – see Table 11.10 (1) and (2), for example. Thus, an overall cumulative chi-squared statistic is defined by

$$\chi^{**2} = \| (\boldsymbol{R}^{*\prime} \otimes \boldsymbol{C}^{*\prime}) z \|^2 = \chi_{11}^{**2} + \cdots + \chi_{a-1,\, b-1}^{**2}. \tag{11.27}$$

This obviously corresponds to $S_{\alpha^*\beta^*}$ of (10.31). The cumulants for the chi-squared approximation are given by

$$\kappa_1 = E\left(\chi^{**2}\right) = (a-1)(b-1) = df, \tag{11.28}$$

$$\kappa_2 = V\left(\chi^{**2}\right) = 2\left\{a-1+2\left(\frac{\lambda_1}{\lambda_2}+\ldots+\frac{\lambda_1+\ldots+\lambda_{a-1}}{\lambda_{a-1}}\right)\right\} \times\left\{b-1+2\left(\frac{\gamma_1}{\gamma_2}+\ldots+\frac{\gamma_1+\ldots+\lambda\gamma_{b-1}}{\gamma_{b-1}}\right)\right\} = 2d^2f, \tag{11.29}$$

and $\kappa_3 = 8\mathrm{tr}\left\{\left(\boldsymbol{R}^{*\prime}\boldsymbol{R}^* \otimes \boldsymbol{C}^{*\prime}\boldsymbol{C}^*\right)^3\right\}$ with

$$\boldsymbol{R}^{*\prime}\boldsymbol{R}^* = \begin{bmatrix} 1 & \sqrt{\lambda_1/\lambda_2} & \sqrt{\lambda_1/\lambda_3} & \cdots & \sqrt{\lambda_1/\lambda_{a-1}} \\ \sqrt{\lambda_1/\lambda_2} & 1 & \sqrt{\lambda_2/\lambda_3} & \cdots & \sqrt{\lambda_2/\lambda_{a-1}} \\ \sqrt{\lambda_1/\lambda_3} & \sqrt{\lambda_2/\lambda_3} & 1 & \cdots & \sqrt{\lambda_3/\lambda_{a-1}} \\ \vdots & \vdots & \vdots & \ddots & \vdots \\ \sqrt{\lambda_1/\lambda_{a-1}} & \sqrt{\lambda_2/\lambda_{a-1}} & \sqrt{\lambda_3/\lambda_{a-1}} & \cdots & 1 \end{bmatrix}.$$

and

$$\boldsymbol{C}^{*\prime}\boldsymbol{C}^* = \begin{bmatrix} 1 & \sqrt{\gamma_1/\gamma_2} & \sqrt{\gamma_1/\gamma_3} & \cdots & \sqrt{\gamma_1/\gamma_{b-1}} \\ \sqrt{\gamma_1/\gamma_2} & 1 & \sqrt{\gamma_2/\gamma_3} & \cdots & \sqrt{\gamma_2/\gamma_{b-1}} \\ \sqrt{\gamma_1/\gamma_3} & \sqrt{\gamma_2/\gamma_3} & 1 & \cdots & \sqrt{\gamma_3/\gamma_{b-1}} \\ \vdots & \vdots & \vdots & \ddots & \vdots \\ \sqrt{\gamma_1/\gamma_{b-1}} & \sqrt{\gamma_2/\gamma_{b-1}} & \sqrt{\gamma_3/\gamma_{b-1}} & \cdots & 1 \end{bmatrix}.$$

where $\lambda_i = U_i/U_i^*$.

11.5.2 Comparing rows

The data of Table 11.8 are taken from a phase II clinical trial to find an optimal dose (Hirotsu, 1992). Table 5.5 employed in Section 5.2.3 (1) is a part of this table. Now the rows represent the dose levels and the columns are the ordered categorical responses in six levels from unfavorable to excellent recovery. This table is characterized by the natural ordering in both rows and columns, and the purpose is to compare the rows based on the response profiles along the ordered columns. Then it seems most appropriate to apply the step change-point contrasts to rows. This implies analyzing Table 11.9 (1) and (2), obtained by separating and collapsing the original table at two cut points in rows. There are several methods for dealing with the ordered columns. We consider three typical ways in the following.

(1) **Max Wil method.** We can apply a Wilcoxon rank sum test to each of Table 11.9 (1) and (2), and take the maximum of the two. The statistic is described as

Table 11.8 Original data from phase II trial for an antibiotic.

Drug	Improvement						Total
	Undesirable	Slightly Undesirable	Not useful	Slightly useful	Useful	Excellent	
Placebo	3	6	37	9	15	1	71
AF3	7	4	33	21	10	1	76
AF6	5	6	21	16	23	6	77
Total	15	16	91	46	48	8	224

Table 11.9 Sub-tables for maximal contrast test for rows.

(1) Collapsing AF3 and AF6

Drug	Improvement						Total
	Undesirable	Slightly Undesirable	Not useful	Slightly useful	Useful	Excellent	
Placebo	3	6	37	9	15	1	71
AF3 & AF6	12	10	54	37	33	7	153
Total	15	16	91	46	48	8	224

(2) Collapsing placebo and AF3

Drug	Improvement						Total
	Undesirable	Slightly Undesirable	Not useful	Slightly useful	Useful	Excellent	
Placebo & AF3	10	10	70	30	25	2	147
AF6	5	6	21	16	23	6	77
Total	15	16	91	46	48	8	224

$$\max \text{Wil} = \max_i \left[\left\{ \boldsymbol{r}^{*\prime}(1,\ldots,i;i+1,\ldots,a) \otimes \boldsymbol{s} \right\} \boldsymbol{z} \right], \ i = 1, 2, \tag{11.30}$$

where $\boldsymbol{s}$ is Wilcoxon's rank score:

$$\boldsymbol{s} = \frac{1}{\sigma_W} \left\{ \sqrt{C_1}\left(\frac{C_1+1}{2} - \frac{N+1}{2}\right), \sqrt{C_2}\left(C_1 + \frac{C_2+1}{2} - \frac{N+1}{2}\right), \ldots, \sqrt{C_b}\left(C_1 + \cdots + C_{b-1} + \frac{C_b+1}{2} - \frac{N+1}{2}\right) \right\}',$$

$$\sigma_W^2 = \sum_{j=1}^{b} \left\{ \sqrt{C_j}\left(C_1 + \cdots + C_{j-1} + \frac{C_j+1}{2} - \frac{N+1}{2}\right) \right\}^2.$$

To calculate the p-value for a moderate sample size, we can apply the normal approximation. Then it is nothing but max acc. $t1$ of a one-way layout with $n_i = R_i$, $i = 1,2,3$ and df for error ∞.

(2) **Max χ^{*2} method.** We can apply also the cumulative chi-squared statistic to each of Table 11.9 (1) and (2), and take the maximum of the two. The statistic is a variation of max acc. χ^2 and described as

$$\max \text{acc.}\, \chi^{*2}\left(\boldsymbol{R}^{*\prime}\right) = \max_i \left\| \left\{ \boldsymbol{r}^{*\prime}(1,\ldots,i;i+1,\ldots,a) \otimes \boldsymbol{C}^{*\prime} \right\} \boldsymbol{z} \right\|^2, \ i = 1, 2. \tag{11.31}$$

The p-value is evaluated approximately as the maximum of the bivariate chi-squared distributions for a moderate sample size.

(3) **Max max χ method.** We can also apply the maximal contrast statistic (max acc. $t1$) to each of Table 11.9 (1) and (2), and take the maximum of the two. The statistic is described as

$$\begin{aligned} \max\max\chi &= \max_i \max_j \left\{ \boldsymbol{r}^{*\prime}(1,\ldots,i;i+1,\ldots,a) \otimes \boldsymbol{c}^{*\prime}(1,\ldots,j;j+1,\ldots,b) \right\} \boldsymbol{z} \\ &= \max_i \max_j \chi_{ij}^{**}, \end{aligned}$$

$$\chi_{ij}^{**} = \left(Y_{ij}^{**} - \frac{U_i V_j}{N}\right) \left(\frac{U_i\, U_i^*\, V_j\, V_j^*}{N^3}\right)^{-1/2}, \tag{11.32}$$

where $Y_{ij}^{**} = \sum_{k \le i} \sum_{l \le j} y_{kl}$, and U_i, U_i^*, V_j, V_j^* are defined in (11.26) and (11.12), respectively. The statistic χ_{ij}^{**} is nothing but the component of χ^{**2} (11.25) in Section 11.5.1. The exact algorithm for the distribution function of max max χ was obtained by Hirotsu (1997). Define a conditional probability

$$F_k\left(\boldsymbol{\chi}_k^{**}\right) = \Pr\left\{ \boldsymbol{\chi}_1^{**} \le c\boldsymbol{j}, \ldots, \boldsymbol{\chi}_k^{**} \le c\boldsymbol{j} \mid \boldsymbol{Y}_k^{**} \right\},$$

where $\boldsymbol{\chi}_k^{**} = \left(\chi_{1k}^{**}, \ldots, \chi_{a-1k}^{**}\right)'$, $\boldsymbol{Y}_k^{**} = \left(Y_{1k}^{**}, \ldots, Y_{ak}^{**}\right)'$ and $\boldsymbol{\chi}_k^{**} \le c\boldsymbol{j}$ means $\chi_{lk}^{**} \le c$ for $l = 1, \ldots, a-1$. Then we have a recursion formula

$$F_{k+1}\left(\chi_{k+1}^{**}\right)=\sum_{Y_k}F_k\left(\chi_k^{**}\right)\times f\left(Y_k^{**}\mid Y_{k+1}^{**}\right),$$

where $f\left(Y_k^{**}\mid Y_{k+1}^{**}\right)$ is a conditional probability of Y_k^{**} given Y_{k+1}^{**}. To be exact, define a matrix

$$A=\begin{bmatrix} 1 & 0 & 0 & \cdots & 0 & 0 \\ -1 & 1 & 0 & \cdots & 0 & 0 \\ 0 & -1 & 1 & \cdots & 0 & 0 \\ \vdots & \vdots & \vdots & \ddots & \vdots & \vdots \\ 0 & 0 & 0 & \cdots & 1 & 0 \\ 0 & 0 & 0 & \cdots & -1 & 1 \end{bmatrix}_{a\times a}.$$

Then $f\left(Y_k^{**}\mid Y_{k+1}^{**}\right)$ is obtained as $f\left(AY_k^{**}\mid AY_{k+1}^{**}\right)$, which is a multivariate hypergeometric distribution given the row total AY_{k+1}^{**} and the column total Y_{ak}^{**}. The p-value is obtained finally by $1-F_b\left(\chi_b^{**}\right)$ (see Hirotsu, 1997 for more details).

Example 11.2. Analysis of Table 11.8

(1) **Overall analysis.** The sub-tables for calculating χ^{**2} are given in Table 11.10 (1) and (2). Then, from (11.27) we obtain $\chi^{**2}=\chi_{11}^{**2}+\cdots+\chi_{25}^{**2}=31.361$. The constants for the chi-squared approximation are obtained from (11.28) and (11.29) as

$$d=1.5125\times 1.2431=1.8802,$$
$$f=(3-1)\times(6-1)/1.8802=5.319.$$

The p-value of χ^{**2} evaluated as $1.88\chi_{5.32}^2$ is 0.0067. This suggests the effects of the active drug, but can give no more details.

(2) **Comparing dose levels**

(a) **Max Wil method.** The Wilcoxon rank test (11.30) applied to each of Table 11.9 (1) and (2) gives $W_1=1.29$ and $W_2=2.76$, respectively. Therefore, max Wil is $W_2=2.76$, with p-value 0.005 (one-sided) and 0.011 (two-sided) by the normal approximation. That is, there is a cut point between AF 3 mg/kg and AF 6 mg/kg, and AF 6 mg/kg is a strongly recommended dose.

(b) **Max χ^{*2} method.** The cumulative chi-squared (11.31) applied to each of Table 11.9 (1) and (2) gives $\chi_1^{**2}=\sum_{j=1}^5\chi_{1j}^{**2}=7.35$ and $\chi_2^{**2}=\sum_{j=1}^5\chi_{2j}^{**2}=24.01$, respectively. Therefore, max acc. $\chi^{*2}(R^{*\prime})$ is $\chi_2^{**2}=24.01$, with p-value 0.005

Table 11.10 Sub-tables for doubly accumulated chi-square χ^{**2}.

(1) Collapsing AF3 and AF6

	Improvement									
Drug	1	2 ~ 6	1, 2	3 ~ 6	1 ~ 3	4 ~ 6	1 ~ 4	5, 6	1 ~ 5	6
Placebo	3	68	9	62	46	25	55	16	70	1
AF3 & AF6	12	141	22	131	76	77	113	40	146	7
	⇓		⇓		⇓		⇓		⇓	
	$\chi_{11}^{**2} = 1.0159$		$\chi_{12}^{**2} = 0.1179$		$\chi_{13}^{**2} = 4.4677$		$\chi_{14}^{**2} = 0.3368$		$\chi_{15}^{**2} = 1.4121$	

(2) Collapsing placebo and AF3

	Improvement									
Drug	1	2 ~ 6	1,2	3 ~ 6	1 ~ 3	4 ~ 6	1 ~ 4	5 ~ 6	1 ~ 5	6
Placebo & AF3	10	137	20	127	90	57	120	27	145	2
AF 6	5	72	11	66	32	45	48	29	71	6
	⇓		⇓		⇓		⇓		⇓	
	$\chi_{21}^{**2} = 0.0078$		$\chi_{22}^{**2} = 0.020$		$\chi_{23}^{**2} = 7.880$		$\chi_{24}^{**2} = 10.033$		$\chi_{25}^{**2} = 6.069$	

(two-sided) and the same conclusion is obtained as for the max Wil method. It should be noted that max χ^{*2} is essentially a two-sided test, since the basic variable is a sum of squares, whereas max Wil and max max χ are useful for both one- and two-sided tests.

(c) **Max max χ method.** The statistic max max χ (11.32) is calculated from Table 11.9 (1) and (2). It is nothing but the square root of $\chi_{24}^{**2} = 10.033$, with p-value 0.0077 (one-sided) and 0.014 (two-sided), and the same conclusion is obtained as for the max Wil method. Although in this example it gives a slightly larger p-value than max Wil and max χ^{*2}, it depends of course on the data. This test is most attractive for its exact p-value and applicability to both one- and two-sided tests. It can further suggest that AF 6 mg is characterized by a high proportion of useful and excellent categories, since a cut point is suggested between columns 4 and 5.

References

Aida, M. and Hirotsu, C. (1983) A method for comparing the multinomial distributions under order constraints and the table of percentiles. *Jap. J. Appl. Statist.* **12**, 101–110 (in Japanese).

Anderson, T. W. (2003) *An introduction to multivariate statistical analysis*, 3rd edn. Wiley, New York.

Cox, D. R. and Snell, E. J. (1981) *Applied statistics*. Chapman & Hall, London.

Greenacre, M. J. (1988) Clustering the rows and columns of a contingency table. *J. Classif.* **5**, 39–51.

Guttman, L. (1971) Measurement as structural theory. *Psychometrika* **36**, 329–347.

Hirotsu, C. (1977) Multiple comparisons and clustering rows in a contingency table. *Quality* **7**, 27–33 (in Japanese).

Hirotsu, C. (1983) Defining the pattern of association in two-way contingency tables. *Biometrika* **70**, 579–590.

Hirotsu, C. (1991) An approach to comparing treatments based on repeated measures. *Biometrika* **78**, 583–594.

Hirotsu, C. (1992) *Analysis of experimental data: Beyond the analysis of variance*. Kyoritsu-shuppan, Tokyo (in Japanese).

Hirotsu, C. (1993) Beyond analysis of variance techniques: Some applications in clinical trials. *Int. Statist. Rev.* **61**, 183–201.

Hirotsu, C. (1997) Two-way change point model and its application. *Austral. J. Statist.* **39**, 205–218.

Hirotsu, C. (2009) Clustering rows and/or columns of a two-way contingency table and a related distribution theory. *Comput. Statist. Data Anal.* **53**, 4508–4515.

Hirotsu, C., Kuriki, S. and Hayter, A. J. (1992) Multiple comparison procedures based on the maximal component of the cumulative chi-squared statistic. *Biometrika* **79**, 381–392.

James, A. T. (1964) Distribution of matrix variates and latent roots derived from normal samples. *Ann. Math. Statist.* **35**, 475–501.

James, A. T. (1968) Calculation of Zonal polynomial coefficients by use of the Laplace–Beltrami operator. *Ann. Math. Statist.* **39**, 1711–1718.

Marcus, R., Peritz, E. and Gabriel, K. R. (1976) On closed testing procedures with special reference to ordered analysis of variance. *Biometrika* **63**, 655–660.

Siotani, M. (1959) The extreme value of the generalized distance of the individual points in the multivariate normal sample. *Ann. Inst. Statist. Math.* **10**, 183–208.

Sugiura, N. (1973) Derivatives of the characteristic root of a symmetric or a Hermitian matrix with two applications in multivariate analysis. *Commun. Statist.* **1**, 393–417.

12

Mixed and Random Effects Model

In factorial experiments, if all factors except the error are fixed effects, we have a fixed effects model. If the factors are all random, except for a general mean, we have a random effects model. If both types of factor are involved in the experiment, we have a mixed effects model. In previous sections fixed effects models were mainly discussed, except for the recovery of inter-block information in the BIBD. However, there are cases where it is better to consider the effects of a factor to be random, and we discuss the basic ideas in this chapter. There is a factor like the variation factor which is dealt with as fixed in the laboratory, but acts as if it were random in extensions to the real world. Therefore, this is a problem of interpretation of data rather than of mathematics.

12.1 One-Way Random Effects Model

12.1.1 Model and parameters

Let us consider an experiment to produce products by a workers using a machine. The yield of the kth experiment by the ith worker is denoted by y_{ik}. Denoting the expected yield of worker i by μ_i and the measurement error by e_{ik}, we have a one-way layout model

$$y_{ik} = \mu_i + e_{ik},\ i = 1, \ldots, a,\ k = 1, \ldots, m.$$

The worker can be a fixed factor but also considered random if they are selected randomly from the population of workers in the factory. Then we are interested in evaluating the variation of the population rather than comparing yields $\mu_1, \ldots, \mu_a$ of the

Advanced Analysis of Variance, First Edition. Chihiro Hirotsu.

selected workers by chance. Therefore, we define the expectation μ and variance σ_α^2 of μ_i in the population. We define the difference

$$\alpha_i = \mu_i - \mu$$

and call it a main effect of worker i. Then we have

$$E(\alpha_i) = 0$$
$$V(\alpha_i) = \sigma_\alpha^2.$$

Finally we reach a model

$$y_{ik} = \mu + \alpha_i + e_{ik},\ i = 1, \ldots, a,\ k = 1, \ldots, m, \tag{12.1}$$

where we assume α_i and e_{ik} are distributed independently of each other as $N(0, \sigma_\alpha^2)$ and $N(0, \sigma^2)$, respectively. If an expert is expected to produce a high yield with smaller variation, we have to assume a model $V(e_{ik}) = \sigma_i^2$, $i = 1, \ldots, a$, but we do not consider the case here. If the assumption of the equality of variance is doubtful, we need to test this also.

12.1.2 Standard form for test and estimation

Similarly to Chapter 2, we can express the model in matrix form as

$$\boldsymbol{y} = [\boldsymbol{j}_n\ X_\alpha] \begin{bmatrix} \mu \\ \boldsymbol{\alpha} \end{bmatrix} + \boldsymbol{e}_n,\ \boldsymbol{\alpha} = (\alpha_1, \ldots, \alpha_a)',$$

where $n = am$, $\boldsymbol{\alpha} \sim N(\boldsymbol{0}, \sigma_\alpha^2 \boldsymbol{I}_a)$, $\boldsymbol{e}_n \sim N(\boldsymbol{0}, \sigma^2 \boldsymbol{I}_n)$, $\boldsymbol{\alpha}$, and $\boldsymbol{e}_n$ are mutually independent, and

$$X_\alpha = \begin{bmatrix} \boldsymbol{j}_m & \boldsymbol{0} & \cdots & \boldsymbol{0} \\ \boldsymbol{0} & \boldsymbol{j}_m & \cdots & \boldsymbol{0} \\ & & \ddots & \\ \boldsymbol{0} & \boldsymbol{0} & \cdots & \boldsymbol{j}_m \end{bmatrix}.$$

Differently from Examples 2.1 and 2.3, we have

$$E(\boldsymbol{y}) = \mu \boldsymbol{j}_n, \tag{12.2}$$

$$V(\boldsymbol{y}) = \sigma_\alpha^2 X_\alpha X_\alpha' + \sigma^2 \boldsymbol{I}_n, \tag{12.3}$$

but the orthonormal transformation $\boldsymbol{M}'$ (2.38) is useful also in this case, and we have a standard form

$$\boldsymbol{z} = \boldsymbol{M}'\boldsymbol{y} = \begin{bmatrix} n^{-1/2} y_{\cdot\cdot} \\ m^{-1/2} \boldsymbol{P}_a' \begin{bmatrix} y_{1\cdot} \\ \vdots \\ y_{a\cdot} \end{bmatrix} \\ \boldsymbol{Q}'\boldsymbol{y} \end{bmatrix} = \begin{bmatrix} n^{1/2}(\mu + \bar{\alpha}_\cdot) \\ m^{1/2} \boldsymbol{P}_a' \boldsymbol{\alpha} \\ \boldsymbol{0} \end{bmatrix} + \boldsymbol{\xi}, \tag{12.4}$$

which is distributed as a multivariate normal distribution with

$$E(z) = \begin{bmatrix} \sqrt{n}\mu \\ \mathbf{0} \\ \mathbf{0} \end{bmatrix},$$

$$V(z) = \begin{bmatrix} m\sigma_\alpha^2 + \sigma^2 & \mathbf{0}' & \mathbf{0}' \\ \mathbf{0} & (m\sigma_\alpha^2 + \sigma^2)\boldsymbol{I}_{a-1} & \mathbf{0} \\ \mathbf{0} & \mathbf{0} & \sigma^2 \boldsymbol{I}_{n-a} \end{bmatrix}. \quad (12.5)$$

This form suggests that all the elements of z are independent and the statistics

$$\boldsymbol{j}_n'\boldsymbol{y} = y_{\cdot\cdot}$$

$$S_\alpha = m^{-1}\|\boldsymbol{P}_a'\boldsymbol{X}_\alpha'\boldsymbol{y}\|^2 = m\sum_i(\bar{y}_{i\cdot} - \bar{y}_{\cdot\cdot})^2$$

$$S_e = \|\boldsymbol{Q}'\boldsymbol{y}\|^2 = \sum_i\sum_k(y_{ik} - \bar{y}_{i\cdot})^2$$

$$= \sum_i\sum_k(y_{ik} - \bar{y}_{\cdot\cdot})^2 - m\sum_i(\bar{y}_{i\cdot} - \bar{y}_{\cdot\cdot})^2$$

are the complete sufficient statistics. These sums of squares are the same as those of the fixed effects model given in Section 6.1. As mentioned there, we first calculate

$$S_T = \sum_i\sum_k {y_{ik}}^2 - y_{\cdot\cdot}^2/n \quad (12.6)$$

and

$$S_\alpha = \sum_i y_{i\cdot}^2/m - y_{\cdot\cdot}^2/n \quad (12.7)$$

and then by subtraction

$$S_e = S_T - S_\alpha. \quad (12.8)$$

By equations (12.4) and (12.5), we have

$$E(S_\alpha) = (a-1)\left(m\sigma_\alpha^2 + \sigma^2\right)$$

$$E(S_e) = (n-a)\,\sigma^2.$$

Therefore,

$$\widehat{\sigma}^2 = S_e/(n-a) \quad (12.9)$$

$$\widehat{\sigma}_\alpha^2 = m^{-1}\{S_\alpha/(a-1) - S_e/(n-a)\} \quad (12.10)$$

are the minimum variance unbiased estimator of σ^2 and σ_α^2 with variances

$$V(\widehat{\sigma}^2) = 2\sigma^4/(n-a), \quad (12.11)$$

and

$$V\left(\widehat{\sigma}_\alpha^2\right) = \frac{2}{m^2}\left\{\frac{\left(m\,\sigma_\alpha^2 + \sigma^2\right)^2}{a-1} + \frac{\sigma^4}{n-a}\right\},$$

respectively.

12.1.3 Problems of negative estimators of variance components

It should be noted, however, that $\widehat{\sigma}_\alpha^2$ (12.10) can take a negative value, whereas it must be positive. Therefore an easy modification

$$\widehat{\sigma}_\alpha^{2\prime} = \max\left(\widehat{\sigma}_\alpha^2, 0\right)$$

is often made. This obviously makes a biased estimator, but the mean squared error is reduced.

The negative best unbiased estimator is a difficult problem, often encountered in the estimation of variance components. We introduce here the idea of Smith and Murray (1984) to justify the negative estimator by regarding a certain variance component as a covariance. For simplicity, we consider a one-way layout model (12.1), where we assume the normal distribution for y_{ik} with covariance structure

$$Cov(y_{ik}, y_{i'k'}) = \begin{cases} \sigma^2 + \theta_a, & i = i',\ k = k', \\ \theta_a, & i = i', k \neq k', \\ 0, & i \neq i'. \end{cases}$$

That is, the σ_α^2 previously defined as the variance of worker α_i is considered as the covariance of measurement errors within the same worker. If index k distinguishes a litter of animals from the same mother, then parameter θ_a expresses the covariance among their characteristic values. It is natural to consider a negative value for θ_a by competition among the animals for nutrition. Then, the non-negative condition for σ_α^2 is relaxed to the condition of positive definiteness of the covariance matrix

$$\sigma^2 \boldsymbol{I}_m + \theta_a \boldsymbol{j}_m \boldsymbol{j}_m'.$$

This condition can still be violated, but the possibility of an inappropriate solution would be considerably reduced.

The confidence interval for σ^2 is easily obtained from the chi-squared distribution of S_e with df $n-a$. For $\widehat{\sigma}_\alpha^2$ the approximate methods are given by Tukey (1951), Moriguti (1954), Bulmer (1957), and Williams (1962). Boardman (1974) compared these methods by Monte Carlo simulation. In particular, Williams' method is preferred in the sense that a true confidence coefficient does not go down below the nominal value.

In contrast, a positive linear combination of the squared normal variables generally follows the chi-squared distribution approximately; the distribution becomes very unstable if negative coefficients are included as in (12.10). Therefore, it would be recommended to estimate $\sigma_\alpha^2 + \sigma^2$ instead of σ_α^2. In this case we have an estimator

$$\widehat{\sigma_\alpha^2 + \sigma^2} = \frac{1}{m}\left(\frac{S_a}{a-1} + \frac{S_e}{a}\right), \tag{12.12}$$

$$V\left(\widehat{\sigma_\alpha^2 + \sigma^2}\right) = \frac{2}{m^2}\left\{\frac{\left(m\,\sigma_\alpha^2 + \sigma^2\right)^2}{a-1} + \frac{(m-1)\sigma^4}{a}\right\}, \tag{12.13}$$

which is more stable. Also, $\sigma_\alpha^2+\sigma^2$ is a variation of one experiment by one worker and easy to interpret.

Meanwhile, the variance of the general mean $\bar{y}_{..}$ is $(m\sigma_\alpha^2+\sigma^2)/n$ from (12.5). This is estimated simply by

$$\left(\widehat{m\sigma_\alpha^2+\sigma^2}\right)/n=S_\alpha/\{n(a-1)\}.$$

Then, it is easy to see that

$$(\bar{y}_{..}-\mu)/[S_\alpha/\{n(a-1)\}]^{1/2}$$

is distributed as a t-distribution with df $a-1$, which leads to a confidence interval for the general mean of

$$\mu\sim\bar{y}_{..}\pm[S_\alpha/\{n(a-1)\}]^{1/2}t_{a-1}(\alpha/2). \tag{12.14}$$

12.1.4 Testing homogeneity of treatment effects

Since the statistics S_α and S_e are distributed as $(m\sigma_\alpha^2+\sigma^2)\chi_{a-1}^2$ and $\sigma^2\chi_{n-a}^2$, the rejection region of the null hypothesis

$$H_0:\sigma_\alpha^2=0$$

is given by the F-test,

$$F=\frac{S_\alpha/(a-1)}{S_e/(n-a)}>F_{a-1,n-a}(\alpha). \tag{12.15}$$

When the null hypothesis fails, the statistic

$$F_{a-1,n-a}=\frac{S_\alpha/\{(a-1)(m\sigma_\alpha^2+\sigma^2)\}}{S_e/\{(n-a)\sigma^2\}} \tag{12.16}$$

is distributed as an F-distribution with df $(a-1,n-a)$. Thus, the power of the test (12.15) is obtained from

$$\Pr\{F>F_{a-1,n-a}(\alpha)\}=\Pr\left\{F_{a-1,n-a}>\frac{\sigma^2}{m\sigma_\alpha^2+\sigma^2}F_{a-1,n-a}(\alpha)\right\}.$$

The power becomes large when m and the ratio σ_α^2/σ^2 are large.

12.1.5 Between and within variance ratio (SN ratio)

Finally, for the confidence interval of the ratio

$$\gamma=\sigma_\alpha^2/\sigma^2 \tag{12.17}$$

we have

$$\Pr\left\{\frac{1}{F_{n-a,\,a-1}(\alpha_1)} \le F_{a-1,n-a} \le F_{a-1,n-a}(\alpha_2)\right\} = 1-\alpha$$

for $F_{a-1,n-a}$ (12.16) and for any (α_1, α_2) satisfying $\alpha_1 \ge 0$, $\alpha_2 \ge 0$, $\alpha_1 + \alpha_2 = \alpha$, which leads to the confidence interval with confidence coefficient $1-\alpha$,

$$\frac{1}{m}\left\{\frac{1}{F_{a-1,n-a}(\alpha_2)} \times \frac{S_\alpha/(a-1)}{S_e/(n-a)} - 1\right\} \le \frac{\sigma_\alpha^2}{\sigma^2} \le \frac{1}{m}\left\{F_{n-a,\,a-1}(\alpha_1) \times \frac{S_\alpha/(a-1)}{S_e/(n-a)} - 1\right\}. \tag{12.18}$$

Usually,

$$\alpha_1 = \alpha_2 = \alpha/2 \text{ or } \alpha_1 = \alpha,\ \alpha_2 = 0$$

is chosen. In the latter case, only the right-hand-side inequality is concerned in (12.18). The ratio σ_α^2/σ^2 is called a signal-to-noise (SN) ratio in engineering.

Example 12.1. Content rate of sulfur in sulfide mineral. Sulfide mineral has been brought to a factory in many freight cars. Six cars are selected at random, and the content rate of sulfur is measured for five samples from each car, as given in Table 12.1 (Moriguti, 1976).

The sums of squares are calculated as follows. First, from (12.6) we have

$$S_T = \left(42.0^2 + 41.8^2 + \cdots + 40.4^2\right) - 1228.6^2/30 = 10.19$$

and from (12.7)

$$S_\alpha = \left(207.0^2 + 207.5^2 + 205.0^2 + 201.9^2 + 205.1^2 + 202.1^2\right)/5 - 1228.6^2/30 = 5.59$$

and then by subtraction

$$S_e = 10.19 - 5.59 = 4.60$$

with df 24. These results are summarized in Table 12.2.

Table 12.1 Content rate of sulfur.

	Sample j						
Car i	1	2	3	4	5	Total	Sum of squares
1	42.0	41.8	40.8	41.4	41.0	207.0	8570.84
2	41.4	41.5	41.1	41.6	41.9	207.5	8611.59
3	41.1	40.8	40.2	41.5	41.4	205.0	8406.10
4	40.5	40.8	39.9	39.7	41.0	201.9	8153.99
5	41.2	40.9	40.7	41.3	41.0	205.1	8413.43
6	40.5	40.3	41.0	39.9	40.4	202.1	8169.51
Total						1228.6	50325.46

Table 12.2 ANOVA table for one-way random effects model.

Factor	Sum of squares	df	Mean sum of squares	F	SN ratio
Treatment	$S_\alpha = 5.59$ (12.7)	5	1.118	$F = 5.84^{**}$ (12.15)	γ (12.17)
Error	$S_e = 4.60$ (12.8)	24	0.192		
Total	$S_T = 10.19$ (12.6)	29			

It should be noted that Table 12.2 is the same as Table 6.1 for a fixed effects model, except for the definition of γ. The observed F ratio is highly significant compared with $F_{5,\,24}(0.01) = 3.90$. Therefore, we proceed to estimate the variance components. The estimate of σ^2 (12.9) is 0.192, as given in Table 12.2, with the estimate of variance

$$V\left(\widehat{\sigma^2}\right) = 2 \times 0.192^2/24 = 3.07 \times 10^{-3},$$

which is obtained by substituting $\widehat{\sigma^2}$ in (12.11).

The estimate of $\sigma_\alpha^2 + \sigma^2$ is, from (12.12),

$$\widehat{\sigma_\alpha^2 + \sigma^2} = 0.378$$

and the estimate of its variance is, from (12.13),

$$\widehat{V}\left(\widehat{\sigma_\alpha^2 + \sigma^2}\right) = \frac{2}{m^2}\left\{\frac{\left(\widehat{m\,\sigma_\alpha^2 + \sigma^2}\right)^2}{a-1} + \frac{(m-1)\left(\widehat{\sigma^2}\right)^2}{a}\right\} = 2.20 \times 10^{-2}.$$

The estimate of the SN ratio is obtained as

$$\widehat{\gamma} = \left(\widehat{\sigma_\alpha^2 + \sigma^2}\right)/\widehat{\sigma^2} - 1 = 0.378/0.192 - 1 = 0.97$$

and the confidence interval at confidence coefficient 0.95 is obtained by putting $\alpha_1 = \alpha_2 = 0.025$ in (12.18) as

$$0.170 \le \gamma \le 7.13.$$

Finally, the confidence interval of the general mean is, from (12.14),

$$\mu \sim 40.95 \pm [5.59/\{30(6-1)\}]^{1/2} t_{6-1}(0.05/2) = 40.95 \pm 0.50.$$

This suggests the range of sulfur content in the arrived lots by many freight cars.

12.2 Two-Way Random Effects Model

12.2.1 Model and parameters

Extending the idea of the previous section, suppose there are many working machines in the factory and b of them are randomly chosen. The a workers work on all b machines m times in random order. Let the yield of the kth experiment by the ith worker on the jth machine be y_{ijk}, the expected yield of the ith worker on the jth machine be μ_{ij}, and an independent random error with expectation zero be e_{ijk}. Then we have a model

$$y_{ijk} = \mu_{ij} + e_{ijk},\ i = 1, \ldots, a; j = 1, \ldots, b; k = 1, \ldots, m. \tag{12.19}$$

Let $\mu_{\alpha,j}$ be the expected yield of machine j over the population of workers and $\mu_{i,\beta}$ the expected yield of worker i over the population of machines. Then the expectation of $\mu_{\alpha,j}$ over the population of machines is denoted by $\mu_{\alpha\beta}$, which is also the expectation of $\mu_{i,\beta}$ over the population of workers and called a general mean. Then

$$\alpha_i = \mu_{i,\beta} - \mu_{\alpha\beta}$$

$$\beta_j = \mu_{\alpha,j} - \mu_{\alpha\beta}$$

$$(\alpha\beta)_{ij} = \mu_{ij} - \mu_{i,\beta} - \mu_{\alpha,j} + \mu_{\alpha\beta}$$

are called the main effects of worker i, machine j, and the interaction between them. The α_i and β_j are independent, since the workers and machines are independently chosen. The covariance between α_i or β_j and $(\alpha\beta)_{ij}$ is shown to be zero, as follows. In the calculation of $E\left\{\alpha_i, (\alpha\beta)_{i'j'}\right\}$, we take the expectation first with respect to the machine, fixing the worker, then we have

$$E\left\{\alpha_i, (\alpha\beta)_{i'j'}\right\} = E\left\{\left(\mu_{i,\beta} - \mu_{\alpha\beta}\right)\left(\mu_{i',\beta} - \mu_{i',\beta} - \mu_{\alpha\beta} + \mu_{\alpha\beta}\right)\right\} = 0$$

and similarly for β_j. We therefore have a model

$$\mu_{ij} = \mu_{\alpha\beta} + \alpha_i + \beta_j + (\alpha\beta)_{ij}$$

where α_i and β_j are independent and their covariances with $(\alpha\beta)_{ij}$ are zero. Further introducing the normality assumption, we have a model in the matrix expression

$$\begin{cases} \boldsymbol{y} = \mu_{\alpha\beta}\,\boldsymbol{j} + \boldsymbol{X}_\alpha\boldsymbol{\alpha} + \boldsymbol{X}_\beta\boldsymbol{\beta} + \boldsymbol{X}_{\alpha\beta}(\boldsymbol{\alpha\beta}) + \boldsymbol{e}, \\ \boldsymbol{\alpha} \sim N\left(\boldsymbol{0}, \sigma_\alpha^2\boldsymbol{I}_a\right), \boldsymbol{\beta} \sim N\left(\boldsymbol{0}, \sigma_\beta^2\boldsymbol{I}_b\right),\ (\boldsymbol{\alpha\beta}) \sim N\left(\boldsymbol{0}, \sigma_{\alpha\beta}^2\boldsymbol{I}_{ab}\right), \boldsymbol{e} \sim N\left(\boldsymbol{0}, \sigma^2\boldsymbol{I}_n\right), n = abm, \\ \boldsymbol{\alpha}, \boldsymbol{\beta},\ (\boldsymbol{\alpha\beta}) \text{ and } \boldsymbol{e} \text{ are mutually independent.} \end{cases} \tag{12.20}$$

The expression of $\boldsymbol{y}$ is formally equivalent to the fixed effects model in Section 10.2, but the distribution theory is different. In this model the expectation of $\boldsymbol{y}$ is $\mu_{\alpha\beta}\,\boldsymbol{j}$ and the variance–covariance matrix is

$$V(y) = \sigma_\alpha^2 X_\alpha X'_\alpha + \sigma_\beta^2 X_\beta X'_\beta + \sigma_{\alpha\beta}^2 X_{\alpha\beta} X'_{\alpha\beta} + \sigma^2 I_n.$$

12.2.2 Standard form for test and estimation

In this case also the orthonormal transformation in Section 10.2 is useful as it is. Using the same notation as (10.9) for $M'y = \left(z_{\mu_{\alpha\beta}}, z'_\alpha, z'_\beta, z'_{\alpha\beta}, z'_e\right)'$, we have

$$E(M'y) = \left(\sqrt{n}\mu_{\alpha\beta}, \mathbf{0}', \mathbf{0}', \mathbf{0}', \mathbf{0}'\right)', \tag{12.21}$$

$$V(M'y) = \begin{bmatrix} \sigma^2 + bm\sigma_\alpha^2 + am\sigma_\beta^2 + m\sigma_{\alpha\beta}^2 & \mathbf{0}' & \mathbf{0}' & \mathbf{0}' & \mathbf{0}' \\ \mathbf{0} & \left(\sigma^2 + bm\sigma_\alpha^2 + m\sigma_{\alpha\beta}^2\right) I_{a-1} & \mathbf{0} & \mathbf{0} & \mathbf{0} \\ \mathbf{0} & \mathbf{0} & \left(\sigma^2 + am\sigma_\beta^2 + m\sigma_{\alpha\beta}^2\right) I_{b-1} & \mathbf{0} & \mathbf{0} \\ \mathbf{0} & \mathbf{0} & \mathbf{0} & \left(\sigma^2 + m\sigma_{\alpha\beta}^2\right) I_{(a-1)(b-1)} & \mathbf{0} \\ \mathbf{0} & \mathbf{0} & \mathbf{0} & \mathbf{0} & \sigma^2 I_{n-ab} \end{bmatrix}. \tag{12.22}$$

That is, $y_{\dots}$, $S_\alpha = \|z_\alpha\|^2, S_\beta = \|z_\beta\|^2$, $S_{\alpha\beta} = \|z_{\alpha\beta}\|^2$, and $S_e = \|Q'y\|^2$ are the complete sufficient statistics and their explicit forms are given in (10.11) ~ (10.14). From the structures of (12.21) and (12.22), the distributions of these sums of squares are easily derived as

$$S_\alpha \sim \left(\sigma^2 + bm\sigma_\alpha^2 + m\sigma_{\alpha\beta}^2\right)\chi_{a-1}^2, \tag{12.23}$$

$$S_\beta \sim \left(\sigma^2 + am\sigma_\beta^2 + m\sigma_{\alpha\beta}^2\right)\chi_{b-1}^2, \tag{12.24}$$

$$S_{\alpha\beta} \sim \left(\sigma^2 + m\sigma_{\alpha\beta}^2\right)\chi_{(a-1)(b-1)}^2, \tag{12.25}$$

$$S_e \sim \sigma^2 \chi_{n-ab}^2. \tag{12.26}$$

We can also easily obtain the minimum variance unbiased estimators of σ_α^2, σ_β^2, $\sigma_{\alpha\beta}^2$, and σ^2 by equating those statistics to their expectations, namely by solving the equation

$$\begin{bmatrix} bm & 0 & m & 1 \\ 0 & am & m & 1 \\ 0 & 0 & m & 1 \\ 0 & 0 & 0 & 1 \end{bmatrix} \begin{bmatrix} \hat{\sigma}_\alpha^2 \\ \hat{\sigma}_\beta^2 \\ \hat{\sigma}_{\alpha\beta}^2 \\ \hat{\sigma}^2 \end{bmatrix} = \begin{bmatrix} S_\alpha/(a-1) \\ S_\beta/(b-1) \\ S_{\alpha\beta}/(a-1)(b-1) \\ S_e/(n-ab) \end{bmatrix}.$$

The explicit forms are as follows:

$$\begin{cases} \hat{\sigma}_\alpha^2 = (bm)^{-1}\left\{\dfrac{S_\alpha}{a-1} - \dfrac{S_{\alpha\beta}}{(a-1)(b-1)}\right\}, \ V(\hat{\sigma}_\alpha^2) = \dfrac{2}{(bm)^2}\left\{\dfrac{\left(\sigma^2 + bm\,\sigma_\alpha^2 + m\,\sigma_{\alpha\beta}^2\right)^2}{a-1} + \dfrac{\left(\sigma^2 + m\,\sigma_{\alpha\beta}^2\right)^2}{(a-1)(b-1)}\right\} \\ \hat{\sigma}_\beta^2 = (am)^{-1}\left\{\dfrac{S_\beta}{b-1} - \dfrac{S_{\alpha\beta}}{(a-1)(b-1)}\right\}, V\left(\hat{\sigma}_\beta^2\right) = \dfrac{2}{(am)^2}\left\{\dfrac{\left(\sigma^2 + am\,\sigma_\beta^2 + m\,\sigma_{\alpha\beta}^2\right)^2}{b-1} + \dfrac{\left(\sigma^2 + m\,\sigma_{\alpha\beta}^2\right)^2}{(a-1)(b-1)}\right\} \\ \hat{\sigma}_{\alpha\beta}^2 = m^{-1}\left\{\dfrac{S_{\alpha\beta}}{(a-1)(b-1)} - \hat{\sigma}^2\right\}, V\left(\hat{\sigma}_{\alpha\beta}^2\right) = \dfrac{2}{m^2}\left\{\dfrac{\left(\sigma^2 + m\,\sigma_{\alpha\beta}^2\right)^2}{(a-1)(b-1)} + \dfrac{\sigma^4}{n-ab}\right\} \\ \hat{\sigma}^2 = S_e/(n-ab), \ \ V(\hat{\sigma}^2) = \dfrac{2\sigma^4}{n-ab} \end{cases}$$

The confidence interval for σ^2 is easily obtained by the chi-squared distribution (12.26) but for other variance components, only the approximate intervals are available. Also, except for $\widehat{\sigma}^2$, negative estimators are possible and one should refer to Smith and Murray (1984) for this situation.

12.2.3 Testing homogeneity of treatment effects

It is easy to construct test procedures on variance components based on the distribution theory (12.23) ~ (12.26). To test

$$H_{\alpha\beta} : \sigma_{\alpha\beta}^2 = 0 \text{ against } K_{\alpha\beta} : \sigma_{\alpha\beta}^2 > 0,$$

the rejection region $R_{\alpha\beta}$ of size α is

$$R_{\alpha\beta} : \frac{S_{\alpha\beta}/\{(a-1)(b-1)\}}{S_e/(n-ab)} > F_{(a-1)(b-1),n-a}(\alpha).$$

To test

$$H_\alpha : \sigma_\alpha^2 = 0 \text{ against } K_\alpha : \sigma_\alpha^2 > 0,$$

the procedure is changed according to $\sigma_{\alpha\beta}^2 = 0$ or $\sigma_{\alpha\beta}^2 \neq 0$, as

$$R_\alpha : \frac{S_\alpha/(a-1)}{S_e/(n-ab)} > F_{a-1,n-a}(\alpha) \ \text{ if } \sigma_{\alpha\beta}^2 = 0,$$

$$R'_\alpha : \frac{S_\alpha/(a-1)}{S_{\alpha\beta}/\{(a-1)(b-1)\}} > F_{a-1,(a-1)(b-1)}(\alpha) \ \text{ if } \sigma_{\alpha\beta}^2 \neq 0$$

If we have a significant result by the rejection region R_α when $\sigma_{\alpha\beta}^2 \neq 0$, it is not clear which of σ_α^2 or $\sigma_{\alpha\beta}^2$ is the cause of it because of the distribution (12.23). When $\sigma_{\alpha\beta}^2 = 0$, $S_{\alpha\beta}$ behaves just like S_e but usually it is not pooled into S_e.

Testing $H_\beta: \sigma_\beta^2 = 0$ is essentially in the same way as H_α. It should be noted here that, as stated in Section 10.2, the test of a factor does not depend on the definition of factors in the fixed effects model, but the tests in this section depend on the definition of these factors. Therefore, one should take care of the interpretation of the main effects when interaction exists.

12.2.4 Easy method for unbalanced two-way random effects model

For the unbalanced two-way data, the sufficient statistics are not complete and uniformly minimum variance unbiased estimators are not available. For various proposals and reviews, one should refer to Searl (1971) and Khuri and Sahai (1985). In contrast, Rao (1971a,b) proposed MINQUE, the minimum norm quadratic unbiased estimator. However, in the unbalanced case the minimal sufficient statistics are not complete and those estimators have no global optimality, except for being unbiased. Therefore, an appropriate analytical method is more strongly desired than a fixed effects model. Under these circumstances, Hirotsu (1968) showed that an easy estimator based on the sum of squares in $x_{ij} = \bar{y}_{ij\cdot}$ and the unbiased variance $\widehat{\sigma}^2$ behaves well if there is no empty cell.

Let $\boldsymbol{x}$ be $(x_{11}, x_{12}, \ldots, x_{ab})'$, then similarly to (12.20) we have

$$\boldsymbol{x} = \mu_{\alpha\beta}\boldsymbol{j}_{ab} + (\boldsymbol{I}_a \otimes \boldsymbol{j}_b)\boldsymbol{\alpha} + (\boldsymbol{j}_a \otimes \boldsymbol{I}_b)\boldsymbol{\beta} + \boldsymbol{I}_{ab}(\boldsymbol{\alpha\beta}) + \bar{\boldsymbol{e}}_{(i)(j)\cdot},$$

$$\boldsymbol{\alpha} \sim N(\boldsymbol{0}, \sigma_\alpha^2 \boldsymbol{I}_a), \boldsymbol{\beta} \sim N\left(\boldsymbol{0}, \sigma_\beta^2 \boldsymbol{I}_b\right), \ (\boldsymbol{\alpha\beta}) \sim N\left(\boldsymbol{0}, \sigma_{\alpha\beta}^2 \boldsymbol{I}_{ab}\right),$$

$$\bar{\boldsymbol{e}}_{(i)(j)\cdot} \sim N\left\{\boldsymbol{0}, \sigma^2 \mathrm{diag}\left(m_{ij}^{-1}\right)\right\},$$

$\boldsymbol{\alpha}$, $\boldsymbol{\beta}$, $(\boldsymbol{\alpha\beta})$, and $\bar{\boldsymbol{e}}$ are mutually independent.

The definition of basic statistics S_e, S_α, S_β, and $S_{\alpha\beta}$ is the same as (10.35) ~ (10.38) in the fixed effects model of Section 10.5.2. In this case again $S_e = \sum_{i=1}^{a}\sum_{j=1}^{b}\sum_{k=1}^{m_{ij}}\left(y_{ijk} - \bar{y}_{ij\cdot}\right)^2$ is distributed as $\sigma^2\chi_{n-ab}^2$ with df $n - ab, n = m_{\cdot\cdot}$ and mutually independent with (S_α, S_β, $S_{\alpha\beta}$). Therefore, $S_e/(n-ab)$ is the best unbiased estimator of σ^2 as usual. To calculate the expectations of S_α, S_β, and $S_{\alpha\beta}$, we note that the variance–covariance matrix of $\boldsymbol{x}$ is simply

$$V(\boldsymbol{x}) = \sigma_\alpha^2\left(\boldsymbol{I}_a \otimes \boldsymbol{j}_b\boldsymbol{j}_b'\right) + \sigma_\beta^2\left(\boldsymbol{j}_a\boldsymbol{j}_a' \otimes \boldsymbol{I}_b\right) + \left(\sigma_{\alpha\beta}^2 + m^{-1}\sigma^2\right)\boldsymbol{I}_{ab} + \sigma^2 \mathrm{diag}\left(m_{ij}^{-1} - m^{-1}\right). \tag{12.27}$$

To calculate $E(S_\alpha)$, we further note that

$$V\left(\bar{\boldsymbol{x}}_{(i)\cdot}\right) = \left\{\sigma_\alpha^2 + b^{-1}\sigma_{\alpha\beta}^2 + (bm)^{-1}\sigma^2\right\}\boldsymbol{I}_a + b^{-1}\sigma_\beta^2\boldsymbol{j}_a\boldsymbol{j}_a' + b^{-2}\sigma^2 \mathrm{diag}\left\{\sum_j\left(m_{ij}^{-1} - m^{-1}\right)\right\},$$

where $\bar{\boldsymbol{x}}_{(i)\cdot} = (\bar{x}_{1\cdot}, \ldots, \bar{x}_{a\cdot})'$. Then we have

$$E(S_\alpha) = \text{tr}\left\{b\boldsymbol{P}_a\boldsymbol{P}_a' V\left(\bar{\boldsymbol{x}}_{(i)\cdot}\right)\right\} = (a-1)\left\{b\sigma_\alpha^2 + \sigma_{\alpha\beta}^2 + m^{-1}\sigma^2\right\}, \tag{12.28}$$

$$\begin{aligned} V(S_\alpha) &= 2\text{tr}\left\{b\boldsymbol{P}_a\boldsymbol{P}_a' V\left(\bar{\boldsymbol{x}}_{(i)\cdot}\right)\right\}^2 \\ &= 2\left[(a-1)\left(b\sigma_\alpha^2 + \sigma_{\alpha\beta}^2 + m^{-1}\sigma^2\right)^2 + f_\alpha\left\{b^{-1}\textstyle\sum_j\left(m_{ij}^{-1} - m^{-1}\right)\right\}\sigma^4\right]. \end{aligned} \tag{12.29}$$

Similarly we have

$$E(S_\beta) = (b-1)\left\{a\sigma_\beta^2 + \sigma_{\alpha\beta}^2 + m^{-1}\sigma^2\right\}, \tag{12.30}$$

$$V(S_\beta) = 2\left[(b-1)\left(a\sigma_\beta^2 + \sigma_{\alpha\beta}^2 + m^{-1}\sigma^2\right)^2 + f_\beta\left\{a^{-1}\textstyle\sum_i\left(m_{ij}^{-1} - m^{-1}\right)\right\}\sigma^4\right]. \tag{12.31}$$

For $S_{\alpha\beta}$ we have, from $V(\boldsymbol{x})$ (12.27),

$$E(S_{\alpha\beta}) = \text{tr}\left\{(\boldsymbol{P}_a \otimes \boldsymbol{P}_b)(\boldsymbol{P}_a \otimes \boldsymbol{P}_b)' \boldsymbol{V}(\boldsymbol{x})\right\} = (a-1)(b-1)\left(\sigma_{\alpha\beta}^2 + m^{-1}\sigma^2\right), \tag{12.32}$$

$$\begin{aligned} V(S_{\alpha\beta}) &= 2\text{tr}\left\{(\boldsymbol{P}_a \otimes \boldsymbol{P}_b)(\boldsymbol{P}_a \otimes \boldsymbol{P}_b)' \boldsymbol{V}(\boldsymbol{x})\right\}^2 \\ &= 2(a-1)(b-1)\left(\sigma_{\alpha\beta}^2 + m^{-1}\sigma^2\right)^2 + f_{\alpha\beta}\left(m_{ij}^{-1} - m^{-1}\right)\sigma^4 \end{aligned} \tag{12.33}$$

The functions f_α, f_β, $f_{\alpha\beta}$ in these equations have been defined in Section 10.5.3.

To test

$$H_{\alpha\beta} : \sigma_{\alpha\beta}^2 = 0 \text{ against } K_{\alpha\beta} : \sigma_{\alpha\beta}^2 > 0$$

at significance level α, an approximate rejection region $R_{\alpha\beta}$,

$$R_{\alpha\beta} : \frac{mS_{\alpha\beta}/\{(a-1)(b-1)\}}{S_e/(n-ab)} > F_{(a-1)(b-1),n-a}(\alpha),$$

is suggested by (12.32). The characteristics of this rejection region are exactly the same as in the fixed effects model in Section 10.5.3. Under the alternative hypothesis $K_{\alpha\beta}$, the power is very close to the balanced case of Section 12.2.3 with hypothetical repetition number m, since the positive $\sigma_{\alpha\beta}^2$ reduces the effects of unbalanced m_{ij}.

To test

$$H_\alpha : \sigma_\alpha^2 = 0 \text{ against } K_\alpha : \sigma_\alpha^2 > 0$$

the procedure is changed according to $\sigma_{\alpha\beta}^2 = 0$ or $\sigma_{\alpha\beta}^2 \neq 0$ by the form of (12.28). For the respective cases, the following rejection regions are suggested:

$$R_\alpha:\ \frac{mS_\alpha/(a-1)}{S_e/(n-ab)} > F_{a-1,n-a}(\alpha)\ \text{if}\ \sigma_{\alpha\beta}^2 = 0$$

$$R'_\alpha:\ \frac{mS_\alpha/(a-1)}{S_{\alpha\beta}/\{(a-1)(b-1)\}} > F_{a-1,(a-1)(b-1)}(\alpha)\ \text{if}\ \sigma_{\alpha\beta}^2 \neq 0$$

The characteristics of the rejection region R_α under the null hypothesis H_α are exactly the same as in the fixed effects model in Section 10.5.3. When $\sigma_{\alpha\beta}^2 = 0$, $S_{\alpha\beta}$ behaves just like S_e but usually it is not pooled into S_e. The test statistic of R'_α is more complicated, since the denominator does not follow the chi-squared distribution. Nevertheless, the size of the rejection region R'_α is very close to the nominal value α, since the positive $\sigma_{\alpha\beta}^2$ reduces again the effects of unbalanced m_{ij}. This tendency becomes more prominent under the alternative hypothesis K_α. Testing $H_\beta: \sigma_\beta^2 = 0$ is discussed similarly to testing H_α. One should refer to Hirotsu (1968) for the details of these properties.

Next we proceed to estimate the variance components by the moment method. We already have the expectation and variance of the sum of squares as in (12.28) ~ (12.33). Then, the unbiased estimators of the variance components are obtained as

$$\hat{\sigma}_\alpha^2 = b^{-1}\left\{\frac{S_\alpha}{a-1} - \frac{S_{\alpha\beta}}{(a-1)(b-1)}\right\},$$

$$\hat{\sigma}_\beta^2 = a^{-1}\left\{\frac{S_\beta}{b-1} - \frac{S_{\alpha\beta}}{(a-1)(b-1)}\right\},$$

$$\hat{\sigma}_{\alpha\beta}^2 = \left\{\frac{S_{\alpha\beta}}{(a-1)(b-1)} - \frac{\hat{\sigma}^2}{m}\right\},$$

$$\hat{\sigma}^2 = S_e/(n-ab),\quad V(\hat{\sigma}^2) = \frac{2\sigma^4}{n-ab}.$$

To calculate the variance of these estimators other than $\hat{\sigma}^2$, we need covariances among the sums of squares in addition to variances. The calculations are tedious but not difficult by expressing the sums of squares in matrix form in $\boldsymbol{x}$. We show only the result in the following (see Hirotsu, 1966 for details):

$$V(\hat{\sigma}_\alpha^2) = \frac{2}{b^2}\left[\frac{b}{a-1}\sigma_\alpha^4 + \frac{b(b-1)}{a-1}\left(\sigma_\alpha^2 + \frac{\sigma_{\alpha\beta}^2}{b-1}\right)^2 + \frac{2b}{m(a-1)}\sigma^2\left(\sigma_\alpha^2 + \frac{\sigma_{\alpha\beta}^2}{b-1}\right)\right.$$
$$\left. + \left\{\frac{b^2}{m^2} - \sum_j\left(\frac{1}{a}\sum_i\frac{1}{m_{ij}}\right)^2 + \frac{a-2}{a}\left(b^2\sum_i\left(\frac{1}{b}\sum_j\frac{1}{m_{ij}}\right)^2 - \sum_i\sum_j\frac{1}{m_{ij}^2}\right)\right\}\left\{\frac{\sigma^2}{(a-1)(b-1)}\right\}^2\right],$$

Table 12.3 Repetition number.

		*D*1 Column			*D*2 Column			*D*3 Column			*D*4 Column			*D*5 Column		
		1	2	3	1	2	3	1	2	3	1	2	3	1	2	3
	1	1	1	1	1	2	3	1	1	1	1	1	4	2	2	2
Row	2	2	2	2	2	3	1	1	2	3	1	4	1	2	2	2
	3	3	3	3	3	1	2	1	3	5	4	1	1	2	2	2

$$V\left(\hat{\sigma}_\beta^2\right) = \frac{2}{a^2}\left[\frac{a}{b-1}\sigma_\beta^4 + \frac{a(a-1)}{b-1}\left(\sigma_\beta^2 + \frac{\sigma_{\alpha\beta}^2}{a-1}\right)^2 + \frac{2a}{m(b-1)}\sigma^2\left(\sigma_\beta^2 + \frac{\sigma_{\alpha\beta}^2}{a-1}\right)\right.$$

$$\left. + \left\{\frac{a^2}{m^2} - \sum_i\left(\frac{1}{b}\sum_j \frac{1}{m_{ij}}\right)^2 + \frac{b-2}{b}\left(a^2\sum_j\left(\frac{1}{a}\sum_i\frac{1}{m_{ij}}\right)^2 - \sum_i\sum_j\frac{1}{m_{ij}^2}\right)\right\}\left\{\frac{\sigma^2}{(a-1)(b-1)}\right\}^2\right]$$

$$V\left(\hat{\sigma}_{\alpha\beta}^2\right) = \frac{2}{(a-1)(b-1)}\left[\sigma_{\alpha\beta}^4 + \frac{2}{m}\sigma_{\alpha\beta}^2\sigma^2 + \right.$$

$$\left.\left\{\frac{(a-2)(b-2)\frac{1}{ab}\sum_i\sum_j\frac{1}{m_{ij}^2} + \frac{a-2}{a}\sum_i\left(\frac{1}{b}\sum_j\frac{1}{m_{ij}}\right)^2 + \frac{b-2}{b}\sum_j\left(\frac{1}{a}\sum_i\frac{1}{m_{ij}}\right)^2 + \frac{1}{m^2}}{(a-1)(b-1)}\right\}\sigma^4\right]$$

$$+\frac{2\sigma^4}{m^2(n-ab)}.$$

These formulae coincide with the formulae in Section 12.2.2 if the m_{ij} are all equal.

Mostafa (1967) proposed purposely unbalanced two-way designs with repetition numbers one or two and without an empty cell to reduce the experimental cost. For the analysis he used the same procedure as proposed by Hirotsu (1966).

Bush and Anderson (1963) evaluated the variances of estimation by the methods G_1 and G_2 of Yates (1934) and G_3 of Henderson (1953) in various unbalanced two-way designs. We compare our easy method with their results in Table 12.4 for those designs given in Table 12.3. We evaluated our method also by the relative variance of Bush and Anderson, which they introduced to define the relative efficiency of various designs to the balanced 6×6 design. The combinations of $\left(\sigma_{\alpha\beta}^2, \sigma_\alpha^2, \sigma_\beta^2, \sigma^2\right)$ are also the same as considered by Bush and Anderson (1963). The easy method is seen to give uniformly minimum variance among compared methods, except for the case of estimating $\sigma_{\alpha\beta}^2$ when $\sigma_{\alpha\beta}^2$ is small. This case is, however, not so important. The easy method for a random effects model is more attractive than the fixed effects model, since the minimal sufficient statistics are not complete and there is no uniformly optimum procedure.

Table 12.4 Comparing the relative variance of several estimators.

$\sigma^2_{\alpha\beta}$	$\sigma^2_{\alpha\beta}$	$\sigma^2_{\alpha\beta}$	σ^2	Design	G_3	G_1	G_2	H	G_3	G_1	G_2	H	G_3	G_1	G_2	H
				D1	4.88	4.88	4.88	4.45	5.54	5.54	4.58	4.45	3.26	3.26	3.26	2.99
				D2	6.12	5.15	4.53	4.45	6.12	5.15	4.53	4.45	3.50	3.15	3.15	2.99
1	1	16	1	D3	6.56	6.25	4.84	4.50	6.56	6.25	4.84	4.50	3.70	3.59	3.59	3.02
				D4	9.42	5.24	4.58	4.52	9.42	5.24	4.58	4.52	4.23	3.20	3.20	3.04
				D5	4.40	4.40	4.40	4.40	4.40	4.40	4.40	4.40	2.94	2.94	2.94	2.94
				D1	1.83	1.83	1.83	1.75	1.80	1.80	1.73	1.74	2.04	2.04	2.04	2.18
				D2	2.23	1.78	1.73	1.74	2.23	1.78	1.73	1.74	3.37	2.10	2.10	2.17
1	1	1	1	D3	3.13	2.01	1.85	1.82	2.13	2.01	1.85	1.82	2.65	2.30	2.30	2.44
				D4	3.27	1.90	1.85	1.85	3.27	1.90	1.85	1.85	6.27	2.50	2.50	2.61
				D5	1.65	1.65	1.65	1.65	1.65	1.65	1.65	1.65	1.84	1.84	1.84	1.84
				D1	1.50	1.50	1.50	1.36	6.49	6.49	5.42	5.31	2.83	2.83	2.83	2.68
				D2	1.40	1.38	1.36	1.36	20.7	6.08	5.37	5.31	5.28	2.78	2.78	2.68
16	0	4	1	D3	1.54	1.55	1.42	1.37	11.7	7.30	5.83	5.54	3.85	3.11	3.11	2.80
				D4	1.51	1.39	1.36	1.37	47.1	6.44	5.68	5.63	9.74	2.93	2.93	2.86
				D5	1.35	1.35	1.35	1.35	5.07	5.07	5.07	5.07	2.55	2.55	2.55	2.55
				D1	1.40	1.40	1.40	1.26	1.33	1.33	1.34	1.35	1.55	1.55	1.55	1.90
				D2	1.32	1.26	1.26	1.26	4.55	1.34	1.34	1.35	52.9	1.67	1.67	1.86
16	2	$\frac{1}{4}$	1	D3	1.41	1.41	1.32	1.27	2.35	1.48	1.42	1.39	16.1	1.81	1.81	2.25
				D4	1.41	1.27	1.27	1.27	9.79	1.39	1.40	1.40	1.38	2.28	2.28	2.56
				D5	1.26	1.26	1.26	1.26	1.30	1.30	1.30	1.30	1.35	1.35	1.35	1.35

12.3 Two-Way Mixed Effects Model

12.3.1 Model and parameters

In the problem of the previous section, let the b machines represent a particular type and their efficiencies be of concern. Then we have a mixed effects model with the machines fixed and the workers random. This situation is most controversial, and various models are proposed, among which we follow the most general model of Scheffé (1959) in this section (see Miller, 1998 for other models). The related topics are the recovery of inter-block information in the incomplete block designs (Chapter 9) and the profile analysis of repeated measurements (Chapter 13).

Similarly to the previous section, let the yield of the kth experiment by the ith worker on the jth machine be y_{ijk}, the expected yield of the ith worker on the jth machine be μ_{ij}, and an independent random error with expectation zero be e_{ijk}. Then we have a model

$$y_{ijk} = \mu_{ij} + e_{ijk},\ i = 1, \ldots, a, j = 1, \ldots, b, k = 1, \ldots, m. \tag{12.34}$$

Let us define a vector

$$\boldsymbol{\mu}_i = (\mu_{i1}, \ldots, \mu_{ib})',\ i = 1, \ldots, a, \tag{12.35}$$

which is regarded as a random sample from the worker population. The expectation and variance–covariance matrix of $\boldsymbol{\mu}_i$ are denoted by

$$\boldsymbol{\mu} = (\mu_1, \ldots, \mu_b)' \text{ and } \boldsymbol{\Omega}_{b\times b} = \{\sigma_{jj'}\}_{b\times b}, \tag{12.36}$$

respectively, where $\boldsymbol{\mu}$ and $\boldsymbol{\Omega}$ are fixed parameters. A general mean is defined by

$$\mu = b^{-1}\textstyle\sum_{j=1}^{b}\mu_j.$$

Then the main effect of the jth machine is defined by

$$\beta_j = \mu_j - \mu.$$

In contrast, the averaged yield of worker i on the b machines is defined by

$$\bar{\mu}_{i\cdot} = b^{-1}\textstyle\sum_{j=1}^{b}\mu_{ij},$$

and the main effect is defined by

$$\alpha_i = \bar{\mu}_{i\cdot} - \mu. \tag{12.37}$$

The effect of worker i on machine j is

$$\mu_{ij} - \mu_j \tag{12.38}$$

and α_i is the average of (12.38). The excess of equation (12.38) over (12.37) is called an interaction, and is expressed by

$$(\alpha\beta)_{ij} = \mu_{ij} - \mu_j - \bar{\mu}_{i.} + \mu.$$

Thus we have a model

$$\mu_{ij} = \mu + \alpha_i + \beta_j + (\alpha\beta)_{ij},\ i = 1, \ldots, a, j = 1, \ldots, b,$$

where α_i and $(\alpha\beta)_{i1}, \ldots, (\alpha\beta)_{ib}$ are random variables not independent of each other. Their covariance structure is determined by $\boldsymbol{\Omega}$ (12.36). In the matrix expression

$$\alpha_i = b^{-1}\boldsymbol{j}'(\boldsymbol{\mu}_i - \boldsymbol{\mu}) = b^{-1}\boldsymbol{j}'\boldsymbol{\mu}_i - \mu,$$

$$(\boldsymbol{\alpha\beta})_i = \begin{bmatrix} (\alpha\beta)_{i1} \\ \vdots \\ (\alpha\beta)_{ib} \end{bmatrix} = (\boldsymbol{I} - b^{-1}\boldsymbol{jj}')\boldsymbol{\mu}_i - \begin{bmatrix} \mu_1 \\ \vdots \\ \mu_b \end{bmatrix} + \mu\boldsymbol{j},$$

and we have

$$V\left(\begin{bmatrix} \alpha_i \\ (\boldsymbol{\alpha\beta})_i \end{bmatrix}\right) = \begin{bmatrix} b^{-1} & b^{-1} & \cdots & b^{-1} \\ 1-b^{-1} & -b^{-1} & \cdots & -b^{-1} \\ & \cdots\cdots\cdots & & \\ -b^{-1} & -b^{-1} & \cdots & 1-b^{-1} \end{bmatrix} \boldsymbol{\Omega} \begin{bmatrix} b^{-1} & b^{-1} & \cdots & b^{-1} \\ 1-b^{-1} & -b^{-1} & \cdots & -b^{-1} \\ & \cdots\cdots\cdots & & \\ -b^{-1} & -b^{-1} & \cdots & 1-b^{-1} \end{bmatrix}'.$$

Therefore we obtain

$$V(\alpha_i) = b^{-2}\textstyle\sum_j \sum_{j'} \sigma_{jj'} = \bar{\sigma}_{..},$$

$$Cov\left\{(\alpha\beta)_{ij}, (\alpha\beta)_{ij'}\right\} = \sigma_{jj'} - \bar{\sigma}_{j.} - \bar{\sigma}_{.j'} + \bar{\sigma}_{..},$$

$$Cov\left\{\alpha_i, (\alpha\beta)_{ij}\right\} = \bar{\sigma}_{j.} - \bar{\sigma}_{..}.$$

Assuming the normal distribution $N(\boldsymbol{\mu}, \boldsymbol{\Omega})$ for $\boldsymbol{\mu}_i$, we finally have a model

$$\begin{cases} \boldsymbol{y} = \mu\boldsymbol{j} + \boldsymbol{X}_\alpha\boldsymbol{\alpha} + \boldsymbol{X}_\beta\boldsymbol{\beta} + \boldsymbol{X}_{\alpha\beta}(\boldsymbol{\alpha\beta}) + \boldsymbol{e} \\ \boldsymbol{\alpha} \sim N(\boldsymbol{0}, \bar{\sigma}_{..}\boldsymbol{I}) \\ (\boldsymbol{\alpha\beta})_i \sim N\left\{\boldsymbol{0}, \left(\sigma_{jj'} - \bar{\sigma}_{j.} - \bar{\sigma}_{.j'} + \bar{\sigma}_{..}\right)_{b\times b}\right\}, \\ \boldsymbol{e} \sim N(\boldsymbol{0}, \sigma^2\boldsymbol{I}_n), \end{cases} \tag{12.39}$$

where the interactions $(\boldsymbol{\alpha\beta})_i$ are mutually independent for different i's, the random error $\boldsymbol{e}$ is independent of other variables $\boldsymbol{\alpha}$ and $(\boldsymbol{\alpha\beta})_i$, and column vectors $\boldsymbol{\alpha}$, $\boldsymbol{\beta}$, and $(\boldsymbol{\alpha\beta})$ are defined as usual. From (12.39) we have

$$
\begin{aligned}
E(y) &= \mu j + X_\beta \beta \\
V(y) &= \bar{\sigma}_{..} X_\alpha X'_\alpha + X_\alpha \{ I_a \otimes (\bar{\sigma}_{1\cdot} - \bar{\sigma}_{..}, \ldots, \bar{\sigma}_{b\cdot} - \bar{\sigma}_{..}) \} X'_{\alpha\beta} \\
&\quad + X_{\alpha\beta} \{ I_a \otimes (\bar{\sigma}_{1\cdot} - \bar{\sigma}_{..}, \ldots, \bar{\sigma}_{b\cdot} - \bar{\sigma}_{..})' \} X'_\alpha \\
&\quad + X_{\alpha\beta} \left\{ I_a \otimes \left(\sigma_{jj'} - \bar{\sigma}_{j\cdot} - \bar{\sigma}_{\cdot j'} + \bar{\sigma}_{..} \right)_{b \times b} \right\} X'_{\alpha\beta} + \sigma^2 I.
\end{aligned}
$$

12.3.2 Standard form for test and estimation

We apply the orthonormal transformation M' of the previous section again to obtain

$$
M'y = z = \left(z_\mu, z'_\alpha, z'_\beta, z'_{\alpha\beta}, z'_e \right)'.
$$

It is easy to see the expectation

$$
E(z) = \begin{bmatrix} \sqrt{n}\mu \\ \mathbf{0} \\ \sqrt{am} P'_b \beta \\ \mathbf{0} \\ \mathbf{0} \end{bmatrix} \tag{12.40}
$$

and after some calculation, the variance is obtained as

$$
V(z) = \begin{bmatrix}
\sigma^2 + bm\bar{\sigma}_{..} & \mathbf{0}' & \sqrt{bm}\,\sigma' P_b & \mathbf{0}' & \mathbf{0}' \\
\mathbf{0} & (\sigma^2 + bm\bar{\sigma}_{..}) I_{a-1} & \mathbf{0} & \sqrt{bm}(I_{a-1} \otimes \sigma' P_b) & \mathbf{0} \\
\sqrt{bm}\, P'_b \sigma & \mathbf{0} & \sigma^2 I_{b-1} + m P'_b \Sigma_{b \times b} P_b & \mathbf{0} & \mathbf{0} \\
\mathbf{0} & \sqrt{bm}\left(I_{a-1} \otimes P'_b \sigma \right) & \mathbf{0} & \begin{matrix} \sigma^2 I_{(a-1)(b-1)} + \\ m\left(I_{a-1} \otimes P'_b \Sigma_{b \times b} P_b \right) \end{matrix} & \mathbf{0} \\
\mathbf{0} & \mathbf{0} & \mathbf{0} & \mathbf{0} & \sigma^2 I_{n-ab}
\end{bmatrix}, \tag{12.41}
$$

where $\sigma = (\bar{\sigma}_{1\cdot}, \ldots, \bar{\sigma}_{b\cdot})'$ and $\Sigma_{b \times b} = \left(\sigma_{jj'} - \bar{\sigma}_{j\cdot} - \bar{\sigma}_{\cdot j'} + \bar{\sigma}_{..} \right)_{b \times b}$.

Now the structure becomes clear, and we proceed to test the effects of factors. By the similarity of the structure of y in (12.20) and (12.39), the sums of squares are the same as in Section 12.2, although their distributions are different.

12.3.3 Null hypothesis $H_{\alpha\beta}$ of interaction and the test statistic

By the form of equation (12.41), the null hypothesis is expressed as

$$
H_{\alpha\beta} : P'_b \Sigma_{b \times b} P_b = \mathbf{0}.
$$

This is equivalent to the equality

$$\sigma_{jj'} = \bar{\sigma}_{j\cdot} + \bar{\sigma}_{\cdot j'} - \bar{\sigma}_{\cdot\cdot},\ j, j' = 1, \ldots, b,$$

or in matrix form

$$\boldsymbol{\Omega}_{b\times b} = \boldsymbol{j}\boldsymbol{\sigma}' + \boldsymbol{\sigma}\boldsymbol{j}' - \bar{\sigma}_{\cdot\cdot}\boldsymbol{j}\boldsymbol{j}' = (\boldsymbol{j}\ \boldsymbol{\sigma})\begin{bmatrix} -\bar{\sigma}_{\cdot\cdot} & 1 \\ 1 & 0 \end{bmatrix}(\boldsymbol{j}\ \boldsymbol{\sigma})'.$$

However, since $\boldsymbol{\Omega}_{b\times b}$ (12.36) is a positive non-negative definite matrix it must be in the form

$$H_{\alpha\beta} : \boldsymbol{\Omega}_{b\times b} = \bar{\sigma}_{\cdot\cdot}\boldsymbol{j}\boldsymbol{j}'. \tag{12.42}$$

Then $\boldsymbol{\mu}_i$ (12.35) is in the form

$$\boldsymbol{\mu}_i = \boldsymbol{\mu} + \alpha_i \boldsymbol{j},$$

where $\boldsymbol{\mu}$ is a fixed vector (12.36) and

$$\alpha_i \sim N(0, \bar{\sigma}_{\cdot\cdot})$$

represents the random part. Thus, the null hypothesis $H_{\alpha\beta}$ implies that the $\boldsymbol{\mu}_i$'s are distributed parallel around the mean vector $\boldsymbol{\mu}$ over the worker population. This model appears as the null model of the profile analysis in Chapter 13.

Under the null hypothesis $H_{\alpha\beta}$, the statistic $S_{\alpha\beta}$ is distributed as $\sigma^2\chi^2_{(a-1)(b-1)}$ and is independent of S_e, which is distributed as $\sigma^2\chi^2_{n-ab}$. When $H_{\alpha\beta}$ fails, $S_{\alpha\beta}$ is still independent of S_e but is not a non-central chi-squared distribution. Since $m\left(\boldsymbol{I}_{a-1}\otimes \boldsymbol{P}_b'\boldsymbol{\Sigma}_{b\times b}\boldsymbol{P}_b\right)$ is a positive matrix, it is statistically larger than under the null hypothesis, similarly to in the fixed effects model. The expectation is, for example,

$$\begin{aligned} \mathrm{E}\left(S_{\alpha\beta}\right) &= \mathrm{tr}\left\{\sigma^2\boldsymbol{I}_{(a-1)(b-1)} + m\left(\boldsymbol{I}\otimes \boldsymbol{P}_b'\boldsymbol{\Sigma}_{b\times b}\boldsymbol{P}_b\right)\right\} \\ &= (a-1)(b-1)\sigma^2 + m(a-1)\textstyle\sum_j\left(\sigma_{jj} - \bar{\sigma}_{\cdot\cdot}\right), \end{aligned}$$

and the second term is positive by definition. Thus,

$$F = \frac{S_{\alpha\beta}/\{(a-1)(b-1)\}}{S_e/(n-ab)}$$

of (10.16) in Section 10.2 is appropriate also as a test statistic in this situation. However, the row- and/or column-wise multiple comparisons would be more useful than an overall F-test. The row-wise multiple comparisons intend to make a group of workers who have similar response pattern to the types of machine. The purpose of the column-wise multiple comparisons will be to select a group of machines of high efficiency or robust against the variation factor. Anyway, the distribution theory under the null hypothesis is the same as the fixed effects model, and the methods of row-wise multiple comparisons in Chapter 10 are applicable as they are. The real examples are given in Chapter 13 for the profile analysis of repeated measurements.

12.3.4 Testing main effects under the null hypothesis $H_{\alpha\beta}$

Under the null hypothesis $H_{\alpha\beta}$, the meaning of the test for main effects

$$H_\alpha : \bar{\sigma}_{..} = \mathbf{0} \qquad (12.43)$$
$$\text{and } \; H_\beta : \boldsymbol{P}_b'\boldsymbol{\mu} = \mathbf{0}$$

is clear. The null hypothesis H_α implies that the random part vanishes and H_β implies

$$H_\beta : \boldsymbol{\mu} = \mu \boldsymbol{j} \qquad (12.44)$$

as usual. Since the variance–covariance matrix under $H_{\alpha\beta}$ is

$$V(z) = \begin{bmatrix} \sigma^2 + bm\bar{\sigma}_{..} & \mathbf{0}' & \mathbf{0}' & \mathbf{0}' & \mathbf{0}' \\ \mathbf{0} & (\sigma^2 + bm\bar{\sigma}_{..})\boldsymbol{I}_{a-1} & \mathbf{0} & \mathbf{0} & \mathbf{0} \\ \mathbf{0} & \mathbf{0} & \sigma^2\boldsymbol{I}_{b-1} & \mathbf{0} & \mathbf{0} \\ \mathbf{0} & \mathbf{0} & \mathbf{0} & \sigma^2\boldsymbol{I}_{(a-1)(b-1)} & \mathbf{0} \\ \mathbf{0} & \mathbf{0} & \mathbf{0} & \mathbf{0} & \sigma^2\boldsymbol{I}_{n-ab} \end{bmatrix}, \qquad (12.45)$$

the F-tests are constructed as usual:

$$R_\alpha : \frac{S_\alpha/(a-1)}{S_e/(n-ab)} > F_{a-1,\, n-ab}(\alpha),$$
$$R_\beta : \frac{S_\beta/(b-1)}{S_e/(n-ab)} > F_{b-1,\, n-ab}(\alpha)$$

12.3.5 Testing main effects H_β when the null hypothesis $H_{\alpha\beta}$ fails

When the null hypothesis $H_{\alpha\beta}$ fails, the test of H_α makes no sense while the test of H_β (12.44) still makes sense. It is testing whether there are differences among the fixed parameter (machines) beyond the variation factor (interaction between worker and machine). However, the result should be combined with the column-wise multiple comparisons of the interaction effects. Usually the rejection region

$$F_\beta = \frac{S_\beta/(b-1)}{S_{\alpha\beta}/\{(a-1)(b-1)\}} > F_{b-1,(a-1)(b-1)}(\alpha)$$

is employed, but F_β does not follow the F-distribution even under H_β, since the variance–covariance matrices of z_β and $z_{\alpha\beta}$ in (12.41) are not idempotent. However, the F-test is applied as a first approximation, since the statistics $S_\beta/(b-1)$ and $S_{\alpha\beta}/\{(a-1)(b-1)\}$ have the same expectation

$$\sigma^2+\frac{m}{b-1}\mathrm{tr}\left(\boldsymbol{P}_b'\boldsymbol{\Sigma}_{b\times b}\boldsymbol{P}_b\right)=\sigma^2+\frac{m}{b-1}\sum_j\left(\sigma_{jj}-\bar{\sigma}_{..}\right)$$

under H_β. On the contrary, when $a \ge b$ an exact test is available based on Hotelling's T^2-statistic.

12.3.6 Exact test of H_β when the null hypothesis $H_{\alpha\beta}$ fails

Definition 12.1. Hotelling's T^2 statistic. Let $\boldsymbol{u}_i$, $i=1,\ldots q$, be distributed independently as a p-variate normal distribution $N(\boldsymbol{\eta}, \boldsymbol{\Xi})$ with $\boldsymbol{\Xi}$ a positive definite matrix, where $q \ge p+1$. Define the usual unbiased estimator of $\boldsymbol{\eta}$ and $\boldsymbol{\Xi}$ by

$$\bar{\boldsymbol{u}}.=\frac{1}{q}\sum_{i=1}^{q}\boldsymbol{u}_i \text{ and } \widehat{\boldsymbol{\Xi}}=\frac{1}{q-1}\sum_{i=1}^{q}(\boldsymbol{u}_i-\bar{\boldsymbol{u}}.)(\boldsymbol{u}_i-\bar{\boldsymbol{u}}.)'.$$

Then

$$T^2=(\bar{\boldsymbol{u}}.-\boldsymbol{\eta})'\left(\frac{1}{q}\widehat{\boldsymbol{\Xi}}\right)^{-1}(\bar{\boldsymbol{u}}.-\boldsymbol{\eta}) \tag{12.46}$$

is called Hotelling's T^2-statistic. The statistic T^2 (12.46) is distributed as a constant times the F-distribution:

$$\frac{(q-1)p}{q-p}F_{p,\,q-p}.$$

When $p=1$, this coincides with the square of the t-statistic. If $\widehat{\boldsymbol{\Xi}}$ is replaced by $\boldsymbol{\Xi}$ in (12.46), the statistic follows a chi-squared distribution with df p. If we discard $\boldsymbol{\eta}$ in (12.46),

$$T'^2=\bar{\boldsymbol{u}}.'\left(\frac{1}{q}\widehat{\boldsymbol{\Xi}}\right)^{-1}\bar{\boldsymbol{u}}.$$

is distributed as a constant times the non-central F-distribution

$$\frac{(q-1)p}{q-p}F_{p,\,q-p,\,\delta}$$

with non-centrality parameter $\delta=q\boldsymbol{\eta}'\boldsymbol{\Xi}^{-1}\boldsymbol{\eta}$.

Now, coming back to the problem of testing H_β, let us define

$$\boldsymbol{u}_i=m^{-1/2}\boldsymbol{P}_b'(y_{i1\cdot},\ldots,y_{ib\cdot})',\ i=1,\ldots,a, \tag{12.47}$$

where $p=b-1$ and $q=a$. Putting model (12.19) into (12.47), we have

$$\boldsymbol{u}_i = m^{1/2}\boldsymbol{P}_b'\left(\boldsymbol{\mu}_i + \begin{bmatrix} \bar{e}_{i1\cdot} \\ \vdots \\ \bar{e}_{ib\cdot} \end{bmatrix}\right),\ i=1,\ldots,a,$$

which implies that the statistics $\boldsymbol{u}_i$, $i=1,\ldots,a$, are distributed independently as

$$N\left(m^{1/2}\boldsymbol{P}_b'\boldsymbol{\mu}, \sigma^2\boldsymbol{I}_{b-1} + m\boldsymbol{P}_b'\boldsymbol{\Omega}\boldsymbol{P}_b\right).$$

That is, if we estimate $\boldsymbol{\Xi}$ by

$$\widehat{\boldsymbol{\Xi}} = \left(\overline{\sigma^2\boldsymbol{I}_{b-1} + m\boldsymbol{P}_b'\boldsymbol{\Omega}\boldsymbol{P}_b}\right) = \frac{1}{a-1}\sum\nolimits_{i=1}^{q}(\boldsymbol{u}_i - \bar{\boldsymbol{u}}_\cdot)(\boldsymbol{u}_i - \bar{\boldsymbol{u}}_\cdot)',$$

then

$$T'^2 = a\bar{\boldsymbol{u}}_\cdot'\widehat{\boldsymbol{\Xi}}^{-1}\bar{\boldsymbol{u}}_\cdot$$

is distributed as the non-central F-distribution

$$\frac{(a-1)(b-1)}{a-(b-1)}F_{b-1,\,a-b+1,\delta},\ \delta = am\left(\boldsymbol{P}_b'\boldsymbol{\mu}\right)'\left(\sigma^2\boldsymbol{I}_{b-1} + m\boldsymbol{P}_b'\boldsymbol{\Omega}\boldsymbol{P}_b\right)^{-1}\boldsymbol{P}_b'\boldsymbol{\mu}.$$

Therefore, for the null hypothesis $H_\beta : \boldsymbol{\mu} = \mu\boldsymbol{j}$ we have an exact rejection region with level α as

$$R:\ \frac{a-b+1}{(a-1)(b-1)}T'^2 > F_{b-1,\,a-b+1}(\alpha).$$

To calculate T'^2, we note that it is rewritten in a form free from $\boldsymbol{P}_b$ as

$$T'^2 = am\begin{bmatrix} \bar{y}_{\cdot1\cdot} - \bar{y}_{\cdots} \\ \vdots \\ \bar{y}_{\cdot b\cdot} - \bar{y}_{\cdots} \end{bmatrix}' \times$$

$$\left[\frac{1}{b}\boldsymbol{j}_b\boldsymbol{j}_b' + \frac{m}{a-1}\sum\nolimits_{i=1}^{a}\left\{\left(\bar{y}_{ij\cdot} - \bar{y}_{i\cdot\cdot} - \bar{y}_{\cdot j\cdot} + \bar{y}_{\cdots}\right)\times\left(\bar{y}_{ij'\cdot} - \bar{y}_{i\cdot\cdot} - \bar{y}_{\cdot j'\cdot} + \bar{y}_{\cdots}\right)\right\}_{b\times b}\right]^{-1} \times \begin{bmatrix} \bar{y}_{\cdot1\cdot} - \bar{y}_{\cdots} \\ \vdots \\ \bar{y}_{\cdot b\cdot} - \bar{y}_{\cdots} \end{bmatrix} \tag{12.48}$$

(see Hirotsu, 1992 for details). The inversion of this type of matrix in (12.48) can be avoided by the following lemma.

Lemma 12.1. For a non-singular matrix $\boldsymbol{A}$, an equation

$$\boldsymbol{a}'\boldsymbol{A}^{-1}\boldsymbol{a} = \frac{|\boldsymbol{A} + \boldsymbol{a}\boldsymbol{a}'|}{|\boldsymbol{A}|} - 1$$

holds.

Proof. It is easy to see that

$$\begin{vmatrix} 1 & \boldsymbol{a}' \\ -\boldsymbol{a} & \boldsymbol{A} \end{vmatrix} = \begin{vmatrix} 1 & \boldsymbol{a}' \\ -\boldsymbol{a} & \boldsymbol{A} + \boldsymbol{a}\boldsymbol{a}' \end{vmatrix} = |\boldsymbol{A} + \boldsymbol{a}\boldsymbol{a}'|.$$

Then

$$\begin{vmatrix} 1 & \boldsymbol{a}' \\ -\boldsymbol{a} & \boldsymbol{A} \end{vmatrix} = \begin{vmatrix} 1 & -\boldsymbol{a}'\boldsymbol{A}^{-1} \\ \boldsymbol{0} & \boldsymbol{I} \end{vmatrix} \times \begin{vmatrix} 1 & \boldsymbol{a}' \\ -\boldsymbol{a} & \boldsymbol{A} \end{vmatrix} = \begin{vmatrix} 1 + \boldsymbol{a}'\boldsymbol{A}^{-1}\boldsymbol{a} & \boldsymbol{0}' \\ -\boldsymbol{a} & \boldsymbol{A} \end{vmatrix} = \left(1 + \boldsymbol{a}'\boldsymbol{A}^{-1}\boldsymbol{a}\right)|\boldsymbol{A}|.$$

By Lemma 12.1, T'^2 is calculated as

$$\frac{\left| \frac{1}{b}\boldsymbol{j}_b\boldsymbol{j}_b' + \frac{m}{a-1}\left\{ \sum_{i=1}^{a} \left(\bar{y}_{ij\cdot} - \bar{y}_{i\cdot\cdot} - \bar{y}_{\cdot j\cdot} + \bar{y}_{\cdots}\right) \times \left(\bar{y}_{ij'\cdot} - \bar{y}_{i\cdot\cdot} - \bar{y}_{\cdot j'\cdot} + \bar{y}_{\cdots}\right) \right\}_{b\times b} + \begin{bmatrix} \bar{y}_{\cdot 1\cdot} - \bar{y}_{\cdots} \\ \vdots \\ \bar{y}_{\cdot b\cdot} - \bar{y}_{\cdots} \end{bmatrix} \begin{bmatrix} \bar{y}_{\cdot 1\cdot} - \bar{y}_{\cdots} \\ \vdots \\ \bar{y}_{\cdot b\cdot} - \bar{y}_{\cdots} \end{bmatrix}' \right|}{\left| \frac{1}{b}\boldsymbol{j}_b\boldsymbol{j}_b' + \frac{m}{a-1} \sum_{i=1}^{a} \begin{bmatrix} \bar{y}_{i1\cdot} - \bar{y}_{i\cdot\cdot} - \bar{y}_{\cdot 1\cdot} + \bar{y}_{\cdots} \\ \vdots \\ \bar{y}_{ib\cdot} - \bar{y}_{i\cdot\cdot} - \bar{y}_{\cdot b\cdot} + \bar{y}_{\cdots} \end{bmatrix} \begin{bmatrix} \bar{y}_{i1\cdot} - \bar{y}_{i\cdot\cdot} - \bar{y}_{\cdot 1\cdot} + \bar{y}_{\cdots} \\ \vdots \\ \bar{y}_{ib\cdot} - \bar{y}_{i\cdot\cdot} - \bar{y}_{\cdot b\cdot} + \bar{y}_{\cdots} \end{bmatrix}' \right|} - am.$$

The estimation of β_j when the interaction is absent is performed in the same way as for a fixed effects model, because of the form of the expectation (12.40) and variance (12.45). On the contrary, if the interaction exists then the row- and/or column-wise multiple comparisons would be more useful than estimating the covariance matrix $\boldsymbol{\Omega}_{b\times b}$.

Finally, the similarity of the model here and the one-way multivariate normal model should be noted. The model (12.34) can be rewritten in matrix form as

$$\boldsymbol{y}_{ik} = \boldsymbol{\mu}_i + \boldsymbol{e}_{ik},$$

where $\boldsymbol{y}_{ik}$ is a b-dimensional random vector. In the multivariate normal model $\boldsymbol{\mu}_i$ is assumed to be distributed as $N(\boldsymbol{\mu}, \boldsymbol{\Omega})$ independently of $\boldsymbol{e}_{ik}$. In contrast, $\boldsymbol{e}_{ik}$ is assumed to be distributed as $N(\boldsymbol{0}, \boldsymbol{\Sigma}_e)$, whereas we assumed $\boldsymbol{\Sigma}_e$ to be $\sigma^2\boldsymbol{I}_b$ in this section. While the likelihood ratio test is usually applied for testing $\boldsymbol{\Omega} = \boldsymbol{0}$ in the multivariate normal model, we partitioned it into two steps, $H_{\alpha\beta}$: $\boldsymbol{\Omega}_{b\times b} = \bar{\sigma}_{\cdot\cdot}\boldsymbol{j}\boldsymbol{j}'$ (12.42) and H_α: $\bar{\sigma}_{\cdot\cdot} = 0$ (12.43), and showed the multiple comparison approach to $H_{\alpha\beta}$. The homogeneity of the components of $\boldsymbol{\mu}$ is also discussed as the hypothesis H_β (12.44). One may consider testing $H_{\alpha\beta}$ and H_α assuming a general error variance $\boldsymbol{\Sigma}_e$, but the distribution theory becomes very complicated.

12.4 General Linear Mixed Effects Model

The random effects models which have been introduced in Sections 12.1 and 12.2 can be classified also as mixed effects models, since they have a general mean as a fixed effect. In this section they are formulated as a general Gaussian linear mixed effects model.

12.4.1 Gaussian linear mixed effects model

The Gaussian linear mixed effects model is defined by

$$\boldsymbol{y}_n = \boldsymbol{X}\boldsymbol{\beta}_p + \boldsymbol{Z}\boldsymbol{U}_q + \boldsymbol{e}_n, \tag{12.49}$$

where $\boldsymbol{X}_{n\times p}$ and $\boldsymbol{Z}_{n\times q}$ are given design matrices, $\boldsymbol{\beta}$ is an unknown parameter vector of fixed effects, $\boldsymbol{U}$ a random vector, and $\boldsymbol{e}$ a random error vector. Random parts $\boldsymbol{U}$ and $\boldsymbol{e}$ are assumed to be distributed as $N(\boldsymbol{0}, \boldsymbol{\Psi})$ and $N(\boldsymbol{0}, \boldsymbol{\Sigma})$, respectively, and mutually independent. The covariance matrices $\boldsymbol{\Psi}$ and $\boldsymbol{\Sigma}$ depend also on some unknown parameters. Then the marginal distribution of $\boldsymbol{y}$ is normal,

$$\boldsymbol{y} \sim N(\boldsymbol{X}\boldsymbol{\beta}, \boldsymbol{\Sigma} + \boldsymbol{Z}\boldsymbol{\Psi}\boldsymbol{Z}'). \tag{12.50}$$

This model can be derived in two stages, where in the first stage $\boldsymbol{y}$ is modeled as a conditional distribution given $\boldsymbol{U}$:

$$f(\boldsymbol{y} \mid \boldsymbol{U} = \boldsymbol{u}) = \frac{1}{(2\pi)^{n/2}|\boldsymbol{\Sigma}|^{1/2}} \exp\left\{ -\frac{1}{2}(\boldsymbol{y} - \boldsymbol{X}\boldsymbol{\beta} - \boldsymbol{Z}\boldsymbol{u})'\boldsymbol{\Sigma}^{-1}(\boldsymbol{y} - \boldsymbol{X}\boldsymbol{\beta} - \boldsymbol{Z}\boldsymbol{u}) \right\}, \tag{12.51}$$

and in the second stage $\boldsymbol{U}$ is modeled by $N(\boldsymbol{0}, \boldsymbol{\Psi})$:

$$f(\boldsymbol{u}) = \frac{1}{(2\pi)^{q/2}|\boldsymbol{\Psi}|^{1/2}} \exp\left\{ -\frac{1}{2}\boldsymbol{u}'\boldsymbol{\Psi}^{-1}\boldsymbol{u} \right\} \tag{12.52}$$

thus leading to

$$\boldsymbol{y} \sim f(\boldsymbol{y} \mid \boldsymbol{U} = \boldsymbol{u}) \times f(\boldsymbol{u}). \tag{12.53}$$

This formulation is called a hierarchical model, and is useful for estimating random effects. Refer to Lee and Nelder (2001) for a general hierarchical model, including the linear mixed effects model here.

Example 12.2. One-way random effects model. We consider for simplicity a balanced model (12.1). It is in the form of (12.49), taking

$$\boldsymbol{X}\boldsymbol{\beta} = \mu\boldsymbol{j}_n, \boldsymbol{U} = \boldsymbol{\alpha}, \boldsymbol{Z} = \boldsymbol{X}_\alpha, \boldsymbol{\Sigma} = \sigma^2\boldsymbol{I}_n \text{ and } \boldsymbol{\Psi} = \sigma_\alpha^2\boldsymbol{I}_a. \tag{12.54}$$

The expectation and variance of $\boldsymbol{y}$ are $E(\boldsymbol{y}) = \mu\boldsymbol{j}_n$ and $V(\boldsymbol{y}) = \sigma_\alpha^2\boldsymbol{X}_\alpha\boldsymbol{X}_\alpha' + \sigma^2\boldsymbol{I}_n$, respectively, as given in (12.2) and (12.3).

Example 12.3. Ramus bone length from Elston and Grizzle (1962). The data of Table 12.5 are the length of the ramus bone for randomly selected boys aged 8 ~ 10 years, taken from Table 2 of Elston and Grizzle (1962), where full data for 20 boys and a detailed analysis by mixed effects model are given. The objective of the study was to establish a normal growth curve for use by orthodontists. For each boy, the bone length was measured four times at age 8, 8.5, 9, and 9.5 years. At the first stage we assume a linear regression model for each boy:

$$\boldsymbol{y}_i = \boldsymbol{X}_i \boldsymbol{\beta}_i + \boldsymbol{e}_i, i = 1, \dots, 5,$$

where

$$\boldsymbol{X}_i = \begin{bmatrix} 1 & 1 & 1 & 1 \\ 8 & 8.5 & 9 & 9.5 \end{bmatrix}', \boldsymbol{\beta}_i = (\beta_{i1}, \beta_{i2})'.$$

Since the boys are selected at random, however, we are interested in the average intercept and slope $\boldsymbol{\beta} = (\beta_1, \beta_2)'$ over the population and the dispersion of $\boldsymbol{\beta}_i$ around the population mean. Therefore, at the second stage $\boldsymbol{\beta}_i$ is modeled by a multivariate normal distribution $N(\boldsymbol{\beta}, \boldsymbol{\Psi})$. Then we have the mixed effects model (12.49) as

$$\boldsymbol{y} = \begin{bmatrix} \boldsymbol{y}_1 \\ \vdots \\ \boldsymbol{y}_5 \end{bmatrix} = (\boldsymbol{j} \otimes \boldsymbol{X})\boldsymbol{\beta} + (\boldsymbol{I} \otimes \boldsymbol{X}) \begin{bmatrix} \boldsymbol{b}_1 \\ \vdots \\ \boldsymbol{b}_5 \end{bmatrix} + \boldsymbol{e},$$

where we put the common design matrix $\boldsymbol{X}_i$ as $\boldsymbol{X}$ and $\boldsymbol{b}_i = \boldsymbol{\beta}_i - \boldsymbol{\beta}$. Assuming independence among the boys, the random part $\boldsymbol{U} = (\boldsymbol{b}_1', \dots, \boldsymbol{b}_5')'$ is modeled by $N\{\boldsymbol{0}, \boldsymbol{I} \otimes \boldsymbol{\Psi}\}$, with $\boldsymbol{\Psi}$ usually modeled by a diagonal matrix. In this case $\boldsymbol{\Sigma}$ may be assumed to be $\sigma^2 \boldsymbol{I}_{20}$. The detailed analysis of the data is given by Elston and Grizzle (1962). This type of modeling is often employed in the profile monitoring field of engineering (see Section 13.1.2).

Table 12.5 Length of the ramus bone (mm).

	Age			
Boy	8	8.5	9	9.5
1	45.1	45.3	46.1	47.2
2	52.5	53.2	53.3	53.7
3	45.0	47.0	47.3	48.3
4	51.2	51.4	51.6	51.9
5	47.2	47.7	48.4	49.5

12.4.2 Estimation of parameters

The parameters in model (12.50) are $\boldsymbol{\beta}$, $\boldsymbol{\Sigma}$ and $\boldsymbol{\Psi}$. The covariance matrices are usually modeled by fewer parameters like as in model (12.54), and denoted by $\boldsymbol{\theta}$ here. Therefore, the parameters to be estimated are $\boldsymbol{\beta}$ and $\boldsymbol{\theta}$. The log likelihood function of the marginal distribution is explicitly

$$\log L(\boldsymbol{\beta}, \boldsymbol{\theta}) = -\frac{1}{2}\log|\boldsymbol{V}(\boldsymbol{\theta})| - \frac{1}{2}(\boldsymbol{y}-\boldsymbol{X\beta})'\boldsymbol{V}^{-1}(\boldsymbol{\theta})(\boldsymbol{y}-\boldsymbol{X\beta}), \qquad (12.55)$$

where $V(\boldsymbol{\theta}) = \boldsymbol{\Sigma} + \boldsymbol{Z\Psi Z}'$. By the derivation of (12.55) with respect to parameter $\boldsymbol{\beta}$, we get an efficient score for $\boldsymbol{\beta}$:

$$\frac{\partial \log L(\boldsymbol{\beta}, \boldsymbol{\theta})}{\partial \boldsymbol{\beta}} = \boldsymbol{X}'\boldsymbol{V}^{-1}(\boldsymbol{y}-\boldsymbol{X\beta}),$$

which leads to the weighted least squares (WLS) estimator $\hat{\boldsymbol{\beta}}$:

$$\boldsymbol{X}'\boldsymbol{V}^{-1}\boldsymbol{X}\hat{\boldsymbol{\beta}} = \boldsymbol{X}'\boldsymbol{V}^{-1}\boldsymbol{y}, \qquad (12.56)$$

However, usually equation (12.56) includes an unknown parameter $\boldsymbol{\theta}$ for variance components. The profile log likelihood for parameter $\boldsymbol{\theta}$ is obtained by substituting $\hat{\boldsymbol{\beta}}$ in (12.55) as

$$\log L(\boldsymbol{\theta}) = -\frac{1}{2}\log|\boldsymbol{V}(\boldsymbol{\theta})| - \frac{1}{2}\left(\boldsymbol{y}-\boldsymbol{X}\hat{\boldsymbol{\beta}}\right)'\boldsymbol{V}^{-1}(\boldsymbol{\theta})\left(\boldsymbol{y}-\boldsymbol{X}\hat{\boldsymbol{\beta}}\right). \qquad (12.57)$$

However, the use of (12.57) suffers from a bias caused by the substitution of the estimate in $\boldsymbol{\beta}$. Therefore, the modified log likelihood

$$\log L^*(\boldsymbol{\theta}) = -\frac{1}{2}\log|\boldsymbol{V}(\boldsymbol{\theta})| - \frac{1}{2}\left(\boldsymbol{y}-\boldsymbol{X}\hat{\boldsymbol{\beta}}\right)'\boldsymbol{V}^{-1}(\boldsymbol{\theta})\left(\boldsymbol{y}-\boldsymbol{X}\hat{\boldsymbol{\beta}}\right) - \frac{1}{2}\log|\boldsymbol{X}'\boldsymbol{V}^{-1}(\boldsymbol{\theta})\boldsymbol{X}| \quad (12.58)$$

is proposed, and the derivation of (12.58) with respect to $\boldsymbol{\theta}$ is set equal to zero. When $\hat{\boldsymbol{\beta}}$ depends on an unknown parameter $\boldsymbol{\theta}$, the equations for $\hat{\boldsymbol{\beta}}$ (12.56) and $\hat{\boldsymbol{\theta}}$ need to be solved iteratively.

The modified log likelihood equals the residual maximum likelihood (REML) of Patterson and Thompson (1971). In the balanced case, REML gives the usual moment estimates of variance components. One should refer to Harville (1977) for a general review of the maximum likelihood approaches to variance component estimation, including REML.

Example 12.4. REML in the balanced one-way random effects model. In this case $E(\boldsymbol{y}) = \mu\boldsymbol{j}_n$ and $V(\boldsymbol{y}) = \sigma_\alpha^2\boldsymbol{X}_\alpha\boldsymbol{X}_\alpha' + \sigma^2\boldsymbol{I}_n$ is a block diagonal matrix, diag $(\boldsymbol{V}_i)$, with ith diagonal matrix

$$\boldsymbol{V}_i = \begin{bmatrix} \sigma^2+\sigma_\alpha^2 & \sigma_\alpha^2 & \cdots & \sigma_\alpha^2 \\ \sigma_\alpha^2 & \sigma^2+\sigma_\alpha^2 & \cdots & \sigma_\alpha^2 \\ & & \ddots & \\ \sigma_\alpha^2 & \sigma_\alpha^2 & \cdots & \sigma^2+\sigma_\alpha^2 \end{bmatrix}_{m\times m}.$$

The parameters to be estimated are therefore μ and $\boldsymbol{\theta}=\left(\sigma^2,\sigma_\alpha^2\right)'$. Now, V_i has an obvious eigenvalue $\sigma^2+m\sigma_\alpha^2$ and σ^2 of multiplicity $m-1$, with respective eigenvectors $\boldsymbol{j}_m$ and $\boldsymbol{P}_a$. Therefore, $\boldsymbol{j}_n$ is an eigenvector of $\boldsymbol{V}(\boldsymbol{y})$ and also $\boldsymbol{V}^{-1}(\boldsymbol{y})$ for eigenvalue $\left(\sigma^2+m\sigma_\alpha^2\right)$ and its inverse, respectively. This implies that the WLS estimator (12.56) is simply $\hat{\mu}=n^{-1}\boldsymbol{j}_n'\boldsymbol{y}=\bar{y}_{\cdot\cdot}$, which is free from unknown parameters σ^2 and σ_α^2. To calculate the terms of (12.58), we note that

$$V_i^{-1}=\frac{1}{\sigma^2}\boldsymbol{I}-\frac{\sigma_\alpha^2}{\sigma^2\left(\sigma^2+m\,\sigma_\alpha^2\right)}\boldsymbol{jj}'.$$

Then the second term is

$$-\frac{1}{2}\left\{\frac{1}{\sigma^2}\sum_{i=1}^{a}\sum_{j=1}^{m}\left(y_{ij}-\bar{y}_{\cdot\cdot}\right)^2-\frac{\sigma_\alpha^2}{\sigma^2\left(\sigma^2+m\,\sigma_\alpha^2\right)}\sum_{i=1}^{a}\left(y_{i\cdot}-m\bar{y}_{\cdot\cdot}\right)^2\right\}=-\frac{1}{2}\left\{\frac{S_T}{\sigma^2}-\frac{m\,\sigma_\alpha^2S_\alpha}{\sigma^2\left(\sigma^2+m\,\sigma_\alpha^2\right)}\right\}$$

$$=-\frac{1}{2}\left\{\frac{S_e}{\sigma^2}-\frac{S_\alpha}{\sigma^2+m\,\sigma_\alpha^2}\right\}$$

The first term is obviously

$$-\frac{1}{2}a\log\left\{\left(\sigma^2+m\sigma_\alpha^2\right)\left(\sigma^2\right)^{m-1}\right\},$$

and the last term is

$$-\frac{1}{2}\log\left\{\frac{n}{\sigma^2}-\frac{mn\,\sigma_\alpha^2}{\sigma^2\left(\sigma^2+m\,\sigma_\alpha^2\right)}\right\}=-\frac{1}{2}\log\left(\frac{n}{\sigma^2+m\,\sigma_\alpha^2}\right).$$

Therefore, the modified log likelihood (12.58) in this case is

$$\log L^*(\boldsymbol{\theta})=-\frac{1}{2}a\log\left\{\left(\sigma^2+m\sigma_\alpha^2\right)\left(\sigma^2\right)^{m-1}\right\}-\frac{1}{2}\left\{\frac{S_e}{\sigma^2}+\frac{S_\alpha}{\sigma^2+m\,\sigma_\alpha^2}\right\}-\frac{1}{2}\log\left(\frac{n}{\sigma^2+m\,\sigma_\alpha^2}\right)$$

and we have partial derivations

$$-2\frac{\partial\log L^*(\boldsymbol{\theta})}{\partial\sigma^2}=\frac{a-1}{\sigma^2+m\,\sigma_\alpha^2}+\frac{a(m-1)}{\sigma^2}-\frac{S_e}{\sigma^4}-\frac{S_\alpha}{\left(\sigma^2+m\,\sigma_\alpha^2\right)^2}, \tag{12.59}$$

$$-2\frac{\partial\log L^*(\boldsymbol{\theta})}{\partial\,\sigma_\alpha^2}=\frac{m(a-1)}{\sigma^2+m\,\sigma_\alpha^2}-\frac{mS_\alpha}{\left(\sigma^2+m\,\sigma_\alpha^2\right)^2}. \tag{12.60}$$

Equating (12.60) to zero, we have at once

$$\widehat{\sigma^2+m\sigma_\alpha^2}=\frac{S_\alpha}{a-1} \tag{12.61}$$

and substituting (12.61) in (12.59) we get

$$\widehat{\sigma}^2=\frac{S_e}{a(m-1)}.$$

These estimators are exactly the same as those obtained in Section 12.1 by a moment method.

12.4.3 Estimation of random effects (BLUP)

The random effects are not model parameters, but it is also of interest to evaluate $\boldsymbol{u}$. By the formulation (12.51) ~ (12.53) of the hierarchical model, we have a log likelihood

$$\begin{aligned}\log L(\boldsymbol{\beta},\boldsymbol{\theta},\boldsymbol{u}) &= \text{const.} - \frac{1}{2}\log|\boldsymbol{\Sigma}| - \frac{1}{2}(\boldsymbol{y}-\boldsymbol{X\beta}-\boldsymbol{Zu})'\boldsymbol{\Sigma}^{-1}(\boldsymbol{y}-\boldsymbol{X\beta}-\boldsymbol{Zu}) \\ &\quad - \frac{1}{2}\log|\boldsymbol{\Psi}(\boldsymbol{\theta})| - \frac{1}{2}\boldsymbol{u}'\boldsymbol{\Psi}^{-1}\boldsymbol{u}.\end{aligned} \tag{12.62}$$

Differentiating (12.62) by $\boldsymbol{\beta}$ and $\boldsymbol{u}$ and equating the resulting equations to zero, we get

$$\begin{bmatrix} \boldsymbol{X}'\boldsymbol{\Sigma}^{-1}\boldsymbol{X} & \boldsymbol{X}'\boldsymbol{\Sigma}^{-1}\boldsymbol{Z} \\ \boldsymbol{Z}'\boldsymbol{\Sigma}^{-1}\boldsymbol{X} & \boldsymbol{Z}'\boldsymbol{\Sigma}^{-1}\boldsymbol{Z}+\boldsymbol{\Psi}^{-1} \end{bmatrix} [\widehat{\boldsymbol{\beta}}\ \widehat{\boldsymbol{u}}] = \begin{bmatrix} \boldsymbol{X}'\boldsymbol{\Sigma}^{-1}\boldsymbol{y} \\ \boldsymbol{Z}'\boldsymbol{\Sigma}^{-1}\boldsymbol{y} \end{bmatrix}. \tag{12.63}$$

This equation can be solved iteratively, starting from the ordinary least squares (OLS) estimator $\widehat{\boldsymbol{\beta}}_0 = (\boldsymbol{X}'\boldsymbol{X})^{-1}\boldsymbol{X}'\boldsymbol{y}$. That is, substituting $\widehat{\boldsymbol{\beta}}_0$ in the second equation of (12.63) we solve it to obtain $\widehat{\boldsymbol{u}}_0$. Then we substitute it in the first equation of (12.63) and solve to obtain the second estimate of $\widehat{\boldsymbol{\beta}}$ and iterate this procedure until convergence is attained. Meanwhile, it is usually necessary to estimate $\widehat{\boldsymbol{\theta}}$ at each stage by REML using $\widehat{\boldsymbol{\beta}}$ at that stage. The estimate $\widehat{\boldsymbol{u}}$ is called the best linear unbiased predictor (BLUP).

Example 12.5. BLUP in the balanced one-way random effects model. In Example 12.4 we already have $\widehat{\boldsymbol{\beta}} = \bar{y}_{\cdot\cdot}$ and the second equation of (12.63) is

$$\left(m\sigma^{-2}\boldsymbol{I}_a + \sigma_\alpha^{-2}\boldsymbol{I}_a\right)\widehat{\boldsymbol{u}} = \sigma^{-2}\left(\boldsymbol{y}_{(i)\cdot} - m\bar{y}_{\cdot}\boldsymbol{j}_n\right),$$

where $\boldsymbol{y}_{(i)\cdot} = (y_{1\cdot}, \ldots, y_{a\cdot})'$. Therefore,

$$\widehat{\boldsymbol{u}} = m\,\frac{\sigma_\alpha^2}{\sigma^2 + m\,\sigma_\alpha^2}\begin{bmatrix} \bar{y}_{1\cdot} - \bar{y}_{\cdot\cdot} \\ \vdots \\ \bar{y}_{a\cdot} - \bar{y}_{\cdot\cdot} \end{bmatrix}$$

and we get the predictor of $\boldsymbol{y}$,

$$\widehat{\boldsymbol{y}} = \boldsymbol{X}\widehat{\boldsymbol{\beta}} + \boldsymbol{Z}\widehat{\boldsymbol{u}} = \bar{y}_{\cdot}\boldsymbol{j}_n + \boldsymbol{X}_\alpha\widehat{\boldsymbol{u}}. \tag{12.64}$$

The (i, j) element of (12.64) is

$$\widehat{y}_{ij} = \bar{y}_{\cdot\cdot} + m\,\frac{\sigma_\alpha^2}{\sigma^2 + m\,\sigma_\alpha^2}(\bar{y}_{i\cdot} - \bar{y}_{\cdot\cdot}) = \{1 - w_i(\gamma)\}\bar{y}_{i\cdot} + w_i(\gamma)\bar{y}_{\cdot\cdot}, \tag{12.65}$$

where $w_i(\gamma) = (1 + m\gamma)^{-1}$ with $\gamma = \sigma_\alpha^2/\sigma^2$ an SN ratio.

The estimate (12.65) is a Stein type shrinkage estimate with shrinkage factor $w_i(\gamma)$, which centers a naïve estimator $\bar{y}_{i\cdot}$ toward the general mean $\bar{y}_{\cdot\cdot}$. Empirical Bayes' interpretation of this type of estimator is given by Laird and Ware (1982) and Madsen and

Thyregod (2011), for example. Of course we need to replace variance parameters by appropriate estimators. In this case we can easily obtain $\widehat{\gamma}$, as given in Example 12.1, without iteration.

References

Boardman, T. J. (1974) Confidence intervals for variance components: A comparative Monte Carlo study. *Biometrics* **30**, 251–262.

Bulmer, M. G. (1957) Approximate confidence limit for components of variance. *Biometrika* **44**, 159–167.

Bush, N. and Anderson, R. L. (1963) A comparison of three different procedures for estimating variance components. *Technometrics* **5**, 421–440.

Elston, R. C. and Grizzle, J. E. (1962) Estimation of time response curves and their confidence bands. *Biometrics* **18**, 148–159.

Harville, D. A. (1977) Maximum likelihood approaches to variance component estimation and to related problems. *J. Amer. Statist. Assoc.* **72**, 320–338.

Henderson, C. R. (1953) Estimation of variance and covariance components. *Biometrics* **9**, 226–252.

Hirotsu, C. (1966) Estimating variance components in a two-way layout with unequal numbers of observations. *Rep. Statist. Appl. Res., JUSE* **13**, 29–34.

Hirotsu, C. (1968) An approximate test for the case of random effects model in a two-way layout with unequal cell frequencies. *Rep. Statist. Appl. Res., JUSE* **15**, 13–26.

Hirotsu, C. (1992) *Analysis of experimental data: Beyond the analysis of variance*. Kyoritsu-shuppan, Tokyo (in Japanese).

Khuri, A. I. and Sahai, H. (1985) Variance components analysis: A selective literature survey. *Int. Statist. Rev.* **53**, 279–300.

Laird, N. M. and Ware, J. H. (1982) Random effects models for longitudinal data. *Biometrics* **38**, 963–974.

Lee, Y. and Nelder, J. A. (2001) Hierarchical generalized linear models: A synthesis of generalized linear models, random effect models and structured dispersions. *Biometrika* **88**, 987–1006.

Madsen, H. and Thyregod, P. (2011) *Introduction to general and generalized linear models*. Chapman & Hall/CRC, London.

Miller, R. G. (1998) *BEYOND ANOVA: Basics of applied statistics*. Chapman & Hall/CRC Texts in Statistical Science, New York.

Moriguti, S. (1954) Confidence limits for a variance component. *Rep. Statist. Appl. Res., JUSE* **3**, 7–19.

Moriguti, S. (ed.) (1976) *Statistical methods*, new edn. Japanese Standards Association, Tokyo (in Japanese).

Mostafa, M. G. (1967) Designs for the simultaneous estimation of functions of variance components from two-way crossed classifications. *Biometrika* **54**, 127–131.

Patterson, H. D. and Thompson, R. (1971) Recovery of inter-block information when block sizes are equal. *Biometrika* **58**, 545–554.

Rao, C. R. (1971a) Estimation of variance and covariance components – MINQUE theory. *J. Mult. Anal.* **1**, 257–275.

Rao, C. R. (1971b) Minimum variance quadratic unbiased estimation of variance components. *J. Mult. Anal.* **1**, 445–456.

Searl, S. R. (1971) Topics in variance component estimation. *Biometrics* **27**, 1–76.

Scheffé, H. (1959) *The analysis of variance*. Wiley, New York.

Smith, D. W. and Murray, L. W. (1984) An alternative to Eisenhart's model II and mixed model in case of negative variance estimates. *J. Amer. Statist. Assoc.* **79**, 145–151.

Tukey, J. W. (1951) Comments in regression. *Biometrics* **27**, 643–657.

Williams, J. S. (1962) A confidence interval for variance components. *Biometrika* **49**, 278–281.

Yates, F. (1934) The analysis of multiple classifications with unequal numbers in the different classes. *J. Amer. Statist. Assoc.* **29**, 51–66.

13

Profile Analysis of Repeated Measurements

13.1 Comparing Treatments Based on Upward or Downward Profiles

13.1.1 Introduction

The data of Table 13.1 are the total cholesterol measurements of 23 subjects every four weeks for six periods (Hirotsu, 1991). Let the repeated measurements of the ith subject assigned to the hth treatment be denoted by $y_{hi} = (y_{hi1}, \ldots, y_{hip})'$, $i = 1, \ldots, n_h; h = 1, \ldots, t$, where t and p are the number of treatments and periods, respectively.

In the example of Table 13.1, $t=2$, $p=6$, $n_1 = 12$, and $n_2 = 11$. The interaction $y_{hij} - \bar{y}_{hi\cdot} - \bar{y}_{h\cdot j} + \bar{y}_{h\cdot\cdot}$ is plotted in Fig. 13.1 for each of active drug $(h=1)$ and placebo $(h=2)$. There is no noticeable difference between the plots at first glance. The thick lines are the averages of the three subgroups – improved, invariant, and deteriorated – obtained later in this chapter.

In comparing the treatments based on these data, the repeated t-tests at each period obviously suffer from the multiplicity problem, enhancing the type I error. There might also be difficulties in combining the different results at six periods into a single consistent conclusion.

Advanced Analysis of Variance, First Edition. Chihiro Hirotsu.

Table 13.1 Total cholesterol amounts.

		Period					
Treatment	Subject	1	2	3	4	5	6
Drug	1	317	280	275	270	274	266
	2	186	189	190	135	197	205
	3	377	395	368	334	338	334
	4	229	258	282	272	264	265
	5	276	310	306	309	300	264
	6	272	250	250	255	228	250
	7	219	210	236	239	242	221
	8	260	245	264	268	317	314
	9	284	256	241	242	243	241
	10	365	304	294	287	311	302
	11	298	321	341	342	357	335
	12	274	245	262	263	235	246
Placebo	13	232	205	244	197	218	233
	14	367	354	358	333	338	355
	15	253	256	247	228	237	235
	16	230	218	245	215	230	207
	17	190	188	212	201	169	179
	18	290	263	291	312	299	279
	19	337	337	383	318	361	341
	20	283	279	277	264	269	271
	21	325	257	288	326	293	275
	22	266	258	253	284	245	263
	23	338	343	307	274	262	309

13.1.2 Popular approaches

The conventional approach assumes a multivariate normal model

$$\boldsymbol{y}_{hi} = \boldsymbol{\mu}_h + \boldsymbol{e}_{hi} \tag{13.1}$$

and compares the mean vectors $\boldsymbol{\mu}_1, \ldots, \boldsymbol{\mu}_t$, assuming that the $\boldsymbol{e}_{hi}$ are independently and identically distributed as $N(\boldsymbol{0}, \boldsymbol{\Omega})$ with serial correlations within subjects. To reduce the number of unknown parameters, a simple model such as an AR model is sometimes assumed for serial correlations (see Ware, 1985, for example). If the covariance matrix $\boldsymbol{\Omega}$ reduces to $\boldsymbol{\Omega} = \left[\sigma_0^2 \boldsymbol{I} + \sigma_1^2 \boldsymbol{j}\boldsymbol{j}'\right]$, then the standard analysis of a split-plot design can be applied, with treatment and period as the first and second-order factor, respectively, and subject as block (Aitkin, 1981; Wallenstein, 1982). This would, however,

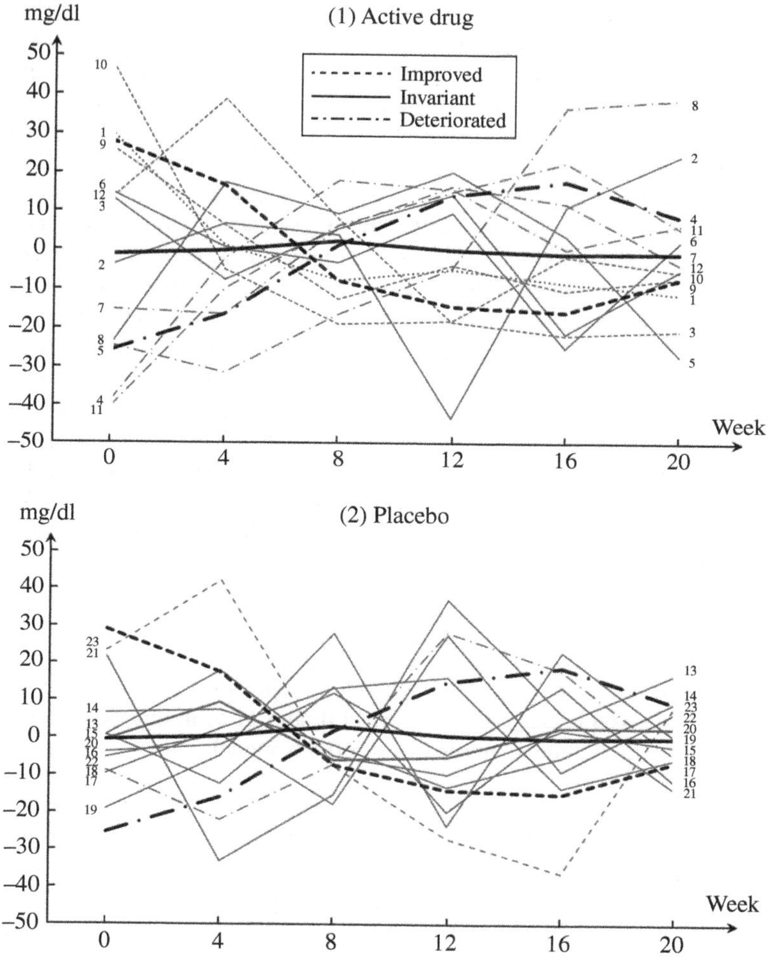

Figure 13.1 Interaction plots for drug and placebo.

rarely be the case, since serial measurements conceptually contradict the assumption of randomization over plots.

The general two-stage model of Section 12.4 permits a more general description of covariance structures through the distribution of subject profiles, and should be an efficient approach if from past experience there is available a reasonable regression model to describe the subject profiles. Starting from a regression model as in Example 12.3, it naturally induces a mixed effects model and provides a within-profile correlation model (see Crowder and Hand, 1990 for additional details).

More recently, profile monitoring has become a popular tool in engineering, where a sequence of measurements of one or more quality characteristics is taken across time or some continuum, producing a curve or functional data that represents the quality of products. This curve is called a profile. In phase I of the monitoring scheme, the historical data are analyzed to see whether they are stable or not. In phase II, the future observations are monitored using the control limits obtained from the stable data set in phase I. An introductory overview of this area is given by Woodall *et al.* (2004). Among many approaches based on the linear regression model, Williams *et al.* (2007) developed a method based on the non-linear parametric model. In contrast, Jensen *et al.* (2008) introduced a linear mixed effects model and proposed a monitoring procedure with T^2 -statistics based on the BLUP of random effects. They consider both cases – balanced and unbalanced – where data points are not necessarily the same for all profiles. Following the simulation result, they conclude that for the balanced data the simple LS approach to the fixed effects model will be sufficient and the linear mixed effects model is recommended as robust for the unbalanced case. Jensen and Birch (2009) extended the mixed effects model to the non-linear profile, and Qiu *et al.* (2010) developed a monitoring procedure based on a non-parametric mixed effects model. Their methods should be very flexible, although large data would be required for the method to work well.

13.1.3 Statistical model and approach

Another type of non-parametric approach is to assume some systematic trend along the time axis, as taken in Sections 6.5, 7.3, 11.4, and 11.5. In this case, a reasonable approach will be to assume a simple mixed effects model

$$\boldsymbol{y}_{hi} = \boldsymbol{\mu}_{hi} + \boldsymbol{e}_{hi}, \tag{13.2}$$

with the $\boldsymbol{\mu}_{hi}$ independently distributed as $N(\boldsymbol{\mu}_h, \boldsymbol{\Omega}_h)$; we can try to incorporate the natural ordering along the time axis in the analysis. The pure measurement errors at different periods are reasonably assumed to be independent of each other, and also mutually independent of the $\boldsymbol{\mu}_{hi}$. Therefore, the $\boldsymbol{e}_{hi}$ are assumed to be independently distributed as $N(\boldsymbol{0}, \sigma^2\boldsymbol{I})$. Given h, this is just Scheffé's (1959) two-way mixed effects model with subject as random factor, period as fixed factor, and without replication. In this case, as in the two-stage model, the covariances among repeated measurements are thought to arise from the inhomogeneity of the individual profiles along the time axis rather than the serial correlations.

Now, if there is no treatment effect, then the expected profile of each subject should be stable over the period, with some random fluctuations around it. In contrast, if a treatment has any effect, individual profiles should change. The effects may, however, not be homogeneous over the subjects, who might therefore be classified into several groups – improved, unchanged, deteriorated, and so on. Actually, some would respond to the placebo, like subject 23 in Fig. 13.1 (2), and some would not respond to the active drug, like subject 8 in Fig. 13.1 (1). The difference between the placebo and the active drug then lies in the proportions of the responder to each drug. This is the conceptual difference between models (13.1) and (13.2). That is, the usual approach of comparing mean vectors – assuming homogeneity within a group – is quite misleading. The proposed approach should be closer to the usual practice of clinical doctors for evaluating a drug. The equality of the covariance matrices should, of course, be tested preliminarily in the classical approach (see Morrison, 1976, for example), but in the proposed approach it shall be the main objective of the experiment. It should also be noted, in comparing covariance matrices, that the usual permutation invariant tests are inappropriate, since we are concerned here with some systematic change in the profile along the time axis.

Now, the proposed approach is to consider the subject as a variation factor and classify them into homogeneous subgroups based on their upward or downward response profiles. The procedure is also regarded as a conditional analysis given each profile. The classification is done by row-wise multiple comparisons of interaction in the two-way table with the subject as row factor, period as ordered column categories, and without replication. The classification naturally induces an ordering among subgroups, and each treatment is then characterized by the distribution of its own subjects over those subgroups. We do not estimate each profile by BLUP, but knowing the class to which each profile is assigned should be useful information.

Renumbering the subjects throughout the treatments, we denote by y_{kj} the measurement of the kth subject at the jth period, $k=1, \ldots, n$, $n=\sum n_h$, for the current example n being equal to $12+11=23$. The basic statistic to measure the difference in profiles between subjects k and k' is given by

$$\|\hat{\boldsymbol{L}}(k;k')\|^2 = \|\left\{(0\cdots 0 2^{-1/2} 0\cdots 0 - 2^{-1/2} 0\cdots 0)\otimes \boldsymbol{P}_p^{*\prime}\right\}\boldsymbol{y}\|^2 = 2^{-1}(\boldsymbol{y}_k-\boldsymbol{y}_{k'})'\boldsymbol{P}_p^{*}\boldsymbol{P}_p^{*\prime}(\boldsymbol{y}_k-\boldsymbol{y}_{k'})$$
$$= \frac{1}{2}\sum_{l=1}^{p-1}\frac{pl}{p-l}\left\{\frac{1}{l}\sum_{j=1}^{l}\left(y_{kj}-y_{k'j}\right)-\left(\bar{y}_{k\cdot}-\bar{y}_{k'\cdot}\right)\right\}^2, \tag{13.3}$$

where $\boldsymbol{y}_k$ is the vector of y_{kj}, $j=1, \ldots, p$, arranged in dictionary order and we are following the notation in Sections 10.3.1 and 10.3.2. Equation (13.3) is obtained by replacing $\boldsymbol{P}_b'$ by $\boldsymbol{P}_p^{*\prime}$ in $\chi^2(l_1;l_2)$ of Section 10.3.2 (1) for detecting upward or downward departure along the time axis. The sums of squares (13.3) calculated for Table 13.1 are given in Table 13.2, where the subjects are rearranged so that subjects with a smaller distance are located closer together. The chi-squared distances between two subgroups and the generalized chi-squared distances are similarly defined to section 10.3.2 (2) and (3) by replacing $\boldsymbol{P}_b'$ by $\boldsymbol{P}_p^{*\prime}$, respectively. The clustering algorithm of Section 11.3.3 is also applicable to this case. Then, the generalized chi-squared distance $\|\left(\boldsymbol{\rho}'\otimes\boldsymbol{P}_p^{*\prime}\right)\boldsymbol{y}\|^2$ is bounded above by the largest eigenvalue W_1^* of the Wishart

Table 13.2 The squared distances (13.3) between two rows.

Row	23	3	1	9	10	6	12	13	14	15	16	17	20	21	22	5	2	19	7	18	4	11	8
23	0	0.7	1.0	0.9	1.2	1.5	2.0	4.4	2.1	2.1	3.7	3.6	2.9	4.9	4.6	4.8	4.3	5.9	7.0	7.7	8.8	11.7	16.7
3		0	1.0	1.1	1.8	1.7	1.8	5.1	2.7	2.1	3.2	3.0	3.1	4.6	4.8	3.4	5.5	5.7	6.6	7.4	8.5	11.5	18.4
1			0	0.03	0.3	0.4	0.5	2.5	1.0	0.8	1.6	1.7	1.3	1.8	2.3	2.6	3.0	3.5	3.9	4.2	6.0	8.1	12.7
9				0	0.4	0.3	0.07	2.2	0.7	0.6	1.4	1.5	1.0	1.7	2.0	2.5	2.6	3.2	3.6	3.8	5.6	7.6	11.9
10					0	1.1	0.4	3.3	1.6	1.7	2.8	3.2	2.2	2.6	3.5	4.7	3.6	5.0	5.4	5.6	8.1	10.1	13.7
6						0	1.4	1.3	0.3	0.3	0.8	0.7	0.5	1.1	1.0	1.7	1.9	2.1	2.2	2.6	3.7	5.5	9.7
12							0	1.3	0.5	0.4	0.6	0.4	0.5	0.9	1.0	1.3	2.2	1.9	1.9	2.2	3.4	5.2	9.7
13								0	0.4	0.8	0.7	1.0	0.4	1.4	0.5	2.2	0.3	0.5	0.8	1.0	1.5	4.3	2.3
14									0	0.1	0.5	0.7	0.1	1.2	0.6	1.7	0.7	1.1	1.5	1.8	2.7	4.0	7.1
15										0	0.3	0.5	0.1	1.2	0.7	1.1	1.1	1.1	1.5	1.9	2.7	4.1	8.0
16											0	0.2	0.2	0.8	0.5	0.6	1.3	0.5	0.7	1.0	1.6	2.8	6.8
17												0	0.4	1.0	0.5	0.4	1.8	0.8	0.8	1.2	1.5	3.0	7.6
20													0	0.9	0.3	1.2	0.7	0.7	1.0	1.2	2.0	3.1	6.4
21														0	0.7	1.7	2.5	1.8	1.1	0.9	2.8	3.6	7.0
22															0	1.2	1.1	0.7	0.4	0.5	1.2	2.0	5.0
5																0	3.2	1.3	1.2	1.7	1.8	3.2	9.1
2																	0	0.9	1.6	1.9	2.3	3.0	4.4
19																		0	0.4	0.7	0.6	1.3	4.3
7																			0	0.09	0.4	0.8	3.8
18																				0	0.7	0.9	3.5
4																					0	0.4	3.5
11																						0	2.0
8																							0

matrix $W\left(\sigma^2 \boldsymbol{P}_p^{*\prime}\boldsymbol{P}_p^*, n-1\right)$ under the null hypothesis of interaction between subjects and periods. A very nice chi-squared approximation for the distribution of the largest eigenvalue of the Wishart matrix has already been obtained in Section 11.4.2 (2).

For the application here the balanced case is of concern, where the largest eigenvalue is large enough compared with the second largest. Therefore, noting that the largest root of $\boldsymbol{P}_p^{*\prime}\boldsymbol{P}_p^*$ is $p/2$, we can employ the first approximation $(p/2)\sigma^2\chi^2_{(1)}$ for the distribution of W_1^*, where $\chi^2_{(1)}$ is explained below. To execute Scheffé type multiple comparisons of subject profiles, however, we need to cancel out the unknown σ^2, since there is no repetition at each cell. Therefore, we introduce a denominator sum of squares which is less affected by a systematic departure among subjects. As one of these we employ

$$\chi^{-2} = \sum_{k=1}^{n}\sum_{j=1}^{p-1}\left\{\frac{y_{kj-1}-y_{kj}-\left(\bar{y}_{\cdot j-1}-\bar{y}_{\cdot j}\right)}{[p/\{j(p-j)\}]^{1/2}}\right\}^2,$$

which was introduced by Hirotsu (1991) as a quadratic form by the generalized inverse matrix of $P_p^*P_p^{*\prime}$, so as to have an inverse characteristic to W_1^*. Then, χ^{-2} is expanded in the form

$$\chi^{-2} = \left\{\tau_1^{-1}\chi^2_{(1)} + \tau_2^{-1}\chi^2_{(2)} + \cdots + \tau_{p-1}^{-1}\chi^2_{(p-1)}\right\}\sigma^2, \tag{13.4}$$

where $\tau_j = p/\{j(j+1)\}$ is the jth eigenvalue of $\boldsymbol{P}_p^{*\prime}\boldsymbol{P}_p^*$ and $\chi^2_{(j)}$ is the chi-squared component of Chebyshev's j th orthogonal polynomial, each with df $n-1$ and mutually independent as in (10.30). This suggests that the χ^{-2} would be appropriate for evaluating σ^2 when some systematic departure-like linear trend exists among subjects. Then, denoting any generalized chi-squared distance by S, the distribution of S/χ^{-2} is asymptotically bounded above by

$$\tau_1\chi^2_{(1)}\Big/\left\{\tau_1^{-1}\chi^2_{(1)} + \tau_2^{-1}\chi^2_{(2)} + \cdots + \tau_{p-1}^{-1}\chi^2_{(p-1)}\right\}.$$

Therefore, the p-value is evaluated by

$$\begin{aligned} p = \Pr\left(S/\chi^{-2} \ge s_0\right) &= \Pr\left\{\frac{\tau_1\chi^2_{(1)}}{\tau_1^{-1}\chi^2_{(1)} + \tau_2^{-1}\chi^2_{(2)} + \cdots + \tau_{p-1}^{-1}\chi^2_{(p-1)}} \ge s_0\right\} \\ &= \Pr\left\{\frac{\left(\tau_1 - s_0\tau_1^{-1}\right)\chi^2_{(1)}}{\tau_2^{-1}\chi^2_{(2)} + \cdots + \tau_{p-1}^{-1}\chi^2_{(p-1)}} \ge s_0\right\}. \end{aligned} \tag{13.5}$$

The denominator of (13.5) is a positive linear combination of the independent chi-squared variables, and is well approximated by a constant times a chi-squared variable $d\chi_f^2$ with constants obtained by adjusting the first two cumulants,

$$df = (n-1)\left(\tau_2^{-1} + \cdots + \tau_{p-1}^{-1}\right),$$

$$d^2f = (n-1)\left(\tau_2^{-2} + \cdots + \tau_{p-1}^{-2}\right).$$

Since the numerator and denominator are mutually independent, we have an F-approximation like

$$p = \Pr\left\{F_{n-1,f} \geq s_0\left(\tau_2^{-1} + \cdots + \tau_{p-1}^{-1}\right) / \left(\tau_1 - s_0\tau_1^{-1}\right)\right\}. \tag{13.6}$$

Thus, we can use (13.6) as a reference value for clustering subjects.

For the data of Table 13.1, the classification into G_1(1, 3, 9, 10, 23), G_2(2, 5, 6, 12, 13, 14, 15, 16, 17, 19, 20, 21, 22), and G_3(4, 7, 8, 11, 18) is obtained at significance level 0.05. The interaction plots for each of these subgroups are given in Fig. 13.2, where the average response pattern is shown by a thick line. The interpretation is now clear: subgroup G_1 is improved, G_2 unchanged, and G_3 deteriorated. The observed distribution of subjects from each of two treatment groups over these three types of subgroup is given in Table 13.3.

It is natural that the placebo is concentrated in the unchanged subgroup. In contrast, the active drug is very strangely distributed – equally on the three types of subgroup, and by no means recommended as good. Table 13.3 suggests also that the difference between the active drug and placebo lies not in the mean profiles, but in the dispersion of profiles around the mean. Therefore, if we had employed the multivariate normal model to compare the mean vectors assuming equality of covariance matrices, we would have failed to detect the difference. Also, both the modified likelihood ratio test (Bartlett, 1937; Anderson, 2003) and Nagao's (1973) invariant test for comparing two Wishart matrices show significance level of approximately 0.25, and are not successful in detecting such a systematic difference in covariance matrices along the time axis.

Finally, the sixth observation of each subject was taken as the post-measurement four weeks after the final dose, and might better be eliminated from the analysis. There is, however, not much difference in doing this; only subject 5 of the treatment group moved from the unchanged subgroup to the deteriorated one.

Example 13.1. From Phase III trial we give another example of comparing two active drugs for lowering cholesterol using a somewhat larger data set taken every four weeks for five periods. In this case a classification into five subgroups is obtained as highly significant. These are summarized in Table 13.4 (1). In this case the control is also an active drug, while the treatment is a well established one, so that the majority of the subjects respond well. The observed distribution of subjects from each treatment over those five classes is given in Table 13.4 (2). Then clearly the treatment drug is shifted in the direction of improved subgroups. The reader is advised to read Hirotsu (1991) for more details.

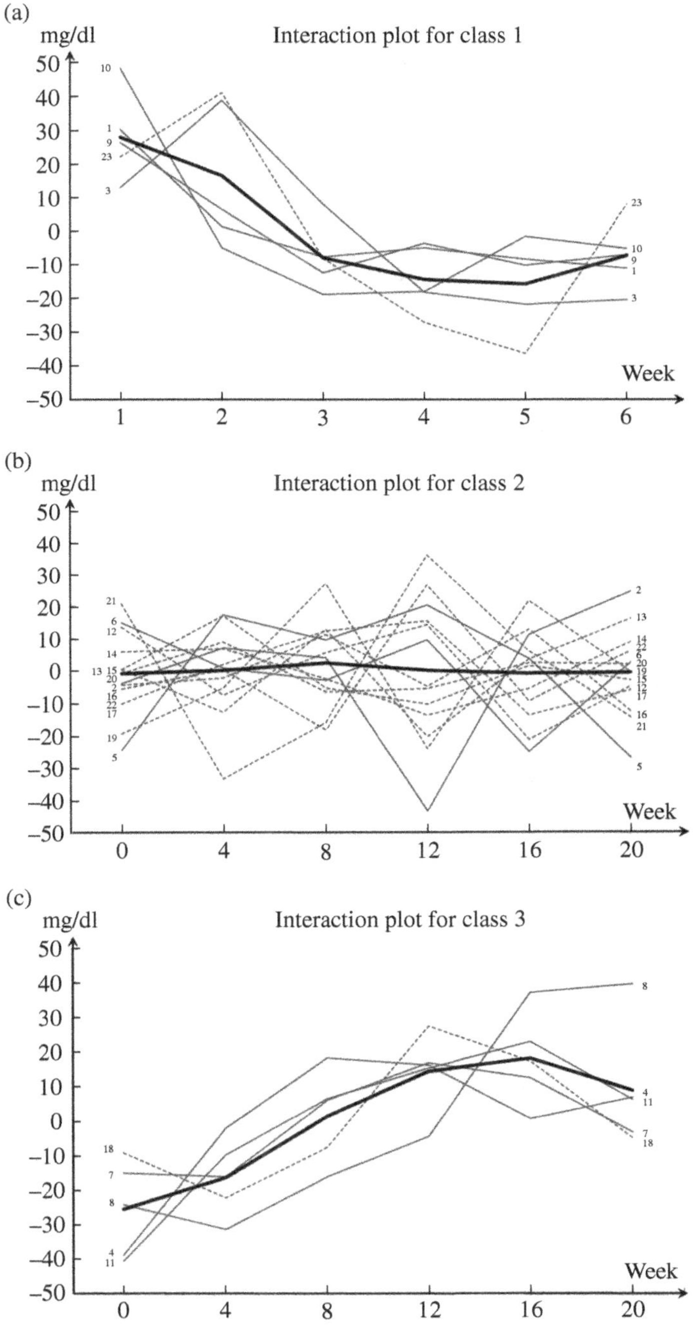

Figure 13.2 Interaction plots for each of three subgroups.

Table 13.3 Observed distribution for active drug and placebo.

	Class		
Treatment	G_1 : Improved	G_2 : Unchanged	G_3 : Deteriorated
Active drug	4	4	4
Placebo	1	9	1

Table 13.4 Summary of the classification into five subgroups.

(1) Estimated interaction patterns with mean profiles in the lower row

	Period					
Subgroup	1	2	3	4	5	Characterization
G_1	79.7	−0.2	−40.9	−34.7	−3.8	Very highly improved
	402.3	292.0	250.3	257.7	289.3	
G_2	22.6	8.6	−1.0	−12.1	−18.1	Highly improved
	296.6	252.2	241.6	231.5	226.3	
G_3	14.3	−5.8	−7.9	−5.7	5.1	Improved
	278.4	227.8	224.7	228.1	239.6	
G_4	−17.0	5.5	10.4	6.8	−5.7	Slightly improved
	260.1	252.2	256.1	253.6	241.9	
G_5	−23.4	−7.7	0.9	12.4	17.7	Unchanged or slightly
	263.4	248.7	256.4	268.9	274.9	deteriorated

(2) Observed distribution for the drug and control

Treatment	G_1	G_2	G_3	G_4	G_5	Total
Drug	3	28	35	14	9	89
Control	0	6	11	33	28	78

13.2 Profile Analysis of 24-Hour Measurements of Blood Pressure

13.2.1 Introduction

There are 48 observations of systolic blood pressure, measured every 30 minutes for 24 hours. These measurements are known to normally go down slightly in the night and a large depression is abnormal, as is a flat pattern or a tendency toward elevation. In the following we call these downward and upward tendencies in the night convex and concave patterns, respectively, since we take the starting point at 3.00 p.m. As a background to these data, it was first pointed out by Miller-Craig *et al.* (1978) that systolic and diastolic blood pressures follow a circadian pattern, with values being lower at night than during the day. Then, an inverse pattern has been reported in subjects with autonomic failure by Mann *et al.* (1983). Kobrin *et al.* (1984) reported that the elderly, who have a higher incidence of cardiovascular disease, may lose this diurnal variation. Following these reports, O'Brien *et al.* (1988) analyzed 24-hour ambulatory blood pressure measurements from 123 hypertensive consecutive patients; they classified these patients into dippers (102 patients, 82.9%) and non-dippers (21 patients, 17.1%), according to whether the difference between mean day-time blood pressure and mean night-time blood pressure showed the normal circadian variation (dipper) or not (non-dipper). Thus, they first concluded that there is a group of patients whose 24-hour blood pressure does not follow the normal circadian pattern and who thus may be at higher risk of cerebrovascular complications. Ohkubo *et al.* (1977) defined further the four classes of extreme dipper (ED), dipper (D), non-dipper (ND), and inverted dipper (ID), finding that the mortality rate is highest in the inverted dipper subgroup. In these works, however, the circadian rhythm has been characterized by several parameters, such as 24-hour mean of measurements, day-time mean, sleep mean, evening acrophase, nocturnal nadir, and amplitude (the last is defined as 50% of the difference between the maximum and minimum of the fitted curve). Although each parameter is useful and informative, the separate analyses of these parameters suffer from the multiplicity problem. Also, another process is required to combine these separate analyses into a single conclusion. Further, some parameters are known to be sensitive to outliers. This, together with the correlations among parameters, makes the simultaneous inferences very complicated. In this section we therefore intend to characterize the 24-hour profile of blood pressure measurements on the whole, with particular interest in their convex and concave patterns.

There have been several attempts to classify subjects based on their time series profiles, such as blood pressure or total cholesterol amount. Among them, a naïve method employed by clinical doctors is to classify subjects according to several prototype patterns. It has been pointed out, however, that there are various difficulties in classifying some hundreds of subjects by sight. Some statistical approaches have also been proposed, including the method proposed in the previous section, which succeeded in classifying subjects according to their upward or downward profile. The 24-hour profile of blood pressure measurements is, however, characterized by coming back approximately to the initial value after 24 hours, and the previous method should not be successful. Also, in this case, we cannot assume the independence of measurements, since they are taken every 30 minutes, whereas in the previous example cholesterol measurements were

Table 13.5 Summary of blood pressure data ($n = 203$).

Data profile	Mean (mmHg)	Standard deviation (mmHg)
Average of 24-hour measurements	137.44	17.47
Minimum of 24-hour measurements	104.07	15.47
Maximum of 24-hour measurements	173.28	22.85
Average of change rate of blood pressure	0.074	0.098

taken every month, allowing for an independence assumption between successive measurements.

13.2.2 Data set and classical approach

The original data set is for 203 elderly subjects taken at the Medical Department of Keio University. It is too large to present here, so we give only a summary of the data in Table 13.5. There we have included the average of the change rate, defined by

$$\Delta = (\text{average in day time} - \text{average in night time}) \div (\text{overall average}),$$

where the day time and night time are tentatively defined as 10 a.m. ~ 8 p.m. and 0 a.m. ~ 6 a.m., respectively. Subjects have been classified as depression, flat, or elevation type, according to $\Delta \geq 0.1$, $-0.1 < \Delta < 0.1$ or $\Delta \leq -0.1$, respectively, and the flat and elevation type at night should be treated (see Shimizu, 1994). The classification was, however, rather arbitrary. We therefore propose a scientific approach to comparing the profile on the whole, rather than based on the somewhat arbitrary amount Δ, and obtain statistically and clinically significant subgroups without prespecifying the number of subgroups.

13.2.3 Statistical model and new approach

(1) **Model and test statistic.** There is apparent serial correlation in the original data. We therefore tried several methods for deleting this, and found that averaging four successive points at every four-point interval could yield an approximately independent sequence (Hirotsu *et al.*, 2003). Averaging four successive points is also useful for smoothing a rather noisy sequence without affecting the systematic trend tendency. We denote the resulting data by y_{kj}, $k = 1, \ldots, n; j = 1, \ldots, p$ and assume the model

$$y_{kj} = \mu_{kj} + e_{kj} \tag{13.7}$$

with normally and independently distributed error e_{ij}. We do not employ any particular parametric model for μ_{kj}, but assume some systematic change in μ_{kj} with respect to the passage of time. This means that the apparently high correlation between the two close measurements is modeled by a smooth systematic change in the mean level, and the casual measurement errors are assumed to be mutually independent in (13.7). Thus we have two components, μ_{kj} and e_{kj}, the former changing slowly and the latter moving rather quickly and independently. The idea is similar to the previous section but in contrast to the monotone profile of the previous section, we intend to catch here those convex and concave systematic changes which are of clinical interest for blood pressure measurements. These patterns can be caught by the contrasts $\boldsymbol{P}_p^{\dagger\prime} = \boldsymbol{D}\left(\boldsymbol{L}_p'\boldsymbol{L}_p\right)^{-1}\boldsymbol{L}_p'$ as developed in Section 6.5.4 (5). So, we modify the squared distance between two subjects (13.3) to

$$\|\hat{\boldsymbol{L}}(k;k')\|^2 = \|\left\{\left(0\cdots 0 2^{-1/2} 0\cdots 0 -2^{-1/2} 0\cdots 0\right)\otimes \boldsymbol{P}_p^{\dagger\prime}\right\}\boldsymbol{y}\|^2 = 2^{-1}(\boldsymbol{y}_k - \boldsymbol{y}_{k'})'\boldsymbol{P}_p^{\dagger}\boldsymbol{P}_p^{\dagger\prime}(\boldsymbol{y}_k - \boldsymbol{y}_{k'})$$

so as to reflect the difference represented by convex and concave patterns. The chi-squared distances between two subgroups and the generalized chi-squared distances are similarly defined to Section 10.3.2 (2) and (3) by replacing $\boldsymbol{P}_b'$ by $\boldsymbol{P}_p^{\dagger\prime}$, respectively. The clustering algorithm of Section 11.3.3 is again applicable to this case. The generalized chi-squared distance is bounded above by the largest eigenvalue $W_1^{\dagger}$ of the Wishart matrix $W\left(\sigma^2\boldsymbol{P}_p^{\dagger\prime}\boldsymbol{P}_p^{\dagger}, n-1\right)$ under the null hypothesis of homogeneity of subjects. It is asymptotically distributed as $\tau_2\sigma^2\chi^2_{(2)}$, where τ_2 and $\chi^2_{(2)}$ are explained below. To cancel out the unknown σ^2 we introduce $S^- = \sum_k \boldsymbol{v}_k\boldsymbol{v}_k'$ instead of χ^{-2}, where $\boldsymbol{v}_k = \text{diag}\left\{\left(\xi_i\delta_i^2\right)^{-1/2}\right\}\boldsymbol{L}_p'(\boldsymbol{y}_k - \bar{\boldsymbol{y}}_.)$ and ξ_i and δ_i are given in Section 6.5.4 (5). This is a quadratic form by the generalized inverse matrix of $\boldsymbol{P}_p^{\dagger}\boldsymbol{P}_p^{\dagger\prime}$ and is simply the weighted sum of squares of the second-order differences in subsequent measurements, and less affected by the systematic change. It is expanded in the form

$$S^- = \left\{\tau_2^{-1}\chi^2_{(2)} + \tau_3^{-1}\chi^2_{(3)} + \cdots + \tau_{p-1}^{-1}\chi^2_{(p-1)}\right\}\sigma^2, \tag{13.8}$$

where $\tau_k = 2p(p+1)/\{(k-1)k(k+1)(k+2)\}$, $k = 2, \ldots, p-1$, is the k th eigenvalue of $P_p^{\dagger\prime}P_p^{\dagger}$ and $\chi^2_{(j)}$ the Chebyshev's chi-squared component for the j th orthogonal polynomial, each with df $n-1$ and mutually independent as in (13.4), with inverse characteristic to $W_1^\dagger$. This equation is similar to that of Section 6.5.4 (5), but differs in the degrees of freedom of the chi-squared components.

(2) **Distribution of test statistic.** By virtue of the expansion (13.8), the p-value of the generalized chi-squared distance is evaluated by formula (13.9) or (13.10):

$$p=\Pr\left(W_1^\dagger/S^- \ge s_0\right)=\Pr\left\{\frac{\tau_2\chi_{(2)}^2}{\tau_2^{-1}\chi_{(2)}^2+\tau_3^{-1}\chi_{(3)}^2+\cdots+\tau_{p-1}^{-1}\chi_{(p-1)}^2}\ge s_0\right\}$$
$$=\Pr\left\{\frac{\left(\tau_2-s_0\tau_2^{-1}\right)\chi_{(2)}^2}{\tau_3^{-1}\chi_{(3)}^2+\cdots+\tau_{t-1}^{-1}\chi_{(p-1)}^2}\ge s_0\right\} \quad (13.9)$$

Now the denominator of (13.9) is a positive linear combination of the independent chi-squared variables, and is well approximated by a constant times the chi-squared variable $d\chi_f^2$, with constants obtained by adjusting the first two cumulants,

$$df=(n-1)\left(\tau_3^{-1}+\cdots+\tau_{p-1}^{-1}\right),$$
$$d^2f=(n-1)\left(\tau_3^{-2}+\cdots+\tau_{p-1}^{-2}\right).$$

Since the numerator and denominator are mutually independent, we finally obtain

$$p=\Pr\left\{F_{n-1,f}\ge s_0\left(\tau_3^{-1}+\cdots+\tau_{p-1}^{-1}\right)/\left(\tau_2-s_0\tau_2^{-1}\right)\right\}, \quad (13.10)$$

where $F_{n-1,f}$ is an F-variable with df $(n-1,f)$.

(3) **Application.** We apply the clustering procedure to the data of $n=203$ and $p=6$ with the reference distribution (13.10). Originally the number of data points was 48, but by averaging four successive points at every four-point interval they are reduced to six points.

First we note that all the classifications beyond $K=2$ are highly significant. However, at $K=3$ the classification fails to detect a flat pattern and the homogeneity within subgroups is not assured. Since the gain in percentage contribution from $K=3$ to $K=4$ is 6.9%, and larger than the 3.0% from $K=4$ to $K=5$, the first choice looks to be $K=4$. Also, the four subgroups are interpreted as moderately convex, slightly convex, flat, and concave, and correspond well to Ohkubo *et al.*'s (1997) classification of extreme dipper (ED), dipper (D), non-dipper (ND), and inverted dipper (ID). In the next step from $K=4$ to $K=5$, however, a highly convex subgroup (UED) of size six is isolated from the moderately convex

Table 13.6 Process of classification.

Significance		Mean profile							Generalized distance	Percentage contribution
K	level	15.00	19.00	23.00	3.00	7.00	11.00	n		
2	1.12E-1	148	138	126	128	139	147	117	9.43E4	50.3
		137	138	136	137	141	132	86		
3	8.99E-5	157	145	128	129	143	154	65	1.24E5	66.3
		137	132	126	129	136	135	102		
		137	143	146	145	146	131	36		
4	1.34E-6	161	146	126	128	144	157	39	1.37E5	73.2
		142	134	125	129	137	142	78		
		138	138	133	133	140	134	62		
		134	139	144	148	145	126	24		
5	1.92E-7	173	152	119	122	136	160	6	1.43E5	76.2
		158	145	129	128	145	155	39		
		141	134	125	129	136	141	74		
		137	137	133	133	141	134	60		
		134	139	144	148	145	126	24		
6	1.99E-7	173	152	119	122	136	160	6	1.43E5	79.3
		158	145	129	128	145	155	39		
		140	132	124	128	136	140	75		
		144	126	128	147	151	129	15		
		137	143	136	132	139	136	47		
		133	141	147	147	145	127	21		

subgroup, while the other subgroups are adjusted only slightly. Thus, the classification of $K=5$ is very natural and easy to interpret, separating UED from ED and giving highly convex (UED), moderately convex (ED), slightly convex (D), flat (ND), and concave (ID) subgroups. The classification of $K=6$ gains us very little in the generalized squared distance over $K=5$, and will not make sense. The process of classification and the final results are shown in Table 13.6 and Fig. 13.3, respectively.

It has been pointed out clinically that the stroke rate is higher in the ND and ID subgroups (O'Brien *et al.*, 1988), and the mortality risk is higher in the ID subgroup. However, no significant difference has been reported between ED and D (Ohkubo *et al.*, 1997). We, however, think that UED might suggest a higher risk than ED, and encourage further clinical research to verify this. The reader is recommended to refer to Hirotsu *et al.* (2003) for a more detailed explanation of the method of this section.

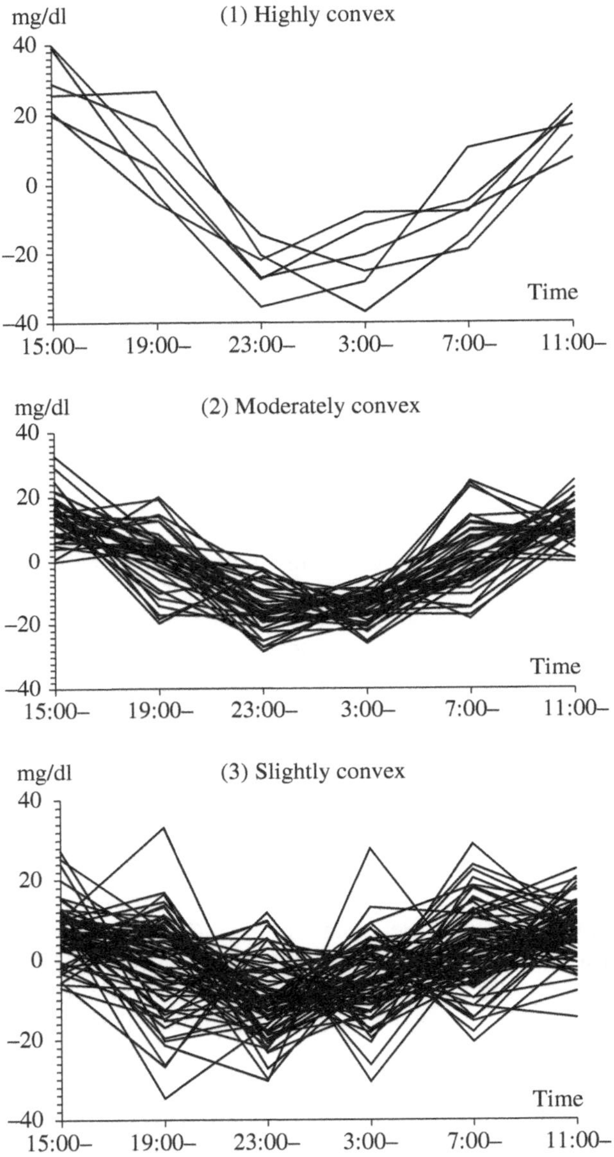

Figure 13.3 Interaction plots of blood pressure for each of five sub-classes.

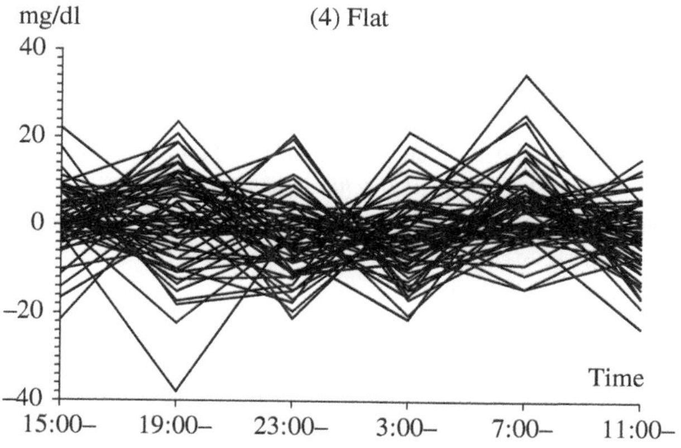

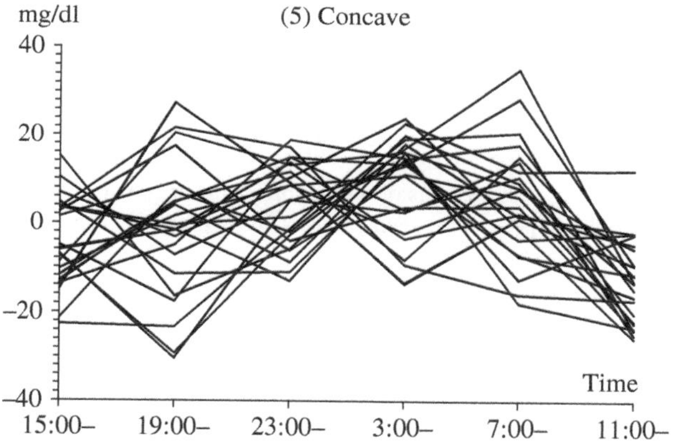

Figure 13.3 (Continued)

References

Aitkin, M. (1981) Regression models for repeated measurements (response to query). *Biometrics* **37**, 831–832.

Anderson, T. W. (2003) *An introduction to multivariate statistical analysis*, 3rd edn. Wiley, New York.

Bartlett, M. S. (1937) Properties of sufficiency and statistical tests. *Proc. R. Soc. Lond. A* **160**, 268–282.

Crowder, M. J. and Hand, D. J. (1990) *Analysis of repeated measures*. Chapman & Hall, London.

Hirotsu, C. (1991) An approach to comparing treatments based on repeated measures. *Biometrika* **78**, 583–594.

Hirotsu, C., Ohta, E., Hirose, N. and Shimizu, K. (2003) Profile analysis of 24-hours measurements of blood pressure. *Biometrics* **59**, 907–915.

Jensen, W. A. and Birch, J. B. (2009) Profile monitoring via nonlinear mixed models. *J. Qual. Technol.* **41**, 18–34.

Jensen, W. A., Birch, J. B. and Woodall, W. H. (2008) Monitoring correlation within linear profiles using mixed models. *J. Qual. Technol.* **40**, 167–183.

Kobrin, I., Oigman, W., Kumar, A., Venture, H. O., Messeri, F. H., Frohlic, E. D. and Dunn, F. G. (1984) Diurnal variation of blood pressure in elderly patients with essential hypertension. *J. Amer. Geriatr. Soc.* **32**, 896–899.

Miller-Craig, M. W., Bishop, C. N. and Raftery, E. B. (1978) Circadian variation of blood pressure. *Lancet* **1**, 795–797.

Morrison, D. F. (1976) *Multivariate statistical methods*, 2nd edn. McGraw-Hill, New York.

Nagao, H. (1973) On some test criteria for covariance matrix. *Ann. Statist.* **1**, 700–709.

O'Brien, E., Sheridan, J. and O'Malley, K. (1988) Dippers and non-dippers. *Lancet* **2**, 397.

Ohkubo, T., Imari, Y. and Tsuji, I. (1997) Relation between nocturnal decline in blood pressure and mortality. *Am. J. Hypertens.* **10**, 1201–1207.

Qiu, P., Zou, C. and Wang, Z. (2010) Nonparametric profile monitoring by mixed effects modeling. *Technometrics* **52**, 265–277.

Scheffé, H. (1959) *The analysis of variance*. Wiley, New York.

Shimizu, K. (1994) Time series analysis of diurnal blood pressure variation – its basis and clinical application to the hypertensive aged subjects. *J. Med., Keio University* **71**, 249–262 (in Japanese).

Wallenstein, S. (1982) Regression models for repeated measurements (reader reaction). *Biometrics* **38**, 489–450.

Ware, J. H. (1985) Linear models for the analysis of longitudinal studies. *Amer. Statist.* **39**, 95–111.

Williams, J. D., Woodall, W. H. and Birch, J. B. (2007) Statistical monitoring of nonlinear product and process quality profiles. *Qual. Reliab. Eng. Int.* **23**, 925–941.

Woodall, W. H., Spitzner, D. J., Montgomery, D. C. and Gupta, S. (2004) Using control charts to monitor process and product quality profiles. *J. Qual. Technol.* **36**, 309–320.

14

Analysis of Three-Way Categorical Data

We denote the frequency data cross-classified according to the three attributes A, B, C by y_{ijk} and the associated cell probabilities by $p_{ijk}, i = 1, \ldots, a, j = 1, \ldots, b, k = 1, \ldots, c$. We denote the vectors of y_{ijk} and p_{ijk} arranged in dictionary order with respect to suffix by $\boldsymbol{y}$ and $\boldsymbol{p}$, respectively. Then there are four different types of sampling scheme to obtain those data.

(1) The y_{ijk} are distributed as an independent Poisson distribution $P_o(\Lambda)$ with mean Λ_{ijk}.

(2) The y_{ijk} are distributed as a multinomial distribution $M(y_{\cdots}, \boldsymbol{p})$ with $p_{\cdots} = 1$.

(3) The two-way categorical data $\boldsymbol{y}_i = (y_{i11}, \ldots, y_{ibc})'$ following a multinomial distribution $M(y_{i\cdot\cdot}, \boldsymbol{p}_i), \boldsymbol{p}_i = (p_{i11}, \ldots, p_{ibc})', i = 1, \ldots, a$, are obtained at each of a levels of a factor A.

(4) The one-way categorical data $\boldsymbol{y}_{ij} = (y_{ij1}, \ldots, y_{ijc})'$ following a multinomial distribution $M(y_{ij\cdot}, \boldsymbol{p}_{ij}), \boldsymbol{p}_{ij} = (p_{ij1}, \ldots, p_{ijc})'$ are obtained at each of $a \times b$ combinations of the levels of factors A and B.

In this section it is convenient to express three factors as roman or italic according to the explanation or response attribute. Model 1 is expressed as $A \times B \times C$ and model 2 is expressed as $\mathrm{A} \times B \times C$, for example. In the following we consider a conditional analysis given the sufficient statistics under the null model, which are the marginal totals.

Advanced Analysis of Variance, First Edition. Chihiro Hirotsu.

Then the Poisson model need not be discussed separately, since it reduces to the multinomial model 1 with $p_{ijk} = \Lambda_{ijk}/\Lambda_{\cdots}$ by considering the conditional distribution given $y_{\cdots}$.

14.1 Analysis of Three-Way Response Data

14.1.1 General theory

In this case the purpose of the analysis is to clarify the association or independence among the three response factors. The simplest model will be

$$p_{ijk} = p_{i\cdot\cdot} \times p_{\cdot j\cdot} \times p_{\cdot\cdot k}, \tag{14.1}$$

expressing the independence of three factors. Also the meaning of the model

$$p_{ijk} = p_{ij\cdot} \times p_{\cdot\cdot k} \tag{14.2}$$

will be straightforward. Then the model

$$p_{ijk} = p_{i\cdot k} \times p_{\cdot jk} / p_{\cdot\cdot k} \tag{14.3}$$

implies the conditional independence of factors A and B given the level of factor C. To cover all these models it is convenient to consider a log linear model of p_{ijk}:

$$\log p_{ijk} = \mu + \alpha_i + \beta_j + \gamma_k + \theta_{ij} + \varphi_{ik} + \tau_{jk} + \omega_{ijk}, \tag{14.4}$$

and test the interaction effects in order of descent, starting from the highest. Model (14.3), for example, corresponds to

$$\log p_{ijk} = \mu + \alpha_i + \beta_j + \gamma_k + \varphi_{ik} + \tau_{jk}. \tag{14.5}$$

This can be shown as follows. If (14.3) holds, it is obvious that $\log p_{ijk}$ is in the form of (14.5). If we assume (14.5), we have

$$p_{i\cdot k} = \exp(\mu + \alpha_i + \gamma_k + \varphi_{ik})\left\{\textstyle\sum_j \exp(\beta_j + \tau_{jk})\right\},$$

$$p_{\cdot jk} = \exp(\mu + \beta_j + \gamma_k + \tau_{jk})\left\{\textstyle\sum_i \exp(\alpha_i + \varphi_{ik})\right\},$$

and $p_{\cdot\cdot k} = \exp(\mu + \gamma_k)\left\{\sum_i \exp(\alpha_i + \varphi_{ik})\right\}\left\{\sum_j \exp(\beta_j + \tau_{jk})\right\}$. Therefore we have

$$\frac{p_{i\cdot k} \times p_{\cdot jk}}{p_{\cdot\cdot k}} = \exp(\mu + \alpha_i + \beta_j + \gamma_k + \varphi_{ik} + \tau_{jk}) = p_{ijk}.$$

It is similarly shown that the models (14.2) and (14.1) correspond to

$$\begin{array}{c}\log p_{ijk} = \mu + \alpha_i + \beta_j + \gamma_k + \theta_{ij}, \\ \log p_{ijk} = \mu + \alpha_i + \beta_j + \gamma_k,\end{array},$$

respectively. We call model (14.4) a saturated model. Then we have further a model

$$\log p_{ijk} = \mu + \alpha_i + \beta_j + \gamma_k + \theta_{ij} + \varphi_{ik} + \tau_{jk} \tag{14.6}$$

between it and model (14.5), and it is in this case that Simpson's paradox occurs by an additive treatment of the data.

For convenience in dealing with the log linear model (14.4), identification conditions like $\sum_j \theta_{ij} = 0$ were introduced by Birch (1963). Plackett (1981) employed the constraint that the parameter with suffix $i = a, j = b$, or $k = c$ is zero on the right-hand side of (14.4), which gives the explicit expression of those parameters in terms of p_{ijk}:

$$\left\{\begin{array}{l} \alpha_{i'} = \log\dfrac{p_{i'bc}}{p_{abc}}, \beta_{j'} = \log\dfrac{p_{aj'c}}{p_{abc}}, \gamma_{k'} = \log\dfrac{p_{abk'}}{p_{abc}} \\ \theta_{i'j'} = \log\dfrac{p_{i'j'c}p_{abc}}{p_{i'bc}p_{aj'c}}, \varphi_{i'k'} = \log\dfrac{p_{i'bk'}p_{abc}}{p_{i'bc}p_{abk'}}, \tau_{j'k'} = \log\dfrac{p_{aj'k'}p_{abc}}{p_{aj'c}p_{abk'}} \\ \omega_{i'j'k'} = \log\dfrac{p_{i'j'k'}p_{i'bc}p_{aj'c}p_{abk'}}{p_{i'j'c}p_{i'bk'}p_{aj'k'}p_{abc}}, \end{array}\right. \tag{14.7}$$

where the prime on the suffix denotes that the range of the suffixes is restricted to $i' = 1, \ldots, a-1; j' = 1, \ldots, b-1$, and $k' = 1, \ldots, c-1$. In contrast, the p_{ijk} are expressed by the new parameters as

$$p_{ijk} = p_{abc}\exp\left(\alpha_i + \beta_j + \gamma_k + \theta_{ij} + \varphi_{ik} + \tau_{jk} + \omega_{ijk}\right),$$

where the parameters with suffix $i = a$, $j = b$, or $k = c$ are zero and p_{abc} is defined so that $p_{\cdots} = 1$.

Both constraints give the same test procedure as long as we follow the strong heredity principle that for testing an interaction θ_{ij}, both parents α_i and β_j should remain in the model. Therefore we start with the test of model (14.6) against the saturated model (14.4). Employing the constraints of Plackett, the likelihood function of the saturated model is

$$\begin{aligned} L = \frac{y_{\cdots}!\, p_{abc}^{y_{\cdots}}}{\Pi_i \Pi_j \Pi_k y_{ijk}!} \exp\Big(& \textstyle\sum_{i'} y_{i'\cdot\cdot}\alpha_{i'} + \sum_{j'} y_{\cdot j'\cdot}\beta_{j'} + \sum_{k'} y_{\cdot\cdot k'}\gamma_{k'} + \sum_{i'}\sum_{j'} y_{i'j'\cdot}\theta_{i'j'} \\ & \textstyle + \sum_{i'}\sum_{k'} y_{i'\cdot k'}\varphi_{i'k'} + \sum_{j'}\sum_{k'} y_{\cdot j'k'}\tau_{j'k'} + \sum_{i'}\sum_{j'}\sum_{k'} y_{i'j'k'}\omega_{i'j'k'}\Big). \end{aligned} \tag{14.8}$$

In the following we mention both the conditional and unconditional tests. The unconditional test based on the MLE is easy to apply and fortunately it often coincides asymptotically with the conditional test. We recommend, however, the conditional test in principle because it gives a similar test. In particular, we recommend an exact conditional test when it is available.

(1) **Test of three-way interaction**

(a) **Conditional test of the null hypothesis H_ω: $\log p_{ijk} = \mu + \alpha_i + \beta_j + \gamma_k + \theta_{ij} + \varphi_{ik} + \tau_{jk}$.** We first test the null hypothesis H_ω (14.6). We consider a conditional analysis given a set of sufficient statistics under H_ω which is obviously $(y_{ij\cdot}, y_{i\cdot k}, y_{\cdot jk})$, where we omit the range of the suffix when it is obvious. Then we have

$$\Pr\{Y_{ijk} = y_{ijk} | y_{ij\cdot}, y_{i\cdot k}, y_{\cdot jk}\} = \frac{1}{C(\omega)} \frac{\exp\left(\sum_{i'}\sum_{j'}\sum_{k'} y_{i'j'k'}\omega_{i'j'k'}\right)}{\Pi_i \Pi_j \Pi_k y_{ijk}!}, \tag{14.9}$$

where $C(\omega)$ is a normalizing constant so that the total probability is unity. In the case of a $2\times2\times2$ table there is only one random variable, which is conveniently specified as Y_{111}. Then the conditional distribution (14.9) reduces to

$$\Pr\{Y_{ijk} = y_{ijk} | y_{ij\cdot}, y_{i\cdot k}, y_{\cdot jk}\} = \frac{1}{C(\omega_{111})}\exp(y_{111}\omega_{111})$$
$$\Big/\left\{\begin{array}{c} y_{111}!(y_{11\cdot} - y_{111})!(y_{1\cdot1} - y_{111})!(y_{1\cdot2} - y_{11\cdot} + y_{111})! \\ \times (y_{\cdot11} - y_{111})!(y_{21\cdot} - y_{\cdot11} + y_{111})!(y_{2\cdot1} - y_{\cdot11} + y_{111})!(y_{2\cdot2} - y_{21\cdot} + y_{\cdot11} - y_{111})! \end{array}\right\}$$

with the aid of Table 14.1. The range of the variable Y_{111} is determined so that all the $2\times2\times2$ cell frequencies in Table 14.1 are non-negative.

In the general case it is very hard to handle the exact distribution, and the normal approximation is usually employed. It should be noted here that the cell probabilities $\{p_{ijk}\}$ are uniquely determined by the given marginal totals $(p_{ij\cdot}, p_{i\cdot k}, p_{\cdot jk})$ and three-factor interaction $\{\omega_{i'j'k'}\}$. They are determined by the iterative scaling procedure,

Table 14.1 Data of $2\times2\times2$ table expressed by y_{111} and two-way marginal totals.

	$i=1$			$i=2$		
J	$k=1$	$k=2$	Total	$k=1$	$k=2$	Total
1	y_{111}	$y_{11\cdot} - y_{111}$	$y_{11\cdot}$	$y_{\cdot11} - y_{111}$	$y_{21\cdot} - y_{\cdot11} + y_{111}$	$y_{21\cdot}$
2	$y_{1\cdot1} - y_{111}$	$y_{1\cdot2} - y_{11\cdot} + y_{111}$	$y_{12\cdot}$	$y_{2\cdot1} - y_{\cdot11} + y_{111}$	$y_{2\cdot2} - y_{21\cdot} + y_{\cdot11} - y_{111}$	$y_{22\cdot}$
Total	$y_{1\cdot1}$	$y_{1\cdot2}$	$y_{1\cdot\cdot}$	$y_{2\cdot1}$	$y_{2\cdot2}$	$y_{2\cdot\cdot}$

adjusting the marginal totals starting from any three-way table $\{q_{ijk}\}$ whose three-way interaction is $\{\omega_{i'j'k'}\}$, such as

$$\begin{cases} q_{i'j'k'} = \exp(\omega_{i'j'k'}), \\ q_{ijk} = 1 \text{ if } (i-a)(j-b)(k-c) = 0, \end{cases} \tag{14.10}$$

for example, see Fienberg (1980) for iterative scaling procedure. Also define the cell frequencies $m_{ijk}(\boldsymbol{\omega})$ that satisfy

$$\begin{cases} m_{ij\cdot}(\boldsymbol{\omega}) = y_{ij\cdot}, \ m_{i\cdot k}(\boldsymbol{\omega}) = y_{i\cdot k}, \ m_{\cdot jk}(\boldsymbol{\omega}) = y_{\cdot jk} \\ \log \dfrac{m_{i'j'k'}(\boldsymbol{\omega}) m_{i'bc}(\boldsymbol{\omega}) m_{aj'c}(\boldsymbol{\omega}) m_{abk'}(\boldsymbol{\omega})}{m_{i'j'c}(\boldsymbol{\omega}) m_{i'bk'}(\boldsymbol{\omega}) m_{aj'k'}(\boldsymbol{\omega}) m_{abc}(\boldsymbol{\omega})} = \omega_{i'j'k'}, \end{cases} \tag{14.11}$$

where $\boldsymbol{\omega}$ is a vector of $\{\omega_{i'j'k'}\}$ arranged in dictionary order. The $m_{ijk}(\boldsymbol{\omega})$ are determined similarly, starting from the q_{ijk} of (14.10) and adjusting two-way marginal totals (14.11). Then the asymptotic distribution of $\{y_{ijk}\}$ is normal,

$$\text{const.exp}\left[-\frac{1}{2} \sum_i \sum_j \sum_k \frac{\{y_{ijk} - m_{ijk}(\boldsymbol{\omega})\}^2}{m_{ijk}(\boldsymbol{\omega})} \right],$$

under the condition

$$\frac{y_{ij\cdot}}{y_{\cdots}} \to \pi_{ij\cdot}, \ \frac{y_{i\cdot k}}{y_{\cdots}} \to \pi_{i\cdot k}, \ \frac{y_{\cdot jk}}{y_{\cdots}} \to \pi_{\cdot jk},$$

where $\{\pi_{ij\cdot}\}$, $\{\pi_{i\cdot k}\}$, and $\{\pi_{\cdot jk}\}$ are the positive numbers that satisfy

$$\sum_j \pi_{ij\cdot} = \sum_k \pi_{i\cdot k}, \ \sum_i \pi_{ij\cdot} = \sum_k \pi_{\cdot jk}, \ \sum_i \pi_{i\cdot k} = \sum_j \pi_{\cdot jk},$$
$$\sum_i \sum_j \pi_{ij\cdot} = 1, \ \sum_i \sum_k \pi_{i\cdot k} = 1, \ \sum_j \sum_k \pi_{\cdot jk} = 1$$

(see Plackett, 1981 for details). This is a degenerated distribution, because of the constraints

$$\sum_i \{y_{ijk} - m_{ijk}(\boldsymbol{\omega})\} = 0, \ \sum_j \{y_{ijk} - m_{ijk}(\boldsymbol{\omega})\} = 0, \ \sum_k \{y_{ijk} - m_{ijk}(\boldsymbol{\omega})\} = 0.$$

Let $\boldsymbol{A}^{*\prime}$ be an $(ab + bc + ca - a - b - c + 1) \times abc$ non-singular matrix, such that $\boldsymbol{A}^{*\prime}\boldsymbol{y}$ forms a set of sufficient statistics under H_ω. It can be obtained, for example, by deleting some of the sufficient statistics $(y_{ij\cdot}, y_{i\cdot k}, y_{\cdot jk})$ so that they become linearly independent of each other. Then, by virtue of equation (6.19), the conditional variance–covariance of $\boldsymbol{y}$ matrix is

$$\boldsymbol{\Omega} - \boldsymbol{\Omega}\boldsymbol{A}^* \left(\boldsymbol{A}^{*\prime}\boldsymbol{\Omega}^{-1}\boldsymbol{A}^*\right)^{-1} \boldsymbol{A}^{*\prime}\boldsymbol{\Omega}, \tag{14.12}$$

where $\boldsymbol{\Omega}$ is a diagonal matrix with $m_{ijk}(\boldsymbol{\omega})$ as its diagonal elements arranged in dictionary order. If we allow for $\boldsymbol{A}^{*\prime}$ being singular, so that $\boldsymbol{A}^{*\prime}\boldsymbol{y}$ represents all of the sufficient statistics $(y_{ij\cdot}, y_{i\cdot k}, y_{\cdot jk})$, we can simply employ a generalized inverse of a matrix in the expression (14.12). An example of the construction of non-singular matrix $\boldsymbol{A}^*$ is given in detail in Section 14.2.2 (1) (a).
Let $\boldsymbol{m}(\boldsymbol{\omega})$ be a vector of $m_{ijk}(\boldsymbol{\omega})$ arranged in dictionary order. Then the quadratic form

$$\chi^2(\boldsymbol{\omega}) = \{\boldsymbol{y} - \boldsymbol{m}(\boldsymbol{\omega})\}'\boldsymbol{\Omega}^{-1}\{\boldsymbol{y} - \boldsymbol{m}(\boldsymbol{\omega})\} = \sum_i \sum_j \sum_k \frac{\{y_{ijk} - m_{ijk}(\boldsymbol{\omega})\}^2}{m_{ijk}(\boldsymbol{\omega})} \tag{14.13}$$

is distributed as a chi-squared distribution with df $(a-1)(b-1)(c-1)$, since

$$\boldsymbol{\Omega}^{-1}\left\{\boldsymbol{\Omega} - \boldsymbol{\Omega}\boldsymbol{A}^*\left(\boldsymbol{A}^{*\prime}\boldsymbol{\Omega}^{-1}\boldsymbol{A}^*\right)^{-1}\boldsymbol{A}^{*\prime}\boldsymbol{\Omega}\right\}$$

is easily verified to be an idempotent matrix with trace equal to $(a-1)(b-1)(c-1)$ (see Lemma 2.1). Under the null hypothesis H_ω the $\chi^2(\boldsymbol{0})$ obtained by putting $\boldsymbol{\omega} = \boldsymbol{0}$ in equation (14.13) is distributed as a chi-squared variable with df $(a-1)(b-1)(c-1)$, where $\boldsymbol{m}(\boldsymbol{0})$ is obtained by the iterative scaling procedure for adjusting two-way marginal totals (14.11), starting from the three-way table $\{q_{ijk}\}$ with $q_{ijk} \equiv 1$ by putting $\omega_{i'j'k'} = 0$ in (14.10).

(b) **Unconditional test of $\boldsymbol{H_\omega}$.** The goodness-of-fit chi-squared statistic is obtained as follows. We go back to the multinomial distribution $M(y_{\cdots}, \boldsymbol{p})$ with $p_{\cdots} = 1$. First, under the saturated model (14.4), the MLEs of the p_{ijk} and m_{ijk} are obviously

$$\hat{p}_{ijk} = y_{ijk}/y_{\cdots} \text{ and } \hat{m}_{ijk} = y_{ijk}, . \tag{14.14}$$

where the hat denotes the MLE. Under the null hypothesis H_ω we maximize the likelihood function (14.8) at $\omega_{i'j'k'} \equiv 0$. Then, after some calculation, we have the likelihood equation

$$\hat{m}_{ij\cdot}(\boldsymbol{0}) = y_{ij\cdot}\,,\ \hat{m}_{i\cdot k}(\boldsymbol{0}) = y_{i\cdot k},\ \hat{m}_{\cdot jk}(\boldsymbol{0}) = y_{\cdot jk},$$
$$\log\frac{\hat{m}_{i'j'k'}(\boldsymbol{0})\hat{m}_{i'bc}(\boldsymbol{0})\hat{m}_{aj'c}(\boldsymbol{0})\hat{m}_{abk'}(\boldsymbol{0})}{\hat{m}_{i'j'c}(\boldsymbol{0})\hat{m}_{i'bk'}(\boldsymbol{0})\hat{m}_{aj'k'}(\boldsymbol{0})\hat{m}_{abc}(\boldsymbol{0})} = 0.$$

That is, the unconditional MLE $\hat{m}_{ijk}(\boldsymbol{0})$ of m_{ijk} coincides with $m_{ijk}(\boldsymbol{0})$ obtained by (14.11) in the previous section. This means that $\chi^2(0)$ (14.13) is nothing but the (unconditional) goodness-of-fit chi-squared statistic.

The likelihood ratio test is then given by

$$S_\omega = 2\log\frac{\hat{L}}{\hat{L}_{\omega=\mathbf{0}}} = 2\sum_i\sum_j\sum_k y_{ijk}\log\frac{y_{ijk}}{m_{ijk}(\mathbf{0})}, \tag{14.15}$$

where $\hat{L}$ and $\hat{L}_{\omega=\mathbf{0}}$ denote the likelihood evaluated at $\hat{m}_{ijk}$ (14.14) and $\hat{m}_{ijk}(\mathbf{0})$, respectively.

(c) **Exact conditional distribution for a $2\times b\times c$ table under H_ω.** An exact conditional null distribution under H_ω has been obtained for a $2\times b\times c$ table in tractable form by Hirotsu *et al.* (2001). In this case the null distribution under H_ω is formally

$$\Pr\{Y_{ijk}=y_{ijk}|y_{ij\cdot},y_{i\cdot k},y_{\cdot jk}\} = C_c^{-1}(y_{ij\cdot},y_{i\cdot k},y_{\cdot jk};i=1,\,2,j=1,\dots,\,b,k=1,\dots,\,c)$$
$$\times\Pi_{i=1}^2\Pi_{j=1}^b\Pi_{k=1}^c(y_{ijk}!)^{-1}. \tag{14.16}$$

Usually, the normalizing constant C_c has not been obtained even for a moderately large table, because of the computational problem. It can, however, be obtained as follows by a similar recursion formula using the Markov property as in Chapter 6.

Lemma 14.1. Exact null distribution under H_ω for $2\times b\times c$ table. Define $\boldsymbol{y}_{1k}=(y_{11k},\dots,y_{1b-1k})'$ and its partial sum $\boldsymbol{Y}_{1k}=\boldsymbol{y}_{11}+\cdots+\boldsymbol{y}_{1k}$, where the capital letters are employed following Chapter 6 to express accumulated statistics instead of the usual terminology for random variables. Then, the factorization of the distribution (14.16) is obtained in terms of $\boldsymbol{Y}_{1k}$ as

$$\Pr\{Y_{ijk}=y_{ijk}|y_{ij\cdot},y_{i\cdot k},y_{\cdot jk}\} = \Pi_{k=1}^{c-1}f_k(\boldsymbol{Y}_{1k}|\boldsymbol{Y}_{1k+1},\boldsymbol{A}^{*\prime}\boldsymbol{y}),$$

$$f_k(\boldsymbol{Y}_{1k}|\boldsymbol{Y}_{1k+1},\boldsymbol{A}^{*\prime}\boldsymbol{y}) = C_{k+1}^{-1}(\boldsymbol{Y}_{1k+1},y_{i\cdot m},y_{\cdot jm};i=12,j=1,\dots,\,b,m=1,\dots,\,k+1)$$
$$\times C_k(\boldsymbol{Y}_{1k},y_{i\cdot m},y_{\cdot jm})\Pi_{i=1}^2\Pi_{j=1}^b\{(Y_{ijk+1}-Y_{ijk})!\}^{-1}, \tag{14.17}$$

where Y_{1jk} is a component of $\boldsymbol{Y}_{1k}$, Y_{2jk} is defined by $\sum_{m=1}^k y_{\cdot jm}-Y_{1jk}$, and $C_k(\boldsymbol{Y}_{1k},y_{i\cdot m},y_{\cdot jm})$ is the summation of $\Pi_{i=1}^2\Pi_{j=1}^b\Pi_{m=1}^k(y_{ijm}!)^{-1}$ with respect to y_{ijm} subject to the condition of fixed marginal totals $\boldsymbol{Y}_{1k},y_{i\cdot m},y_{\cdot jm};i=12,j=1,\dots,$ $b,m=1,\dots,\,k$. It is determined by the recursion formula

$$C_k\left(\boldsymbol{Y}_{1k}, y_{i\cdot m}, y_{\cdot jm}\right) = \sum_{\boldsymbol{Y}_{1k-1}} C_{k-1}\left(\boldsymbol{Y}_{1k-1}, y_{i\cdot m}, y_{\cdot jm}\right) \Pi_{i=1}^{2} \Pi_{j=1}^{b} \left\{\left(Y_{ijk} - Y_{ijk-1}\right)!\right\}^{-1},\ k = 2, \ldots, c,$$

starting from

$$C_1\left(\boldsymbol{Y}_{11}, y_{i\cdot m}, y_{\cdot jm}\right) = \Pi_{i=1}^{2} \Pi_{j=1}^{b} \left\{\left(Y_{ij1}\right)!\right\}^{-1},$$

and coincides with C_c of (14.16) at $k = c$.

Proof. In the log linear model approach of the three-way contingency table, the model under H_ω is most controversial and an exact distribution is not known. It makes the problem difficult that an amalgamation invariance does not hold; that is, for the collapsed table with respect to k, the null model H_ω no longer holds (Darroch, 1974). However, since sub-table invariance holds, the probability function up to $m = 1,\ldots k$ is given by

$$C_{k+1}^{-1}\left(\boldsymbol{Y}_{1k+1}, y_{i\cdot m}, y_{\cdot jm}; i = 12, j = 1, \ldots, b, m = 1, \ldots, k+1\right) \Pi_{i=1}^{2} \Pi_{j=1}^{b} \Pi_{m=1}^{k+1} \left(y_{ijm}!\right)^{-1},$$

and it is factorized into a product of two probability functions,

$$C_k^{-1} \Pi_{i=1}^{2} \Pi_{j=1}^{b} \Pi_{m=1}^{k} \left(y_{ijm}!\right)^{-1} \times C_{k+1}^{-1} C_k \Pi_{i=1}^{2} \Pi_{j=1}^{b} \left\{\left(Y_{ijk+1} - Y_{ijk}\right)!\right\}^{-1}. \qquad (14.18)$$

The latter part of (14.18) is nothing but the conditional distribution of (14.17). It should be noted that the random variable $\boldsymbol{Y}_{1k}$ is included also in $C_k\left(\boldsymbol{Y}_{1k}, y_{i\cdot m}, y_{\cdot jm}\right)$, the normalizing constant one step before.

The result of Lemma 14.1 is utilized in Section 14.2.2 (2) (a).

(2) Test of two-way interaction

(a) Conditional test of the null hypothesis H_τ: $\log p_{ijk} = \mu + \alpha_i + \beta_j + \gamma_k + \theta_{ij} + \varphi_{ik}$ assuming H_ω. Next we consider a test of the null hypothesis $H_\tau : \tau_{jk} = 0$, assuming H_ω. The sufficient statistics under $H_\omega \cap H_\tau$ are $\left(y_{i'\cdot\cdot}, y_{\cdot j'\cdot}, y_{\cdot\cdot k'}, y_{i'j'\cdot}, y_{i'\cdot k'}\right)$, or simply $\left(y_{ij\cdot}, y_{i\cdot k}\right)$ in a redundant expression, and the sufficient statistics for the parameter $\tau_{j'k'}$ are $y_{\cdot j'k'}$. Therefore, the inference should be based on $y_{\cdot j'k'}$ conditionally on $\left(y_{ij\cdot}, y_{i\cdot k}\right)$. For each i the distribution $MH\left(y_{ijk} \mid y_{ij\cdot}, y_{i\cdot k}\right)$ of y_{ijk} under $H_\tau \cap H_\omega$ given $\left(y_{ij\cdot}, y_{i\cdot k}\right)$ is a multivariate hypergeometric distribution,

$$\frac{\left(\Pi_{j=1}^{b} y_{ij\cdot}!\right)\left(\Pi_{k=1}^{c} y_{i\cdot k}!\right)}{y_{i\cdot\cdot}!\left(\Pi_{k=1}^{c}\Pi_{j=1}^{b} y_{ijk}!\right)},\ i=1,\ \ldots,\ a.$$

The expectation and covariance of this distribution are therefore easily obtained as follows:

$$E\left(Y_{ijk}\mid y_{ij\cdot}, y_{i\cdot k}\right)=y_{ij\cdot}\,y_{i\cdot k}/y_{i\cdot\cdot}$$

$$Cov\left(Y_{i_1j_1k_1}, Y_{i_2j_2k_2}\mid y_{ij\cdot}, y_{i\cdot k}\right)=\begin{cases}\dfrac{y_{ij_1\cdot}\left(\delta_{j_1j_2}y_{i\cdot\cdot}-y_{ij_2\cdot}\right)y_{i\cdot k_1}\left(\delta_{k_1k_2}y_{i\cdot\cdot}-y_{i\cdot k_2}\right)}{\left\{y_{i\cdot\cdot}^2\left(y_{i\cdot\cdot}-1\right)\right\}}, & i_1=i_2=i,\\ 0, \text{ otherwise.}\end{cases}$$

Then, the expectation and covariance of $Y_{\cdot jk}$ are obtained as

$$E\left(Y_{\cdot jk}\mid y_{ij\cdot}, y_{i\cdot k}\right)=\sum_i y_{ij\cdot}\,y_{i\cdot k}/y_{i\cdot\cdot},$$

$$Cov\left(Y_{\cdot j_1k_1}, Y_{\cdot j_2k_2}\mid y_{ij\cdot}, y_{i\cdot k}\right)=\sum_i\frac{y_{ij_1\cdot}\left(\delta_{j_1j_2}y_{i\cdot\cdot}-y_{ij_2\cdot}\right)y_{i\cdot k_1}\left(\delta_{k_1k_2}y_{i\cdot\cdot}-y_{i\cdot k_2}\right)}{\left\{y_{i\cdot\cdot}^2\left(y_{i\cdot\cdot}-1\right)\right\}}.$$

Let $\boldsymbol{t}$ be a vector of $y_{\cdot j'k'}$,

$$\boldsymbol{t}=\left(y_{\cdot 11},\ \ldots, y_{\cdot 1c-1},\ y_{\cdot 21},\ \ldots, y_{\cdot 2c-1},\ \ldots, y_{\cdot b-1c-1}\right)', \tag{14.19}$$

and $E(\boldsymbol{t})$ and $V(\boldsymbol{t})$ be the expectation and variance–covariance matrix of $\boldsymbol{t}$. Then the quadratic form of $\boldsymbol{t}$,

$$\{\boldsymbol{t}-E(\boldsymbol{t})\}'V^{-1}(\boldsymbol{t})\{\boldsymbol{t}-E(\boldsymbol{t})\}, \tag{14.20}$$

is nothing but the efficient score test of Birch (1965), which is asymptotically distributed as chi-squared with df $f_\tau=(b-1)(c-1)$. The chi-squared approximation of the statistic (14.20) is good for moderately large $y_{ij\cdot}$ and $y_{i\cdot k}$, because of the summation with respect to i.

(b) **Unconditional test of $\boldsymbol{H_\tau}$.** The unconditional MLE of the cell frequencies $m_{ijk}(\mathbf{0}, \mathbf{0})$ under $H_\omega \cap H_\tau$ satisfies the equation

$$m_{i\cdot k}(\mathbf{0}, \mathbf{0})=y_{i\cdot k},\ m_{ij\cdot}(\mathbf{0}, \mathbf{0})=y_{ij\cdot}, \log\frac{m_{ij'k'}(\mathbf{0}, \mathbf{0})m_{ibc}(\mathbf{0}, \mathbf{0})}{m_{ij'c}(\mathbf{0}, \mathbf{0})m_{ibk'}(\mathbf{0}, \mathbf{0})}=0.$$

This equation is solved at once, giving

$$m_{ijk}(\mathbf{0}, \mathbf{0})=y_{ij\cdot}y_{i\cdot k}/y_{i\cdot\cdot}.$$

Therefore, the likelihood ratio test is obtained as

$$S_\tau=2\log\frac{\hat{L}_{\omega=0}}{\hat{L}_{\tau=0,\ \omega=0}}=2\sum_i\sum_j\sum_k y_{ijk}\log\frac{m_{ijk}(\mathbf{0})}{m_{ijk}(\mathbf{0}, \mathbf{0})}. \tag{14.21}$$

This statistic is asymptotically distributed as chi-squared with df $f_\tau = (b-1)(c-1)$ under $H_\omega \cap H_\tau$ and the condition $y_{ij\cdot}/y_{\cdots} \to \pi_{ij\cdot}$, $y_{i\cdot k}/y_{\cdots} \to \pi_{i\cdot k}$. In contrast, a more convenient expression of S_τ is obtained as

$$S_\tau = 2\sum_i \sum_j \sum_k m_{ijk}(\mathbf{0}) \log \frac{m_{ijk}(\mathbf{0})}{m_{ijk}(\mathbf{0}, \mathbf{0})}.$$

The goodness-of-fit chi-squared asymptotically equivalent to S_τ is

$$\chi_\tau^2 = \sum_i \sum_j \sum_k \frac{\{m_{ijk}(\mathbf{0}) - m_{ijk}(\mathbf{0}, \mathbf{0})\}^2}{m_{ijk}(\mathbf{0}, \mathbf{0})}.$$

(c) **Unconditional test of the null hypothesis H_φ:log $p_{ijk} = \mu + \alpha_i + \beta_j + \gamma_k + \theta_{ij}$ assuming $H_\omega \cap H_\tau$.** This test is equivalent to the test of the null hypothesis

$$p_{i\cdot k} = p_{i\cdot\cdot} \times p_{\cdot\cdot k}$$

in the collapsed two-way table of $\{y_{i\cdot k}\}$. Therefore, the goodness-of-fit chi-squared and the likelihood ratio test are given by

$$\chi_\varphi^2 = \sum_i \sum_k \frac{(y_{i\cdot k} - y_{i\cdot\cdot} y_{\cdot\cdot k}/y_{\cdots})^2}{y_{i\cdot\cdot} y_{\cdot\cdot k}/y_{\cdots}}$$

$$\text{and } S_\varphi = 2\sum_i \sum_k y_{i\cdot k} \log \frac{y_{i\cdot k}}{y_{i\cdot\cdot} y_{\cdot\cdot k}/y_{\cdots}}$$

$$= 2\left(\sum_i \sum_k y_{i\cdot k} \log y_{i\cdot k} - \sum_i y_{i\cdot\cdot} \log y_{i\cdot\cdot} - \sum_k y_{\cdot\cdot k} \log y_{\cdot\cdot k} + y_{\cdots} \log y_{\cdots}\right). \quad (14.22)$$

This is also obtained as follows. The MLE under $H_\omega \cap H_\tau$ was $m_{ijk}(\mathbf{0}, \mathbf{0}) = y_{ij\cdot} y_{i\cdot k}/y_{i\cdot\cdot}$ and the MLE under $H_\omega \cap H_\tau \cap H_\varphi$ is easily obtained as $m_{ijk}(\mathbf{0}, \mathbf{0}, \mathbf{0}) = y_{ij\cdot} y_{\cdot\cdot k}/y_{\cdots}$. Therefore, twice the log likelihood is

$$S_\varphi = 2\sum_i \sum_j \sum_k y_{ijk} \log \frac{m_{ijk}(\mathbf{0}, \mathbf{0})}{m_{ijk}(\mathbf{0}, \mathbf{0}, \mathbf{0})}$$

$$= 2\sum_i \sum_j \sum_k m_{ijk}(\mathbf{0}, \mathbf{0}) \log \frac{m_{ijk}(\mathbf{0}, \mathbf{0})}{m_{ijk}(\mathbf{0}, \mathbf{0}, \mathbf{0})}$$

$$= 2\sum_i \sum_j \sum_k y_{ijk} \log \frac{y_{ij\cdot} y_{i\cdot k}/y_{i\cdot\cdot}}{y_{ij\cdot} y_{\cdot\cdot k}/y_{\cdots}} = (14.22)$$

(d) **Unconditional test of the null hypothesis H_θ: log$p_{ijk} = \mu + \alpha_i + \beta_j + \gamma_k$ assuming $H_\omega \cap H_\tau \cap H_\varphi$.** This test is equivalent to the test of independence

$$p_{ij\cdot}=p_{i\cdot\cdot}\times p_{\cdot j\cdot}$$

in the collapsed two-way table $\{y_{ij\cdot}\}$. Therefore, the goodness-of-fit chi-squared and the likelihood ratio test are given by

$$\chi_\theta^2=\sum_i\sum_j\frac{(y_{ij\cdot}-y_{i\cdot\cdot}y_{\cdot j\cdot}/y_{\cdots})^2}{y_{i\cdot\cdot}y_{\cdot j\cdot}/y_{\cdots}}$$

$$\text{and } S_\theta=2\sum_i\sum_j y_{ij\cdot}\log\frac{y_{ij\cdot}}{y_{i\cdot\cdot}y_{\cdot j\cdot}/y_{\cdots}} \tag{14.23}$$

$$=2\left(\sum_i\sum_j y_{ij\cdot}\log y_{ij\cdot}-\sum_i y_{i\cdot\cdot}\log y_{i\cdot\cdot}-\sum_j y_{\cdot j\cdot}\log y_{\cdot j\cdot}+y_{\cdots}\log y_{\cdots}\right).$$

This is also obtained as follows. The MLE under $H_\omega\cap H_\tau\cap H_\varphi\cap H_\theta$ is obviously $m_{ijk}(\mathbf{0},\mathbf{0},\mathbf{0},\mathbf{0})=y_{i\cdot\cdot}y_{\cdot j\cdot}y_{\cdot\cdot k}/y_{\cdots}^2$. Therefore,

$$S_\theta=2\sum_i\sum_j\sum_k y_{ijk}\log\frac{m_{ijk}(\mathbf{0},\mathbf{0},\mathbf{0})}{m_{ijk}(\mathbf{0},\mathbf{0},\mathbf{0},\mathbf{0})}$$

$$=2\sum_i\sum_j\sum_k m_{ijk}(\mathbf{0},\mathbf{0},\mathbf{0})\log\frac{m_{ijk}(\mathbf{0},\mathbf{0},\mathbf{0})}{m_{ijk}(\mathbf{0},\mathbf{0},\mathbf{0},\mathbf{0})}$$

$$=2\sum_i\sum_j y_{ij\cdot}\log\frac{y_{\cdots}y_{ij\cdot}}{y_{i\cdot\cdot}y_{\cdot j\cdot}}=(14.23)$$

The series of test statistics S_ω (14.15), S_τ (14.21), S_φ (14.22), and S_θ (14.23) is the factorization of the test statistic of the likelihood ratio test of overall independence,

$$p_{ijk}=p_{i\cdot\cdot}\times p_{\cdot j\cdot}\times p_{\cdot\cdot k},$$

against the saturated model. The following equation is easily verified:

$$2\sum_i\sum_j\sum_k y_{ijk}\log\frac{y_{ijk}}{y_{i\cdot\cdot}y_{\cdot j\cdot}y_{\cdot\cdot k}/y_{\cdots}^2}=2\sum_i\sum_j\sum_k y_{ijk}\log\frac{y_{ijk}}{m_{ijk}(\mathbf{0})}$$
$$+2\sum_i\sum_j\sum_k m_{ijk}(\mathbf{0})\log\frac{m_{ijk}(\mathbf{0})}{m_{ijk}(\mathbf{0},\mathbf{0})}+2\sum_i\sum_j\sum_k m_{ijk}(\mathbf{0},\mathbf{0})\log\frac{m_{ijk}(\mathbf{0},\mathbf{0})}{m_{ijk}(\mathbf{0},\mathbf{0},\mathbf{0})}$$
$$+2\sum_i\sum_j\sum_k m_{ijk}(\mathbf{0},\mathbf{0},\mathbf{0})\log\frac{m_{ijk}(\mathbf{0},\mathbf{0},\mathbf{0})}{m_{ijk}(\mathbf{0},\mathbf{0},\mathbf{0},\mathbf{0})}.$$

14.1.2 Cumulative chi-squared statistics for the ordered categorical responses

(1) **Order-restricted inference of the null hypothesis H_ω**

(a) **Assuming simple order effects for all the row, column, and layer categories.** First, from (14.13) and (14.12) we note that

$$\chi^2_\omega(\mathbf{0}) = \{\mathbf{y}-\mathbf{m}(\mathbf{0})\}'\boldsymbol{\Omega}^{-1}\{\mathbf{y}-\mathbf{m}(\mathbf{0})\} = \textstyle\sum_i\sum_j\sum_k \dfrac{\{y_{ijk}-m_{ijk}(\mathbf{0})\}^2}{m_{ijk}(\mathbf{0})}$$

is the quadratic form of $(\mathbf{P}'_a\otimes\mathbf{P}'_b\otimes\mathbf{P}'_c)\{\mathbf{y}-\mathbf{m}(\mathbf{0})\}$ by the inverse of its variance–covariance matrix

$$(\mathbf{P}'_a\otimes\mathbf{P}'_b\otimes\mathbf{P}'_c)\left\{\boldsymbol{\Omega}-\boldsymbol{\Omega}\mathbf{A}^*\left(\mathbf{A}^{*\prime}\boldsymbol{\Omega}^{-1}\mathbf{A}^*\right)^{-1}\mathbf{A}^{*\prime}\boldsymbol{\Omega}\right\}(\mathbf{P}_a\otimes\mathbf{P}_b\otimes\mathbf{P}_c)$$
$$=\left\{(\mathbf{P}'_a\otimes\mathbf{P}'_b\otimes\mathbf{P}'_c)\boldsymbol{\Omega}^{-1}(\mathbf{P}_a\otimes\mathbf{P}_b\otimes\mathbf{P}_c)\right\}^{-1}$$

This is easily verified, since the two-way marginal totals of $\mathbf{y}$ and $\mathbf{m}(\mathbf{0})$ coincide. For the assumed order effects we introduce the accumulated statistics

$$(\mathbf{P}^{*\prime}_a\otimes\mathbf{P}^{*\prime}_b\otimes\mathbf{P}^{*\prime}_c)\{\mathbf{y}-\mathbf{m}(\mathbf{0})\}$$

and the sum of squares of its standardized components χ^{***2}. The matrix $\mathbf{P}^{*\prime}_a$ is defined in Section 6.5.3 (2) (b), regarding the cumulative chi-squared χ^{*2} in a one-way layout. Let the diagonal elements of

$$\mathbf{V}^{***} = (\mathbf{P}^{*\prime}_a\otimes\mathbf{P}^{*\prime}_b\otimes\mathbf{P}^{*\prime}_c)\left\{\boldsymbol{\Omega}-\boldsymbol{\Omega}\mathbf{A}^*\left(\mathbf{A}^{*\prime}\boldsymbol{\Omega}^{-1}\mathbf{A}^*\right)^{-1}\mathbf{A}^{*\prime}\boldsymbol{\Omega}\right\}(\mathbf{P}^*_a\otimes\mathbf{P}^*_b\otimes\mathbf{P}^*_c) \qquad (14.24)$$

be $v_{i'j'k'}$, and $\mathbf{D}(v_{i'j'k'}) = \mathrm{diag}\{v_{i'j'k'}\}$ a diagonal matrix of $v_{i'j'k'}$ arranged in dictionary order. Then the cumulative chi-squared statistic is defined by

$$\chi^{***2}_\omega = \|\mathbf{D}^{-1/2}(v_{i'j'k'})(\mathbf{P}^{*\prime}_a\otimes\mathbf{P}^{*\prime}_b\otimes\mathbf{P}^{*\prime}_c)\{\mathbf{y}-\mathbf{m}(\mathbf{0})\}\|^2,$$

where $\mathbf{D}^{-1/2}(v_{i'j'k'})$ is a diagonal matrix of $v^{-1/2}_{i'j'k'}$. The constants for the chi-squared approximation $d\chi^2_f$ are obtained as usual by

$$\kappa_1 = E(\chi^{***2}_\omega) = \mathrm{tr}\{\mathbf{D}^{-1}(v_{i'j'k'})\mathbf{V}^{***}\} = (a-1)(b-1)(c-1) = df, \qquad (14.25)$$

$$\kappa_2 = V(\chi^{***2}_\omega) = 2\mathrm{tr}\left[\{\mathbf{D}^{-1}(v_{i'j'k'})\mathbf{V}^{***}\}^2\right] = 2d^2f. \qquad (14.26)$$

(b) **Assuming simple order effects for the row and column categories.** For the assumed order effects we introduce the two-way accumulated statistic

$$(P_a^{*\prime} \otimes P_b^{*\prime} \otimes P_c^{\prime})\{y - m(0)\}.$$

Let $V_{i'j'}$, $i' = 1, \ldots, a-1; j' = 1, \ldots, b-1$ be a variance–covariance matrix

$$\begin{aligned} V_{i'j'} = &\{r^{*\prime}(1,\ldots,i';i'+1,\ldots,a),\otimes,c^{*\prime}(1,\ldots,j';j'+1,\ldots,b)\otimes P_c^{\prime}\} \\ &\times\left\{\Omega - \Omega A^*\left(A^{*\prime}\Omega^{-1}A^*\right)^{-1}A^{*\prime}\Omega\right\} \\ &\times\{r^*(1,\ldots,i';i'+1,\ldots,a)\otimes c^*(1,\ldots,j';j'+1,\ldots,b)\otimes P_c\}, \end{aligned}$$

where $r^{*\prime}$ and $c^{*\prime}$ are defined as the rows of $P_a^{*\prime}$ and $P_b^{*\prime}$ in Sections 10.4.1 (3) and 10.4.2 (2), respectively. That is, each $V_{i'j'}$ is a principal $(c-1)\times(c-1)$ sub-matrix of

$$V^{**} = (P_a^{*\prime}\otimes P_b^{*\prime}\otimes P_c^{\prime})\left\{\Omega - \Omega A^*\left(A^{*\prime}\Omega^{-1}A^*\right)^{-1}A^{*\prime}\Omega\right\}(P_a^*\otimes P_b^*\otimes P_c).$$

Then the two-way cumulative chi-squared is defined by

$$\chi_\omega^{**2} = \left[(P_a^{*\prime}\otimes P_b^{*\prime}\otimes P_c^{\prime})\{y-m(0)\}\right]^{\prime} D^{-1}(V_{i'j'})\left[(P_a^{*\prime}\otimes P_b^{*\prime}\otimes P_c^{\prime})\{y-m(0)\}\right],$$

where $D(V_{i'j'})$ is a block diagonal matrix with $V_{i'j'}$ as its diagonal elements in dictionary order. The constants for the chi-squared approximation for the statistic χ^{**2} are given by

$$\begin{aligned} \kappa_1 &= E(\chi_\omega^{**2}) = \mathrm{tr}\{D^{-1}(V_{i'j'})V^{**}\} = (a-1)(b-1)(c-1) = df \\ \kappa_2 &= V(\chi_\omega^{**2}) = 2\mathrm{tr}\left[\{D^{-1}(V_{i'j'})V^{**}\}^2\right] = 2d^2f. \end{aligned},$$

(c) **Assuming simple order effects only for the row categories.** For the assumed order effects, we introduce the cumulative sum statistic

$$(P_a^{*\prime}\otimes P_b^{\prime}\otimes P_c^{\prime})\{y-m(0)\}.$$

Let $V_{i'}$ be a variance–covariance matrix

$$\begin{aligned} V_{i'} = &\{r^{*\prime}(1,\ldots,i';i'+1,\ldots,a)\otimes P_b^{\prime}\otimes P_c^{\prime}\}\left\{\Omega - \Omega A^*\left(A^{*\prime}\Omega^{-1}A^*\right)^{-1}A^{*\prime}\Omega\right\} \\ &\times\{r^*(1,\ldots,i';i'+1,\ldots,a)\otimes P_b\otimes P_c\}, \end{aligned}$$

which is a principal $\{(b-1)(c-1)\}\times\{(b-1)(c-1)\}$ sub-matrix of

$$V^* = (P_a^{*\prime}\otimes P_b^{\prime}\otimes P_c^{\prime})\left\{\Omega - \Omega A^*\left(A^{*\prime}\Omega^{-1}A^*\right)^{-1}A^{*\prime}\Omega\right\}(P_a^*\otimes P_b\otimes P_c).$$

Then the cumulative chi-squared is defined by

$$\chi_{\omega}^{*2} = \left[\left(\boldsymbol{P}_a^{*\prime}\otimes\boldsymbol{P}_b^{\prime}\otimes\boldsymbol{P}_c^{\prime}\right)\{\boldsymbol{y}-\boldsymbol{m}(\boldsymbol{0})\}\right]^{\prime}\boldsymbol{D}^{-1}(\boldsymbol{V}_{i^{\prime}})\left[\left(\boldsymbol{P}_a^{*}\otimes\boldsymbol{P}_b\otimes\boldsymbol{P}_c\right)\{\boldsymbol{y}-\boldsymbol{m}(\boldsymbol{0})\}\right],$$

where $\boldsymbol{D}(\boldsymbol{V}_{i^{\prime}})$ is a block diagonal matrix with $\boldsymbol{V}_{i^{\prime}}$ as its diagonal elements in dictionary order. The constants for the chi-squared approximation for the statistic χ^{*2} are given by

$$\kappa_1 = E\left(\chi_{\omega}^{*2}\right) = \mathrm{tr}\left\{\boldsymbol{D}^{-1}(\boldsymbol{V}_{i^{\prime}})\boldsymbol{V}^*\right\} = (a-1)(b-1)(c-1) = df,$$
$$\kappa_2 = V\left(\chi_{\omega}^{*2}\right) = 2\mathrm{tr}\left[\left\{\boldsymbol{D}^{-1}(\boldsymbol{V}_{i^{\prime}})\boldsymbol{V}^*\right\}^2\right] = 2d^2f.$$

(2) **Order-restricted inference of the null hypothesis $\boldsymbol{H}_\tau$ assuming $\boldsymbol{H}_\omega$.** Birch's test is easy to extend to a cumulative chi-squared type test, since it is an efficient score test. We can simply define the accumulated statistics according to the ordered column and/or layer categories.

(a) **Assuming simple order effects for both column and layer categories.** We follow the notation of Section 14.1.1 (2) (a) and to define the cumulative sum statistics we introduce a lower triangular matrix of unities:

$$\boldsymbol{T}_l^{\prime} = \begin{bmatrix} 1 & 0 & 0 & 0 & \cdots & 0 & 0 & 0 \\ 1 & 1 & 0 & 0 & \cdots & 0 & 0 & 0 \\ 1 & 1 & 1 & 0 & \cdots & 0 & 0 & 0 \\ & & & & \vdots & & & \\ 1 & 1 & 1 & 1 & \cdots & 1 & 1 & 1 \end{bmatrix}_{l\times l}.$$

Then the two-way accumulated statistics are defined by

$$\boldsymbol{t}^{**} = (\boldsymbol{T}_{b-1}\otimes\boldsymbol{T}_{c-1})^{\prime}\{\boldsymbol{t}-E(\boldsymbol{t})\}, \tag{14.27}$$

where $\boldsymbol{t}$ is defined in (14.19). Define the variance–covariance matrix of $\boldsymbol{t}^{**}$ by

$$\boldsymbol{W}^{**} = (\boldsymbol{T}_{b-1}\otimes\boldsymbol{T}_{c-1})^{\prime}V(\boldsymbol{t})(\boldsymbol{T}_{b-1}\otimes\boldsymbol{T}_{c-1})$$

and its diagonal elements by $w_{j^{\prime}k^{\prime}}$, where $V(\boldsymbol{t})$ is defined in Section 14.1.1 (2) (a). Further, define a diagonal matrix $\boldsymbol{D}(w_{j^{\prime}k^{\prime}}) = \mathrm{diag}\{w_{j^{\prime}k^{\prime}}\}$ by diagonal elements $w_{j^{\prime}k^{\prime}}$ arranged in dictionary order. Then the cumulative chi-squared statistic is defined by

$$\chi_{\tau}^{**2} = \|\,\boldsymbol{D}^{-1/2}(w_{j^{\prime}k^{\prime}})\boldsymbol{t}^{**}\,\|^2.$$

The constants for the chi-squared approximation for the statistic χ_{τ}^{**2} are given by

$$\kappa_1 = E\left(\chi_\tau^{**2}\right) = \mathrm{tr}\left\{\boldsymbol{D}^{-1}\left(w_{j'k'}\right)\boldsymbol{W}^{**}\right\} = (b-1)(c-1) = df, \tag{14.28}$$

$$\kappa_2 = V\left(\chi_\tau^{**2}\right) = 2\mathrm{tr}\left[\left\{\boldsymbol{D}^{-1}\left(w_{j'k'}\right)\boldsymbol{W}^{**}\right\}^2\right] = 2d^2f. \tag{14.29}$$

(b) **Assuming simple order effects only for the column categories.** The accumulated statistics are defined by

$$\boldsymbol{t}^* = \left(\boldsymbol{T}'_{b-1}\otimes\boldsymbol{I}_{c-1}\right)\boldsymbol{t}.$$

Define the variance–covariance matrix of $\boldsymbol{t}^*$ by

$$\boldsymbol{W}^* = \left(\boldsymbol{T}'_{b-1}\otimes\boldsymbol{I}_{c-1}\right)V(\boldsymbol{t})\left(\boldsymbol{T}_{b-1}\otimes\boldsymbol{I}_{c-1}\right)$$

and its principal $(c-1)\times(c-1)$ sub-matrix by $\boldsymbol{W}_{j'}, j' = 1, \ldots, b-1$. Define a block diagonal matrix $\boldsymbol{D}\left(\boldsymbol{W}_{j'}\right)$ with diagonal elements $\boldsymbol{W}_{j'}$ arranged in dictionary order. Then the cumulative chi-squared statistic and the constants for the chi-squared approximation are given by

$$\chi_\tau^{*2} = \boldsymbol{t}^{*\prime}\boldsymbol{D}^{-1}\left(\boldsymbol{W}_{j'}\right)\boldsymbol{t}^*,$$

$$\kappa_1 = E\left(\chi_\tau^{*2}\right) = \mathrm{tr}\left\{\boldsymbol{D}^{-1}\left(\boldsymbol{W}_{j'}\right)\boldsymbol{W}^*\right\} = (b-1)(c-1) = df,$$

$$\kappa_2 = V\left(\chi_\tau^{*2}\right) = 2\mathrm{tr}\left[\left\{\boldsymbol{D}^{-1}\left(\boldsymbol{W}_{j'}\right)\boldsymbol{W}^*\right\}^2\right] = 2d^2f.$$

There are various variations of max acc. t type test statistics, and some of them are given later in the examples of this chapter.

14.2 One-Way Experiment with Two-Way Categorical Responses

14.2.1 General theory

A multinomial distribution $M\left(y_{i\cdot\cdot}, p_{ijk} \mid p_{i\cdot\cdot} = 1\right)$ is assumed at each level of the factor A: $\Pr\left\{Y_{ijk} = y_{ijk}, j = 1, \ldots, b; k = 1, \ldots, c\right\} = y_{i\cdot\cdot}!\Pi_j\Pi_k\left(p_{ijk}^{y_{ijk}}/y_{ijk}!\right)$. Therefore, the total probability function is

$$\Pr\left\{Y_{ijk} = y_{ijk}, i = 1, \ldots, a; j = 1, \ldots, b; k = 1, \ldots, c\right\} = \Pi_i\left\{y_{i\cdot\cdot}!\Pi_j\Pi_k\left(p_{ijk}^{y_{ijk}}/y_{ijk}!\right)\right\}.$$

Employing Plackett's identification condition we have the likelihood function

$$L = \frac{\Pi_i y_{i..}!\, p_{ibc}^{y_{i..}}}{\Pi_i \Pi_j \Pi_k y_{ijk}!} \exp\Big(\sum_{j'} y_{\cdot j' \cdot} \beta_{j'} + \sum_{k'} y_{\cdot\cdot k'} \gamma_{k'} + \sum_{i'} \sum_{j'} y_{i'j'\cdot} \theta_{i'j'}$$

$$+ \sum_{i'} \sum_{k'} y_{i'\cdot k'} \varphi_{i'k'} + \sum_{j'} \sum_{k'} y_{\cdot j'k'} \tau_{j'k'} + \sum_{i'} \sum_{j'} \sum_{k'} y_{i'j'k'} \omega_{i'j'k'} \Big), \quad (14.30)$$

where it should be noted that p_{abc} has been replaced by p_{ibc} and the parameter $\alpha_{i'}$ deleted in (14.8). The definition of the other parameters is the same as in equation (14.7). In this case the p_{ijk} are expressed by the new parameters as

$$p_{ijk} = p_{ibc} \exp\left(\beta_j + \gamma_k + \theta_{ij} + \varphi_{ik} + \tau_{jk} + \omega_{ijk} \right),$$

where

$$p_{ibc} = \Big\{ 1 + \sum_{j'} \exp\left(\beta_{j'} + \theta_{ij'} \right) + \sum_{k'} \exp(\gamma_{k'} + \varphi_{ik'})$$

$$+ \sum_{j'} \sum_{k'} \exp\left(\beta_{j'} + \gamma_{k'} + \theta_{ij'} + \varphi_{ik'} + \tau_{j'k'} + \omega_{ij'k'} \right) \Big\}^{-1}$$

so that $p_{i..} = 1$. It should be noted that the number of new parameters is $a(bc-1)$ and they are interpreted as follows. In the ith two-way table the interaction between the factors B and C is expressed as

$$\log \frac{p_{ij'k'} p_{ibc}}{p_{ij'c} p_{ibk'}}, j' = 1, \ldots, b-1, k' = 1, \ldots, c-1.$$

The hypothesis that the interactions are equivalent for all levels i is expressed as the null hypothesis

$$H_\omega : \omega_{i'j'k'} = \log \frac{p_{i'j'k'} p_{i'bc}}{p_{i'j'c} p_{i'bk'}} - \log \frac{p_{aj'k'} p_{abc}}{p_{aj'c} p_{abk'}} \equiv 0, i' = 1, \ldots, a-1, j' = 1, \ldots, b-1, k' = 1, \ldots, c-1.$$

If the null hypothesis H_ω is not rejected, we are interested in testing the null hypothesis of effects of A on B and C by

$$H_\theta : \theta_{i'j'} = 0, i' = 1, \ldots, a-1, j' = 1, \ldots, b-1,$$

and

$$H_\varphi : \varphi_{i'k'} = 0, i' = 1, \ldots, a-1, k' = 1, \ldots, c-1$$

assuming H_ω. This is the case of Section 14.1.1 (2), and Birch's test is available in addition to the likelihood ratio and the goodness-of-fit chi-squared tests. Because of the similarities of the likelihood function, all the test procedures are found to be the same as in the previous section and the difference exists only in the interpretation.

If both hypotheses H_θ and H_φ are rejected, then the independence between factors B and C is tested by

$$H_\tau : \tau_{j'k'} = 0, j' = 1, \ldots, b-1; k' = 1, \ldots, c-1,$$

assuming H_ω. If H_τ is not rejected, then the recommended model is

$$H_\omega \cap H_\tau : p_{ijk} = p_{ij\cdot} \times p_{i\cdot k}.$$

That is, the factors B and C are independent at each level of i but the effects of A on B and C are different for each i and estimated by

$$\hat{p}_{ij\cdot} = \frac{y_{ij\cdot}}{y_{i\cdot\cdot}}, \hat{p}_{i\cdot k} = \frac{y_{i\cdot k}}{y_{i\cdot\cdot}}.$$

Other possible cases are as follows:

$$H_\omega \cap H_\theta : p_{ijk} = p_{i\cdot k} \times a^{-1} p_{\cdot jk} / \left(a^{-1} p_{\cdot\cdot k}\right) \tag{14.31}$$

$$H_\omega \cap H_\varphi : p_{ijk} = p_{ij\cdot} \times a^{-1} p_{\cdot jk} / \left(a^{-1} p_{\cdot j\cdot}\right) \tag{14.32}$$

$$H_\omega \cap H_\theta \cap H_\tau : p_{ijk} = p_{i\cdot k} \times a^{-1} p_{\cdot j\cdot} \tag{14.33}$$

$$H_\omega \cap H_\varphi \cap H_\tau : p_{ijk} = p_{ij\cdot} \times a^{-1} p_{\cdot\cdot k} \tag{14.34}$$

$$H_\omega \cap H_\tau \cap H_\varphi \cap H_\theta : p_{ijk} = a^{-1} p_{\cdot j\cdot} \times a^{-1} p_{\cdot\cdot k}. \tag{14.35}$$

Models (14.31) and (14.32) are a little strange, since the factor A affects only one of B or C, while B and C are dependent on each other. Model (14.33) is easy to interpret, where the factors B and C are independent and the factor A affects only C. Therefore, the occurrence probability of $p_{\cdot j\cdot}$ is estimated by the total observations, while $p_{i\cdot k}$ is estimated only by the observation at level i. That is, we have

$$a^{-1}\hat{p}_{\cdot j\cdot} = \frac{y_{\cdot j\cdot}}{y_{\cdot\cdot\cdot}}, \hat{p}_{i\cdot k} = \frac{y_{i\cdot k}}{y_{i\cdot\cdot}}.$$

Similarly, under $H_\omega \cap H_\tau \cap H_\varphi$ (14.34) the factors B and C are independent and the factor A affects only B; the occurrence probabilities are estimated by

$$\hat{p}_{ij\cdot} = \frac{y_{ij\cdot}}{y_{i\cdot\cdot}}, a^{-1}\hat{p}_{\cdot\cdot k} = \frac{y_{\cdot\cdot k}}{y_{\cdot\cdot\cdot}}.$$

Finally, under $H_\omega \cap H_\tau \cap H_\varphi \cap H_\theta$ (14.35), the equation

$$p_{ijk} = a^{-1} p_{\cdot j\cdot} \times a^{-1} p_{\cdot\cdot k}$$

holds. That is, the factors B and C are independent and there is no effect of A on B and C. The estimates are given by

$$a^{-1}\hat{p}_{\cdot j\cdot} = \frac{y_{\cdot j\cdot}}{y_{\cdot\cdot\cdot}}, a^{-1}\hat{p}_{\cdot\cdot k} = \frac{y_{\cdot\cdot k}}{y_{\cdot\cdot\cdot}}.$$

In this section the cell frequencies are estimated by $y_{i\cdot\cdot}\hat{p}_{ijk}$, differently from the $y_{\cdots}\hat{p}_{ijk}$ of the previous section. Therefore, we have estimates of the cell frequencies for the respective models as

$$H_\omega \cap H_\tau \to y_{i\cdot\cdot}\hat{p}_{ijk} = y_{i\cdot\cdot} \times \frac{y_{ij\cdot}}{y_{i\cdot\cdot}} \times \frac{y_{i\cdot k}}{y_{i\cdot\cdot}} = \frac{y_{ij\cdot}y_{i\cdot k}}{y_{i\cdot\cdot}},$$

$$H_\omega \cap H_\theta \cap H_\tau \to y_{i\cdot\cdot}\hat{p}_{ijk} = y_{i\cdot\cdot} \times \frac{y_{i\cdot k}y_{\cdot j\cdot}}{y_{i\cdot\cdot}\,y_{\cdots}} = \frac{y_{i\cdot k}y_{\cdot j\cdot}}{y_{\cdots}},$$

$$H_\omega \cap H_\varphi \cap H_\tau \to y_{i\cdot\cdot}\hat{p}_{ijk} = y_{i\cdot\cdot} \times \frac{y_{ij\cdot}\,y_{\cdot\cdot k}}{y_{i\cdot\cdot}\,y_{\cdots}} = \frac{y_{ij\cdot}y_{\cdot\cdot k}}{y_{\cdots}},$$

$$H_\omega \cap H_\tau \cap H_\varphi \cap H_\theta \to y_{i\cdot\cdot}\hat{p}_{ijk} = y_{i\cdot\cdot} \times \frac{y_{\cdot j\cdot}}{y_{\cdots}} \times \frac{y_{\cdot\cdot k}}{y_{\cdots}} = \frac{y_{i\cdot\cdot}y_{\cdot j\cdot}y_{\cdot\cdot k}}{y_{\cdots}^2}.$$

As a result, these estimators are equivalent to those obtained in previous sections under the respective models.

14.2.2 Applications

(1) **Application of the cumulative chi-squared statistics.** The data of Table 14.2 are the number of cancer patients cross-classified by age A (four levels, $i = 1, \ldots, 4$), metastasis of cancer B (two levels, $j = 1, 2$), and saturation C (three levels, $k = 1, 2, 3$) (Hirotsu, 1992). They are characterized by the natural ordering in age and saturation. This is the case of A×B×C, but the test statistics are the same as in the case $A \times B \times C$. Assuming a log linear model (14.30), we begin by testing H_ω, the effects of age on metastasis and saturation.

(a) **Test of the null hypothesis $\boldsymbol{H_\omega}$.** This is the case of 14.1.2 (1) (b), with natural ordering in C instead of B. However, since the factor B is binary, we can also apply the formula of (a) for χ_ω^{***2}, taking $\boldsymbol{P}_b^{*\prime} = \left(-2^{1/2}, 2^{1/2}\right)$ and that would be easier. The expected cell frequencies $\boldsymbol{m}(\mathbf{0})$ under H_ω are obtained by the iterative scaling procedure as in Table 14.3. There is standard software for this, but it is also easy to calculate.

The conditional variance–covariance matrix of $\boldsymbol{y}$ given the sufficient statistics $\left(y_{ij\cdot}, y_{i\cdot k}, y_{\cdot jk}\right)$ is

$$\boldsymbol{\Omega} - \boldsymbol{\Omega}\boldsymbol{A}^*\left(\boldsymbol{A}^{*\prime}\boldsymbol{\Omega}^{-1}\boldsymbol{A}^*\right)^{-1}\boldsymbol{A}^{*\prime}\boldsymbol{\Omega} \tag{14.36}$$

from (14.12), where $\boldsymbol{\Omega}$ is a diagonal matrix with $m_{ijk}(\mathbf{0})$ as its diagonal elements arranged in dictionary order. The matrix $\boldsymbol{A}^*$ is determined so as to produce the sufficient statistics. To make it full rank, we employ the same idea which derived equation (10.49). That is, we express the log linear model under H_ω in matrix form as

Table 14.2 Cancer patients cross-classified by levels of age i, metastasis j, and saturation k.

	$i=1$			$i=2$			$i=3$			$i=4$		
j	$k=1$	$k=2$	$k=3$	$k=1$	$k=2$	$k=3$	$k=1$	$k=2$	$k=3$	$k=1$	$k=2$	$k=3$
1	9	5	12	10	9	15	6	13	11	8	5	9
2	4	3	5	5	9	16	2	9	18	4	3	9
Total	13	8	17	15	18	31	8	22	29	12	8	18

Table 14.3 Expected cell frequencies under H_{ω}.

	$i=1$			$i=2$			$i=3$			$i=4$		
j	$k=1$	$k=2$	$k=3$	$k=1$	$k=2$	$k=3$	$k=1$	$k=2$	$k=3$	$k=1$	$k=2$	$k=3$
1	10.02	5.54	10.44	9.70	9.91	14.39	5.08	11.83	13.09	8.20	4.72	9.08
2	2.98	2.46	6.56	5.30	8.09	16.61	2.92	10.17	15.91	3.80	3.28	8.92
Total	13	8	17	15	18	31	8	22	29	12	8	18

$$\log p_{ijk} = \boldsymbol{A}^* \begin{bmatrix} \mu \\ \boldsymbol{\alpha}^* \\ \boldsymbol{\beta}^* \\ \boldsymbol{\gamma}^* \\ \boldsymbol{\theta}^* \\ \boldsymbol{\varphi}^* \\ \boldsymbol{\tau}^* \end{bmatrix} = \left[\boldsymbol{j}\ \boldsymbol{X}_\alpha^*\ \boldsymbol{X}_\beta^*\ \boldsymbol{X}_\gamma^*\ \boldsymbol{X}_\theta^*\ \boldsymbol{X}_\varphi^*\ \boldsymbol{X}_\tau^*\right] \begin{bmatrix} \mu \\ \boldsymbol{\alpha}^* \\ \boldsymbol{\beta}^* \\ \boldsymbol{\gamma}^* \\ \boldsymbol{\theta}^* \\ \boldsymbol{\varphi}^* \\ \boldsymbol{\tau}^* \end{bmatrix}, \tag{14.37}$$

where $\boldsymbol{\alpha}^* = (\alpha_1, \alpha_2, \alpha_3)'$, $\boldsymbol{\beta}^* = (\beta_1)'$, $\boldsymbol{\gamma}^* = (\gamma_1, \gamma_2)'$, $\boldsymbol{\theta}^* = (\theta_{11}, \theta_{21}, \theta_{31})'$, $\boldsymbol{\varphi}^* = (\varphi_{11}, \varphi_{12}, \varphi_{21}, \varphi_{22}, \varphi_{31}, \varphi_{32})'$, $\boldsymbol{\tau}^* = (\tau_{11}, \tau_{12})'$. Equation (14.37) defines a full rank coefficient matrix $\boldsymbol{A}^*$, which should be substituted in (14.36). The calculation is simple matrix multiplications but too big (24×24) to present here. Therefore, we give only the cumulative statistics

$$\boldsymbol{x} = \left(\boldsymbol{P}_a^{*\prime} \otimes \boldsymbol{P}_b^{*\prime} \otimes \boldsymbol{P}_c^{*\prime}\right)\{\boldsymbol{y} - \boldsymbol{m}(\boldsymbol{0})\} = \begin{bmatrix} 2.040 \\ 3.120 \\ 1.247 \\ 3.759 \\ -0.400 \\ 0.160 \end{bmatrix}$$

and its variance–covariance matrix

$$V(\boldsymbol{x}) = \boldsymbol{V}^{***} = \begin{bmatrix} 5.201 & 3.579 & 2.344 & 1.604 & 1.388 & 0.970 \\ 3.579 & 6.786 & 1.582 & 3.479 & 0.970 & 1.693 \\ 2.344 & 1.582 & 4.911 & 3.112 & 2.403 & 1.681 \\ 1.604 & 3.479 & 3.112 & 8.254 & 1.681 & 2.933 \\ 1.388 & 0.970 & 2.403 & 1.681 & 4.163 & 2.911 \\ 0.970 & 1.693 & 1.681 & 2.933 & 2.911 & 5.080 \end{bmatrix} \tag{14.38}$$

by (14.24). It should be noted that the conditional variance (14.36) depends on the choice of A^* but $V(\boldsymbol{x})$ does not depend on it after multiplying $\left(\boldsymbol{P}_a^{*\prime} \otimes \boldsymbol{P}_b^{*\prime} \otimes \boldsymbol{P}_c^{*\prime}\right)$. Then the diagonal matrix $\boldsymbol{D}(v_{i'j'k'}) = \mathrm{diag}(v_{i'j'k'})$ is formed by the diagonal elements of (14.38) and the test statistic is formed as the sum of squares of the standardized accumulated statistics

$$\chi_\omega^{***2} = \chi_\omega^{**2} = \frac{2.040^2}{5.201} + \frac{3.120^2}{6.786} + \frac{1.247^2}{4.911} + \frac{3.759^2}{8.254} + \frac{(-0.400)^2}{4.163} + \frac{0.160^2}{5.080} = 4.307.$$

The constants for the chi-squared approximation are obtained by (14.25) and (14.26) as

$$df = E\left(\chi_{\omega}^{***2}\right) = \mathrm{tr}\left\{\boldsymbol{D}^{-1}\left(v_{i'j'k'}\right)\boldsymbol{V}^{***}\right\} = (a-1)(b-1)(c-1) = 6,$$

$$2d^2f = 2\mathrm{tr}\left\{\boldsymbol{D}^{-1}\left(v_{i'j'k'}\right)\boldsymbol{V}^{***}\right\}^2 = 21.972.$$

giving $d = 1.831, f = 3.277$, and a non-significant p-value 0.266 for $\chi_{\omega}^{***2} = 4.307$. Therefore, we proceed to test H_{θ}, the effects of age on metastasis based on $y_{i'j'\cdot}$ assuming H_{ω}.

(b) **Test of H_{θ} assuming H_{ω}.** This is the pattern of 14.1.2 (2) (b), replacing H_{τ} by H_{θ}. We can, however, apply the formula of 14.1.2 (2) (a) for χ_{τ}^{**2} again and it would be easier. The two-way accumulated statistics are defined by

$$\boldsymbol{t}^{**} = \left(\boldsymbol{T}_3' \otimes \boldsymbol{T}_1'\right)\{\boldsymbol{t} - E(\boldsymbol{t})\}, \tag{14.39}$$

where $\boldsymbol{t}$ is a vector of $y_{i'j'\cdot}$ and it is explicitly $\boldsymbol{t} = (y_{11\cdot}, y_{21\cdot}, y_{31\cdot})' = (26, 34, 30)'$ in this case. The conditional expectation and covariance of y_{ijk} given $y_{i\cdot k}, y_{\cdot jk}$ are as follows:

$$E\left(y_{ijk} \mid y_{i\cdot k}, y_{\cdot jk}\right) = y_{i\cdot k}y_{\cdot jk}/y_{\cdot\cdot k}, \tag{14.40}$$

$$Cov\left(y_{i_1j_1k_1}, y_{i_2j_2k_2} \mid y_{i\cdot k}, y_{\cdot jk}\right) = \begin{cases} \dfrac{y_{i_1\cdot k}\left(\delta_{i_1i_2}y_{\cdot\cdot k} - y_{i_2\cdot k}\right)y_{\cdot j_1k}\left(\delta_{j_1j_2}y_{\cdot\cdot k} - y_{\cdot j_2k}\right)}{\left\{y_{\cdot\cdot k}^2\left(y_{\cdot\cdot k} - 1\right)\right\}}, & k_1 = k_2 = k, \\ 0, \text{ otherwise.} \end{cases} \tag{14.41}$$

We define the vector of expectation $\boldsymbol{\mu}_{\theta_0} = E\left(\boldsymbol{y} \mid y_{i\cdot k}, y_{\cdot jk}\right)$ and the variance–covariance matrix $\boldsymbol{V}_{\theta_0} = V\left(\boldsymbol{y} \mid y_{i\cdot k}, y_{\cdot jk}\right)$ based on (14.40) and (14.41). They are too big to present here, but there is no difficulty in the calculation. Then $E(\boldsymbol{t})$ and $V(\boldsymbol{t})$ are calculated as

$$E(\boldsymbol{t}) = \boldsymbol{X}_{\theta}^{*\prime}\boldsymbol{\mu}_{\theta_0} = \begin{bmatrix} 21.919 \\ 35.935 \\ 32.419 \end{bmatrix}$$

and

$$V(\boldsymbol{t}) = \boldsymbol{X}_{\theta}^{*\prime}\boldsymbol{V}_{\theta_0}\boldsymbol{X}_{\theta}^{*} = \begin{bmatrix} 7.316 & -2.934 & -2.570 \\ -2.934 & 10.584 & -4.703 \\ -2.570 & -4.703 & 9.883 \end{bmatrix}.$$

Therefore we have from (14.39)

$$t^{**} = (T_3' \otimes T_1')\{t - E(t)\} = \begin{bmatrix} 4.081 \\ 2.145 \\ -0.273 \end{bmatrix},$$

and the variance–covariance matrix of t^{**} is obtained as

$$W^{**} = (T_3' \otimes T_1') V(t) (T_3 \otimes T_1) = \begin{bmatrix} 7.316 & 4.382 & 1.812 \\ 4.382 & 12.032 & 4.760 \\ 1.812 & 4.760 & 7.370 \end{bmatrix}.$$

The cumulative chi-squared statistic is obtained by the sum of squares of the standardized elements of t^{**}, and is found to be

$$\chi_\theta^{**2} = \frac{(4.081)^2}{7.316} + \frac{(2.145)^2}{12.032} + \frac{(-0.273)^2}{7.370} = 2.669.$$

Applying formulae (14.28) and (14.29) for the chi-squared approximation, we obtain

$$df = (4-1)(2-1) = 3$$
$$2d^2f = 2\mathrm{tr}\left[\{D^{-1}(w_{j'k'}) W^{**}\}^2\right] = 8.138,$$

where $D(w_{j'k'})$ is a diagonal matrix with diagonal elements 7.316, 12.032, 7.370. Thus, we have $d = 1.356$ and $f = 2.212$. The p-value of $\chi_\theta^{**2} = 2.669$ is 0.30 and not significant at significance level 0.05.

Similarly we can test H_φ, the effects of age on saturation based on $y_{i' \cdot k'}$, assuming H_ω conditionally given $y_{ij\cdot}$, and $y_{\cdot jk}$. In this case $t = (y_{1\cdot 1}, y_{1\cdot 2}, y_{2\cdot 1}, y_{2\cdot 2}, y_{3\cdot 1}, y_{3\cdot 2})'$ $= (13, 8, 15, 18, 8, 22)'$ and

$$t^{**} = (T_3' \otimes T_2')\{t - E(t)\} = \begin{bmatrix} 3.270 \\ -2.739 \\ -0.190 \\ 0.010 \\ -5.839 \\ 5.429 \end{bmatrix}.$$

Then, similarly to testing H_θ, we obtain $\chi_\varphi^{**2} = 4.349$, $d = 1.74$, $f = 3.44$, and the related p-value 0.284. There is no evidence of the effect of age on saturation. We may also test H_φ by the collapsed two-way table $\{y_{i\cdot k}\}$, assuming $H_\omega \cap H_\theta$. As a conclusion, the effects of age on metastasis and saturation are not observed.

Table 14.4 Table of metastasis and saturation obtained by collapsing age levels.

j	$k=1$	$k=2$	$k=3$	Total
1	33	32	47	112
2	15	24	48	87
Total	48	56	95	199

Finally, we test the independence between metastasis and saturation throughout age by the collapsed two-way table $\{y_{\cdot jk}\}$ given in Table 14.4.

The cumulative chi-squared statistic for the collapsed 2×3 table is calculated as follows:

$$\chi^2(1;2,3)=\frac{199(33\times 72-15\times 79)^2}{48\times 151\times 112\times 87}=3.997,$$

$$\chi^2(1,2;3)=\frac{199(65\times 48-39\times 47)^2}{104\times 95\times 112\times 87}=3.424,$$

$$\chi^{*2}=3.585+3.424=7.421.$$

The constants for the chi-squared approximation are $d=1.29$ and$f=1.55$. Since 1.29 $\chi^2_{1.55}(0.05)=6.48$, the observed value of the cumulative chi-squared is significant at level 0.05. In conclusion, there is a positive association between metastasis and saturation but the age effect on them is not observed.

(2) **Application of max acc. *t* type statistic to the analysis of the association between disease and alleles, with particular interest in haplotype analysis.** To analyze the association between disease and allele frequencies at highly polymorphic loci based on the data as shown in Table 14.5, some statistical tests for a $2\times b$ contingency table have been employed. Among them, Sham and Curtis (1995) compared four statistical tests by simulation, including a goodness-of-fit chi-squared test and multiple comparisons of one cell at a time against the others. For the latter test, Hirotsu *et al.* (2001) proposed an exact algorithm called a max one-to-others χ^2 test. This test is a simple modification of max acc. $t1$ of Section 5.2.3, and obviously appropriate if there is only one susceptibility allele in the locus. They do not, however, take into account the natural ordering in the number of CA repeats, whereas an abnormal extension of CAG repeats has been reported associated with Huntington's disease. Of course, differently from CAG repeats, the CA repeats do not correspond to the actual amino acid arrangement, but still are considered to reflect some indications of the disease. In consideration of the natural ordering, the max acc. $t1$ of Section 5.2.3 (2) would be an appropriate approach. In Hirotsu *et al.* (2001), a combined test of these two chi-squares is also proposed when there is no prior information to specify one of

Table 14.5 Allele frequencies at the D20S95 locus.

	97	99	101	103	105	107	109	111	113	115	Total
Schizophrenia	4	30	25	10	17	90	34	37	4	1	252
Normal	5	18	28	6	41	67	5	34	5	5	214
Total	9	48	53	16	58	157	39	71	9	6	466

Table 14.6 p-Values of the proposed methods.

	Statistics	p-Values
Max one-to-others χ^2	18.78 (109 vs. others)	0.000059
Max acc. $t1$	6.59	0.075732
Combined maximal statistic	18.78 (109 vs. others)	0.000113

the alternatives. While the approximate p-values may be obtained easily by Monte Carlo simulation, it is very important to obtain exact p-values in the context of genome scans for linkage disequilibrium, where the adjustment for multiplicity of tests requires us to seek extremely small p-values, not easily estimated by simulation.

The exact p-values of the proposed methods are shown in Table 14.6. Actually, in the research we made simultaneous analyses for 34 loci on the 19, 20, 21, and 22 chromosomes and it was decided beforehand to apply the combined method with adjustment of the p-values for the number of analyses by Bonferroni inequality. The obtained p-values multiplied by 34 for the combined test amount only to 0.0038, inviting further research focused on that locus using independent samples.

In contrast, the simultaneous analyses of the two closely linked loci in a chromosome have been called a haplotype analysis. As an example, Sham and Curtis (1995) analyzed the bivariate allele frequencies of DXS548 (192 ~ 206) and FRAXAC2 (143 ~ 169) as if they were from a $2 \times bc$ two-way table. However, the frequency data are presented in a $2 \times b \times c$ contingency table as shown in Table 14.7. Then it is obvious that we need the analysis of three-way interactions to distinguish two cases as shown in Table 14.8. In Table 14.8 (1) the probability model $p_{ijk} = p_{ij\cdot} \times p_{i\cdot k} / p_{i\cdot\cdot}$ holds, suggesting that the singularities of row 2 and column 3 are associated with the disease without interaction effects of the row and column on the disease. In other words, the row and column are conditionally independent given $i = 1$ (normal) or 2 (disease). On the contrary, Table 14.8 (2) suggests the interaction effect pointing out the singularity of the (2, 3) cell associated with the disease, and in this case the separate analyses of marginal tables collapsing rows or columns are quite

Table 14.7 Haplotype allele frequencies at two loci in fragile X and normal chromosome.

DXS548	FRAXAC2										
	k										
j	1	2	3	4	5	6	7	8	9	10	Total
					Fragile $X (i=1)$						
1	0*	0*	0*	0*	0*	0*	0*	0*	0*	0*	0
2	1*	3	0	0	9	16	12	0*	0*	0*	41
3	0*	5	9	1	11	1	3	0*	0*	0*	30
4	0*	0	0*	0*	0*	0*	1	0*	0*	0*	1
5	0*	0*	0*	0*	0*	0	1	0*	0*	0*	1
6	0*	1	0*	1	9	3	14	0*	0*	0*	28
7	0*	0*	0*	0*	1	0*	0	0*	0*	0*	1
Total	1	9	9	2	30	20	31	0	0	0	102
					Normal $(i=2)$						
1	0*	0*	0*	0*	0*	2*	0*	0*	0*	0*	2
2	1*	7	5	1	17	67	7	4*	1*	0*	110
3	0*	4	6	0	3	8	1	0*	0*	0*	22
4	0*	1	0*	0*	0*	0*	0	0*	0*	0*	1
5	0*	0*	0*	0*	0*	1	1	0*	0*	0*	2
6	0*	2	0*	0	3	6	2	0*	0*	1*	14
7	0*	0*	0*	0*	0	0*	2	0*	0*	0*	2
Total	1	14	11	1	23	84	13	4	1	1	153

Table 14.8 Configuration of p_{ijk} (normalizing constant omitted).

Normal $(i=1)$							Disease $(i=2)$					
Locus 1		Locus 2 k					Locus 1	Locus 2 k				
	j	1	2	3	4	5	j	1	2	3	4	5
(1)	1	1	1	1	1	1	1	1	1	2	1	1
	2	1	1	1	1	1	2	2	2	4	2	2
	3	1	1	1	1	1	3	1	1	2	1	1
	4	1	1	1	1	1	4	1	1	2	1	1
(2)	1	1	1	1	1	1	1	1	1	1	1	1
	2	1	1	1	1	1	2	1	1	4	1	1
	3	1	1	1	1	1	3	1	1	1	1	1
	4	1	1	1	1	1	4	1	1	1	1	1

misleading. The existence of a three-way interaction suggests more strongly the potential candidate gene associated with the disease affecting simultaneously the bivariate allele frequencies. We therefore recommend the analysis of the three-way interaction first for the haplotype analysis. In particular, we propose max one-to-others χ^2 and two-way max acc. $t1$.

(a) **Two-way max acc. *t1* test.** We assume a log linear model

$$\log p_{ijk} = \mu + \alpha_i + \beta_j + \gamma_k + \theta_{ij} + \varphi_{ik} + \tau_{jk} + \omega_{ijk}.$$

and test the null hypothesis H_ω against the alternative hypothesis $H_{\omega 1}$:

$$\begin{cases} \omega_{1jk} - \omega_{1j-1k} - \omega_{1jk-1} + \omega_{1j-1k-1} \geq 0, j=2, \ldots, b; k=2, \ldots, c, \\ \text{or } \omega_{1jk} - \omega_{1j-1k} - \omega_{1jk-1} + \omega_{1j-1k-1} \leq 0, j=2, \ldots, b; k=2, \ldots, c, \\ \text{with at least one inequality strong.} \end{cases} \tag{14.42}$$

It is easy to see that $H_{\omega 1}$ implies that the relative occurrence probability for disease against normal is increasing or decreasing as j and k are increasing. Then, by the complete class Lemma 6.2, the appropriate test should be increasing in every $\left|Y_{1jk}^{**} - \hat{Y}_{1jk}^{**}\right|, j=1, \ldots, b-1, k=1, \ldots, c-1$, where $Y_{1jk}^{**} = \sum_{l=1}^{j} \sum_{m=1}^{k} y_{1lm}$ is the two-way accumulated statistic up to j and k, $\hat{Y}_{1jk}^{**}$ the MLE of the accumulated cell frequency under H_ω, and a function of the two-way marginal totals $\boldsymbol{A}^{*\prime}\boldsymbol{y}$. Then the most natural test would be based on the maximum of

$$\chi_{jk}^{**2} = \frac{\left(Y_{1jk}^{**} - \hat{Y}_{1jk}^{**}\right)^2}{V_{jk}}, j=1, \ldots, b-1; k=1, \ldots, c-1,$$

with V_{jk} the appropriate sum of elements of the variance matrix (14.12), where $\boldsymbol{\Omega}$ is a diagonal matrix with $\hat{y}_{ijk} = \hat{m}_{ijk}(\boldsymbol{0}) = m_{ijk}(\boldsymbol{0})$ as diagonal elements arranged in dictionary order. The V_{jk} is explicitly given by $\boldsymbol{l}_{jk}'\left\{\boldsymbol{\Omega} - \boldsymbol{\Omega}\boldsymbol{A}^*\left(\boldsymbol{A}^{*\prime}\boldsymbol{\Omega}^{-1}\boldsymbol{A}^*\right)^{-1}\boldsymbol{A}^{*\prime}\boldsymbol{\Omega}\right\}\boldsymbol{l}_{jk}$, with $\boldsymbol{l}_{jk} = \boldsymbol{l}_1(2) \otimes \boldsymbol{l}_j(b) \otimes \boldsymbol{l}_k(c)$, where $\boldsymbol{l}_i(a)$ denotes an a-dimensional column vector with the first i elements unity and the rest zero.

For the exact algorithm of the p-value, define the conditional probability

$$F_k\left(\boldsymbol{Y}_{1k}^{**}\right) = \Pr\left(\chi_{lm}^{**2} < c, l=1, \ldots, b-1; m=1, \ldots, k \middle| \boldsymbol{Y}_{1k}^{**}, \boldsymbol{A}^{*\prime}\boldsymbol{y}\right), \tag{14.43}$$

where $\boldsymbol{Y}_{1k}^{**} = \left(Y_{11k}^{**}, \ldots, Y_{1b-1k}^{**}\right)'$. Then we have a recursion formula

$$F_{k+1}\left(\boldsymbol{Y}_{1k+1}^{**}\right) = \sum_{\boldsymbol{Y}_{1k}^{**}} F_k\left(\boldsymbol{Y}_{1k}^{**}\right) \times f_k\left(\boldsymbol{Y}_{1k}^{**} | \boldsymbol{Y}_{1k+1}^{**}, \boldsymbol{A}^{*\prime}\boldsymbol{y}\right) \tag{14.44}$$

where the summation is with respect to $\boldsymbol{Y}_{1k}^{**}$. The conditional distribution $f_k\left(\boldsymbol{Y}_{1k}^{**} | \boldsymbol{Y}_{1k+1}^{**}, \boldsymbol{A}^{*\prime}\boldsymbol{y}\right)$ is obtained similarly to the formula (14.17), just by noting that y_{ijk} is expressed by the doubly accumulated statistics as

$$y_{ijk} = Y_{ijk}^{**} - Y_{ij-1k}^{**} - Y_{ijk-1}^{**} + Y_{ij-1k-1}^{**}.$$

That is, replacing $y_{ijk+1} = Y_{ijk+1} - Y_{ijk}$ of (14.17) by $y_{ijk+1} = Y_{ijk+1}^{**} - Y_{ij-1k+1}^{**} - Y_{ijk}^{**} + Y_{ij-1k}^{**}$. In executing the summation with respect to $Y_{1jk}^{**}, j = 1, \ldots, b-1$, in (14.44), it should be noted that it is defined only in the region

$$L\left(Y_{1jk+1}^{**}, Y_{1j+1k}^{**}, Y_{1j+1k+1}^{**}, y_{\cdot jk}\right) \le Y_{1jk}^{**} \le \mathrm{U}\left(Y_{1jk+1}^{**}, Y_{1j+1k}^{**}, Y_{1j+1k+1}^{**}, y_{\cdot jk}\right),$$

$$L = \max\left(0, Y_{1j+1k}^{**} - Y_{1j+1k+1}^{**} + Y_{1jk+1}^{**}, Y_{1j+1k}^{**} - \sum_{m=1}^{k} y_{\cdot j+1m}, Y_{1jk+1}^{**} - \sum_{l=1}^{j} y_{\cdot lk+1}\right),$$

$$U = \min\left(Y_{1j+1k}^{**}, Y_{1jk+1}^{**}, \sum_{l=1}^{j}\sum_{m=1}^{k} y_{\cdot lm}, Y_{1jk+1}^{**} + Y_{1j+1k}^{**} - Y_{1j+1k+1}^{**} + y_{\cdot j+1k+1}\right).$$

L and U are obtained from Table 14.9, which is essentially the same as Table 14.1 so that every entry of the table expressed by Y_{1jk}^{**} should be positive.

(b) **Max one-to-others chi-squared test.** We assume a log linear model

$$\log p_{ijk} = \mu + \alpha_i + \beta_j + \gamma_k + \theta_{ij} + \varphi_{ik} + \tau_{jk} + \omega_{ijk}.$$

and test the null hypothesis H_ω against the alternative hypothesis

$$H_{\omega 2} : \omega_{ijk} = 0 \text{ for all but one cell } (i, j, k).$$

The test statistic should naturally be

$$\max \chi_{jk}^2, j = 1, \ldots, b; k = 1, \ldots, c,$$

where

$$\chi_{jk}^2 = \left(y_{1jk} - \hat{y}_{1jk}\right)^2 / v_{1jk}$$
$$= \left(y_{2jk} - \hat{y}_{2jk}\right)^2 / v_{2jk}$$

and v_{ijk} is a conditional variance of $y_{ijk} - \hat{y}_{ijk}$ given as the diagonal element of (14.12). Then, expressing y_{1jk} by $Y_{1jk}^{**} - Y_{1j-1k}^{**} - Y_{1jk-1}^{**} + Y_{1j-1k-1}^{**}$, the recursion formula for max χ_{jk}^{**2} can be applied as it is for $\max \chi_{jk}^2$ just by replacing χ_{lm}^{**2} by χ_{lm}^2 based on

Table 14.9 Cell frequencies of pooled $2 \times 2 \times 2$ table expressed in the function of Y^{**}_{1jk}.

Rows	Columns		Total
	$1 \cdots k$	$k+1$	
		$i=1$	
$1, \ldots, j$	Y^{**}_{1jk}	$Y^{**}_{1jk+1} - Y^{**}_{1jk}$	Y^{**}_{1jk+1}
$j+1$	$Y^{**}_{1j+1k} - Y^{**}_{1jk}$	$Y^{**}_{1j+1k+1} - Y^{**}_{1j+1k} - Y^{**}_{1jk+1} + Y^{**}_{1jk}$	$Y^{**}_{1j+1k+1} - Y^{**}_{1jk+1}$
Total	Y^{**}_{1j+1k}	$Y^{**}_{1j+1k+1} - Y^{**}_{1j+1k}$	$Y^{**}_{1j+1k+1}$
		$i=2$	
$1, \ldots, j$	$\sum_{l=1}^{j}\sum_{m=1}^{k} y_{\cdot lm} - Y^{**}_{1jk}$	$\sum_{l=1}^{j} y_{\cdot lk+1} - Y^{**}_{1jk+1} + Y^{**}_{1jk}$	$\sum_{l=1}^{j}\sum_{m=1}^{k+1} y_{\cdot lm} - Y^{**}_{1jk+1}$
$j+1$	$\sum_{m=1}^{k} y_{\cdot j+1m} - Y^{**}_{1j+1k} + Y^{**}_{1jk}$	$y_{\cdot j+1k+1} - Y^{**}_{1j+1k+1} + Y^{**}_{1j+1k} + Y^{**}_{1jk+1} - Y^{**}_{1jk}$	$\sum_{m=1}^{k+1} y_{\cdot j+1m} - Y^{**}_{1j+1k+1} + Y^{**}_{1jk+1}$
Total	$\sum_{l=1}^{j+1}\sum_{m=1}^{k} y_{\cdot lm} - Y^{**}_{1j+1k}$	$\sum_{l=1}^{j+1} y_{\cdot lk+1} - Y^{**}_{1j+1k+1} + Y^{**}_{1j+1k}$	$\sum_{l=1}^{j+1}\sum_{m=1}^{k+1} y_{\cdot lm} - Y^{**}_{1j+1k+1}$

Table 14.10 The values of statistics and their p-values.

	Statistics		p-values
Max one-to-others χ^2	4.799	(7, 7)	0.4878
Max acc. $t1$	4.844	(1–6, 1–5)	0.3747
Combined maximal statistic	4.844	(1–6, 1–5)	0.5344

$Y_{1lm}^{**} - Y_{1l-1m}^{**} - Y_{1lm-1}^{**} + Y_{1l-1m-1}^{**}$, $l = 1, \ldots, b; m = 1, \ldots, k$ in (14.43). The results shown in Table 14.10 suggest that there is no three-way interaction and the separate association analyses for the two loci, FRAXAC2 and DXS548, will be accepted. Then, in the locus DXS548 a highly significant chi-squared component appears in each of the two test statistics of the previous section.

In executing the analysis, it should be noted that the three-way table, table 14.7, is very sparse, obliging some of the y_{ijk} to be fixed when the two-way marginal totals $y_{ij\cdot}, y_{i\cdot k}, y_{\cdot jk}$ are fixed. For example, $y_{11k}, k = 1, \ldots, 10$ must be all zeros, since $y_{11\cdot} = 0$, and y_{21k} must be 0 for $k = 1, 2, 3, 4, 5, 7, 8, 9, 10$, since $y_{\cdot 1k} = 0$ except at $k = 6$. Then, y_{216} must be 2 since $y_{\cdot 16} = 2$ and y_{116} is already known to be zero. By these considerations it is known that all the cells marked $*$ in Table 14.7 are fixed in this particular data set. Then, this information must be included in calculating the variance matrix of (14.12). This can be done either by adding the column to $\boldsymbol{A}^*$, which is composed of zero elements except for the one unit element corresponding to the fixed cell, or equivalently by eliminating those fixed cells in advance. Refer to Hirotsu *et al.* (2001) for more details.

14.3 Two-Way Experiment with One-Way Categorical Responses

14.3.1 General theory

A multinomial distribution $M(y_{ij\cdot}, p_{ijk} \mid p_{ij\cdot} = 1)$ is assumed at each combination of factors A and B: $\Pr\{Y_{ijk} = y_{ijk},\ k = 1, \ldots, c\} = y_{ij\cdot}!\Pi_k \left(p_{ijk}^{y_{ijk}} / y_{ijk}!\right)$. Therefore, the total probability function is

$$\Pr\{Y_{ijk} = y_{ijk}, i = 1, \ldots, a, j = 1, \ldots, b; k = 1, \ldots, c\} = \Pi_i \Pi_j \left\{ y_{ij\cdot}!\Pi_k \left(p_{ijk}^{y_{ijk}} / y_{ijk}!\right)\right\}.$$

Assuming a log linear model and employing Plackett's identification condition, we again have the likelihood function

$$L = \frac{\Pi_i \Pi_j y_{ij\cdot}!\, p_{ijc}^{y_{ij\cdot}}}{\Pi_i \Pi_j \Pi_k y_{ijk}!} \exp\left(\sum_{k'} y_{\cdot\cdot k'} \gamma_{k'} + \sum_{i'} \sum_{k'} y_{i'\cdot k'} \varphi_{i'k'} + \sum_{j'} \sum_{k'} y_{\cdot j'k'} \tau_{j'k'} + \sum_{i'} \sum_{j'} \sum_{k'} y_{i'j'k'} \omega_{i'j'k'}\right),$$

where it should be noted that p_{ibc} is replaced by p_{ijc} and the parameters not related to suffix k' are deleted in (14.30). In this case, the p_{ijk}'s are expressed by the new parameters as

$$p_{ijk} = p_{ijc} \exp\left(\gamma_k + \varphi_{ik} + \tau_{jk} + \omega_{ijk}\right),$$

where $p_{ijc} = \left\{1 + \sum_{k'} \exp\left(\gamma_{k'} + \varphi_{ik'} + \tau_{jk'} + \omega_{ijk'}\right)\right\}^{-1}$, so that $p_{ij\cdot} = 1$. It should be noted that the number of new parameters is $ab(c-1)$, and they are interpreted as follows.

We are basically interested in comparing ab multinomial distributions. This reduces to testing whether the functions log $\left(p_{ijk'}/p_{ijc}\right)$ are equivalent or not for all combinations of (i, j). Then, the interaction between the two factors A and B is expressed by

$$\log \frac{p_{i'j'k'}}{p_{i'j'c}} - \log \frac{p_{i'bk'}}{p_{i'bc}} - \log \frac{p_{aj'k'}}{p_{aj'c}} + \log \frac{p_{abk'}}{p_{abc}} = \omega_{i'j'k'}.$$

If it does not exist, namely under the null hypothesis $H_\omega : \omega_{i'j'k'} = 0$, the main effects of A on C are homogeneous, irrespective of the level j of factor B and expressed by

$$\log \frac{p_{i'1k'}}{p_{i'1c}} - \log \frac{p_{a1k'}}{p_{a1c}} = \cdots = \log \frac{p_{i'bk'}}{p_{i'bc}} - \log \frac{p_{abk'}}{p_{abc}} = \varphi_{i'k'}.$$

Similarly, the main effects of B on C are homogeneous, irrespective of the level i of factor A and expressed by

$$\log \frac{p_{1j'k'}}{p_{1j'c}} - \log \frac{p_{1bk'}}{p_{1bc}} = \cdots = \log \frac{p_{aj'k'}}{p_{aj'c}} - \log \frac{p_{abk'}}{p_{abc}} = \tau_{j'k'}.$$

Again, by the similarity of the likelihood function, the test of $H_\omega : \omega_{i'j'k'} = 0$, the test of $H_\tau : \tau_{j'k'} = 0$ assuming H_ω, and the test of $H_\varphi : \varphi_{i'k'} = 0$ assuming $H_\omega \cap H_\tau$ are the same as given in Section 14.1. The interpretations of the respective models are as follows.

Under $H_\omega \cap H_\tau$ the equation

$$p_{ijk} = b^{-1} p_{i\cdot k}$$

holds. That is, for the categorical response C, only the main effect of A exists. Its estimate is given by

$$b^{-1} \hat{p}_{i\cdot k} = y_{i\cdot k} / y_{i\cdot\cdot}.$$

Under $H_{\omega} \cap H_{\tau} \cap H_{\varphi}$ the equation

$$p_{ijk} = (ab)^{-1} p_{\cdot\cdot k}$$

holds. That is, there is no effect of A and B on C, and the occurrence probability is estimated by

$$(ab)^{-1} \hat{p}_{\cdot\cdot k} = y_{\cdot\cdot k} / y_{\cdot\cdot\cdot}.$$

In this section the cell frequencies are estimated by $y_{ij\cdot} \hat{p}_{ijk}$. For the respective models, they are

$$H_{\omega} \cap H_{\tau} \rightarrow \hat{m}_{ijk} = y_{ij\cdot} \hat{p}_{ijk} = y_{ij\cdot} \times \frac{y_{i\cdot k}}{y_{i\cdot\cdot}}$$

$$H_{\omega} \cap H_{\tau} \cap H_{\varphi} \rightarrow \hat{m}_{ijk} = y_{ij\cdot} \hat{p}_{ijk} = y_{ij\cdot} \times y_{\cdot\cdot k} / y_{\cdot\cdot\cdot}$$

These are formally equivalent to those obtained in previous sections.

The modeling in this section is equivalent to that of the logit linear model if the factor C takes two levels with $p_{ij1} + p_{ij2} = 1$. In this case we can put

$$p_{ij1} = p_{ij},\ p_{ij2} = 1 - p_{ij}$$

and from (14.4) we have

$$\log\left(\frac{p_{ij}}{1 - p_{ij}}\right) = \log p_{ij1} - \log p_{ij2} = \gamma_1 + \varphi_{i1} + \tau_{j1} + \omega_{ij1}. \tag{14.45}$$

We can delete the common suffix 1 in (14.45) and then the equation is nothing but a logit two-way ANOVA model. Further, if $b = 2$ (so in the case of $a \times 2 \times 2$ table),

$$\log \frac{p_{i11}}{p_{i12}} - \log \frac{p_{i21}}{p_{i22}} = \log \frac{p_{i11} p_{i22}}{p_{i12} p_{i21}} = \tau_{11} + \omega_{i11}$$

is the odds ratio of the ith 2×2 table and H_{ω} is nothing but the hypothesis of the homogeneity of the odds ratios through $i = 1, \ldots, a$. This example is given in Section 14.3.2 (2) (a) and (b).

14.3.2 Applications

(1) **Reanalysis of Table 5.10 from a clinical trial.** The data of Table 5.10 are from a typical phase III clinical trial comparing a new drug against an active control in the infectious disease of respiratory organs. The effectiveness of a drug in this field is very sensitive to the existence of pseudomonas, and the results are classified according to its detected case or not. Therefore it is necessary first to check the interaction of pseudomonas vs. drug on the effectiveness by testing the null hypothesis H_{ω}. This is essentially the same approach to testing the constancy of the differences of efficacy rates

between two drugs over the two classes of pseudomonas detected or not in Section 5.3.4, but we are taking here an approach based on the multiplicative model. If H_ω is not rejected, we can proceed to testing the overall effects of drugs through the two classes, where we still need to take care of Simpson's paradox. Classically, the Breslow-Day test has been employed for the interaction test, but it is asymptotically equivalent to the likelihood ratio test of Section 14.1.1 (1) (b). If H_ω is not rejected, Mantel-Haenszel's test is well known for testing the overall effects of drugs in this field. However, this is nothing but Birch's test (14.20) of Section 14.1.1 (2) (a), and is again asymptotically equivalent to the likelihood ratio test of (b). We therefore give only the likelihood ratio test approach here.

(a) **Test of the null hypothesis H_ω.** First we need to calculate the MLE $\hat{m}_{ijk}(\mathbf{0})$ under H_ω by an iterative scaling procedure as explained in Section 14.1.1 (1) (a), and obtained as in Table 14.11. It should be noted that all the two-way marginal totals are retained.

By inserting $\hat{m}_{ijk}(\mathbf{0})$ in (14.15) we get $S_\omega = 1.243$. As a chi-squared variable with df 1, this is not significant at level 0.05.

Before testing the drug effects τ_{jk}, we test the null hypotheses H_θ and H_φ. The test of H_θ is to verify the balance of the pseudomonas detected case or not between the test and control drugs. The test of H_φ is to verify if the pseudomonas affects the effectiveness through both drugs.

(b) **Test of the null hypothesis H_θ assuming H_ω.** This is the case of 14.1.1 (2) (b), replacing H_τ by H_θ. The test statistic is given by (14.21) as

$$S_\theta = 2\sum_i\sum_j\sum_k\left\{y_{ijk}\log m_{ijk}(\mathbf{0}) - y_{ijk}\log m_{ijk}(\mathbf{0},\mathbf{0})\right\}. \quad (14.46)$$

Table 14.11 Maximum likelihood estimates $\hat{m}_{ijk}(\mathbf{0})$ under H_ω.

Pseudomonas	Drug	Effectiveness		
i	j	$k=1\,(-)$	$k=2+(+)$	Total
1 (Detected)	1 (Active)	16.3692	19.6039	36
	2 (Control)	11.6039	11.3961	23
Total		28	31	59
2 (No)	1 (Active)	5.6039	21.3961	27
	2 (Control)	10.3961	32.6039	43
Total		16	54	70
Total	1 (Active)	22	41	63
	2 (Control)	22	44	66

Since the first term of (14.46) has already been calculated for S_ω, only the second term needs to be calculated. In this case the MLE under $H_\omega \cap H_\theta$ is given by

$$m_{ijk}(\mathbf{0}, \mathbf{0}) = y_{i\cdot k} y_{\cdot jk} / y_{\cdot\cdot k}. \tag{14.47}$$

Substituting (14.47) in (14.46), the test statistic is obtained as $S_\theta = 6.719$. This value is highly significant as the chi-squared variable with df 1. That is, the pseudomonas detected case is more often seen in the test drug than in the control.

(c) **Test of the null hypothesis H_φ assuming H_ω.** In this case the MLE under $H_\omega \cap H_\varphi$ is given by

$$m_{ijk}(\mathbf{0}, \mathbf{0}) = y_{ij\cdot} y_{\cdot jk} / y_{\cdot j \cdot}. \tag{14.48}$$

Substituting (14.48) in (14.46), the test statistic is obtained as $S_\varphi = 8.888$ with the associated p -value 0.003 as the chi-squared variable with df 1. That is, the pseudomonas detected case shows a lower success rate.

Since both of the null hypotheses H_θ and H_φ are rejected, this is a possible case of Simpson's paradox. Therefore, we test the drug effects H_τ assuming only H_ω.

(d) **Test of the null hypothesis H_τ assuming H_ω.** This is exactly the case of Section 14.1.1 (2) (b), and $S_\tau = 0.247$ is obtained. Its p-value is by no means significant. In conclusion, there is no significant difference between the test and control drugs after adjusting for the effects of pseudomonas. This result is similar to the result of Example 5.12.

(2) **Test of a trend in odds ratios of $2 \times 2 \times 2$ table**

(a) **Exact analysis.** The data of Table 14.12 are from a case–control study on breast cancer caused by hormone replacement therapy (HRT) (Sala *et al.*, 2000). Long-term hormone therapy for menopausal disorders has been pointed out to raise the risk of breast cancer. Therefore, Sala *et al.* carried out a case–control study on the effects of duration of therapy and menopausal status at the start of therapy, where high and low-risk mammographic patterns are taken as case and control, respectively.

For the data we assume a multivariate generalized hypergeometric distribution

$$\Pr\{Y_{ijk} = y_{ijk} | y_{ij\cdot}, y_{i\cdot k}, y_{\cdot jk}\} = \frac{1}{C(\omega)} \frac{\exp\left(\sum_{i'} y_{i'11} \omega_{i'11}\right)}{\Pi_i \Pi_j \Pi_k y_{ijk}!},$$

Table 14.12 Odds ratio for high-risk mammographic patterns according to HRT use starting time within HRT duration categories.

Duration of therapy (yr)	Case–control	Status at start of therapy		
i	j	$k=1$ (before)	$k=2$ (after)	Odds ratio
1 (<1)	1 (Case)	8	5	1.07
	2 (Control)	9	6	
2 (1 ~ 4)	1 (Case)	22	20	1.70
	2 (Control)	11	17	
3 (5 ≤)	1 (Case)	18	11	9.82
	2 (Control)	2	12	

which is a special case of (14.9) with $b=c=2$. Then we are interested in verifying an increased risk of cancer for the case, according to the duration of therapy. Therefore, we set a one-sided monotone hypothesis

$$H_\omega : \omega_{111} \leq \cdots \leq \omega_{a-111}. \tag{14.49}$$

This can be dealt with as a special case of (14.42), since the second equation is equivalent to $\omega_{111} \leq \cdots \leq \omega_{11c-1}$ when $a=b=2$ and coincides with the situation of this example. Therefore, an exact one-sided test is available based on $Y^*_{i11} - \hat{Y}^*_{i11}$, where Y^*_{i11} is an accumulated statistic with respect to suffix i. The exact p-value is 0.05 one-sided, suggesting an increased risk of cancer according to the duration of HRT with the case starting therapy before menopause.

The complexity of calculation is almost the same with max acc. $t1$ for the binomial data in Chapter 7, except for the range of variables, which is determined with the aid of Table 14.1. Ohta *et al.* (2003) developed a method specific to an $a \times 2 \times 2$ table, and made detailed power comparisons with the restricted likelihood test approach of El Barmi (1997). The powers are rather similar, but the method based on maximal accumulated statistics is much easier to handle. The exact algorithms for the power and also the confidence region of a change-point are also easily obtained by applying the methods of Chapter 8. Thus, an ordered $a \times 2 \times 2$ table is an important special case of example in Section 14.2.2 (2) (a).

(b) **Useful normal approximation for large data.** The data of Table 14.13 are from Ashford and Sowden (1970), and were used by El Barmi (1997) to illustrate the restricted maximum likelihood approach to the trend test of the odds ratios. For the data, we are interested in the effects of age on breathlessness and wheezing. Therefore, this is an example of Section 14.2 but again the one-sided monotone hypothesis (14.49) is of interest, similarly to part (a) of this section. However, an exact test is very time-consuming, since each entry is a little too large in Table 14.13. On the contrary, a very nice normal approximation is available in this case. First, the conditional

Table 14.13 Coal miners classified by breathlessness, wheezing, and age.

Age	Breathlessness	Wheeze	
i	j	$k=1$ (Yes)	$k=2$ (No)
1 (20–24)	1 (Yes)	9	7
	2 (No)	95	1841
2 (25–29)	1 (Yes)	23	9
	2 (No)	105	1654
3 (30–34)	1 (Yes)	54	19
	2 (No)	177	1863
4 (35–39)	1 (Yes)	121	48
	2 (No)	257	2357
5 (40–44)	1 (Yes)	169	54
	2 (No)	273	1778
6 (45–49)	1 (Yes)	269	88
	2 (No)	324	1712
7 (50–54)	1 (Yes)	404	117
	2 (No)	245	1324
8 (55–59)	1 (Yes)	406	152
	2 (No)	225	967
9 (60–64)	1 (Yes)	372	106
	2 (No)	132	526

expectation of $Y_{i11}^* - \hat{Y}_{i11}^*$ given all the two-way marginal totals is zero, since the MLE $\hat{Y}_{i11}^*$ under $H_{\omega 0}$ coincides with the conditional expectation of Y_{i11}^*. Usually, the calculation of the conditional variance based on (14.12) is a little complicated. However, in this case an explicit form of it is obtained as

$$Cov\left(Y_{i11}^*, Y_{i'11}^*\right) = M_i M_{i'}^* / M_a \text{ for } i \le i',$$

where $M_i = \sum_{l=1}^{i} \left\{ m_{l11}^{-1}(\mathbf{0}) + m_{l12}^{-1}(\mathbf{0}) + m_{l21}^{-1}(\mathbf{0}) + m_{l21}^{-1}(\mathbf{0}) \right\}^{-1}$,

$$M_i^* = \sum_{l=i+1}^{a} \left\{ m_{l11}^{-1}(\mathbf{0}) + m_{l12}^{-1}(\mathbf{0}) + m_{l21}^{-1}(\mathbf{0}) + m_{l21}^{-1}(\mathbf{0}) \right\}^{-1},$$

and the correlation matrix of Y_{i11}^* for $i = 1, \ldots, a-1$ is exactly in the form of (6.28) for max acc. $t1$ with $N_i = M_i$ and $N_i^* = M_i^*$. Therefore, an algorithm for max acc. $t1$ can be applied to the standardized statistic $\left(Y_{i11}^* - \hat{Y}_{i11}^*\right)/M_i^{1/2}$ as it is. Thus, we obtain the p-value 0.00015, which is almost the same as obtained by El Barmi. Agresti (2012) also discuss some ordinal effects of age on breathlessness and wheezing.

References

Agresti, A. (2012) *Categorical data analysis*, 3rd edn. Wiley, New York.

Ashford, J. R. and Sowden, R. D. (1970) Multivariate probit analysis. *Biometrics* **26**, 535–546.

Birch, M. W. (1963) Maximum likelihood in three-way contingency tables. *J. Roy. Statist. Soc. B* **25**, 220–233.

Birch, M. W. (1965) The detection of partial association II: The general case. *J. Roy. Statist. Soc. B* **27**, 111–124.

Darroch, J. N. (1974) Multiplicative and additive interaction in contingency tables. *Biometrika* **61**, 207–214.

El Barmi, H. (1997) Testing for or against a trend in odds ratios. *Commun. Statist.: Theor. Meth.* **26**, 1877–1891.

Fienberg, S. E. (1980) *The analysis of cross classified data*. MIT Press, Boston, MA.

Hirotsu, C. (1992) *Analysis of experimental data: Beyond analysis of variance*. Kyoritsu-shuppann, Tokyo (in Japanese).

Hirotsu, C., Aoki, S., Inada, T. and Kitao, Y. (2001) An exact test for the association between the disease and alleles at highly polymorphic loci with particular interest in the haplotype analysis. *Biometrics* **57**, 769–778.

Ohta, E., Aoki, S. and Hirotsu, C. (2003) Evaluating the max accumulated χ^2 in the analysis of K ordered odds ratio parameter. *Jap. J. Appl. Statist.* **32**, 107–126 (in Japanese).

Plackett, R. L. (1981) *The analysis of categorical data*. Griffin, London.

Sala, E., Warren, R., Duffy, S., Lube, R. and Day, N. (2000) High-risk mammographic parenchymal patterns, hormone replacement therapy and other risk factors: A case–control study. *Int. J. Epidemiol.* **29**, 629–636.

Sham, P. C. and Curtis, D. (1995) Monte Carlo tests for associations between disease and alleles at highly polymorphic loci. *Ann. Human Genet.* **59**, 97–105.

15

Design and Analysis of Experiments by Orthogonal Arrays

In the early stage of experiments in a factory it is usual to have to consider many factors simultaneously. It is, however, impossible to perform a multi-way factorial experiment because of its size. It is also unnecessary since higher-order interaction effects are usually small compared with main and two-way interaction effects, and also difficult to interpret even if they exist. It should therefore be better to consider it as a sort of noise at first. Under these circumstances, an orthogonal array is often employed in a factory to collect the necessary information quickly and efficiently.

15.1 Experiments by Orthogonal Array

15.1.1 Orthogonal array

An experiment was planned for seven factors, each with two levels, as shown in Table 15.1. Then, every row of Table 15.2 corresponds to an experiment of the total $n = 16$ experiments for seven factors. The entries ± 1 of the table imply the level of factors; in experiment #1, the levels of all seven factors are set at 1, for example.

If we employ all combinations of the seven factors, that is if we take a seven-way layout, then the number of experiments should be 2^7 without replication. In contrast, the experiments in Table 15.2 are composed of 2^4 experiments and called 2^{7-3} or

Advanced Analysis of Variance, First Edition. Chihiro Hirotsu.

Table 15.1 Factors for improving the fixing time of special aluminum printing.

Factor	Level 1	Level 2
A: Concentration of caustic soda	5.0 kg/m^2	6.0 kg/m^2
B: Temperature of caustic soda	75°C	80°C
C: Time of caustic soda	30 s	40 s
D: Concentration of nitric acid	2.0 kg/m^2	3.0 kg/m^2
E: Material for printing roller	hard gum	soft gum
F: Amount of ink	large	small
G: Drying temperature	170° C	180° C

Table 15.2 Design of 16 experiments.

Experiment #	*A*	*B*	*C*	*D*	*E*	*F*	*G*
1	1	1	1	1	1	1	1
2	1	1	1	−1	−1	−1	−1
3	1	1	−1	1	−1	1	−1
4	1	1	−1	−1	1	−1	1
5	1	−1	1	−1	1	1	−1
6	1	−1	1	1	−1	−1	1
7	1	−1	−1	−1	−1	1	1
8	1	−1	−1	1	1	−1	−1
9	−1	1	1	1	1	1	−1
10	−1	1	1	−1	−1	−1	1
11	−1	1	−1	1	−1	1	1
12	−1	1	−1	−1	1	−1	−1
13	−1	−1	1	−1	1	1	1
14	−1	−1	1	1	−1	−1	−1
15	−1	−1	−1	−1	−1	1	−1
16	−1	−1	−1	1	1	−1	1

one-eighth experiment. The statistical model of this experiment can be expressed in linear form as

$$\mathbf{y} = \begin{bmatrix} y_1 \\ y_2 \\ \vdots \\ y_{16} \end{bmatrix} = \begin{bmatrix} \mu+\alpha_1+\beta_1+\theta_1+\delta_1+\epsilon_1+\varphi_1+\gamma_1+e_1 \\ \mu+\alpha_1+\beta_1+\theta_1+\delta_2+\epsilon_2+\varphi_2+\gamma_2+e_2 \\ \vdots \\ \mu+\alpha_2+\beta_2+\theta_2+\delta_1+\epsilon_1+\varphi_2+\gamma_1+e_{16} \end{bmatrix}. \tag{15.1}$$

Then, a standard linear statistical inference can be applied to estimate the main effects of seven factors and the variance $\sigma^2$of the error e_i. Equivalently, we can employ a linear model taking the entries ± 1 of Table 15.2 as coefficients:

$$\boldsymbol{y} = \begin{bmatrix} y_1 \\ y_2 \\ \vdots \\ y_{16} \end{bmatrix} = \begin{bmatrix} \mu+\alpha+\beta+\theta+\delta+\epsilon+\varphi+\gamma+e_1 \\ \mu+\alpha+\beta+\theta-\delta-\epsilon-\varphi-\gamma+e_2 \\ \vdots \\ \mu-\alpha-\beta-\theta+\delta+\varepsilon-\varphi+\gamma+e_{16} \end{bmatrix} = \boldsymbol{X}\boldsymbol{\vartheta},$$

where $\boldsymbol{\vartheta} = (\mu, \alpha, \beta, \theta, \delta, \epsilon, \varphi, \gamma)'$ is an unknown parameter vector. This is equivalent to imposing identification conditions on the parameters in (15.1) like $\alpha_1 + \alpha_2 = 0$, $\beta_1 + \beta_2 = 0$, and so on. Then, the design matrix $\boldsymbol{X}$ is obviously orthogonal and the estimator

$$\widehat{\boldsymbol{\vartheta}} = (\boldsymbol{X}'\boldsymbol{X})^{-1}\boldsymbol{X}'\boldsymbol{y} = 16^{-1}\boldsymbol{X}'\boldsymbol{y}$$

is obtained at once. Each row of $\boldsymbol{X}'\boldsymbol{y}$ represents the difference between the sums of y_i corresponding to level 1 and 2, respectively. The degrees of freedom for treatment effects are $1 \times 7 = 7$, and therefore the degrees of freedom for the error are $16 - 7 - 1 = 8$. If we employ the 2^7 experiments, the degrees of freedom amount to $128 - 1 = 127$ and should be too large for the necessary estimation problem. If we need to estimate all the interaction effects, the required degrees of freedom are exactly

$$\binom{7}{1} + \binom{7}{2} + \cdots + \binom{7}{7} = 2^7 - 1.$$

However, as stated first, the three- and higher-way interactions are usually small and better to be neglected, at least at an early stage of experiments. This is called the sparsity principle.

In the design of Table 15.2 there are still some degrees of freedom for two-way interaction effects. The contrast to extract the interaction effects is obtained by the products of signs of parent factors in each experiment. The contrast for the interaction $(\alpha\beta)$ is, for example, obtained as $(1, 1, 1, 1, -1, -1, -1, -1, -1, -1, -1, -1, 1, 1, 1, 1)$ by the product of the signs of columns A and B. This vector is orthogonal to every column of $\boldsymbol{X}$, and therefore the interaction $(\alpha\beta)$ is estimable.

Alternatively, the contrast for interaction $(\beta\varphi)$ is $(1, -1, 1, -1, -1, 1, -1, 1, 1, -1, 1, -1, -1, 1, -1, 1)$ by the products of the signs of columns B and F, which coincides with the column of D. Namely, the effects of D and $B \times F$ are confounded and cannot be estimated separately. Therefore, the design given in Table 15.2 is inappropriate for estimating interaction $(\beta\varphi)$.

In the case of $n = 2^4 = 16$ there are 15 orthogonal vectors composed of ± 1, which are shown as columns of Table 15.3. This table is called an orthogonal array and expressed as $L_{16}(2^{15})$, where L is the initial of the Latin square. For two-level

Table 15.3 Orthogonal array $L_{16}(2^{15})$.

Row number	Column number 1	2	3	4	5	6	7	8	9	10	11	12	13	14	15	Data
1	1	1	1	1	1	1	1	1	1	1	1	1	1	1	1	5.9
2	1	1	1	1	1	1	1	−1	−1	−1	−1	−1	−1	−1	−1	8.2
3	1	1	1	−1	−1	−1	−1	1	1	1	1	−1	−1	−1	−1	5.7
4	1	1	1	−1	−1	−1	−1	−1	−1	−1	−1	1	1	1	1	4.7
5	1	−1	−1	1	1	−1	−1	1	1	−1	−1	1	1	−1	−1	5.0
6	1	−1	−1	1	1	−1	−1	−1	−1	1	1	−1	−1	1	1	3.3
7	1	−1	−1	−1	−1	1	1	1	1	−1	−1	−1	−1	1	1	3.7
8	1	−1	−1	−1	−1	1	1	−1	−1	1	1	1	1	−1	−1	5.9
9	−1	1	−1	1	−1	1	−1	1	−1	1	−1	1	−1	1	−1	4.9
10	−1	1	−1	1	−1	1	−1	−1	1	−1	1	−1	1	−1	1	4.5
11	−1	1	−1	−1	1	−1	1	1	−1	1	−1	−1	1	−1	1	4.6
12	−1	1	−1	−1	1	−1	1	−1	1	−1	1	1	−1	1	−1	10.7
13	−1	−1	1	1	−1	−1	1	1	−1	−1	1	1	−1	−1	1	4.4
14	−1	−1	1	1	−1	−1	1	−1	1	1	−1	−1	1	1	−1	8.5
15	−1	−1	1	−1	1	1	−1	1	−1	−1	1	−1	1	1	−1	2.1
16	−1	−1	1	−1	1	1	−1	−1	1	1	−1	1	−1	−1	1	1.0
Factor	*A*	*B*	*A* × *B*	*C*		*B* × *C*	*F* × *G*	*F*		*D*	*A* × *D*	*E*			*G*	

Table 15.4 Column number where the interaction of two specified columns appears.

Column	1	2	3	4	5	6	7	8	9	10	11	12	13	14	15
1	*	3	2	5	4	7	6	9	8	11	10	13	12	15	14
2	3	*	1	6	7	4	5	10	11	8	9	14	15	12	13
3	2	1	*	7	6	5	4	11	10	9	8	15	14	13	12
4	5	6	7	*	1	2	3	12	13	14	15	8	9	10	11
5	4	7	6	1	*	3	2	13	12	15	14	9	8	11	10
6	7	4	5	2	3	*	1	14	15	12	13	10	11	8	9
7	6	5	4	3	2	1	*	15	14	13	12	11	10	9	8
8	9	10	11	12	13	14	15	*	1	2	3	4	5	6	7
9	8	11	10	13	12	15	14	1	*	3	2	5	4	7	6
10	11	8	9	14	15	12	13	2	3	*	1	6	7	4	5
11	10	9	8	15	14	13	12	3	2	1	*	7	6	5	4
12	13	14	15	8	9	10	11	4	5	6	7	*	1	2	3
13	12	15	14	9	8	11	10	5	4	7	6	1	*	3	2
14	15	12	13	10	11	8	9	6	7	4	5	2	3	*	1
15	14	13	12	11	10	9	8	7	6	5	4	3	2	1	*

experiments, the orthogonal arrays $L_8(2^7)$, $L_{32}(2^{31})$, and $L_{64}(2^{63})$ are employed very often in the factories. There are also orthogonal arrays such as $L_9(3^4)$, $L_{27}(3^{13})$, and $L_{81}(3^{40})$ for three-level experiments (Taguchi, 1962), which are omitted here.

The design of Table 15.2 allocates factors A ~ G to columns 1, 2, 4, 10, 12, 8, 15 of Table 15.3. In choosing the allocation, it is necessary to take care that the expected interaction effects are not confounded with the main effects. To see the pattern of confounding, Table 15.4 is useful. It suggests, for example, that the interaction between column 2 (B) and 8 (F) appears at column 10, and thus the effects of $B \times F$ are confounded with the main effect D.

In the aluminum experiment of Table 15.1, the experimenter wished to evaluate the interaction $A \times B$, $B \times C$, $A \times D$, and $F \times G$. In $L_{16}(2^{15})$ there are still eight degrees of freedom left, after taking seven degrees of freedom for the main effects. Therefore, it still seems possible to estimate the four 2 by 2 interaction effects of interest, each with df 1. It seems very complicated to find such a design by trial and error referring to Table 15.4. However, with the aid of an interaction diagram we can obtain such a design very easily.

15.1.2 Planning experiments by interaction diagram

Example 15.1. Planning an experiment for seven factors of Table 15.1 with interest in the two-way interactions $A \times B$, $B \times C$, $A \times D$, and $F \times G$. The required degrees of freedom for the factors of interest are 1×7 for main effects and 1×4 for interaction effects, adding to 11. Therefore, the first choice of the orthogonal array will be $L_{16}(2^{15})$ with 15 degrees of freedom for factors. The required pattern of the interaction diagram is expressed in Fig. 15.1, whose meaning is self-explanatory. There are six patterns of such diagram available for $L_{16}(2^{15})$, as shown in Fig. 15.2 (Taguchi, 1962).

Then, the diagram of Fig. 15.1 can be obtained as part of diagram (2) of Fig. 15.2, as shown in Fig. 15.3. This is nothing but the allocation of Table 15.3 and the rest of the columns (5, 9, 13, 14) are used for estimating errors.

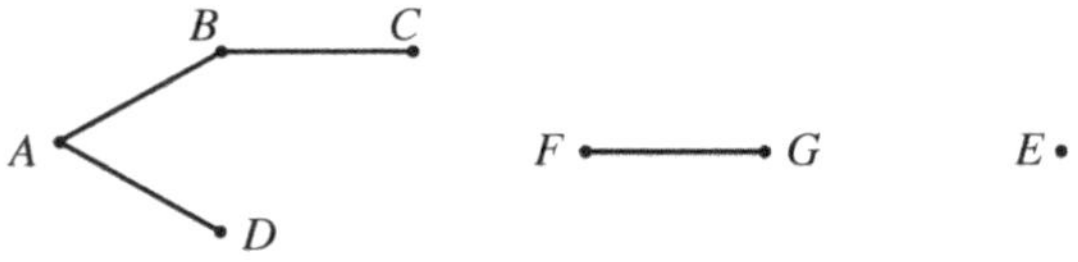

Figure 15.1 The required interaction diagram.

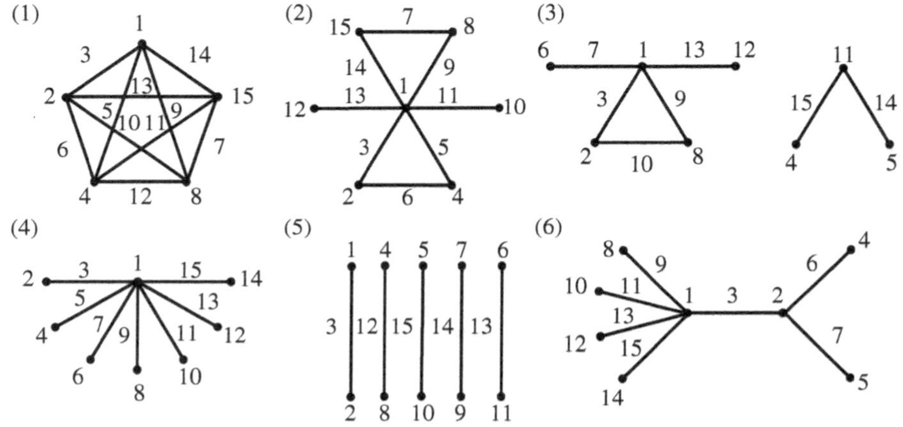

Figure 15.2 The interaction diagrams for $L_{16}(2^{15})$.

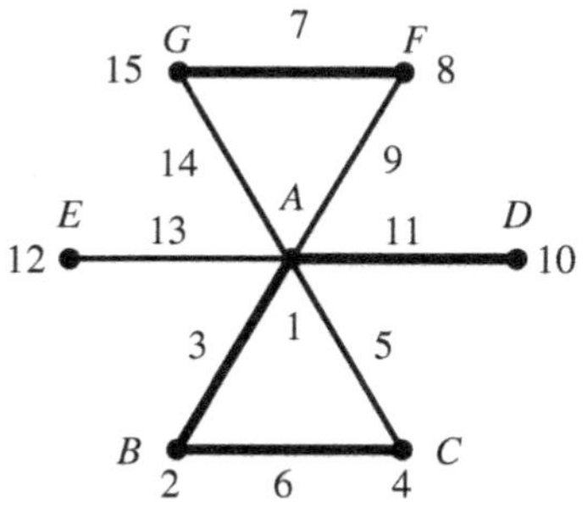

Figure 15.3 The required pattern taken on Fig. 15.2 (2).

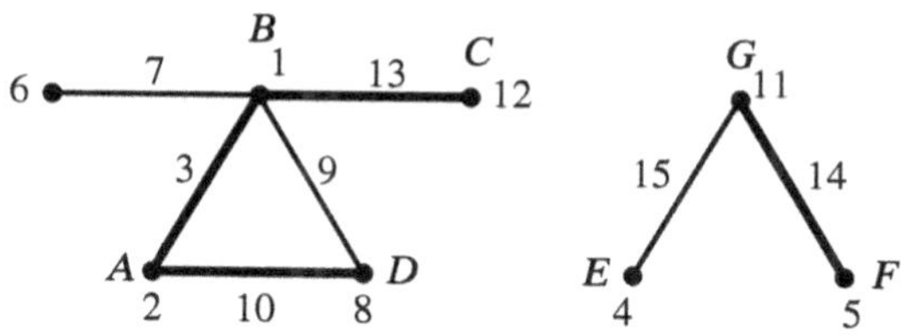

Figure 15.4 The required pattern taken on Fig. 15.2 (3).

The interaction diagram Fig. 15.2 (3) can also be applied as shown in Fig. 15.4, while Fig. 15.2 (6) is not useful for this problem.

Finally, in executing the experiments by the allocation of Table 15.3, all 16 experiments are randomized. Further, the two levels of each factor are assigned randomly to ± 1.

15.1.3 Analysis of experiments from an orthogonal array

The statistical model for the experiments of Table 15.3 is expressed as

$$y = \begin{bmatrix} y_1 \\ y_2 \\ \vdots \\ y_{16} \end{bmatrix} = \begin{bmatrix} \mu+\alpha+\beta+\theta+\delta+\epsilon+\varphi+\gamma+(\alpha\beta)+(\beta\theta)+(\varphi\gamma)+(\alpha\delta)+e_1 \\ \mu+\alpha+\beta+\theta-\delta-\epsilon-\varphi-\gamma+(\alpha\beta)+(\beta\theta)+(\varphi\gamma)-(\alpha\delta)+e_2 \\ \vdots \\ \mu-\alpha-\beta-\theta+\delta+\varepsilon-\varphi+\gamma+(\alpha\beta)+(\beta\theta)-(\varphi\gamma)-(\alpha\delta)+e_{16} \end{bmatrix}. \quad (15.2)$$

Then the standard analysis of a linear statistical model can be applied. In contrast, some particular formulae for the two-level orthogonal array are also available. Obviously the sum of squares of each factor is obtained from the allocated columns as

$$S = \left(\sum_1 y_i - \sum_{-1} y_i\right)^2/n,$$

where n is the number of experiments, and $\sum_1 y_i$ and $\sum_{-1} y_i$ are the sum of data corresponding to the sign 1 or −1 of the allocated column, respectively. The residual sum of squares is obtained by subtracting the sum of squares of factors from the total sum of squares S_T. It is obtained also from the columns not assigned to the factors. Of course, S_T coincides with the sum of squares of all $n-1$ columns.

Example 15.2. Analysis of experiments of Table 15.3 from Moriguti (1989). The analysis of variance table is obtained as Table 15.5. The calculations of the sums of squares for the factors are shown in the table. The total sum of squares is obtained as

Table 15.5 Analysis of variance table for the aluminum experiment.

Factor	Column	Σ_1	Σ_{-1}	$\Sigma_1 - \Sigma_{-1}$	Square	S	df	Unbiased variance	F
A	1	42.4	40.7	1.7	2.89	0.18	1	0.18	0.16
B	2	49.2	33.9	15.3	234.09	14.63	1	14.63	13.3^*
C	4	44.7	38.4	6.3	39.69	2.48	1	2.48	2.2
D	10	39.8	43.3	−3.5	12.25	0.77	1	0.77	0.7
E	12	42.5	40.6	1.9	3.61	0.23	1	0.23	0.2
F	8	36.3	46.8	−10.5	110.25	6.89	1	6.89	6.2
G	15	32.1	51.0	−18.9	357.21	22.33	1	22.33	20.2^*
$A \times B$	3	40.5	42.6	−2.1	4.41	0.28	1	0.28	0.25
$B \times C$	6	36.2	46.9	−10.7	114.49	7.16	1	7.16	6.5
$A \times D$	11	42.5	40.6	1.9	3.61	0.23	1	0.23	0.2
$F \times G$	7	51.9	31.2	20.7	428.49	26.78	1	26.78	24.3^{**}
Total						81.96	11		
Error	(5 9 13 14)					$S_e = 4.39$	4	$\hat{\sigma}^2 = 1.10$	
Total						$S_T = 86.35$	15		

$$S_T = \sum_i y_i^2 - y_.^2/n = 517.95 - 83.1^2/16 = 86.35,$$

with df 15, which is the sum of squares for all 15 columns. Then, the sum of squares for the error is

$$S_e = 86.35 - 81.96 = 4.39$$

with df $15 - 11 = 4$ and the unbiased variance of error is

$$\hat{\sigma}^2 = 4.39/4 = 1.10.$$

The F-ratios are obtained by dividing the unbiased variance of factors by that of the error. By Table 15.5, the factors B, G, and $F \times G$ are statistically significant. Then, the proposed model is

$$y_{ijkl} = \mu + \beta_i + \varphi_j + \gamma_k + (\varphi\gamma)_{jk} + e_{ijkl} \tag{15.3}$$

recovering the parent factor F. This is a model of a three-way layout with repetitions and one two-way interaction $F \times G$. The factor B has no interaction with the other factors and can be analyzed separately. The effect of B is estimated by

$$\widehat{\beta_1 - \beta_2} = 49.2/8 - 33.9/8 = 1.91$$

and its variance is estimated by

$$\left(\frac{1}{8} + \frac{1}{8}\right)\hat{\sigma}^2 = 0.275.$$

Then, the confidence interval with confidence coefficient 0.95 is obtained as

$$\beta_1 - \beta_2 \sim 1.91 \pm 0.275^{1/2} t_4(0.05/2) = 1.91 \pm 1.46,$$

and the suggested level will be B_2 since the shorter time is preferable.

Since an interaction is observed between factors F and G, we are interested in comparing four combinations $(FG)_{11} \sim (FG)_{22}$. The sum of squares for F and G is obtained as the sum of squares of the columns 8, 15, and 7, amounting to 56.00. The averages of those four combinations have already been given in Fig. 1.1. It is seen that the combination $(FG)_{22}$ shows a high response against the others. Actually, the contribution of the contrast $(-1, -1, -1, 1)$ is

$$\frac{\{\bar{y}_{\cdot 22} - 3^{-1}(\bar{y}_{\cdot 11} + \bar{y}_{\cdot 12} + \bar{y}_{\cdot 21})\}^2}{(1 + 3^{-2} \times 3) \times 4^{-1}} = 52.52$$

and explains 94% of the total $S_{FG} = 6.89 + 22.33 + 26.78 = 56.00$. This contrast is significant almost at level 0.01 by Scheffé type multiple comparisons. Among the other combinations, the first choice will be $(FG)_{21}$.

It is interesting to estimate the expected mean for the suggested levels of B_2, F_2, and G_1. It is obtained by a standard analysis of the linear model (15.3), but more simply it is obtained as

$$\hat{\mu}_{221} = \left\{\widehat{\mu + \beta_2 + \varphi_2 + \gamma_1 + (\varphi\gamma)_{21}}\right\}$$
$$= \left(\widehat{\mu + \beta_2}\right) + \left\{\widehat{\mu + \varphi_2 + \gamma_1 + (\varphi\gamma)_{21}}\right\} - \hat{\mu} \quad (15.4)$$
$$= 33.9/8 + 13.5/4 - 83.1/16 = 2.42\,(\mathrm{s}).$$

The meaning of calculation of (15.4) is self-explanatory. There are two formulae for the variance of the estimator (15.4). Both formulae give the effective repetition number n_e to obtain the variance of (15.4) in the form of σ^2/n_e.

Ina's formula:

$$n_e = \frac{\text{Total number of experiments}}{\text{Total number of degrees of freedom for factors including a general mean}}$$
$$= \frac{16}{1+1+1+1+1} = \frac{16}{5} = 3.2.$$

Taguchi's formula:

$$n_e = \{\text{Sum of inverses of the number of repetitions of terms in (15.4)}\}^{-1}$$
$$= \left(\frac{1}{8} + \frac{1}{4} - \frac{1}{16}\right)^{-1} = \left(\frac{2+4-1}{16}\right)^{-1} = \frac{16}{5} = 3.2.$$

It should be noted that the negative sign in the estimating equation (15.4) corresponds to the minus in the summation in Taguchi's formula. Finally, the confidence interval with confidence coefficient 0.95 for μ_{221} is obtained as

$$\mu_{221} \sim 2.42 \pm \sqrt{1.10/3.2}\,t_4(0.025)$$
$$= 2.42 \pm 1.63 \quad (15.5)$$

In Example 15.2 we reached a preferable result (15.5) for the selected factors $(BFG)_{221}$. However, it should be noted that experiments by orthogonal array are usually performed to screen out the abundant possible factors suggested by the workers actively engaged in the production process, and to find a few promising factors to improve the quality of the products. Therefore, the analytical result inevitably suffers from a large selection bias. There is a simulation result of testing the large observed sum of squares by the small observed sum of squares in the $L_{16}(2^{15})$ experiment with no real factor, which suggests that the probability of declaring some columns significant at the level 0.05 amounts to 0.30. Therefore, a confirmatory experiment is certainly necessary for the selected factors, extending the number of levels of each factor.

Finally, there are various devices for an extended use of orthogonal arrays. They can be used for a split-plot design, not just for completely randomized experiments. There are methods to allocate the factors with three or four levels to a two-level orthogonal array. In contrast, the factors with two levels can be allocated to a three-level orthogonal array. For these extensions, refer to Taguchi (1962).

15.2 Ordered Categorical Responses in a Highly Fractional Experiment

The data of Table 15.6 reported by Taguchi and Wu (1980) are the result of an arc-welding experiment to find the important factors that affect the workability of an arc-welded section between two steel plates. Workability is the degree of difficulty in welding the two steel plates together, which was judged in three levels: 1 (easy), 2 (normal), and 3 (difficult). It was a 2^{9-5} experiment for nine factors ($A \sim I$) with four interactions ($A \times G, A \times H, A \times C, G \times H$) of interest. Taguchi and Wu reported three significant factors D, F, and G by the so-called accumulation analysis. However, the experiment is so highly fractional, with many factors of interest, that the conclusion should suffer from large selection bias. Further, the analysis is based on the collapsed data for the respective factors, so that inevitably it suffers from Simpson's paradox. It is essentially difficult to give a proper analytical method for highly fractional categorical data like this. Therefore, it might be inevitable to make some crude analysis first, although this cannot certainly be a conclusion. One should refer also to Wu and Hamada (1990), and comments regarding the analysis of this experiment. In this book we take the approach of conditional analysis based on the cumulative efficient score for the ordered categorical data, like in this example. However, such a conditional analysis is not applicable to the original data, since it is so highly fractional.

Table 15.6 Design and workability data for arc-welding experiment.

Experiment #	A	B	C	D	E	F	G	H	I	Workability
1	1	1	1	1	1	1	1	1	1	2
2	1	1	2	2	2	2	1	1	2	2
3	1	2	2	1	1	1	1	2	1	2
4	1	2	1	2	2	2	1	2	2	1
5	1	2	2	1	1	2	2	1	2	1
6	1	2	1	2	2	1	2	1	1	3
7	1	1	1	1	1	2	2	2	2	1
8	1	1	2	2	2	1	2	2	1	3
9	2	2	2	1	2	1	1	1	2	2
10	2	2	1	2	1	2	1	1	1	2
11	2	1	1	1	2	1	1	2	2	2
12	2	1	2	2	1	2	1	2	1	2
13	2	1	1	1	2	2	2	1	1	1
14	2	1	2	2	1	1	2	1	2	3
15	2	2	2	1	2	2	2	2	1	2
16	2	2	1	2	1	1	2	2	2	2

Table 15.7 Summarized data with respect to factors D, F, and G.

Factor levels			Workability			
D	F	G	1	2	3	Total
1	1	1	0	4	0	4
2	2	1	1	3	0	4
1	2	2	3	1	0	4
2	1	2	0	1	3	4
Total			4	9	3	16

We therefore apply the method to the summarized data (Table 15.7) with respect to factors D, F, and G. We assume a multinomial distribution and a log linear model for the cell probability:

$$\log p_{ijkl} = \mu_{ijk} + \delta_{il} + \varphi_{jl} + \gamma_{kl} + \rho_l,$$

where the suffix i, j and k are for the factors D, F and G each with two levels, and ρ_l, $l = 1, 2, 3$, is the parameter for the response category, with number of levels three. It should be noted that it is only a 2^{3-1} experiment, even for the summarized data, and only half of the combinations are realized. Further, it contains so many zero cells, making conditional analysis impossible for factors D and G. We therefore deal only with the null hypothesis $H_\varphi : \varphi_{jl} = 0$ of factor F.

To derive the sufficient statistics under H_φ we drop the factors ρ_l, δ_{il} and γ_{kl} when $(i-2)(k-2)(l-3) = 0$, while we leave four combinations of μ_{ijk} as they are. Then, the linearly independent sufficient statistics are obtained as

$$(y_{111\cdot}, y_{221\cdot}, y_{122\cdot}, y_{212\cdot}, y_{1\cdot\cdot 1}, y_{1\cdot\cdot 2}, y_{\cdot\cdot 11}, y_{\cdot\cdot 12}, y_{\cdot\cdot\cdot 1}, y_{\cdot\cdot\cdot 2}). \tag{15.6}$$

It should be noted here that $y_{1\cdot\cdot 1}$ is actually $y_{1111} + y_{1221}$, since this is a half experiment. Then, starting from a table with all entries unity, we obtain the maximum likelihood estimate $\hat{y}_{ijkl}$ by the iterative scaling procedure to keep the sets of sufficient statistics $(y_{111\cdot}, y_{221\cdot}, y_{122\cdot}, y_{212\cdot})$ for μ_{ijk}, $(y_{1\cdot\cdot 1}, y_{1\cdot\cdot 2})$ for δ_{il}, $(y_{\cdot\cdot 11}, y_{\cdot\cdot 12})$ for γ_{kl}, and $(y_{\cdot\cdot\cdot 1}, y_{\cdot\cdot\cdot 2})$ for ρ_l. The result is given in Table 15.8 (1).

The variance of the estimated cell frequencies is obtained in the same form as (14.12), by expressing (15.6) as $\boldsymbol{A}^{*\prime}\boldsymbol{y}$ with $\boldsymbol{\Omega}$ a diagonal matrix of $\hat{y}_{ijkl}$. For the calculation of (14.12), a detailed explanation is given in Section 14.2.2 (1) (a). The standardized residual of each cell is shown in Table 15.8 (2). There is only one linearly independent residual, so that the degree of freedom for testing H_φ is unity. The standardized residual of Table 15.8 (2) will suggest weak evidence of the effect of F.

As a worked example, we consider the data of Table 15.9, changing Table 15.7 slightly. The marginal tables and their cumulative chi-squared statistics are shown in Table 15.10, suggesting some effects for F and D.

Table 15.8 Summarized data with respect to the factors *D*, *F*, and *G*.

Factor levels			Workability			
D	*F*	*G*	1	2	3	Total
			(1) Estimated cell frequencies			
1	1	1	0.5529	3.4471	0	4.0000
2	2	1	0.4471	3.5529	0	4.0000
1	2	2	2.4471	1.5529	0	4.0000
2	1	2	0.5529	0.4471	3	4.0000
	Total		4.0000	9.0000	3	
			(2) Standardized residuals			
1	1	1	−1.723	1.723	0	
2	2	1	1.723	−1.723	0	
1	2	2	1.723	−1.723	0	
2	1	2	−1.723	1.723	0	

Table 15.9 Worked example.

Factor levels			Workability			
D	*F*	*G*	1	2	3	Total
1	1	1	1	2	1	4
2	2	1	1	3	0	4
1	2	2	3	1	0	4
2	1	2	0	1	3	4
	Total		5	7	4	16

Table 15.10 The marginal tables for the worked example.

Marginal table 1					Marginal table 2					Marginal table 3				
	Workability					Workability					Workability			
Factor *D*	1	2	3	Total	Factor *F*	1	2	3	Total	Factor *G*	1	2	3	Total
1	4	3	1	8	1	1	3	4	8	1	2	5	1	8
2	1	4	3	8	2	4	4	0	8	2	3	2	3	8
Total	5	7	4	16	Total	5	7	4	16	Total	5	7	4	16
$\chi^{*2} = 3.952$					$\chi^{*2} = 7.952$					$\chi^{*2} = 1.624$				

Table 15.11 Estimated cell frequencies under H_φ.

Factor levels			Workability			
D	F	G	1	2	3	Total
1	1	1	1.6020	2.1465	0.2515	4.0000
2	2	1	0.3980	2.8535	0.7485	4.0000
1	2	2	2.3980	0.8535	0.7485	4.0000
2	1	2	0.6020	1.1465	2.2515	4.0000
	Total		5.0000	7.0000	4.0000	16.0000

Now, the worked example is only slightly different from the welding experiment, but there are still two, one, and one degrees of freedom for the conditional analyses of factors F, D, and G, respectively.

To test the null hypothesis of the factor F, the conditional MLEs of the cell frequencies are calculated as in Table 15.11. From these fitted values we can calculate the conditional expectation of the sufficient statistics $\boldsymbol{s} = (y_{\cdot 1 \cdot 1}, y_{\cdot 1 \cdot 2})'$ for $\boldsymbol{\varphi} = (\varphi_{11}, \varphi_{12})'$ as

$$E(\boldsymbol{s}) = \begin{bmatrix} 1.6020 + 0.6020 \\ 2.1465 + 1.1465 \end{bmatrix} = \begin{bmatrix} 2.2040 \\ 3.2930 \end{bmatrix}.$$

The variance–covariance matrix is similarly obtained via (14.12) again as

$$V(\boldsymbol{s}) = \begin{bmatrix} 0.55148 & -0.39300 \\ -0.39300 & 0.681956 \end{bmatrix}.$$

The variance–covariance matrix of the accumulated statistics is easily obtained as

$$\boldsymbol{V}^* = \begin{bmatrix} 1 & 0 \\ 1 & 1 \end{bmatrix} V(\boldsymbol{s}) \begin{bmatrix} 1 & 1 \\ 0 & 1 \end{bmatrix} = \begin{bmatrix} 0.55148 & 0.15848 \\ 0.15848 & 0.44744 \end{bmatrix}. \tag{15.7}$$

Then we obtain the cumulative chi-squared

$$\chi^{*2}(\text{conditional}) = \frac{(1 - 2.2040)^2}{0.55148} + \frac{\{1 + 3 - (2.2040 + 3.2930)\}^2}{0.4474} = 7.637^*.$$

The correlation matrix of the components of χ^{*2} is easily obtained from (15.7) as

$$\boldsymbol{C}^{*\prime}\boldsymbol{C}^* = \begin{bmatrix} 1 & 0.3190 \\ 0.3190 & 1 \end{bmatrix}.$$

Then the constants for the chi-squared approximation are obtained from

$$df = \mathrm{tr}\left(C^{*\prime}C^{*}\right) = 2,\ 2d^2f = 2\mathrm{tr}\left(C^{*\prime}C^{*}\right)^2 = 4.4070$$

as $d = 1.1018$ and $f = 1.8152$. The upper 0.05 point is calculated as $1.1018 \times 5.628 = 6.20$. Therefore, the crude analysis of the marginal table 2 of Table 15.10 will be supported.

The conditional standardized residuals under the null hypothesis of factor D, conditional on the factors F and G, are all ± 1.249 excluding zero residuals. Hence, a somewhat large χ^{*2} value in the marginal table 1 of Table 15.10 should better be considered as spurious. Similarly, the conditional standardized residuals under the null hypothesis of factor G, conditional on the factors D and F, are ± 0.2334, suggesting that the effects of factor G are negligible (see Hirotsu, 1990 for details).

15.3 Optimality of an Orthogonal Array

We generalize the optimality of the weighing experiment with balance in Section 1.1. Now the problem is to estimate the weight of p materials $\boldsymbol{\mu} = (\mu_1 \ldots, \mu_p)'$ by n measurements with a balance. In an experiment we can put each material on the left or right plate, or there is a choice not to weigh the material in the experiment. This is formulated mathematically by considering a variable x_{ij} which takes 1, -1, or 0 according to whether the jth material is weighed on the left or right plate, or not weighed at the ith experiment. Then, introducing an $n \times p$ design matrix $\boldsymbol{X} = [x_{ij}]$, the statistical model is expressed in linear form as

$$\boldsymbol{y} = \boldsymbol{X}\boldsymbol{\mu} + \boldsymbol{e}, \qquad (15.8)$$

where the error $\boldsymbol{e}$ is assumed to be uncorrelated and of equal variance, namely

$$V(\boldsymbol{e}) = \sigma^2 \boldsymbol{I}.$$

The problem is to select an optimal design $\boldsymbol{X}$ of weighing. There are several definitions of optimality. First, the rank of $\boldsymbol{X}$ must be p for the unbiased estimator of $\boldsymbol{\mu}$ for all components to be available. Then, $\widehat{\boldsymbol{\mu}} = (\boldsymbol{X}'\boldsymbol{X})^{-1}\boldsymbol{y}$ is an unbiased estimator of $\boldsymbol{\mu}$ with variance

$$V(\widehat{\boldsymbol{\mu}}) = \boldsymbol{M}^{-1}\sigma^2,$$

where $\boldsymbol{M} = \boldsymbol{X}'\boldsymbol{X}$ is Fisher's information matrix. If there are two designs $\boldsymbol{X}_1$ and $\boldsymbol{X}_2$, and supposing $\boldsymbol{X}_1'\boldsymbol{X}_1 - \boldsymbol{X}_2'\boldsymbol{X}_2$ to be semi-positive definite, then the design $\boldsymbol{X}_1$ is strongly recommended since for any parameter $\boldsymbol{l}'\boldsymbol{\mu}$, the variance of the estimator $\boldsymbol{l}'\widehat{\boldsymbol{\mu}}$ by $\boldsymbol{X}_1$ is always less than or equal to that by $\boldsymbol{X}_2$. However, it is rarely the case that such an optimal design $\boldsymbol{X}_1$ exists. Therefore, usually the following four criteria are used, where $\lambda_j, j = 1, \ldots, p$ are the eigenvalues of $\boldsymbol{M}$.

1. D-optimum. Maximize the determinant $|\boldsymbol{M}| = \Pi\lambda_j$. It is equivalent to minimizing the generalized variance $|V(\hat{\boldsymbol{\mu}})|$. D is the initial of determinant.

2. A-optimum. Minimize tr $\left(\boldsymbol{M}^{-1}\right) = \sum \lambda_j^{-1}$. It is equivalent to minimizing the average of the variance of $\hat{\mu}_j$. A is the initial of average.

3. E-optimum. Maximize the minimum of $\lambda_j, j = 1, \ldots, p$. It is equivalent to minimizing the maximum variance of the standardized linear combination $\boldsymbol{l}'\hat{\boldsymbol{\mu}}$, $\boldsymbol{l}'\boldsymbol{l} = 1$. E is the initial of eigenvalue.

4. Mini-max criterion. Minimize the maximum diagonal element of $\boldsymbol{M}^{-1}$. It is equivalent to minimizing the maximum variance of $\hat{\mu}_j, j = 1, \ldots, p$.

An orthogonal array of the two levels satisfies all criteria (1) ~ (4) simultaneously. We start by stating Hotelling's theory.

Theorem 15.1. Hotelling's theorem (1944)

(1) By n weighings of p materials with a balance, $V\left(\hat{\mu}_j\right) \geq \sigma^2/n, j = 1, \ldots, p$. That is, the variance cannot be made smaller than σ^2/n for each material.

(2) The necessary and sufficient condition for $V\left(\hat{\mu}_j\right) = \sigma^2/n$ for some j is $x_{ij} = 1$ or -1 for all i and $\sum_i x_{ij}x_{ij'} = 0$ for all $j'(\neq j)$.

Proof. Without any loss of generality, we can discuss $\hat{\mu}_1$ assuming $j = 1$. Let the first column of $\boldsymbol{X}$ (15.8) be $\boldsymbol{x}_1$ and the rest of the columns $\boldsymbol{X}_2$, so that $\boldsymbol{X} = [\boldsymbol{x}_1\ \boldsymbol{X}_2]$. Then, by simple algebra we have

$$V(\hat{\mu}_1) = \left\{\boldsymbol{x}_1'\boldsymbol{x}_1 - \boldsymbol{x}_1'\boldsymbol{X}_2\left(\boldsymbol{X}_2'\boldsymbol{X}_2\right)^{-1}\boldsymbol{X}_2'\boldsymbol{x}_1\right\}^{-1}\sigma^2, \tag{15.9}$$

where an obvious modification is necessary in the case of rank $(\boldsymbol{X}) < p$. Now in (15.9) it is obvious that

$$\boldsymbol{x}_1'\boldsymbol{x}_1 \leq n,$$
$$\boldsymbol{x}_1'\boldsymbol{X}_2\left(\boldsymbol{X}_2'\boldsymbol{X}_2\right)^{-1}\boldsymbol{X}_2'\boldsymbol{x}_1 \geq 0.$$

This immediately proves the first part of Theorem 15.1. Then $V(\hat{\mu}_1)$ is minimized when

$$\boldsymbol{x}_1'\boldsymbol{x}_1 = n \tag{15.10}$$

and

$$\boldsymbol{x}_1'\boldsymbol{X}_2\left(\boldsymbol{X}_2'\boldsymbol{X}_2\right)^{-1}\boldsymbol{X}_2'\boldsymbol{x}_1 = 0. \tag{15.11}$$

Equation (15.10) holds if and only if $x_{i1} = 1$ or -1 and not 0 for all i; that means for material j to be weighed every time, either on the left or right plate. Equation (15.11) holds if and only if material 1 is weighed the same number of times on the same or opposite plates to other materials $\mu_j, j=2, \ldots, p$. This is an orthogonality relationship between two columns of $\boldsymbol{X}$, and implies the latter part of the necessary and sufficient condition of Theorem 15.1.

In the case of $n=4n'$ the design matrix $\boldsymbol{X}$ composed of any p columns of a Hadamard matrix gives $\boldsymbol{X}'\boldsymbol{X} = n\boldsymbol{I}_p$ and satisfies the necessary and sufficient condition of Theorem 15.1 for each column. Therefore, it satisfies optimality criteria (2) and (4). Further, in the weighing problem we have obviously

$$\sum \lambda_j = \mathrm{tr}\,(\boldsymbol{X}'\boldsymbol{X}) \le pn.$$

Therefore we have

$$\Pi \lambda_j \le \left(\sum \lambda_j / p\right)^p \le n^p. \tag{15.12}$$

However, the design matrix from a Hadamard matrix satisfies the equality of equation (15.12) showing D-optimality. Then it is also E-optimal since it satisfies the equality of equation (15.13),

$$\min \lambda_j \le \sum \lambda_j / p \le n. \tag{15.13}$$

Definition 15.1. Hadamard matrix. 1. A square matrix composed of ± 1, each row of which is orthogonal to other rows. A Hadamard matrix of order n satisfies $H'H = nI_n$.

References

Hirotsu, C. (1990) Discussion on a critical look at accumulation analysis and related methods. *Technometrics* **32**, 133–136.

Hotelling, H. (1944) Some improvements in weighing and other experimental techniques. *Ann. Math. Statist.* **15**, 297–305.

Moriguti, S. (1989) *Statistical methods*, new edn. Japanese Standards Association, Tokyo (in Japanese).

Taguchi, G. (1962) *Design of experiments* (1), (2). Maruzen, Tokyo (in Japanese).

Taguchi, G. and Wu, Y. (1980) *Introduction to off-line quality control*. Central Quality Control Association, Nagoya, Japan.

Wu, C. F. J. and Hamada, M. (1990) A critical look at accumulation analysis and related methods. *Technometrics* **32**, 119–130.

Appendix

Table A Upper percentiles $t_{\alpha}(a, f)$ of max acc. $t1$.

$\alpha = 0.01$

f	a = 3	4	5	6	7	8
5	3.900	4.203	4.39	4.56	4.70	4.78
10	3.115	3.309	3.44	3.54	3.62	3.68
15	2.908	3.075	3.19	3.27	3.34	3.39
20	2.813	2.968	3.07	3.15	3.21	3.26
25	2.758	2.907	3.01	3.08	3.14	3.19
30	2.723	2.867	2.97	3.04	3.09	3.14
40	2.680	2.819	2.91	2.98	3.04	3.08
60	2.638	2.772	2.86	2.93	2.98	3.02
120	2.598	2.726	2.81	2.88	2.92	2.96
∞	2.558	2.682	2.764	2.824	2.871	2.909

(*continued overleaf*)

Advanced Analysis of Variance, First Edition. Chihiro Hirotsu.

Table A (*Continued*)

$\alpha = 0.05$

	a					
f	3	4	5	6	7	8
5	2.441	2.676	2.84	2.95	3.05	3.12
10	2.151	2.333	2.46	2.54	2.62	2.67
15	2.067	2.235	2.35	2.43	2.49	2.54
20	2.027	2.188	2.30	2.37	2.43	2.48
25	2.004	2.161	2.26	2.34	2.40	2.45
30	1.989	2.143	2.24	2.32	2.38	2.43
40	1.970	2.121	2.22	2.29	2.35	2.40
60	1.952	2.100	2.20	2.27	2.32	2.37
120	1.934	2.079	2.18	2.24	2.30	2.34
∞	1.916	2.058	2.151	2.219	2.271	2.314

$\alpha = 0.10$

	a					
f	3	4	5	6	7	8
5	1.873	2.089	2.23	2.34	2.43	2.49
10	1.713	1.894	2.02	2.10	2.17	2.23
15	1.666	1.836	1.95	2.03	2.10	2.15
20	1.643	1.808	1.92	2.00	2.06	2.11
25	1.629	1.792	1.90	1.98	2.04	2.09
30	1.620	1.781	1.89	1.96	2.02	2.07
40	1.609	1.768	1.87	1.95	2.01	2.05
60	1.598	1.755	1.86	1.93	1.99	2.04
120	1.588	1.742	1.84	1.92	1.97	2.02
∞	1.577	1.730	1.829	1.901	1.956	2.001

Table B Upper percentiles of max acc. $\chi^2(\boldsymbol{P}_a^{*\prime})$.

$\alpha = 0.01$

	b	2			4			6			8		
	a	3	4	5	3	4	5	3	4	5	3	4	5
m	2	17.74	15.96	14.97	21.12	20.10	19.51	24.31	23.63	23.22	27.51	27.01	26.71
	3	11.47	11.46	11.46	16.33	16.48	16.58	20.12	20.38	20.56	23.60	23.94	24.18
	4	10.04	10.33	10.53	15.03	15.44	15.72	18.90	19.40	19.75	22.41	22.99	23.38
	5	9.405	9.822	10.10	14.42	14.95	15.31	18.32	18.93	19.35	21.84	22.53	22.99
	10	8.468	9.044	9.435	13.48	14.17	14.64	17.39	18.19	18.75	20.94	21.82	22.14
	∞	7.808	8.480	8.944	12.77	13.58	14.14	16.69	17.59	18.21	20.21	21.19	21.86

$\alpha = 0.05$

	b	2			4			6			8		
	a	3	4	5	3	4	5	3	4	5	3	4	5
m	2	8.195	8.256	8.287	12.83	13.00	13.10	16.34	16.62	16.79	19.55	19.91	20.14
	3	6.262	6.695	6.971	10.86	11.39	11.74	14.41	15.04	15.45	17.63	18.34	18.80
	4	5.753	6.261	6.595	10.27	10.91	11.32	13.81	14.54	15.02	17.02	17.83	18.37
	5	5.520	6.058	6.416	9.995	10.67	11.12	13.52	14.30	14.81	16.73	17.58	18.16
	10	5.159	5.739	6.134	9.550	10.29	10.79	13.05	13.90	14.48	16.24	17.18	17.75
	∞	4.894	5.500	5.919	9.210	9.994	10.53	12.69	13.58	14.20	15.86	16.85	17.52

(continued overleaf)

Table B (*Continued*)

$\alpha = 0.10$

	b	2			4			6			8		
	a	3	4	5	3	4	5	3	4	5	3	4	5
	2	5.440	5.806	6.002	9.841	10.29	10.56	13.21	13.75	14.09	16.28	16.90	17.30
	3	4.440	4.934	5.255	8.654	9.288	9.702	11.98	12.72	13.20	15.03	15.85	16.39
m	4	4.161	4.685	5.034	8.292	8.976	9.429	11.60	12.39	12.92	14.62	15.51	16.09
	5	4.031	4.566	4.927	8.117	8.824	9.296	11.41	12.23	12.77	14.42	15.34	15.94
	10	3.826	4.378	4.756	7.835	8.575	9.076	11.09	11.96	12.54	14.09	15.05	15.67
	∞	3.672	4.235	4.625	7.615	8.382	8.904	10.85	11.75	12.35	13.83	14.83	15.50

Table C Upper percentiles of the largest eigenvalue of Wishart matrix.

$$W\{\boldsymbol{I}_{\min(a-1,\,b-1)},\ \max(a-1,b-1)\}$$

Upper column for $\alpha=0.05$ and lower column for $\alpha=0.01$

	$b-1$											
$a-1$	1	2	3	4	5	6	7	8	9	10	15	20
2	5.99	8.59	10.74	12.68	14.49	16.21	17.88	19.49	21.06	22.61	29.97	36.94
	9.21	12.16	14.57	16.73	18.73	20.64	22.47	24.23	25.95	27.63	35.60	43.08
3	7.82	10.74	13.11	15.24	17.21	19.09	20.88	22.62	24.31	25.96	33.80	41.18
	11.35	14.57	17.18	19.50	21.65	23.69	25.64	27.52	29.34	31.12	39.51	47.38
4	9.45	12.68	15.24	17.52	19.63	21.62	23.53	25.37	27.15	28.90	37.13	44.84
	13.28	16.73	19.50	21.96	24.24	26.38	28.43	30.41	32.32	34.18	42.95	51.10
5	11.07	14.49	17.21	19.63	21.86	23.95	25.96	27.88	29.75	31.58	40.14	48.14
	15.09	18.73	21.65	24.24	26.62	28.86	31.00	33.05	35.04	36.98	46.05	54.48
6	12.59	16.21	19.09	21.62	23.95	26.14	28.23	30.24	32.18	34.07	42.96	51.22
	16.81	20.64	23.69	26.38	28.86	31.19	33.40	35.53	37.59	39.59	48.96	57.64
7	14.07	17.88	20.88	23.53	25.96	28.23	30.40	32.48	34.50	36.45	45.62	54.10
	18.48	22.47	25.64	28.43	31.00	33.40	35.69	37.89	40.01	42.07	51.71	60.60
8	15.51	19.49	22.62	25.37	27.88	30.24	32.48	34.63	36.70	38.72	48.15	56.86
	20.09	24.23	27.52	30.40	33.05	35.53	37.89	40.15	42.33	44.45	54.33	63.42
10	18.31	22.61	25.96	28.90	31.58	34.07	36.45	38.72	40.91	43.04	52.94	62.04
	23.21	27.63	31.12	34.18	36.98	39.59	42.07	44.45	46.74	48.95	59.28	68.76

Index

Advanced Analysis of Variance, First Edition. Chihiro Hirotsu.